FOUNDATIONS OF MATHEMATICAL ANALYSIS

RICHARD JOHNSONBAUGH
DePaul University
Chicago, Illinois

W. E. PFAFFENBERGER
University of Victoria
Victoria, British Columbia, Canada

DOVER PUBLICATIONS, INC.
Mineola, New York

Bibliographical Note

This Dover edition, first published in 2010, is an unabridged republication of the slightly corrected 2002 Dover edition of the work originally published in 1981 by Marcel Dekker, Inc., New York.

Library of Congress Cataloging-in-Publication Data

Johnsonbaugh, Richard, 1941–
  Foundations of mathematical analysis / Richard Johnsonbaugh and W. E. Pfaffenberger. — Dover ed.
    p. cm.
  Originally published: New York : Dekker, 1981.
  Includes bibliographical references and index.
  ISBN-13: 978-0-486-47766-4
  ISBN-10: 0-486-47766-5
  1. Mathematical analysis—Foundations. I. Pfaffenberger, W. E. II. Title.

QA299.8.J63 2010
515—dc22
                                                                                2009043354

Manufactured in the United States by Courier Corporation
47766503   2013
www.doverpublications.com

Preface

This book evolved from a one-year Advanced Calculus course that we have given during the last decade. Our audiences have included junior and senior majors and honors students, and, on occasion, gifted sophomores.

The material is logically self-contained; that is, all of our results are proved and are ultimately based on the axioms for the real numbers. We do not use results from other sources, except for a few results from linear algebra which are summarized in a brief appendix. Thus, theoretically, no prerequisites are necessary to understand this material. Realistically, the prerequisite is some mathematical maturity such as one might acquire by taking calculus and, perhaps, linear algebra.

Our intent is to teach students the tools of modern analysis as it relates to further study in mathematics, especially statistics, numerical analysis, differential equations, mathematical analysis, and functional analysis.

It is our belief that the key to a sound foundation for the study of analysis lies in an understanding of the limit concept. Thus, after initial chapters on sets and the real number system, we introduce the limit concept using numerical sequences and series (Chapters IV and V). This is followed by Chapter VI on the limit of a function. We then move to the general setting of metric spaces (Chapter VII). Chapter VIII is a review of differential calculus. Chapter IX gives a detailed introduction to the theory of Riemann-Stieltjes integration. We then turn to the study of sequences and series of functions (Chapters X and XI), Fourier series (Chapter XII), the Riesz representation theorem (Chapter XIII), and the Lebesgue integral (Chapter XIV). The first seven chapters could be used for a one-term course on the Concept of Limit.

Because we believe that an essential part of learning mathematics is doing mathematics, we have included over 750 exercises, some containing several parts, of varying degrees of difficulty. Hints and solutions to selected exercises, indicated by an asterisk, are given at the back of the book.

We would like to thank our colleagues, Dr. Rosalind Reichard, who taught this course from a preliminary version and gave us useful information, and Dr. Keith Rose, who read the manuscript and offered valuable criticism. Thanks also to our many students who studied this material and offered suggestions, and especially Mr. James Africh, who worked nearly every exercise and made many helpful comments. Our thanks also go to the secretarial staff at the University of Victoria, who over the years typed various versions of the manuscript. Of course, we assume joint responsibility for the book's strengths and weaknesses, and we welcome comment.

Richard Johnsonbaugh
W. E. Pfaffenberger

Preface to the Dover Edition

Foundations of Mathematical Analysis covers real analysis—from naïve set theory and the axioms for the real numbers to the Lebesgue integral, with sequences and series, metric spaces, the Riemann-Stieltjes integral, inner product spaces, Fourier series, Tauberian theorems, the Riesz representation theorem, and a brief discussion of Hilbert spaces in between.

The book evolved from a one-year Advanced Calculus course that both authors taught during the late 1960s and throughout the 1970s. The audience included junior and senior majors and honors students, and, on occasion, gifted sophomores. Professor Pfaffenberger has continued to teach this course. (Professor Johnsonbaugh moved to computer science.) The intervening years have confirmed the importance of real analysis. Analysis is a core subject in mathematics and is a prerequisite for further study in mathematics. Analysis is also fundamental to many related fields such as statistics. Several of Professor Pfaffenberger's students have completed doctorates at distinguished institutions (e.g., Princeton, Harvard, Berkeley, Cambridge), and many have specialized in analysis.

Because we believe that an essential part of learning mathematics is doing mathematics, we have included over 750 exercises, some containing several parts, of varying degrees of difficulty. Hints and solutions to selected exercises, indicated by an asterisk, are given in the back of the book.

We will maintain a World Wide Web site for this reprint that contains additional problems, errata, and other supplementary material.

About This Book

The material is logically self-contained; that is, all of our results are proved and are ultimately based on the axioms for the real numbers. We do not use results from other sources except for a few results from linear algebra that are summarized in a brief appendix. Thus, logically, no prerequisites are necessary to understand this material. Realistically, the prerequisite is some mathematical maturity such as one might acquire by taking calculus and, perhaps, linear algebra.

It is our belief that understanding the limit concept is the key to a sound foundation for the study of analysis. Thus, after initial chapters on sets and the real number system (Chapters I–III), we introduce the limit concept using numerical sequences and series (Chapters IV and V). Chapter VI then covers the limit of a function. Theorem 16.6 in Chapter IV contains an interesting proof that $\{(1+1/n)^n\}$ is increasing and convergent. A discussion of double series (Section 29) leads to a quick proof that a rearrangement of an absolutely convergent series converges and the sum of the rearranged series is equal to the sum of the original series (Theorem 29.7). Double series are also used to discuss the Cauchy product of two series (Theorem 29.9) and to give some results on power series (Corollary 29.10 and Theorem 29.11).

In Chapter VII, we move to the general setting of metric spaces. Results include the Bolzano-Weierstrass characterization of a compact metric space (Section 43) and the Baire category theorem (Section 47). The first seven chapters could be used for a one-term course on the Concept of Limit.

After a review of differential calculus (Chapter VIII), Chapter IX gives a detailed introduction to the theory of Riemann-Stieltjes integration. We discuss measure zero (Section 57) and give a necessary and sufficient condition for the existence of the Riemann integral (Section 58).

We then turn to the study of sequences and series of functions (Chapter X). We provide applications to power series (Section 63) and Abel's limit theorems (Section 64). We also discuss summability methods and Tauberian theorems (Section 65).

Chapter XI discusses the exponential, logarithm, and trigonometric functions. The exponential, sine, and cosine functions are defined by power series, after which their standard properties are derived. The other trigonometric functions are defined in terms of the sine and cosine functions, and the logarithm function is defined as the inverse of the exponential function.

Inner product spaces and Fourier series are the topics for Chapter XII. Included is a discussion of Cesàro summability (Section 77) and Hardy's Tauberian theorem (Theorem 79.1) with an application to Fourier series (Theorem 79.3).

Chapter XIII develops normed linear spaces and proves the following version of the Riesz representation theorem (Theorem 84.1), where $C[a, b]$ is the set of continuous functions on $[a, b]$: If T is a continuous linear functional on $C[a, b]$, then there exists a function of bounded variation α on $[a, b]$ such that

$$T(f) = \int_a^b f \, d\alpha \qquad \text{for all } f \in C[a, b].$$

Furthermore, the norm of T is equal to the total variation of α on $[a, b]$.

The last chapter (Chapter XIV) studies the Lebesgue integral. Topics include measurable functions (Section 87), integration on positive measure spaces (Section 88), and Lebesgue measure (Sections 89 and 90). The concluding section (Section 91) introduces Hilbert spaces and proves the Riesz-Fischer theorem (Theorem 91.9).

The book concludes with an appendix on vector spaces, a list of references, and hints to selected exercises. The appendix summarizes some of the key definitions and theorems concerning vector spaces that are used in the book.

Examples

The book contains nearly 100 worked examples. These examples clarify the theory, show students how to develop proofs, demonstrate applications of the theory, elucidate proofs, and help motivate the material.

World Wide Web Site

The World Wide Web site

http://condor.depaul.edu/~rjohnson

contains

- Expanded explanations of particular topics and further information on analysis.
- Additional exercises.
- An errata list.

Acknowledgments

We thank our former colleagues Rosalind Reichard and Keith Rose, who offered valuable comments on early drafts of the book. Thanks also to our many students, who studied this material and contributed suggestions—in particular, we thank James Africh, who worked most of the exercises. Our thanks also go to the secretarial staff at the University of Victoria, who typed many drafts of the book. In particular, we are grateful to Yvonne Leeming for typing the first camera-ready version of the book. Finally, we thank John W. Grafton, Senior Reprint Editor, Dover Publications, for proposing this edition of the book and for his assistance in producing it.

R.J.
W.E.P.

Contents

I

Sets and Functions

In this brief chapter we will summarize some of the fundamental notation and definitions which will be used throughout the text.

1. Sets

As a starting point we describe what we mean by a *set*. We will not attempt to define the term *set*, but we demonstrate the use of the term by examples. The term *set* is used (roughly) to describe any collection of objects. For example, the set whose objects are the positive integers 1, 2, 3 may be denoted

$$\{1, 2, 3\}$$

If a set consists of a finite number of objects, we may denote the set by listing its objects. If a set consists of infinitely many objects such a listing is impossible. In this case we may describe the set by naming a property common to all the objects in the set. For example, to describe the set of positive integers, we use the notation

$$\{x \mid x \text{ is a positive integer}\}$$

This notation is read "the set of all x such that x is a positive integer." (The bar "|" is read "such that.") In general, to describe the set of all objects having a particular property P, we write

$$\{x \mid x \text{ has property } P\}$$

If x is an object in a set A, we write

$$x \in A$$

and say that x is an *element* (or a *point*) of A. If x is not an element of the set A, we write

$$x \notin A$$

For example, if $\mathbf{Z}$ denotes the set of integers, then $1 \in \mathbf{Z}$, but $\frac{1}{2} \notin \mathbf{Z}$.

We shall say two sets A and B are *equal* and write $A = B$ if A and B have the same elements. Thus $A = B$ if whenever $x \in A$, then $x \in B$, and whenever $x \in B$, then $x \in A$.

Definition 1.1   If A and B are sets, the *union* of A and B is the set

$$A \cup B = \{x \mid x \in A \text{ or } x \in B\}$$

The *intersection* of A and B is the set

$$A \cap B = \{x \mid x \in A \text{ and } x \in B\}$$

For example, if $A = \{1, 2, 3\}$ and $B = \{2, 3, 4\}$, then

$$A \cup B = \{1, 2, 3, 4\}$$
$$A \cap B = \{2, 3\}$$

We may extend the definitions of union and intersection to more than two sets as follows. If $\mathscr{A}$ is a collection of sets, we define

$$\cup \mathscr{A} = \{x \mid x \in A \text{ for some } A \in \mathscr{A}\}$$
$$\cap \mathscr{A} = \{x \mid x \in \mathscr{A} \text{ for every } A \in \mathscr{A}\}$$

In case the family $\mathscr{A}$ is indexed by the positive integers, $\mathscr{A} = \{A_1, A_2, A_3, \ldots\}$, we write

$$\cup \mathscr{A} = \bigcup_{n=1}^{\infty} A_n \qquad \cap \mathscr{A} = \bigcap_{n=1}^{\infty} A_n$$

For example, if

$$\mathscr{A} = \{A_1, A_2, \ldots\}$$

where $A_1 = \{1\}, \quad A_2 = \{1, 2\}, \ldots, \quad A_n = \{1, 2, \ldots, n\}, \ldots$

then $$\cap \mathscr{A} = \bigcap_{n=1}^{\infty} A_n = \{1\} \qquad \cup \mathscr{A} = \bigcup_{n=1}^{\infty} A_n$$

is the set of positive integers.

Definition 1.2   The *empty set* is the set with no elements and is denoted $\varnothing$.

Definition 1.3   If every element of the set A is an element of the set B, we write $A \subset B$ and say that A is *contained in* B or that A is a *subset* of B.

We write $B \supset A$ if $A \subset B$ and say that B *contains* A.

We note that according to Definition 1.3, $A \subset B$ allows the possibility that $A = B$. It follows immediately from our definitions that $A = B$ if and only if $A \subset B$ and $B \subset A$. We call A a *proper subset* of B if $A \subset B$ and $A \neq B$.

Definition 1.4   If A and B are sets, the *difference* of A and B is the set

$$A \setminus B = \{x \mid x \in A \text{ and } x \notin B\}$$

For example, if $A = \{1, 2, 3\}$ and $B = \{2, 3, 4\}$, then $A \setminus B = \{1\}$.

If we are working with sets all of which are subsets of some particular set U, we sometimes say that U is the *universe* in which we are working. For example, if we are working with sets of integers, we could agree that the universe U is the set $\mathbf{Z}$ of all integers.

Definition 1.5   If we are working in a fixed universe U, and $A \subset U$, we write

$$A' = U \setminus A$$

The set $U \setminus A$ is called the *complement* of A *relative to* U (or just simply the *complement* of A if the universe U is understood).

For example, if U is the set of real numbers, we would state that the complement of the set of rational numbers is the set of irrational numbers.

We now prove two theorems which are known as *De Morgan's laws*. These theorems can often be used to convert a statement about unions into a statement about intersections and vice versa.

Theorem 1.6   If A and B are subsets of a universe U, then

$$(A \cup B)' = A' \cap B' \qquad (A \cap B)' = A' \cup B'$$

Proof.   We prove only the first equation leaving the second as an exercise.
Let $x \in (A \cup B)'$. Then $x \notin A \cup B$, and hence $x \notin A$ and $x \notin B$. Thus $x \in A'$ and $x \in B'$ which implies $x \in A' \cap B'$. Therefore,

$$(A \cup B)' \subset A' \cap B'$$

Next suppose $x \in A' \cap B'$. Then $x \in A'$ and $x \in B'$, and hence $x \notin A$ and $x \notin B$. Thus $x \notin A \cup B$, which implies $x \in (A \cup B)'$. Therefore,

$$A' \cap B' \subset (A \cup B)'$$

which completes the proof.   ■

In a similar way one may prove the following theorem.

Theorem 1.7 Let $\mathscr{A}$ be a collection of subsets of a universe U, and let

$$\mathscr{A}' = \{A' \mid A \in \mathscr{A}\}$$

Then $\qquad\qquad\qquad (\cup\mathscr{A})' = \cap\mathscr{A}' \qquad (\cap\mathscr{A})' = \cup\mathscr{A}'$

Exercises

1.1 Let $U = \{1, 2, 3, 4, 5, 6, 7, 8, 9, 10\}$
$A = \{1, 2, 3, 4\}$
$B = \{1, 3, 5, 7, 9\}$
Compute the following sets:
(a) $A \cup B$
(b) $A \cap B$
(c) $A\backslash B$
(d) $B\backslash A$
(e) A'
(f) B'

1.2 Prove the second equation of Theorem 1.6.

1.3 Prove Theorem 1.7.
In exercises 4 to 13, A, B, and C are all subsets of a universe U.

1.4 Prove: $A \subset \varnothing$ if and only if $A = \varnothing$

1.5 Prove: If $A \subset B$ and $B \subset C$, then $A \subset C$

1.6 Prove: $A \cup B = B \cup A$

1.7 Prove: $A \cup (B \cup C) = (A \cup B) \cup C$

1.8 Prove: $A \cap (B \cup C) = (A \cap B) \cup (A \cap C)$

1.9 Prove: $A \cup (B \cap C) = (A \cup B) \cap (A \cup C)$

1.10 Prove: $(A')' = A$

1.11 Prove: If $A \subset B$, then $B' \subset A'$

1.12 Prove: $A \cap A' = \varnothing$

1.13 Prove: $A \not\subset B$ if and only if $A \cap B' \neq \varnothing$ ($\not\subset$ means "is not a subset of")

1.14 Let U be a universe. Let A be a subset of U and let $\mathscr{A}$ be a collection of subsets of U. Prove:
(a) $A \cap (\cup\mathscr{A}) = \cup\{A \cap X \mid X \in \mathscr{A}\}$
(b) $A \cup (\cap\mathscr{A}) = \cap\{A \cup X \mid X \in \mathscr{A}\}$

2. Functions

The concept of a *function* is central to all of mathematics, and in this section we will give a precise definition of a function and prove several properties of functions. We begin with the concept of an *ordered pair*.

Definition 2.1 The *ordered pair* of elements a and b, written (a, b), is the set

$$(a, b) = \{\{a\}, \{a, b\}\}$$

a is called the *first element* of (a, b) and b is called the *second element* of (a, b).

The crucial property of an ordered pair is stated in the next theorem. The theorem states that two ordered pairs are identical if and only if they have the same first elements and the same second elements.

Theorem 2.2 Let (a, b) and (c, d) be ordered pairs. Then $(a, b) = (c, d)$ if and only if $a = c$ and $b = d$.

Proof. Suppose $a = c$ and $b = d$. Then $\{a\} = \{c\}$ and $\{a, b\} = \{c, d\}$, and therefore $(a, b) = (c, d)$.

Conversely, suppose $(a, b) = (c, d)$. Then

$$\{\{a\}, \{a, b\}\} = \{\{c\}, \{c, d\}\} \tag{2.1}$$

First, we consider the case that $a = b$. Now

$$(a, b) = \{\{a\}, \{a, b\}\} = \{\{a\}, \{a\}\} = \{\{a\}\}$$

and so
$$\{\{c\}, \{c, d\}\} = \{\{a\}\}$$

Thus $\{a\} = \{c\} = \{c, d\}$, and therefore $a = b = c = d$.

Now suppose $a \neq b$. From equation (2.1) we see that either $\{c\} = \{a\}$ or $\{c\} = \{a, b\}$. Since $a \neq b$, we must have $\{c\} = \{a\}$, which implies that $a = c$. Again using equation (2.1), we have that either $\{a, b\} = \{c\}$ or $\{a, b\} = \{c, d\}$. If $\{a, b\} = \{c\}$, then $a = b = c$, which is not the case, and thus $\{a, b\} = \{c, d\}$. It follows that $b = c$ or $b = d$, but if $b = c$, we would have the contradiction $b = c = a$. Therefore $b = d$, and we have established the theorem. ∎

Definition 2.3 If X and Y are sets, the *Cartesian product* of X and Y denoted $X \times Y$ is the set

$$X \times Y = \{(x, y) \mid x \in X \text{ and } y \in Y\}$$

Definition 2.4 Let X and Y be sets. A *function* from X into Y is a subset f of $X \times Y$ satisfying

(i) If (x, y) and (x, y') belong to f, then $y = y'$.
(ii) If $x \in X$, then $(x, y) \in f$ for some $y \in Y$.

If f is a function from X into Y, we will write $f : X \to Y$.

The crucial property of a function from X into Y is that with *each* [2.4(ii)] element x in X there is associated a *unique* [2.4(i)] element y in Y.

Definition 2.5 Let f be a function from X into Y. Let $A \subset X$ and $B \subset Y$.

(i) X is called the *domain* of f. Y is called the *codomain* of f.

(ii) If $(x, y) \in f$, we write $y = f(x)$ and call y the *(direct) image* of x under f.

(iii) The *range* of f is the set

$$\{ f(x) \mid x \in X \}$$

(iv) The *image* of A under f is the set $f(A) = \{ f(x) \mid x \in A \}$.

(v) The *inverse image* of B under f is the set

$$f^{-1}(B) = \{ x \mid f(x) \in B \}$$

(vi) f is *onto* Y if $f(X) = Y$.

(vii) f is *one to one* if $f(x) = f(x')$ always implies that $x = x'$ for $x, x' \in X$.

If f is a one-to-one function from X into Y, we may define the *inverse function* to f, denoted f^{-1}, from the range of f onto X by the rule

$$(y, x) \in f^{-1} \text{ if and only if } (x, y) \in f$$

Notice that the *function* f^{-1} is defined only if f is one to one but that $f^{-1}(B)$ is defined for an arbitrary function f and for all sets $B \subset Y$.

If $g \colon X \to Y$ and $f \colon Y \to Z$, we define the *composition* $f \circ g \colon X \to Z$ by the rule

$$(f \circ g)(x) = f(g(x)) \qquad \text{for each } x \in X$$

If $f \colon X \to Y$ and $A \subset X$, we define the function

$$f \mid A = \{ (x, y) \in f \mid x \in A \}$$

$f \mid A$ is called the *restriction* of f to A, and f is said to an *extension* of $f \mid A$ to X.

We illustrate the preceding definitions with an example.

Example. Let

$$f = \{(1, 1), (2, 1), (3, 4)\}, \quad A = \{1, 2\}, \quad B = \{1\}$$

The domain of f is $\{1, 2, 3\}$ and the range of f is $\{1, 4\}$. The image of A under f is the set $f(A) = \{1\}$. The inverse image of B under f is the set $f^{-1}(B) = \{1, 2\}$. If we let $Y = \{1, 4\}$, f is onto Y. The function f is not one to one. The restriction of f to A is the function

$$f \mid A = \{(1, 1), (2, 1)\}$$

Let $\qquad\qquad\qquad\quad g = \{(1, 1), (2, 3)\}$

Then g is one to one and the inverse function to g is the function

$$g^{-1} = \{(1, 1), (3, 2)\}$$

The composition $f \circ g$ is the function

$$f \circ g = \{(1, 1), (2, 4)\}$$

The next theorem establishes several properties of inverse and direct images.

Theorem 2.6 Let f be a function from X into Y. Let $\mathscr{A}$ be a collection of subsets of X, and let $\mathscr{C}$ be a collection of subsets of Y. Let $C \subset Y$.
Then

(i) $f(\cup\mathscr{A}) = \cup\{f(A) \mid A \in \mathscr{A}\}$

(ii) $f^{-1}(\cup\mathscr{C}) = \cup\{f^{-1}(C) \mid C \in \mathscr{C}\}$

(iii) $f^{-1}(\cap\mathscr{C}) = \cap\{f^{-1}(C) \mid C \in \mathscr{C}\}$

(iv) $f^{-1}(C') = [f^{-1}(C)]'$

Proof. We prove part (ii) only. Let $x \in f^{-1}(\cup\mathscr{C})$. Then $f(x) \in \cup\mathscr{C}$, and thus $f(x) \in C_1$ for some $C_1 \in \mathscr{C}$. Thus $x \in f^{-1}(C_1)$, and therefore $x \in \cup\{f^{-1}(C): C \in \mathscr{C}\}$. We have proved that

$$f^{-1}(\cup\mathscr{C}) \subset \cup\{f^{-1}(C) \mid C \in \mathscr{C}\}$$

Now let $x \in \cup\{f^{-1}(C): C \in \mathscr{C}\}$, then $x \in f^{-1}(C_1)$ for some C_1 in $\mathscr{C}$. Therefore $f(x) \in C_1$, and hence $f(x) \in \cup\mathscr{C}$. It follows that $x \in f^{-1}(\cup\mathscr{C})$. We have proved that

$$\cup\{f^{-1}(C) \mid C \in \mathscr{C}\} \subset f^{-1}(\cup\mathscr{C})$$

and part (ii) now follows. ∎

In case $\mathscr{A} = \{A, B\}$ and $\mathscr{C} = \{C, D\}$ the conclusions of Theorem 2.6 may be written

$$f(A \cup B) = f(A) \cup f(B)$$
$$f^{-1}(C \cup D) = f^{-1}(C) \cup f^{-1}(D)$$
$$f^{-1}(C \cap D) = f^{-1}(C) \cap f^{-1}(D)$$

It is not true that in general one has $f(A \cap B) = f(A) \cap f(B)$. (Verify.)

Exercises

2.1 Let $g = \{(1, 2), (2, 2), (3, 1), (4, 4)\}$
$f = \{(1, 5), (2, 7), (3, 9), (4, 17)\}$
$A = \{1, 2\}$
Compute
(a) The domain of g

    (b)   The range of g

    (c) $g(A)$

    (d) $g^{-1}(A)$

    (e) $f \circ g$

    (f) f^{-1}

2.2 Let $h\colon X \to Y$, $g\colon Y \to Z$, and $f\colon Z \to W$. Prove that $(f \circ g) \circ h = f \circ (g \circ h)$.

2.3 Let $f\colon X \to Y$ and $A \subset X$ and $B \subset Y$. Prove

    (a) $f(f^{-1}(B)) \subset B$

    (b) $A \subset f^{-1}(f(A))$

2.4 Prove Theorem 2.6 (i), (iii), and (iv).

2.5 Let $g\colon X \to Y$ and $f\colon Y \to Z$.

    (a) Prove that if g and f are one-to-one functions, then $f \circ g$ is a one-to-one function and $(f \circ g)^{-1} = g^{-1} \circ f^{-1}$.

    (b) Prove that if g and f are onto functions, then $f \circ g$ is an onto function.

    (c) Prove that if $A \subset Z$, then $(f \circ g)^{-1}(A) = g^{-1}(f^{-1}(A))$.

2.6 Let $f\colon X \to Y$. Prove that f is a one-to-one function if and only if

$$f(A \cap B) = f(A) \cap f(B)$$

for all subsets A and B of X.

II

The Real Number System

The central topic of study in this text is the real number system. We will define the real numbers by specifying which axioms or rules the real numbers are assumed to satisfy. In an appropriate theory of sets, one may construct a number system which satisfies these axioms [see Kelley (1955) and Landau (1960)]. It can be shown that these axioms determine the real numbers (see Exercise 7.9).

3. The Algebraic Axioms of the Real Numbers

We begin with a definition.

Definition 3.1 A *binary operation* on a set X is a function from $X \times X$ into X.

Intuitively, a binary operation on a set X is a rule which associates with each ordered pair of elements of X a unique element of X. Binary operations are often written $+$, $\cdot$, or $*$, and the value of the function at an ordered pair (x, y) is usually written $x + y$, $x \cdot y$, or $x*y$. The particular binary operations with which we shall be immediately concerned are addition and multiplication of real numbers. Addition or multiplication of real numbers associates with each ordered pair of real numbers (a, b) another real number, namely the sum of a and b or the product of a and b.

Definition 3.2 The *real numbers* **R** is a set of objects satisfying Axioms 1 to 13 as listed in the following:

Axiom 1.   There is a binary operation called *addition* and denoted $+$ such that if x and y are real numbers, $x + y$ is a real number.

Axiom 2.   Addition is associative.

$$(x + y) + z = x + (y + z)$$

for all $x, y, z \in \mathbf{R}$.

Axiom 3.   Addition is commutative.

$$x + y = y + x$$

for all $x, y \in \mathbf{R}$.

Axiom 4.   An additive identity exists. There exists a real number denoted 0 which satisfies

$$x + 0 = x = 0 + x$$

for all $x \in \mathbf{R}$.

Axiom 5.   Additive inverses exist. For each $x \in \mathbf{R}$, there exists $y \in \mathbf{R}$ such that

$$x + y = 0 = y + x$$

The number y of Axiom 5 may be shown to be unique, and it is denoted $-x$. We define $x - z$ as $x + (-z)$ for all x and z in $\mathbf{R}$.

Any mathematical system which satisfies axioms 1 to 5 is called a *commutative* (or *abelian*) *group*. Using axioms 1 to 5 one may establish the usual rules for addition of real numbers. We give one example in the next theorem and several other additive properties are given as exercises.

Theorem 3.3   The additive identity of axiom 4 is unique, that is, if there exists $0' \in \mathbf{R}$ such that $x + 0' = x$ for all $x \in \mathbf{R}$, then $0 = 0'$.

Proof.   Suppose there exists $0' \in \mathbf{R}$ such that $x + 0' = x$ for all $x \in \mathbf{R}$. Then $0 + 0' = 0$. On the other hand, from axiom 4 we have $0' + 0 = 0'$. Since addition is commutative (axiom 3),

$$0 = 0 + 0' = 0' + 0 = 0' \qquad \blacksquare$$

We continue by giving the axioms for multiplication of real numbers.

Axiom 6.   There is a binary operation called *multiplication* and denoted $\cdot$ such that if x and y are real numbers, then $x \cdot y$ (or xy) is a real number.

Axiom 7. Multiplication is associative.

$$(xy)z = x(yz)$$

for all $x,y,z \in \mathbf{R}$.

Axiom 8. Multiplication is cummutative.

$$xy = yx$$

for all $x,y \in \mathbf{R}$.

Axiom 9. A multiplicative identity exists. There exists a real number, different from 0, denoted 1 which satisfies

$$x \cdot 1 = x = 1 \cdot x$$

for all x in $\mathbf{R}$.

Axiom 10. Multiplicative inverses exist for nonzero real numbers. For any $x \in \mathbf{R}$ with $x \neq 0$, there exists $y \in \mathbf{R}$ such that

$$xy = 1 = yx$$

The next axiom links the operations of addition and multiplication.

Axiom 11. Multiplication distributes over addition.

$$x(y + z) = xy + xz \qquad (y + z)x = yx + zx$$

for all $x,y,z \in \mathbf{R}$.

As in the case of addition one may show that 1 and multiplicative inverses are unique. The multiplicative inverse of a nonzero real number x is denoted x^{-1} or $1/x$. We define $x/y = x(y^{-1})$ if $y \neq 0$. Any mathematical system satisfying axioms 1 to 11 is called a *field*. Using axioms 1 to 11 one may establish all of the well-known algebraic properties of the real numbers. We give one example; others are given in the exercises. In subsequent sections we will assume that all of the well-known algebraic properties of the real numbers have been verified from the axioms. The interested reader may consult Landau (1960) to see how this might be accomplished.

Theorem 3.4 $x \cdot 0 = 0$ for all x in $\mathbf{R}$.

Proof. $x \cdot (0 + 0) = x \cdot 0$ by axiom 4. On the other hand $x \cdot (0 + 0) = x \cdot 0 + x \cdot 0$ by axiom 11. Thus $x \cdot 0 = x \cdot 0 + x \cdot 0$. Now

$$x \cdot 0 + [-(x \cdot 0)] = (x \cdot 0 + x \cdot 0) + [-(x \cdot 0)]$$

Again using axiom 5 and axiom 2, we have

$$0 = x \cdot 0 + 0$$

and so by axiom 4 we have $0 = x \cdot 0$. ∎

Exercises

3.1 Prove that the additive inverse of axiom 5 is unique.

3.2 Prove that
  (a) $(x + y) + (z + w) = (x + (y + z)) + w$
        for all $x, y, z, w \in \mathbf{R}$,
  (b) The sum $x + y + z + w$ is independent of the manner in which the parentheses are inserted.

3.3 Prove that $-(-x) = x$ for all $x \in \mathbf{R}$.

3.4 Prove that $-(x + y) = -x - y$ for all $x, y \in \mathbf{R}$.

3.5 Let $x, y \in \mathbf{R}$. Prove that $xy = 0$ if and only if $x = 0$ or $y = 0$.

3.6 Let $x, y \in \mathbf{R}$. Prove that if $xy = xz$ and $x \neq 0$, then $y = z$.

3.7 Prove that $-(xy) = x(-y) = (-x)y$ for all $x, y \in \mathbf{R}$.

3.8 Prove that $(-1)x = -x$ for all $x \in \mathbf{R}$.

4. The Order Axiom of the Real Numbers

In this section we give the order axiom of the real numbers and derive several useful results.

Axiom 12.   There is a subset P of $\mathbf{R}$ called the *positive real numbers* satisfying

 (i) If x and y are in P, then $x + y$ and xy are in P.
 (ii) If x is in $\mathbf{R}$, exactly one of the following statements is true:

$$x \in P \quad \text{or} \quad x = 0 \quad \text{or} \quad -x \in P.$$

Using axiom 12 we can define the usual notation for order.

Definition 4.1 Let x and y be real numbers.

 (i) x is *negative* if $-x$ is positive.
 (ii) $x > y$ means $x - y$ is positive.
 (iii) $x \geq y$ means $x > y$ or $x = y$.
 (iv) $x < y$ means $y > x$.
 (v) $x \leq y$ means $y \geq x$.

The inequality $x > y$ ($x < y$) is read "x is greater (less) than y" and the inequality $x \geq y$ ($x \leq y$) is read "x is greater (less) than or equal to y." Several order properties of the real numbers are given in the next theorem.

Theorem 4.2

 (i) $1 > 0$.
 (ii) If $x > y$ and $y > z$, then $x > z$, $x,y,z \in \mathbf{R}$.
 (iii) If $x > y$, then $x + z > y + z$, $x,y,z \in \mathbf{R}$.
 (iv) If $x > y$ and $z > 0$, then $xz > yz$, $x,y,z \in \mathbf{R}$.
 (v) If $x > y$ and $z < 0$, then $xz < yz$, $x,y,z \in \mathbf{R}$.

Proof. We prove parts (i) and (ii) leaving the others as exercises. By axiom 12, exactly one of the following statements is true

$$1 \in P \quad \text{or} \quad 1 = 0 \quad \text{or} \quad -1 \in P$$

By axiom 9, $1 \neq 0$. Suppose $-1 \in P$. Then by axiom 12, $(-1)(-1) = 1 \in P$ (see Exercise 3.8) and hence $1 \in P$ and $-1 \in P$ which contradicts axiom 12. Therefore $1 \in P$ and (i) holds.

If $x > y$ and $y > z$, then $x - y$, $y - z \in P$ by Definition 4.1. By axiom 12, $x - z = (x - y) + (y - z) \in P$ and thus $x > z$. ∎

We will employ the following notation throughout the remainder of the text.

Definition 4.3   Let $a,b \in \mathbf{R}$ with $a < b$. We define

$$(a, b) = \{x \in \mathbf{R} \mid a < x < b\}$$
$$[a, b] = \{x \in \mathbf{R} \mid a \leq x \leq b\}$$
$$[a, b) = \{x \in \mathbf{R} \mid a \leq x < b\}$$
$$(a, b] = \{x \in \mathbf{R} \mid a < x \leq b\}$$
$$(-\infty, a) = \{x \in \mathbf{R} \mid x < a\}$$
$$(-\infty, a] = \{x \in \mathbf{R} \mid x \leq a\}$$
$$[a, \infty) = \{x \in \mathbf{R} \mid a \leq x\}$$
$$(a, \infty) = \{x \in \mathbf{R} \mid a < x\}$$
$$(-\infty, \infty) = \mathbf{R}$$

We call (c, d) an *open interval*; $[c, d]$ a *closed interval*; and either $[c, d)$ or $(c, d]$ a *half-open interval*, where c or d are possibly $\pm\infty$.

We conclude this section by defining the absolute value of a real number and deriving several results concerning absolute value.

Definition 4.4   Let x be a real number. We define

$$|x| = \begin{cases} x & \text{if } x \geq 0 \\ -x & \text{if } x < 0 \end{cases}$$

We call $|x|$ the *absolute value* of x.

Theorem 4.5

(i) Let $\varepsilon > 0$. Then $|x| < \varepsilon$ if and only if $-\varepsilon < x < \varepsilon$ and $|x| \leq \varepsilon$ if and only if $-\varepsilon \leq x \leq \varepsilon$.
(ii) $x \leq |x|$ for all $x \in \mathbf{R}$.
(iii) $|xy| = |x|\,|y|$ for all $x, y \in \mathbf{R}$.
(iv) $|x + y| \leq |x| + |y|$ for all $x, y \in \mathbf{R}$.

Proof.   We prove only (ii) and (iv) leaving the others as exercises.

If $x \geq 0$, then $x = |x| \leq |x|$. If $x < 0$, then $x = -|x| < |x|$. In either case $x \leq |x|$.

If $x + y \geq 0$, then $|x + y| = x + y \leq |x| + |y|$ by (ii). If $x + y < 0$, then $|x + y| = -(x + y) = -x - y \leq |x| + |y|$ by (ii). ∎

Exercises

4.1   Prove parts (iii), (iv), and (v) of Theorem 4.2.
4.2   Prove parts (i) and (iii) of Theorem 4.5.
4.3   Prove that if $x \leq y$ and $y \leq x$, then $x = y$, $x, y \in \mathbf{R}$.
4.4   Prove that if $xy > 0$, then either $x > 0$ and $y > 0$ or $x < 0$ and $y < 0$, $x, y \in \mathbf{R}$.
4.5   Prove that if $x > 0$, then $1/x > 0$, $x \in \mathbf{R}$.
4.6   Prove that $x^2 > 0$ for all $x \in \mathbf{R}$ with $x \neq 0$.
4.7   Prove that $x^2 + y^2 \geq 2xy$ for all $x, y \in \mathbf{R}$.
4.8   Prove that if $0 < xy$ and $x < y$, then $1/y < 1/x$, $x, y \in \mathbf{R}$.
4.9   Prove that if $x \leq y + \varepsilon$ for every $\varepsilon > 0$, then $x \leq y$, $x, y \in \mathbf{R}$.
4.10 Prove that if $x + y > z$ and $w > x$, then $w + y > z$, $x, y, z, w \in \mathbf{R}$.

5. The Least-Upper-Bound Axiom

Before giving the final axiom for the real numbers we must make some definitions.

Definition 5.1   A nonempty subset X of $\mathbf{R}$ is said to be *bounded above (below)* if there exists a real number a such that $x \le a\ (x \ge a)$ for all $x \in X$. The number a is called an *upper (lower) bound* for X.

Definition 5.2 Let X be a nonempty subset of **R**.

A number a in **R** is said to be a *least upper bound* for X if

(i) a is an upper bound for X.

(ii) If b is an upper bound for X, then $a \le b$.

A number a in **R** is said to be a *greatest lower bound* for X if

(i) a is a lower bound for X.

(ii) If b is a lower bound for X, then $b \le a$.

Part (ii) of Definition 5.2 concerning least upper bounds may be equivalently stated as follows: if $b < a$, then b is not an upper bound for X. By Definition 5.1, this last statement is equivalent to

(ii′) If $b < a$, there exists $x \in X$ such that $b < x$.

In proving that a is a least upper bound for a set X, it is sometimes easier to use (ii′) rather than (ii). Part (ii) of Definition 5.2 concerning greatest lower bounds may be rephrased in a similar way.

The next theorem states that if a set has a least upper bound, it is unique.

Theorem 5.3 Let X be a subset of **R**. If a and b are least upper bounds for X, then $a = b$.

Proof. By Definition 5.2, since a is a least upper bound for X and b is an upper bound for X, we have $a \le b$. Similarly, $b \le a$, and thus $a = b$. ∎

Theorem 5.3 (and ·the corresponding result for greatest lower bounds) allows us to speak of *the* least upper bound (or *the* greatest lower bound) and justifies the following notation. If a is the least upper bound of a set X, we let $a = \text{lub } X$. Similarly, if b is the greatest lower bound of a set X, we let $b = \text{glb } X$. If X is finite set, we also denote lub X by max X, and we denote glb X by min X.

For example,

$$\text{lub } (0, 1) = 1 = \text{lub } [0, 1]$$

and

$$\text{glb } (0, 1) = 0 = \text{glb } [0, 1]$$

The least upper bound of a set X is also called the *supremum* of X, and the greatest lower bound of a set X is also called the *infimum* of X. The notation lub $X = \sup X$ and glb $X = \inf X$ is also used.

Axiom 13. A nonempty subset of real numbers which is bounded above has a least upper bound.

Figure 5.1

Axiom 13 is called the *completeness axiom* for the real numbers. The real numbers are complete in the sense that there are no "holes" in the real line. Informally, if there were a hole in the real line (see Figure 5.1), the set of numbers to the left of the hole would have no least upper bound.

The least-upper-bound axiom is the basis of many deep theorems in analysis concerning the real numbers. Many theorems about the real numbers (for example, Theorems 16.2 and 18.1) which involve the existence of a number with special properties ultimately rest on this axiom.

Using axiom 13, we are able to prove the existence of greatest lower bounds.

Theorem 5.4 A nonempty subset of real numbers which is bounded below has a greatest lower bound.

Proof. Let X be a nonempty subset of real numbers which is bounded below, and let Y be the set of lower bounds for X. Let $c \in X$. Then $y \leq c$ for $y \in Y$. Thus Y is bounded above, and by the least-upper-bound axiom Y has a least upper bound a. We will show that a is the greatest lower bound of X.

Let $x \in X$. Then $y \leq x$ for all $y \in Y$, and thus x is an upper bound for Y. Since a is the least upper bound of Y, we have $a \leq x$. Therefore, a is a lower bound for X.

Let b any lower bound for X. Then $b \in Y$, and hence $b \leq a$. By Definition 5.2, a is the greatest lower bound of X. ∎

Exercises

5.1 Let X be a set of real numbers with least upper bound a. Prove that if $\varepsilon > 0$, there exists $x \in X$ such that $a - \varepsilon < x \leq a$.

5.2 Prove that the greatest lower bound of a set of real numbers is unique.

5.3 Give another proof of Theorem 5.4 by considering the least upper bound of the set

$$-X = \{-x \mid x \in X\}$$

5.4 Show that if X is a nonempty subset of real numbers which is bounded above then

$$\text{lub } X = -\text{glb}\,(-X).$$

5.5 Let X and Y be nonempty subsets of real numbers such that $X \subset Y$ and Y is bounded above. Prove that

$$\text{lub } X \leq \text{lub } Y$$

5.6 Let X be a set of real numbers with least upper bound a. Let $t \geq 0$. Prove that ta is the least upper bound of the set

$$tX = \{tx \mid x \in X\}$$

State and prove an analogous result if $t < 0$.

5.7 Let X and Y be sets of real numbers with least upper bounds a and b, respectively. Prove that $a + b$ is the least upper bound of the set

$$X + Y = \{x + y \mid x \in X, y \in Y\}$$

5.8* If f is a real-valued function on (a, b) and $c \in (a, b)$, we say that f is *strictly increasing at* c if there exists $\delta > 0$ such that if $c - \delta < x < c$, then $f(x) < f(c)$ and if $c < x < c + \delta$, then $f(c) < f(x)$. We say that f is *strictly increasing on* (a, b) if whenever $x, y \in (a, b)$ with $x < y$, we have $f(x) < f(y)$. Prove that if f is strictly increasing at each point of (a, b), then f is strictly increasing on (a, b).

6. The Set of Positive Integers

We now isolate that subset of the real numbers known as the *positive integers*. We will define the positive integers as the smallest subset **P** of **R** having the properties that $1 \in \mathbf{P}$ and if $n \in \mathbf{P}$, then $n + 1 \in \mathbf{P}$.

Definition 6.1 A subset X of **R** is said to be a *successor set*

(i) If $1 \in X$,
(ii) If $n \in X$, then $n + 1 \in X$.

Since **R** itself is a successor set, successor sets exist.

Lemma 6.2 If $\mathscr{A}$ is any nonempty collection of successor sets, then $\cap \mathscr{A}$ is a successor set.

Proof. Since $1 \in A$ for every $A \in \mathscr{A}$, $1 \in \cap \mathscr{A}$.

Suppose $n \in \cap \mathscr{A}$. Then $n \in A$ for every $A \in \mathscr{A}$. Since every set A in $\mathscr{A}$ is a successor set, $n + 1 \in A$ for every $A \in \mathscr{A}$, and thus $n + 1 \in \cap \mathscr{A}$. ∎

Definition 6.3 The set **P** of *positive integers* is the intersection of the family of all successor sets.

By Lemma 6.2, **P** is a successor set. **P** is the smallest successor set in the following sense: if X is a successor set, $\mathbf{P} \subset X$. This is important but fairly

---

*The asterisk indicates that the entry is discussed in the Hints to Selected Exercises at the back of the book.

easy to see. Let $x \in \mathbf{P}$. Then x belongs to every successor set, and in particular, x belongs to X.

The theorem of mathematical induction follows almost immediately from Definition 6.3.

Theorem 6.4 (Mathematical Induction) Suppose that for each positive integer n, we have a statement $S(n)$. Suppose also that

(i) $S(1)$ is true.
(ii) If $S(n)$ is true, then $S(n + 1)$ is true.

Then $S(n)$ is true for every positive integer n.

Proof. Let $G = \{n \in \mathbf{P} \mid S(n)$ is true$\}$. Then $G \subset \mathbf{P}$. On the other hand, $1 \in G$, and if $n \in G$, then $n + 1 \in G$ by (ii). Thus G is a successor set, and so $\mathbf{P} \subset G$. Therefore $G = \mathbf{P}$. ∎

As an example of a proof by induction we will establish the following theorem.

Theorem 6.5 If n is a positive integer, then $n \geq 1$.

Proof. Let $S(n)$ be the statement "$n \geq 1$." Since $1 \geq 1$, $S(1)$ is true. If $n \geq 1$, then $n + 1 > n \geq 1$; so if $S(n)$ is true, then $S(n + 1)$ is true. Therefore $S(n)$ is true for every positive integer n by Theorem 6.4. ∎

Theorem 6.6 If $m, n \in \mathbf{P}$, then $m + n \in \mathbf{P}$.

Proof. Let $S(m)$ be the statement "$m + n \in \mathbf{P}$ for all $n \in \mathbf{P}$." The term $1 + n \in \mathbf{P}$ for all $n \in \mathbf{P}$ since $\mathbf{P}$ is a successor set, and so $S(1)$ is true. Next assume $m + n \in \mathbf{P}$ for all $n \in \mathbf{P}$. Then $(m + 1) + n = (m + n) + 1 \in \mathbf{P}$ for all $n \in \mathbf{P}$ since $\mathbf{P}$ is a successor set ($m + n \in \mathbf{P}$ implies $(m + n) + 1 \in \mathbf{P}$). Thus if $S(m)$ is true, then $S(m + 1)$ is true; so by Theorem 6.4, $S(m)$ is true for all $m \in \mathbf{P}$. ∎

The next theorem we will prove is the well-ordering theorem for the positive integers (Theorem 6.10). The proof of this theorem depends on the fact that if n is a positive integer, there is no positive integer between n and $n + 1$ (Lemma 6.9), and in order to prove this result we must establish two lemmas.

Lemma 6.7 If $n \in \mathbf{P}$, then either $n - 1 = 0$ or $n - 1 \in \mathbf{P}$.

Proof. Let

$$G = \{n \in \mathbf{P} \mid n - 1 = 0 \text{ or } n - 1 \in \mathbf{P}\}.$$

We show that G is a successor set. Clearly $1 \in G$. Suppose $n \in G$. Then $(n + 1) - 1 = n \in \mathbf{P}$; so $n + 1 \in G$. Therefore $G = \mathbf{P}$. (Why?) ∎

Lemma 6.8   If $m,n \in \mathbf{P}$ and $m < n$, then $n - m \in \mathbf{P}$.

Proof.   We proceed by induction on m letting $S(m)$ be the statement of the lemma.

Suppose $1 < n$, $n \in \mathbf{P}$. By Lemma 6.7, either $n - 1 = 0$ or $n - 1 \in \mathbf{P}$. Since $n \neq 1$, $n - 1 \in \mathbf{P}$.

Assume that if $m,n \in \mathbf{P}$ and $m < n$, then $n - m \in \mathbf{P}$. Suppose $m + 1 < n$, $m,n \in \mathbf{P}$. By Lemma 6.7 and Theorem 6.5, $n - 1 \in \mathbf{P}$. Since $m < n - 1$, by induction we see that $(n - 1) - m \in \mathbf{P}$. Thus $n - (m + 1) \in \mathbf{P}$, and we have completed the inductive step. ∎

Lemma 6.9   Let n be a positive integer. No positive integer m satisfies the inequality $n < m < n + 1$.

Proof.   Suppose there exists $m \in \mathbf{P}$ such that $n < m < n + 1$. By Lemma 6.8, $m - n \in \mathbf{P}$. On the other hand, since $m < n + 1$, we have $m - n < 1$, which contradicts Theorem 6.5. ∎

We are now ready to prove the well-ordering theorem.

Theorem 6.10 (Well-Ordering Theorem)   If X is a nonvoid subset of the positive integers, then X contains a least element; that is, there exists $a \in X$ such that $a \leq x$ for all $x \in X$.

Proof.   We use induction on n, letting $S(n)$ be the statement: "If $n \in X$, then X contains a least element."

If $1 \in X$, then 1 is the least element of X by Theorem 6.5.

Now assume $S(n)$ is true and suppose $n + 1 \in X$. Since $S(n)$ is true, $X \cup \{n\}$ contains a least element m. If $m \in X$, then m is the least element of X. If $m \notin X$, then $m = n$ and $n \leq x$ for all $x \in X$. Since $n \notin X$, we have $n + 1 \leq x$ for all $x \in X$ by Lemma 6.9. In this case $n + 1$ is the least element of X. ∎

Theorem 6.11   The set of positive integers is not bounded above.

Proof.   If $\mathbf{P}$ is bounded above, then by the least-upper-bound axiom, $\mathbf{P}$ has a least upper bound a. Since $a - 1$ is not an upper bound for $\mathbf{P}$, there exists $n \in \mathbf{P}$ such that $a - 1 < n$. But then $a < n + 1$, which contradicts the assumption that a is an upper bound for $\mathbf{P}$. ∎

Corollary 6.12   The set of real numbers is Archimedean ordered; that is, if a and b are positive real numbers, there exists a positive integer n such that $a < nb$.

Proof. Since **P** is not bounded above, there exists $n \in \mathbf{P}$ such that $a/b < n$. Thus $a < nb$. ∎

Corollary 6.13 If ε is a positive real number, there exists a positive integer N such that $1/N < \varepsilon$.

Proof. Take $a = 1$, $b = \varepsilon$ in Corollary 6.12. ∎

Exercises

6.1 Prove the formula $1 + 2 + 3 + \cdots + n = n(n + 1)/2$.

6.2 Prove that if m and n are positive integers, then $m \cdot n$ is a positive integer.

6.3 Prove the binomial theorem: If a and b are real numbers and n is a positive integer, then

$$(a + b)^n = \binom{n}{0}a^n + \binom{n}{1}a^{n-1}b + \binom{n}{2}a^{n-2}b^2 + \cdots + \binom{n}{n-1}ab^{n-1} + \binom{n}{n}b^n$$

where $\binom{n}{k} = \dfrac{n!}{k!(n-k)!}$.

6.4 Prove that if X is a nonempty subset of positive integers which is bounded above, then X contains a greatest element.

6.5 Where is the flaw in the following "proof" by induction that any two positive integers are equal?
Theorem Let k and m be positive integers. If n is the maximum of k and m, then $n = k = m$.
Proof. Let $S(n)$ be the statement of the theorem. If $1 = \max\{k, m\}$, then $k = m = 1$ by Theorem 6.5. Suppose $n + 1 = \max\{k, m\}$. Then $n = \max\{k - 1, m - 1\}$ so by induction $n = k - 1 = m - 1$, and thus $n + 1 = k = m$. ∎

6.6* Prove the second form of mathematical induction: Suppose that for each positive integer n, we have a statement $S(n)$. Suppose also that
(a) $S(1)$ is true.
(b) If $S(k)$ is true for all $k < n$, then $S(n)$ is true.
Then $S(n)$ is true for all positive integers n.

6.7 Deduce Theorem 6.4 from Theorem 6.10.

6.8 Deduce Lemma 6.9 from Theorem 6.10.

7. Integers, Rationals, and Exponents

In this section we will define the integers, rational numbers, and rational exponents and collect some miscellaneous results.

Definition 7.1   The set of *integers*, denoted **Z**, is the set

$$\{0\} \cup \mathbf{P} \cup -\mathbf{P}$$

where $-\mathbf{P} = \{-n \mid n \in \mathbf{P}\}$.

It can be shown that **Z** is a commutative group (that is, **Z** satisfies axioms 1 to 5). We may define the rational numbers in terms of the integers.

Definition 7.2   The set of *rational numbers*, denoted **Q**, is the set

$$\left\{ \frac{p}{q} \mid p,q \in \mathbf{Z} \text{ and } q \neq 0 \right\}$$

It can be shown that **Q** is a field (that is, **Q** satisfies axioms 1 to 11). However we will shortly prove that $\mathbf{Q} \neq \mathbf{R}$.

It is convenient at this point to define x^n, where $x \in \mathbf{R}$ and $n \in \mathbf{Z}$.

Definition 7.3   Let $x \in \mathbf{R}$. We define

$$x^1 = x \qquad \text{and} \qquad x^{n+1} = x \cdot x^n, \qquad n \in \mathbf{P}$$

If $x \neq 0$, we define

$$x^0 = 1 \qquad \text{and} \qquad x^{-n} = \frac{1}{x^n}, \qquad n \in \mathbf{P}$$

From this definition one can deduce the laws of integer exponents (see Exercise 7.5).

We next show that there is no rational number x such that $x^2 = 2$ (Theorem 7.4), but there is a *real* number x such that $x^2 = 2$ (Theorem 7.5). These results show that **Q** is a proper subset of **R**. The fact that $\sqrt{2}$ is not rational is one reason for extending **Q** to the larger set **R**. A real number which is not rational is said to be *irrational*.

Theorem 7.4   There is no rational number r satisfying $r^2 = 2$.

Proof.   Suppose there exists $r \in \mathbf{Q}$ satisfying $r^2 = 2$. Then $r = p/q$ where $p,q \in \mathbf{Z}$ and $q \neq 0$. We may assume that not both p and q are even. Then $p^2 = 2q^2$. This implies that p^2, and hence also p is even. Thus $p = 2n$ for some $n \in \mathbf{Z}$. Now $4n^2 = 2q^2$; so $2n^2 = q^2$, and hence q is also even. This contradiction establishes the theorem.   ∎

We now prove the existence of nth roots in the set of real numbers. The idea of the proof is that the nth root of a should be the largest real number b such that $b^n \leq a$. The least-upper-bound axiom will guarantee the existence of such a number.

Theorem 7.5   If a is a nonnegative real number and n is a positive integer, there exists a real number $b \geq 0$ such that $b^n = a$.

Proof. Let

$$X = \{x \in \mathbf{R} \mid x \geq 0 \text{ and } x^n \leq a\}$$

X is nonempty since $0 \in X$.

We argue by contradiction to show that X is bounded above by $a + 1$. Suppose there exists $x \in X$ such that $x > a + 1$. Then

$$a \geq x^n > (a + 1)^n = \sum_{k=0}^{n} \binom{n}{k} a^k \geq na$$

which is impossible; so X is bounded above. By the least-upper-bound axiom, X has a least upper bound b. We will prove that $b^n = a$. Either $b^n < a$, $b^n > a$, or $b^n = a$. We show that the first two possibilities cannot occur.

Suppose $b^n < a$, and let $\delta = a - b^n$. Choose positive integers $m_0, \ldots,$ m_{n-1} such that

$$\binom{n}{k} b^k \frac{1}{m_k^{n-k}} < \frac{\delta}{n}, \qquad k = 0, 1, \ldots, n-1$$

Let $m = \max \{m_0, \ldots, m_{n-1}\}$. Then

$$\left(b + \frac{1}{m}\right)^n = \sum_{k=0}^{n} \binom{n}{k} b^k \frac{1}{m^{n-k}}$$

$$= \sum_{k=0}^{n-1} \binom{n}{k} b^k \frac{1}{m^{n-k}} + b^n$$

$$< \sum_{k=0}^{n-1} \frac{\delta}{n} + b^n$$

$$= \delta + b^n = a$$

Therefore $b + 1/m \in X$, but $b < b + 1/m$ which is impossible; thus $a \leq b^n$. Similarly, one shows that $a < b^n$ is false, and therefore $a = b^n$. ∎

It can be shown that the number b in Theorem 7.5 is unique (Exercise 7.8).

Corollary 7.6 If a is a real number and n is an odd positive integer, there exists a real number b such that $b^n = a$.

Proof. By Theorem 7.5, there exists a real number c such that $c^n = |a|$. If $a \geq 0$, $c^n = a$, and we have the desired conclusion. If $a < 0$, let $b = -c$. Then $b^n = -c^n = a$, and again we have the conclusion. ∎

We are now in a position to define x^r, where x is real and r is rational. We will conclude our definitions of exponents in Section 17 when we define x^y, where x and y are real numbers and $x > 0$.

Definition 7.7 Let x be a nonnegative real number, and let n be a positive integer. We define $x^{1/n}$ to be the nonnegative real number y such that $y^n = x$.

If x is a real number and n is an odd positive integer, we define $x^{1/n}$ to be the real number y such that $y^n = x$.

If x is a real number and n is a positive integer, we define

$$x^{-1/n} = \frac{1}{x^{1/n}}$$

provided $x^{1/n} \neq 0$ and $x^{1/n}$ is defined.

If x is a real number and r is a rational number, $r = p/q$, where $p, q \in \mathbf{Z}$ and r is expressed in lowest terms, we define

$$x^r = (x^{1/q})^p$$

whenever $x^{1/q}$ is defined.

From this definition one can deduce the laws of rational exponents (see Exercise 7.6).

We close this chapter with two theorems which show that there are many rational and irrational numbers. The proof of the first theorem is illustrated in Figure 7.1.

Theorem 7.8 If a and b are real numbers with $a < b$, there exists a rational number r such that $a < r < b$.

Proof. First we consider the case that $b > 0$. Choose a positive integer n such that $b - a > 1/n$. Let m be the least positive integer such that $b \leq m/n$. If $m = 1$, clearly $(m - 1)/n < b$. If $m > 1$, $m - 1$ is a positive integer less than m, and thus $(m - 1)/n < b$.

Since $a - b < -1/n$ and $b \leq m/n$, we have

$$a = b + (a - b) < \frac{m}{n} - \frac{1}{n} = \frac{m - 1}{n}$$

Thus the rational number $r = (m - 1)/n$ satisfies $a < r < b$. We have established the theorem in case $b > 0$.

Suppose now that a and b are arbitrary real numbers with $a < b$. Choose a positive integer n such that $b + n > 0$. By the special case just established, there exists a rational number r' such that $a + n < r' < b + n$. Thus $r = r' - n$ is a rational number satisfying $a < r < b$. ∎

Figure 7.1

Lemma 7.9 The sum of a rational number and an irrational number is an irrational number.

Proof. Suppose the lemma is false. Then there exists a rational number r and an irrational number s such that $t = r + s$ is a rational number. Then $s = t - r$ is the difference of two rational numbers and is therefore rational. This is a contradiction. ∎

Theorem 7.10 If a and b are real numbers with $a < b$, then there is an irrational number s such that $a < s < b$.

Proof. By Theorem 7.8, there exists a rational number r_1 such that $a - \sqrt{2} < r_1 < b - \sqrt{2}$. By Theorem 7.4 and Lemma 7.9, $r = r_1 + \sqrt{2}$ is an irrational number. Since $a < r < b$ we have completed the proof. ∎

In Chapter III we will give a very different proof of Theorem 7.10.

Exercises

7.1 Prove that $\mathbf{Z}$ is a commutative group.

7.2 Prove that $\mathbf{Q}$ is a field.

7.3 Prove that there is no rational number r satisfying $r^2 = 3$.

7.4 Prove that there is no rational number r satisfying $r^3 = 2$.

7.5 Prove the laws of integer exponents:

    (a) $x^{m+n} = x^m x^n$, $x \in \mathbf{R}$, $x \neq 0$, $m, n \in \mathbf{Z}$

    (b) $x^n = \dfrac{1}{x^{-n}}$, $x \in \mathbf{R}$, $x \neq 0$, $n \in \mathbf{Z}$

    (c) $(xy)^n = x^n y^n$, $x, y \in \mathbf{R}$, $x \neq 0 \neq y$, $n \in \mathbf{Z}$

    (d) $(x^m)^n = x^{mn}$, $x \in \mathbf{R}$, $x \neq 0$, $m, n \in \mathbf{Z}$

    (e) $\left(\dfrac{x}{y}\right)^n = \dfrac{x^n}{y^n}$, $x, y \in \mathbf{R}$, $x \neq 0 \neq y$, $n \in \mathbf{Z}$

    (f) If $0 < x < y$, then $x^n < y^n$, $n \in \mathbf{P}$

    (g) If $n < m$ and $x > 1$, then $x^n < x^m$, $n, m \in \mathbf{Z}$.

7.6 (a) Do Exercise 7.5 with x and y positive and m and n rational numbers.

    (b) Discuss Exercise 7.5, where x and y are arbitrary real numbers and m and n are rational numbers.

7.7 Complete the proof of Theorem 7.5 by showing that $b^n > a$ is impossible.

7.8 Prove that if $a > 0$ and n is a positive integer there exists a unique $b > 0$ such that $b^n = a$.

7.9* Prove that the set of real numbers as defined by axioms 1 to 13 is unique in the following sense: If $\mathbf{R}'$ is a set which satisfies axioms 1 to 13, there exists a one-to-one function f from $\mathbf{R}$ onto $\mathbf{R}'$ which satisfies

$$f(x + y) = f(x) + f(y) \qquad \text{for all } x, y \in \mathbf{R}$$
$$f(xy) = f(x)f(y) \qquad \text{for all } x, y \in \mathbf{R}$$
$$f(x) < f(y) \text{ if and only if } x < y \qquad \text{for } x, y \in \mathbf{R}$$

7.10 Prove that no equilateral triangle in the plane can have all vertices with rational coordinates.

III

Set Equivalence

When do two (possibly infinite) sets have the same number of elements? In this chapter we will give a definition (Definition 8.1) which answers this question. We will show that according to our definition there are as many positive integers as there are rational numbers, but that there are more real numbers than positive integers.

8. Definitions and Examples

Suppose we are given two finite sets X and Y and are asked to compare their sizes. One way to do this is to count the number of elements in each set and compare the results. An alternative method is to pair off the elements in X with those in Y.

$$X = \{a, b, c, d\}$$
$$\updownarrow \ \updownarrow \ \updownarrow \ \updownarrow$$
$$Y = \{1, 2, 3, 4\}$$

This second method generalizes to infinite sets, and thus we make the following definition.

Definition 8.1 Let X and Y be nonempty sets. We say that X is *equivalent* to Y and write $X \approx Y$ if there exists a one-to-one function from X onto Y.

Obviously $X \approx X$. (Why?) If $X \approx Y$, then $Y \approx X$. (Why?) The relation $\approx$ is also transitive. If $X \approx Y$ and $Y \approx Z$, then $X \approx Z$. (Why?)

Sets are divided into two categories as specified in the following definition.

Definition 8.2 A set X is *finite* if either $X = \varnothing$ or $X \approx \{1, 2, \ldots, n\}$ for some positive integer n. A set which is not finite is said to be *infinite*. A set X is *countable* if either X is finite or $X \approx \mathbf{P}$. A set X is *uncountable* if X is not countable.

If a set X is countable and infinite there is a one-to-one function f from $\mathbf{P}$ onto X. Thus the elements of X may be listed

$$f(1), \quad f(2), \quad f(3), \quad \ldots$$

or as we sometimes write

$$f_1, \quad f_2, \quad f_3, \quad \ldots$$

. The set of even positive integers is countable, since the function f defined by $f(n) = 2n$ is a one-to-one function from $\mathbf{P}$ onto $\{2, 4, 6, \ldots\}$. The correspondence or listing is

$$
\begin{array}{cccc}
2 & 4 & 6 & 8 \\
\updownarrow & \updownarrow & \updownarrow & \updownarrow \quad \cdots \\
1 & 2 & 3 & 4
\end{array}
$$

The set of all integers is countable since the function f defined by

$$
f(n) = \begin{cases} -\dfrac{(n-1)}{2} & \text{if } n \text{ is odd} \\[2ex] \dfrac{n}{2} & \text{if } n \text{ is even} \end{cases}
$$

is a one-to-one function from $\mathbf{P}$ onto $\mathbf{Z}$. The correspondence or listing is

$$
\begin{array}{ccccc}
0 & 1 & -1 & 2 & -2 \\
\updownarrow & \updownarrow & \updownarrow & \updownarrow & \updownarrow \quad \cdots \\
1 & 2 & 3 & 4 & 5
\end{array}
$$

To prove that a set X is countable from the definition, one must exhibit a listing like those given above (a one-to-one function from $\mathbf{P}$ onto X). In view of the remarks preceding Definition 8.2, one may also show that a set X is countable by showing that X is equivalent to a set which is known to be countable. To prove that a set X is uncountable, one would have to show that there is no one-to-one function from $\mathbf{P}$ onto X. The usual method is to assume the existence of such a function and deduce a contradiction. A remarkable fact is that uncountable sets exist. We will see later that the real numbers are uncountable. At this point we will give a different example of an uncountable set.

Example 8.3 A real sequence is a function from $\mathbf{P}$ into $\mathbf{R}$. In this example we will denote a real sequence as

$$(a_1, a_2, \ldots) \qquad a_n \in \mathbf{R}, n = 1, 2, \ldots$$

Let

$$X = \{(a_1, a_2, \ldots) \mid a_n \in \{0, 1\}, n = 1, 2, \ldots\}$$

Thus X is the set of all real sequences which assume only the values 0 or 1. We show that X is uncountable. If X is countable, the elements of X may be listed:

$$1 \leftrightarrow (a_1^{(1)}, a_2^{(1)}, a_3^{(1)}, \ldots)$$
$$2 \leftrightarrow (a_1^{(2)}, a_2^{(2)}, a_3^{(2)}, \ldots)$$
$$\vdots$$

We will now define a sequence which is not in the list, thus deducing a contradiction. Let

$$b_n = \begin{cases} 0 & \text{if } a_n^{(n)} = 1 \\ 1 & \text{if } a_n^{(n)} = 0, n = 1, 2, \ldots \end{cases}$$

Then $(b_1, b_2, \ldots)$ is a sequence in X which is not in the list, and hence X is uncountable.

If X is a set, we let $P(X)$ denote the collection of all subsets of X. [$P(X)$ is called the *power set* of X.]

Theorem 8.4   Let X be a set. Then $X \not\approx P(X)$.

Proof.   If $X = \varnothing$, then $P(X) = \{\varnothing\}$; so $X \not\approx P(X)$. Therefore, suppose $X \neq \varnothing$ and $X \approx P(X)$. Then there exists a one-to-one function f from X onto $P(X)$. We now define a subset Y of X which is not in the range of f thus deducing a contradiction. Let

$$Y = \{x \in X \mid x \notin f(x)\}$$

If $Y \in f(X)$, there exists $y \in X$ such that $f(y) = Y$. Either $y \in Y$ or $y \notin Y$, and we examine the two possibilities.

1. If $y \in Y$, then $y \notin f(y)$. Thus $y \notin f(y) = Y$. The first case is impossible.
2. If $y \notin Y$, then $y \in f(y)$. Thus $y \in f(y) = Y$. The second case is impossible.

Thus $Y \notin f(X)$, and we have the desired contradiction. ∎

Corollary 8.5 $P(\mathbf{P})$ is uncountable.

Exercises

8.1 Prove that the set of even integers is countable.

8.2 Suppose that X and Y are equivalent sets. Prove that $P(X) \approx P(Y)$.

8.3 Prove that every infinite set has a countably infinite subset.

8.4* Prove that a set X is infinite if and only if X is equivalent to a proper subset of itself.

8.5 Let X be a set. Let S be the set of all functions from X into $\{0, 1\}$. (Such functions are called *characteristic functions*.) Prove that $S \approx P(X)$. How does this exercise connect Example 8.3 and Corollary 8.5?

9. Countable and Uncountable Sets

The following result should not be too surprising.

Theorem 9.1 Any subset of the positive integers is countable.

Proof. Let X be a subset of **P**. If X is finite, there is nothing to prove; so assume X is infinite. We define a function f from **P** to X as follows: Let $f(1)$ be the least element in X. Let $f(2)$ be the least element of $X \backslash \{f(1)\}$. Continuing in this way, having defined $f(1), f(2), \ldots, f(n)$, we let $f(n + 1)$ be the least element of $X \backslash \{f(1), f(2), \ldots, f(n)\}$. It is easy to verify that f is a one-to-one function from **P** onto X, and therefore X is countable. ■

Corollary 9.2 Any subset of a countable set is countable.

The next theorem shows that the one to one requirement need not be checked in proving that a set is countable.

Theorem 9.3 Let X be a nonempty set, then X is countable if and only if there exists a function f from **P** onto X.

Proof. If X is countable and infinite, there is nothing to prove. If X is finite, then $X \approx \{1, 2, \ldots, n\}$ for some $n \in \textbf{P}$; so $X = \{x_1, x_2, \ldots, x_n\}$. Define $f(i) = x_i$ for $i = 1, 2, \ldots, n$ and $f(j) = x_n$ for all $j > n$. Then f is a function from **P** onto X.

Suppose f is a function from **P** onto a set X. If X is finite, there is nothing to prove; so assume X is infinite. We define a function g from X onto a subset of **P** by the rule

$$g(x) = \min \{f^{-1}(x)\}, \quad \text{for each } x \in X$$

Then g is a one-to-one (check this) function from X onto $g(X)$. $g(X)$ is an infinite subset of $\mathbf{P}$; hence by Theorem 9.1, $g(X)$ is countable. Thus $X \approx g(X) \approx \mathbf{P}$ and X is countable. ∎

Suppose A_1, A_2, ... are countably infinite sets. We may list the elements of A_n as

$$a_1^{(n)}, \quad a_2^{(n)}, \quad a_3^{(n)}, \quad \ldots$$

We now find that we may list the elements of $\bigcup_{n=1}^{\infty} A_n$:

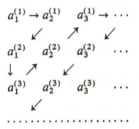

The arrows indicate that

$$1 \to a_1^{(1)}$$
$$2 \to a_2^{(1)}$$
$$3 \to a_1^{(2)}$$
$$\vdots$$

We have described a function from $\mathbf{P}$ onto $\bigcup_{n=1}^{\infty} A_n$; so by Theorem 9.3, $\bigcup_{n=1}^{\infty} A_n$ is countable. This result is often summarized by saying that a countable union of countable sets is countable. We will now give another (tricky) proof of this theorem.

Lemma 9.4   The set $\mathbf{P} \times \mathbf{P}$ is countable.

Proof.   The function $f: \mathbf{P} \times \mathbf{P} \to \mathbf{P}$ defined by

$$f(n, m) = 2^n 3^m$$

is one to one, since factorization into primes is unique. Thus $\mathbf{P} \times \mathbf{P}$ is equivalent to a subset of $\mathbf{P}$. By Theorem 9.1, $\mathbf{P} \times \mathbf{P}$ is countable. ∎

Theorem 9.5   Let A_1, A_2, ... be a countable family of countable sets. Then

$$\bigcup_{n=1}^{\infty} A_n$$

is a countable set.

Proof. Since A_n is countable, we may list the elements of A_n

$$A_n: a_1^{(n)}, \quad a_2^{(n)}, \quad a_3^{(n)}, \quad \ldots$$

(In case A_n is finite, $A_n = \{a_1^{(n)}, \ldots, a_m^{(n)}\}$, we let $a_i^{(n)} = a_m^{(n)}$ if $i \geq m$. If $A_n = \varnothing$, redefine $A_n = A_m$ for some $A_m \neq \varnothing$. If all $A_m = \varnothing$ then $\bigcup_{m=1}^{\infty} A_m = \varnothing$ is countable.)

The function f defined by

$$f(n, m) = a_n^{(m)}$$

maps $\mathbf{P} \times \mathbf{P}$ onto $\bigcup_{n=1}^{\infty} A_n$. By Lemma 9.4, $\mathbf{P} \approx \mathbf{P} \times \mathbf{P}$, and it follows that we may map $\mathbf{P}$ onto $\bigcup_{n=1}^{\infty} A_n$. By Theorem 9.3 $\bigcup_{n=1}^{\infty} A_n$ is countable. ∎

Corollary 9.6 If A and B are countable sets, then $A \cup B$ is a countable set.

Corollary 9.7 The set $\mathbf{Q}$ of rational numbers is countable.

Proof. For each positive integer n, let

$$A_n = \left\{ \ldots, \frac{-2}{n}, \frac{-1}{n}, \frac{0}{n}, \frac{1}{n}, \frac{2}{n}, \ldots \right\}$$

Then A_n is countable for each positive integer n. (Why?) By Theorem 9.5, $\mathbf{Q} = \bigcup_{n=1}^{\infty} A_n$ is countable. ∎

Theorem 9.8 Any interval of $\mathbf{R}$ is uncountable; in particular, $\mathbf{R}$ is uncountable.

Proof. Let I be an interval of $\mathbf{R}$, and suppose I is countable. Then the elements of I may be listed

$$r_1, \quad r_2, \quad r_3, \quad \ldots$$

Choose $a_1, b_1 \in \mathbf{R}$ such that $r_1 \notin (a_1, b_1) \subset I$. Choose $a_2, b_2 \in \mathbf{R}$ such that $a_1 < a_2 < b_2 < b_1$ and $r_2 \notin (a_2, b_2)$. Continue in this way so that having chosen $a_n, b_n \in \mathbf{R}$, we choose $a_{n+1}, b_{n+1} \in \mathbf{R}$ such that $a_n < a_{n+1} < b_{n+1} < b_n$ and $r_{n+1} \notin (a_{n+1}, b_{n+1})$.

The set $\{a_1, a_2, \ldots\}$ is bounded above by b; so the least upper bound r of this set exists. It follows that $a_n < r$ for all $n \in \mathbf{P}$. On the other hand, $r < b_n$ for all $n \in \mathbf{P}$ for if $b_m \leq r$ for some $m \in \mathbf{P}$, we would have

$$a_n < b_{m+1} < b_m \leq r$$

for all $n \in \mathbf{P}$. Thus b_{m+1} would be an upper bound for $\{a_1, a_2, \ldots\}$, but since $b_{m+1} < r$, this is impossible. Therefore $r \in (a_n, b_n)$ for all $n \in \mathbf{P}$. It follows that $r \neq r_n$ for every $n \in \mathbf{P}$ and $r \in I$. This contradiction establishes the theorem. ∎

Corollary 9.9   Irrational numbers exist.

Proof.   If every real number were rational, **R** would be countable by Corollary 9.7. This statement contradicts Theorem 9.8. ■

The proof of Corollary 9.9 caused considerable consternation when first given by Cantor, for it proves the existence of irrational numbers without exhibiting a *single* irrational number!

Corollary 9.10   The set of irrational numbers is uncountable.

Proof.   Suppose **Q**′ is countable. By Corollary 9.7, **Q** is countable; so by Corollary 9.6, **R** = **Q** ∪ **Q**′ is countable. This contradicts Theorem 9.8. ■

We are now in a position to give another proof of Theorem 7.10.

Theorem 9.11 (Revisited)   If a and b are real numbers with $a < b$, then there is an irrational number s such that $a < s < b$.

Proof.   If $(a, b) \subset$ **Q**, then by Corollaries 9.7 and 9.2, (a, b) is countable, and this contradicts Theorem 9.8. ■

Exercises

9.1   Prove Corollary 9.2 (give all the details).

9.2   When is the function described before Lemma 9.4 one to one?

9.3   Let A and B be sets. Prove that if A is countable and B is uncountable, then $A \cup B$ is uncountable.

9.4   Let A and B be sets such that $A \subset B$. Prove that if A is uncountable, then B is uncountable.

9.5   Deduce the fact that **P** × **P** is countable from Theorem 9.5.

9.6   Prove that if A and B are countable sets, then $A \times B$ is countable.

9.7*   Prove that the set of all polynomials of degree n with rational coefficients is countable.

9.8*   Prove that the set of all polynomials with rational coefficients is countable.

9.9   A real number r is said to be *algebraic* if r is a root of a polynomial with rational coefficients. If r is not algebraic, r is said to be *transcendental*. (It is known that e and π are transcendental.)

   (a)   Prove that every rational number and every root of a positive rational number are algebraic numbers.

   (b)   Prove that r is an algebraic number if and only if r is a root of a polynomial with integer coefficients.

(c)* Prove that the set of algebraic numbers is countable. Deduce that transcendental numbers exist.

9.10* Prove that the plane is *not* the union of a countable family of straight lines.

9.11 Prove that the collection of finite subsets of **P** is countable. Deduce that the collection of infinite subsets of **P** is uncountable.

IV

Sequences of Real Numbers

Analysis is concerned in one form or another with limits. We begin our study of real analysis with the study of limits of real sequences.

10. Limit of a Sequence

Let X be a set. A sequence of elements of X is a "list" of elements from the set X. In this chapter we will be concerned with sequences of real numbers. We first make the notion of sequence precise.

Definition 10.1 Let X be a set. A *sequence* of elements of X is a function from the set of positive integers into X.

In particular, a *real sequence* (or *sequence of real numbers*) is a function from **P** into **R**. The usual notation for a real sequence is $\{a_n\}_{n=1}^{\infty}$, where a denotes the function from **P** into **R** and a_n is the value of the function at the positive integer n. The notations $a_1, a_2, \ldots$ and $\{a_n\}$ are also used to denote a real sequence. The number a_n is called the nth *term* of the sequence $\{a_n\}_{n=1}^{\infty}$.

A sequence may be defined by giving an explicit formula for the nth term. For example, the formula

$$a_n = \frac{1}{n}$$

defines the sequence whose value at the positive integer n is $1/n$. The first three terms of this sequence are

$$a_1 = 1, \quad a_2 = \tfrac{1}{2}, \quad a_3 = \tfrac{1}{3}$$

A sequence may also be defined inductively. Thus equations

$$a_1 = 0, \quad a_2 = 1, \quad a_{n+2} = \frac{a_n + a_{n+1}}{2}, \qquad n = 1, 2, \ldots$$

define the sequence whose first six terms are

$$0, \quad 1, \quad \tfrac{1}{2}, \quad \tfrac{3}{4}, \quad \tfrac{5}{8}, \quad \tfrac{11}{16}$$

Consider the sequence $\{a_n\}_{n=1}^{\infty}$ whose nth term is defined by the formula

$$a_n = \frac{n}{n+1}$$

The first four terms of this sequence are

$$\tfrac{1}{2}, \quad \tfrac{2}{3}, \quad \tfrac{3}{4}, \quad \tfrac{4}{5}$$

The terms corresponding to $n = 100, 101, 102$ are

$$\tfrac{100}{101}, \quad \tfrac{101}{102}, \quad \tfrac{102}{103}$$

which are close to 1. For example, $\frac{100}{101}$ differs from 1 by only $\frac{1}{101}$. It is clear that $n/(n + 1)$ is "close to" 1 "for all large positive integers n." For this reason we say that the sequence $\{n/(n + 1)\}_{n=1}^{\infty}$ has limit 1. In general, we say that a sequence $\{a_n\}_{n=1}^{\infty}$ has limit L if a_n is "close to" L "for all large positive integers n." To define the limit of a sequence, we need to make the concepts "close to" and "for all large positive integers n" precise.

Since $|a_n - L|$ is the distance between a_n and L, we would agree that a_n is "close to" L if $|a_n - L|$ is small. How small? Given any positive number ε (no matter how small) we wish to make $|a_n - L| < \varepsilon$ for all large positive integers n. If we can do this, we can make a_n as close to L as we wish simply by making ε small enough. "For all large positive integers n" means "for all n greater than (or equal to) some fixed positive integer N." Thus the condition we are seeking is that if $\varepsilon > 0$, there exists a positive integer N such that $|a_n - L| < \varepsilon$, for all $n \geq N$. The positive integer N may depend on the ε which we are given. We would expect that as ε is taken smaller, we would have to choose N larger.

Definition 10.2 Let $\{a_n\}_{n=1}^{\infty}$ be a sequence of real numbers. We say that $\{a_n\}_{n=1}^{\infty}$ has *limit* $L \in \mathbf{R}$ if for every $\varepsilon > 0$, there exists a positive integer N, such that if $n \geq N$, then

$$|a_n - L| < \varepsilon$$

Let us return to the sequence $\{n/(n + 1)\}_{n=1}^{\infty}$, and prove that $\{n/(n + 1)\}_{n=1}^{\infty}$ has limit 1. If $\varepsilon > 0$, we are required to show that for some positive integer N

$$\left| \frac{n}{n+1} - 1 \right| < \varepsilon$$

if $n \geq N$. Now

$$\left| \frac{n}{n+1} - 1 \right| = \frac{1}{n+1}$$

and hence if we choose a positive integer N such that $1/(N+1) < \varepsilon$, then if $n \geq N$,

$$\left| \frac{n}{n+1} - 1 \right| = \frac{1}{n+1} \leq \frac{1}{N+1} < \varepsilon$$

and therefore $\{n/(n+1)\}_{n=1}^{\infty}$ has limit 1.

In exactly the same way, one can show that the sequence $\{1/n\}_{n=1}^{\infty}$, has limit 0.

Since the inequality $|a_n - L| < \varepsilon$ in Definition 10.2 is equivalent to $L - \varepsilon < a_n < L + \varepsilon$, we may restate Definition 10.2 as follows. The sequence $\{a_n\}_{n=1}^{\infty}$ has limit L if for every $\varepsilon > 0$, there exists a positive integer N such that if $n \geq N$, then a_n is in the open interval $(L - \varepsilon, L + \varepsilon)$. See Figure 10.1. Figure 10.1 provides a geometric interpretation of the statement "a_n is close to L for all large positive integers n."

The next theorem allows us to speak of *the* limit of a sequence. The proof of this theorem is illustrated in Figure 10.2.

Theorem 10.3 The limit of a sequence is unique. That is, if a sequence $\{a_n\}_{n=1}^{\infty}$ has limit L and limit L', then $L = L'$.

Proof. We argue by contradiction. Suppose $L \neq L'$. We assume that $L < L'$. (The argument is similar if $L' < L$.) Now $\varepsilon = (L' - L)/2 > 0$; so there exists a positive integer N_1 such that if $n \geq N_1$, then

$$|a_n - L| < \frac{L' - L}{2}$$

There exists a positive integer N_2 such that if $n \geq N_2$, then

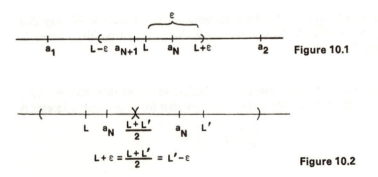

Figure 10.1

Figure 10.2

$$|a_n - L'| < \frac{L' - L}{2}$$

If $N = \max \{N_1, N_2\}$, then

$$a_N - L < \frac{L' - L}{2} \quad \text{and} \quad -\frac{L' - L}{2} < a_N - L'$$

and therefore

$$\frac{L' + L}{2} < a_N < \frac{L' + L}{2}$$

which is impossible. ∎

If L is *the* limit of a sequence $\{a_n\}$, we write

$$\lim_{n \to \infty} a_n = L$$

which is read "the limit of a_n as n approaches infinity is L."

Let $\{a_n\}_{n=1}^{\infty}$ be a sequence such that $\lim_{n \to \infty} a_n = L$, and let $\varepsilon > 0$. Then there exists a positive integer N such that if $n \geq N$, then $|a_n - L| < \varepsilon$. The inequality $|a_n - L| \geq \varepsilon$ can hold for at most finitely many positive integers n (namely $n = 1, 2, \ldots, N - 1$). Thus a sequence $\{a_n\}_{n=1}^{\infty}$ has limit L if and only if for every $\varepsilon > 0$, $|a_n - L| \geq \varepsilon$ for finitely many positive integers n. Therefore, a sequence $\{a_n\}_{n=1}^{\infty}$ does not have limit L if for some $\varepsilon > 0$, $|a_n - L| \geq \varepsilon$ for infinitely many positive integers n. As an example, we prove that the sequence $\{n\}_{n=1}^{\infty}$ does not have limit L for any real number L. Let $\varepsilon = 1$. Since $|n - L| \geq 1$ for infinitely many positive integers n, $\{n\}_{n=1}^{\infty}$ does not converge to L for any real number L.

Exercises

In Exercises 10.1 to 10.8 prove the limit using Definition 10.2.

10.1 $\displaystyle \lim_{n \to \infty} \frac{1}{n} = 0$

10.2 $\displaystyle \lim_{n \to \infty} \frac{1}{2n} = 0$

10.3 $\displaystyle \lim_{n \to \infty} \frac{1}{n + 2} = 0$

10.4 $\displaystyle \lim_{n \to \infty} \left(2 - \frac{1}{n}\right) = 2$

10.5 $\displaystyle \lim_{n \to \infty} \frac{n}{n + 2} = 1$

10.6 $\displaystyle \lim_{n \to \infty} \frac{2n}{n + 2} = 2$

10.7 $\displaystyle\lim_{n\to\infty}\frac{(-1)^n}{n}=0$

10.8 $\displaystyle\lim_{n\to\infty}\frac{1}{n^2}=0$

10.9 Prove that the sequence $\{(-1)^n\}_{n=1}^{\infty}$ has no limit.

10.10 Prove that the sequence $\{(n+(1/n))\}_{n=1}^{\infty}$ has no limit.

10.11 Let $\{a_n\}_{n=1}^{\infty}$ be a sequence with limit L. Suppose that $\{b_n\}_{n=1}^{\infty}$ is a sequence such that for some positive integer N we have

$$b_n = a_n \qquad \text{for every } n \geq N.$$

Prove that $\displaystyle\lim_{n\to\infty} b_n = L$.

10.12 Let $\{a_n\}_{n=1}^{\infty}$ be a sequence such that $\lim_{n\to\infty} a_n = L$. Prove that

$$\lim_{n\to\infty}|a_n| = |L|$$

10.13 What is wrong with the following "proof" of Theorem 10.3? Suppose the sequence $\{a_n\}_{n=1}^{\infty}$ has limits L and L'. Then

$$L = \lim_{n\to\infty} a_n \quad \text{and} \quad L' = \lim_{n\to\infty} a_n$$

so $L = \displaystyle\lim_{n\to\infty} a_n = L'$ ■

11. Subsequences

Consider the sequence $\{a_n\}_{n=1}^{\infty}$ defined by $a_n = 1/n$ which begins

$$1, \quad \tfrac{1}{2}, \quad \tfrac{1}{3}, \quad \tfrac{1}{4}, \quad \tfrac{1}{5}, \quad \cdots$$

If we were to cross out every other term our sequence would become

$$1, \quad \tfrac{1}{3}, \quad \tfrac{1}{5}, \quad \tfrac{1}{7}, \quad \cdots$$

The resulting sequence is called a *subsequence* of the original sequence. The first term of this subsequence is the first term of the original sequence; the second term of this subsequence is the third term of the original sequence; the third term of this subsequence is the fifth term of the original sequence; etc. Thus the subsequence is defined by

$$1 \to 1 \to a_1 = 1$$
$$2 \to 3 \to a_3 = \tfrac{1}{3}$$
$$3 \to 5 \to a_5 = \tfrac{1}{5}$$
$$\cdots\cdots\cdots\cdots\cdots$$
$$n \to 2n-1 \to a_{2n-1} = \frac{1}{2n-1}$$
$$\cdots\cdots\cdots\cdots\cdots\cdots\cdots\cdots$$

This subsequence is defined by composing the functions $n \to 2n - 1$ and $n \to a_n$. We now define subsequences of arbitrary sequences.

Definition 11.1 Let $\{a_n\}_{n=1}^{\infty}$ be a sequence. Let f be a strictly increasing function from **P** into **P**. The sequence $\{a_{f(n)}\}_{n=1}^{\infty}$ is called a *subsequence* of the sequence $\{a_n\}_{n=1}^{\infty}$. [The function f is *strictly increasing* if $f(m) < f(n)$ whenever $m < n$.]

In our example, the function f of Definition 11.1 is defined by $f(n) = 2n - 1$. The function f of Definition 11.1 specifies which terms of the original sequence we are to keep to form the subsequence.

We now prove the "obvious" fact that if a sequence $\{a_n\}_{n=1}^{\infty}$ has limit L, then any subsequence of $\{a_n\}_{n=1}^{\infty}$ also has limit L.

Theorem 11.2 Let $\{a_n\}_{n=1}^{\infty}$ be a sequence with limit L. Then any subsequence of $\{a_n\}_{n=1}^{\infty}$ has limit L.

Proof. Let $\{a_{f(n)}\}_{n=1}^{\infty}$ be a subsequence of $\{a_n\}_{n=1}^{\infty}$. Let $\varepsilon > 0$. There exists a positive integer N such that if $n \geq N$, then

$$|a_n - L| < \varepsilon$$

If $n \geq N$, then $f(n) \geq f(N) \geq N$ (verify), and therefore

$$|a_{f(n)} - L| < \varepsilon \qquad \blacksquare$$

Examples. The sequences $\{1/2^n\}_{n=1}^{\infty}$ and $\{1/n!\}_{n=1}^{\infty}$ are subsequences of $\{1/n\}_{n=1}^{\infty}$ and $\lim_{n \to \infty} 1/n = 0$, thus

$$\lim_{n \to \infty} \frac{1}{2^n} = 0 = \lim_{n \to \infty} \frac{1}{n!}$$

The sequence $\{(-1)^n\}_{n=1}^{\infty}$ has no limit because the subsequence $\{(-1)^{2n}\}_{n=1}^{\infty}$ has limit 1 and the subsequence $\{(-1)^{2n-1}\}_{n=1}^{\infty}$ has limit -1.

We have used the notation $\{a_{f(n)}\}_{n=1}^{\infty}$ to denote a subsequence of a sequence $\{a_n\}_{n=1}^{\infty}$. The standard (and more confusing) notation for a subsequence of $\{a_n\}_{n=1}^{\infty}$ is $\{a_{n_k}\}_{k=1}^{\infty}$. Here, n denotes the function f, and k denotes an element of the domain **P**.

Exercises

In Exercises 11.1 to 11.4 a sequence is given and then a subsequence of the given sequence. Find a formula for the function f of Definition 11.1; give a formula for the sequence and a formula for the subsequence.

11.1 $1, 2, 3, 4, \ldots$
          $2, 4, 6, 8, \ldots$

11.2 $1, \frac{1}{2}, \frac{1}{3}, \frac{1}{4}, \ldots$
          $\frac{1}{2}, \frac{1}{4}, \frac{1}{8}, \frac{1}{16}, \ldots$

11.3 $1, 2, 3, 4, 5, \ldots$
          $1, 3, 6, 10, 15, \ldots$

11.4 $1, \frac{1}{2}, \frac{3}{4}, \frac{5}{8}, \frac{11}{16}, \frac{21}{32}, \ldots$
          $\frac{1}{2}, \frac{5}{8}, \frac{21}{32}, \ldots$

11.5   Let f be as in Definition 11.1. Prove that $f(n) \geq n$ for every $n \in \mathbf{P}$.

11.6   Use Theorem 11.2 to prove the following limits.

$$\lim_{n \to \infty} \frac{1}{n^2} = 0 \qquad \lim_{n \to \infty} \frac{1}{n+2} = 0$$

11.7   Prove that the sequence $\{a_n\}_{n=1}^{\infty}$ defined by

$$a_n = \begin{cases} 1 & \text{if } n \text{ is even} \\ 0 & \text{if } n \text{ is odd} \end{cases}$$

has no limit.

11.8   Let $\{a_n\}_{n=1}^{\infty}$ be a sequence. Suppose that

$$\lim_{n \to \infty} a_{2n} = L = \lim_{n \to \infty} a_{2n-1}$$

Prove that $\lim_{n \to \infty} a_n = L$.

11.9   Let $\{a_n\}_{n=1}^{\infty}$ be a sequence which has a finite range. Prove that there is a subsequence of $\{a_n\}_{n=1}^{\infty}$ which has a limit.

11.10 Let S_1 and S_2 be two subsequences of a sequence $\{a_n\}_{n=1}^{\infty}$. How would one distinguish S_1 and S_2 using the two notations for subsequences discussed at the end of this section?

11.11 Prove that a subsequence of a subsequence is a subsequence. Discuss the notation for a subsequence of a subsequence.

11.12 Prove that the set of subsequences of $\{1/n\}_{n=1}^{\infty}$ is uncountable.

12. The Algebra of Limits

If a sequence $\{a_n\}$ has a limit, we will say that $\{a_n\}$ *converges*. If $\{a_n\}$ has no limit, we will say that $\{a_n\}$ *diverges*. In this section we will show that sums, products, and quotients of convergent sequences are convergent.

Theorem 12.1   If $a_n = L$ for all $n \in \mathbf{P}$, then $\lim_{n \to \infty} a_n = L$.

Proof.   Let $\varepsilon > 0$. If $N = 1$, then

$$|a_n - L| = 0 < \varepsilon$$

for all $n \geq N$.   ∎

Theorem 12.2 Let $\{a_n\}$ and $\{b_n\}$ be sequences such that $\lim_{n\to\infty} a_n = L$ and $\lim_{n\to\infty} b_n = M$. Then

$$\lim_{n\to\infty} (a_n + b_n) = L + M$$

Proof. Let $\varepsilon > 0$. There exists a positive integer N_1 such that if $n \geq N_1$, then

$$|a_n - L| < \frac{\varepsilon}{2}$$

There exists a positive integer N_2 such that if $n \geq N_2$, then

$$|b_n - M| < \frac{\varepsilon}{2}$$

If $n \geq \max\{N_1, N_2\}$, then

$$|(a_n + b_n) - (L + M)| \leq |a_n - L| + |b_n - M| < \frac{\varepsilon}{2} + \frac{\varepsilon}{2} = \varepsilon \qquad \blacksquare$$

Theorem 12.2 states two facts about sums of convergent sequences. If $\{a_n\}$ and $\{b_n\}$ are convergent sequences, then the sequence $\{a_n + b_n\}$ converges, and moreover, if $\{a_n\}$ and $\{b_n\}$ have limits L and M, respectively, then $\{a_n + b_n\}$ has limit $L + M$. Analogous remarks apply to Theorems 12.3, 12.4, 12.6, and 12.9.

Example. In Section 11 we showed that

$$\lim_{n\to\infty} \frac{1}{2^n} = 0 = \lim_{n\to\infty} \frac{1}{n!}$$

By Theorem 12.2,

$$\lim_{n\to\infty} \frac{1}{2^n} + \frac{1}{n!} = 0$$

Theorem 12.3 Let $\{a_n\}$ be a sequence such that $\lim_{n\to\infty} a_n = L$. If c is any real number, then

$$\lim_{n\to\infty} ca_n = cL$$

Proof. If $c = 0$, the conclusion follows from Theorem 12.1, so assume $c \neq 0$. Let $\varepsilon > 0$. There exists a positive integer N such that if $n \geq N$, then

$$|a_n - L| < \frac{\varepsilon}{|c|}$$

If $n \geq N$, then

$$|ca_n - cL| = |c||a_n - L| < |c|\frac{\varepsilon}{|c|} = \varepsilon \quad \blacksquare$$

Corollary 12.4 Let $\{a_n\}$ and $\{b_n\}$ be sequences such that $\lim_{n \to \infty} a_n = L$ and $\lim_{n \to \infty} b_n = M$. Then

$$\lim_{n \to \infty} (a_n - b_n) = L - M$$

Proof. By Theorems 12.2 and 12.3,

$$\lim_{n \to \infty} (a_n - b_n) = \lim_{n \to \infty} [a_n + (-1)b_n] = \lim_{n \to \infty} a_n + \lim_{n \to \infty} (-1)b_n$$
$$= L + (-1)M = L - M \quad \blacksquare$$

Before proving that the limit of a product is the product of the limits, we prove a lemma which is a special case.

Lemma 12.5 Let $\{a_n\}$ and $\{b_n\}$ be sequences such that

$$\lim_{n \to \infty} a_n = 0 = \lim_{n \to \infty} b_n$$

Then $\lim_{n \to \infty} a_n b_n = 0$.

Proof. Let $\varepsilon > 0$. There exists a positive integer N_1 such that if $n \geq N_1$, then

$$|a_n| < 1$$

There exists a positive integer N_2 such that if $n \geq N_2$, then

$$|b_n| < \varepsilon$$

If $n \geq \max \{N_1, N_2\}$, then

$$|a_n b_n| < \varepsilon \quad \blacksquare$$

Theorem 12.6 Let $\{a_n\}$ and $\{b_n\}$ be sequences such that $\lim_{n \to \infty} a_n = L$ and $\lim_{n \to \infty} b_n = M$. Then $\lim_{n \to \infty} a_n b_n = LM$.

Proof. We write

$$a_n b_n = (a_n - L)(b_n - M) + a_n M + L b_n - LM$$

By Theorem 12.1 and Corollary 12.4,

$$\lim_{n \to \infty} (a_n - L) = 0 = \lim_{n \to \infty} (b_n - M)$$

and using Lemma 12.5, we have

$$\lim_{n \to \infty} (a_n - L)(b_n - M) = 0$$

By Theorem 12.3,

$$\lim_{n \to \infty} a_n M = LM = \lim_{n \to \infty} Lb_n$$

By Theorem 12.2,

$$\lim_{n \to \infty} a_n b_n = \lim_{n \to \infty} [(a_n - L)(b_n - M) + a_n M + Lb_n - LM]$$

$$= \lim_{n \to \infty} (a_n - L)(b_n - M) + \lim_{n \to \infty} a_n M + \lim_{n \to \infty} Lb_n + \lim_{n \to \infty} (-LM)$$

$$= 0 + LM + LM - LM = LM \qquad \blacksquare$$

Corollary 12.7 Let $\{a_n\}$ be a sequence such that $\lim_{n \to \infty} a_n = L$, and let k be a positive integer. Then $\lim_{n \to \infty} a_n^k = L^k$.

Proof. The proof follows from Theorem 12.6 by induction on k. $\blacksquare$

As a preliminary to the quotient theorem for limits, we prove the following lemma.

Lemma 12.8 Let $\{a_n\}$ be a sequence such that $\lim_{n \to \infty} a_n = L \neq 0$. Then $a_n \neq 0$ for all but finitely many positive integers n, and

$$\lim_{n \to \infty} \frac{1}{a_n} = \frac{1}{L}$$

Proof. If $\varepsilon = |L|/2$; there exists a positive integer N_1 such that if $n \geq N_1$, then

$$|a_n - L| < \frac{|L|}{2}$$

Thus if $n \geq N_1$,

$$|L| \leq |L - a_n| + |a_n| < \frac{|L|}{2} + |a_n|$$

and

$$\frac{|L|}{2} < |a_n| \qquad (n \geq N_1)$$

Thus $a_n \neq 0$ if $n \geq N_1$.

Let $\varepsilon > 0$. There exists a positive integer N_2 such that if $n \geq N_2$, then

$$|a_n - L| < \frac{|L|^2 \varepsilon}{2}$$

If $n \geq \max\{N_1, N_2\}$, then

$$\left|\frac{1}{a_n} - \frac{1}{L}\right| = \left|\frac{a_n - L}{a_n L}\right|$$

$$= \frac{|a_n - L|}{|a_n||L|}$$

$$< \frac{|L|^2 \varepsilon}{2} \frac{2}{|L|} \frac{1}{|L|} = \varepsilon \qquad \blacksquare$$

Obviously the existence of the limit of a sequence and the value of the limit (if it exists) are not affected by altering or deleting finitely many elements of the sequence. Lemma 12.8 illustrates this situation in that we delete from the sequence $\{1/a_n\}$ those terms for which $a_n = 0$.

Theorem 12.9 Let $\{a_n\}$ and $\{b_n\}$ be sequences such that $\lim_{n\to\infty} a_n = L$ and $\lim_{n\to\infty} b_n = M \neq 0$. Then

$$\lim_{n\to\infty} \frac{a_n}{b_n} = \frac{L}{M}$$

Proof. By Lemma 12.8,

$$\lim_{n\to\infty} \frac{1}{b_n} = \frac{1}{M}$$

Thus, $\lim_{n\to\infty} \frac{a_n}{b_n} = \lim_{n\to\infty} a_n \cdot \frac{1}{b_n} = L\frac{1}{M} = \frac{L}{M}$

by Theorem 12.6. $\blacksquare$

Example. We use the preceding theorems to calculate

$$\lim_{n\to\infty} \frac{2n}{n+2}$$

$$\lim_{n\to\infty} \frac{2n}{n+2} = \lim_{n\to\infty} \frac{2}{1 + 2/n} = \frac{\lim_{n\to\infty} 2}{\lim_{n\to\infty}(1 + (2/n))} \qquad \text{(Theorem 12.9)}$$

$$= \frac{\lim_{n\to\infty} 2}{\lim_{n\to\infty} 1 + \lim_{n\to\infty}(2/n)} \qquad \text{(Theorem 12.2)}$$

$$= \frac{\lim_{n\to\infty} 2}{\lim_{n\to\infty} 1 + 2\lim_{n\to\infty}(1/n)} \qquad \text{(Theorem 12.3)}$$

$$= \frac{2}{1 + 2\cdot 0} = 2$$

Exercises

In Exercises 12.1 to 12.4 compute the limit of the given sequence by using the theorems of Section 12 as in the example at the end of Section 12.

12.1 $\lim\limits_{n \to \infty} \dfrac{2}{n+2}$

12.2 $\lim\limits_{n \to \infty} \dfrac{2n^2 + n + 3}{n^2 + 1}$

12.3 $\lim\limits_{n \to \infty} \dfrac{2n^2}{n^3 + 3}$

12.4 $\lim\limits_{n \to \infty} \sqrt{n+1} - \sqrt{n}$

12.5 Give the details of the inductive proof of Corollary 12.7.

12.6 If $a_n \geq 0$ for $n = 1, 2, \ldots$ and $\lim_{n \to \infty} a_n = L$, prove that

$$\lim_{n \to \infty} \sqrt{a_n} = \sqrt{L}$$

12.7 If $\lim_{n \to \infty} (a_{n+1}/a_n) = 0$, find $\lim_{n \to \infty} a_n$.

13. Bounded Sequences

We begin with a definition.

Definition 13.1 We say that a sequence $\{a_n\}$ is *bounded above* (*below*) if there exists a number M such that $a_n \leq M$ $(a_n \geq M)$ for every positive integer n. We say that $\{a_n\}$ is *bounded* if $\{a_n\}$ is bounded both above and below.

It is easy to see that a sequence $\{a_n\}$ is bounded if and only if there exists a positive number M such that $|a_n| \leq M$ for every positive integer n.

We next show that a convergent sequence is bounded.

Theorem 13.2 If $\{a_n\}$ is a convergent sequence, then $\{a_n\}$ is bounded.

Proof. Suppose $\lim_{n \to \infty} a_r = L$. If $\varepsilon = 1$, there exists a positive integer N such that if $n \geq N$, then

$$|a_n - L| < 1$$

Thus, if $n \geq N$, we have

$$|a_n| \leq |a_n - L| + |L| < 1 + |L|$$

If $n < N$, then

$$|a_n| \leq \max \{|a_1|, |a_2|, \ldots, |a_{N-1}|\}$$

Thus for any positive integer n,

$$|a_n| \leq M$$

if we let

$$M = \max \{|a_1|, |a_2|, \ldots, |a_{N-1}|, 1 + |L|\}$$ ∎

It follows from Theorem 13.2, that the sequence $\{n\}_{n=1}^{\infty}$ diverges since the set of positive integers is not bounded.

The sequence $\{(-1)^n\}_{n=1}^{\infty}$ is bounded and divergent and thus the converse of Theorem 13.2 does not hold. We will show later (Theorem 16.2) that a bounded *monotone* sequence is convergent.

Our next theorem generalizes Lemma 12.5.

Theorem 13.3   Let $\{a_n\}$ and $\{b_n\}$ be sequences such that $\{a_n\}$ is bounded and $\lim_{n \to \infty} b_n = 0$. Then $\lim_{n \to \infty} a_n b_n = 0$.

Proof.   There exists a positive number M such that $|a_n| \le M$ for every positive integer n. Let $\varepsilon > 0$. There exists a positive integer N such that if $n \ge N$, then

$$|b_n| < \varepsilon/M$$

If $n \ge N$, then

$$|a_n b_n| < M\frac{\varepsilon}{M} = \varepsilon$$ ∎

Exercises

13.1   Prove that if $\{a_n\}$ and $\{b_n\}$ are bounded sequences and c is a real number, then $\{ca_n\}$, $\{a_n + b_n\}$, and $\{a_n b_n\}$ are bounded sequences.

13.2   Let $\{a_n\}$ be a sequence with limit 0. Prove that
$$\lim_{n \to \infty} (-1)^n a_n = 0$$

13.3   Give an example of sequences $\{a_n\}$ and $\{b_n\}$ such that $\{a_n\}$ is bounded and $\{b_n\}$ is convergent, but $\{a_n + b_n\}$ and $\{a_n b_n\}$ are divergent.

13.4   Let $a_n = 1 \cdot 3 \cdot 5 \cdots (2n-1)/[2 \cdot 4 \cdot 6 \cdots (2n)]$. Prove that $\{a_n\}$ is a bounded sequence.

14.   Further Limit Theorems

In this brief section we collect some miscellaneous theorems which are often useful in proving limits.

Lemma 14.1   Let $\{a_n\}$ be a sequence such that $0 \le a_n$ for every positive integer n and $L = \lim_{n \to \infty} a_n$. Then $0 \le L$.

Proof. We argue by contradiction. Suppose $L < 0$. There exists a positive integer N such that if $n \geq N$, then

$$|a_n - L| < \frac{-L}{2}$$

Thus

$$a_N - L < \frac{-L}{2}$$

and therefore

$$a_N < \frac{L}{2} < 0$$

which is a contradiction. ∎

Theorem 14.2 Let $\{a_n\}$ and $\{b_n\}$ be sequences such that $\lim_{n \to \infty} a_n = L$ and $\lim_{n \to \infty} b_n = M$. If $a_n \leq b_n$ for every positive integer n, then $L \leq M$.

Proof. Since $0 \leq b_n - a_n$ for all $n \in \mathbf{P}$, we may use Theorems 12.4 and 14.1 to conclude that $0 \leq M - L$. ∎

A special case of Theorem 14.2 is often used with either $\{a_n\}$ or $\{b_n\}$ taken to be a constant sequence. For example, if $a_n \leq M$ for all $n \in \mathbf{P}$ and $\lim_{n \to \infty} a_n = L$, then $L \leq M$.

Theorem 14.3 (Squeeze Theorem) Let $\{a_n\}$, $\{b_n\}$, and $\{c_n\}$ be sequences such that

$$a_n \leq b_n \leq c_n$$

for every positive integer n. If

$$\lim_{n \to \infty} a_n = L = \lim_{n \to \infty} c_n$$

then

$$\lim_{n \to \infty} b_n = L$$

Proof. Let $\varepsilon > 0$. There exists a positive integer N_1 such that if $n \geq N_1$, then

$$L - \varepsilon < a_n < L + \varepsilon$$

There exists a positive integer N_2 such that if $n \geq N_2$, then

$$L - \varepsilon < c_n < L + \varepsilon$$

Thus if $n \geq \max \{N_1, N_2\}$, we have

$$L - \varepsilon < a_n \leq b_n \leq c_n \leq L + \varepsilon \qquad \blacksquare$$

Example. Since $0 < 1/2^n < 1/n$ and $\lim_{n \to \infty} 1/n = 0$, by Theorem 14.3, $\lim_{n \to \infty} 1/2^n = 0$.

Exercises

14.1 Use the squeeze theorem to prove that $\lim_{n \to \infty} (1/n!) = 0$.

14.2 Why can't Theorem 14.3 be deduced directly from Theorem 14.2?

14.3 Prove the following theorem. Let $\{a_n\}$, $\{b_n\}$, and $\{c_n\}$ be sequences such that for some positive integer N

$$a_n \leq b_n \leq c_n$$

for all $n \geq N$. If

$$\lim_{n \to \infty} a_n = L = \lim_{n \to \infty} c_n$$

then

$$\lim_{n \to \infty} b_n = L$$

15. Divergent Sequences

We begin with a definition.

Definition 15.1 Let $\{a_n\}$ be a sequence. We say that $\{a_n\}$ *diverges to infinity* (*minus infinity*) and write $\lim_{n \to \infty} a_n = \infty$ ($\lim_{n \to \infty} a_n = -\infty$) if for every real number M, there exists a positive integer N such that if $n \geq N$, then $a_n > M$ ($a_n < M$).

A sequence which diverges to infinity (or to minus infinity) is unbounded and hence divergent because of Theorem 13.2. It is possible to prove analogues of the theorems given in Sections 11, 12, and 14 for sequences which diverge to infinity (or minus infinity). We will give two examples and list several others as exercises.

Theorem 15.2 Let $\{a_n\}$ and $\{b_n\}$ be sequences such that $\lim_{n \to \infty} a_n = \infty = \lim_{n \to \infty} b_n$. Then $\lim_{n \to \infty} (a_n + b_n) = \infty$.

Proof. Let M be a real number. There exists a positive integer N_1 such that if $n \geq N_1$, then

$$\frac{M}{2} < a_n$$

There exists a positive integer N_2 such that if $n \geq N_2$, then

$$\frac{M}{2} < b_n$$

If $n \geq \max \{N_1, N_2\}$, then

$$M = \frac{M}{2} + \frac{M}{2} < a_n + b_n \qquad \blacksquare$$

Theorem 15.3 (Squeeze Theorem)   Let $\{a_n\}$ and $\{b_n\}$ be sequences such that $a_n \leq b_n$ for every positive integer n. If $\lim_{n \to \infty} b_n = -\infty$, then $\lim_{n \to \infty} a_n = -\infty$.

Proof.   Let M be a real number. There exists a positive integer N such that if $n \geq N$, then

$$b_n < M$$

Thus if $n \geq N$, we have

$$a_n \leq b_n < M$$ ∎

Exercises

15.1   Let $\{a_n\}$ be a sequence such that $\lim_{n \to \infty} a_n = \infty$. Prove that any subsequence of $\{a_n\}$ diverges to infinity.

15.2   Let $\{a_n\}$ and $\{b_n\}$ be sequences such that $\{a_n\}$ diverges to infinity and $\{b_n\}$ is bounded. Prove that $\{a_n + b_n\}$ diverges to infinity.

15.3   Prove that the sequence $\{a_n\}$ diverges to infinity if and only if $\{-a_n\}$ diverges to minus infinity.

15.4   Let $\{a_n\}$ and $\{b_n\}$ be sequences such that $\lim_{n \to \infty} a_n = \infty = \lim_{n \to \infty} b_n$. Prove that $\lim_{n \to \infty} a_n b_n = \infty$.

15.5   State and prove an analogue of Exercise 15.4 for sequences $\{a_n\}$ and $\{b_n\}$ which diverge to minus infinity.

15.6   Let $\{a_n\}$ be a sequence. Prove that $\{a_n\}$ diverges to infinity if and only if there exists a positive integer N such that $0 < a_n$ for all $n \geq N$ and $\lim_{n \to \infty} (1/a_n) = 0$.

15.7   What can be said about a divergent sequence which diverges to neither ∞ nor $-\infty$?

16.   Monotone Sequences and the Number *e*

In order to prove that a sequence $\{a_n\}$ is convergent using Definition 10.2, we must know the value of the limit of $\{a_n\}$ in advance. In this section we will prove a theorem which will allow us to prove that certain sequences (monotone sequences) are convergent without knowing the value of the limit in advance. In Section 19 we will prove a general condition for convergence.

Definition 16.1   Let $\{a_n\}$ be a sequence. We say that $\{a_n\}$ is *increasing* (*decreasing*) if $a_n \leq a_{n+1}$ ($a_n \geq a_{n+1}$) for every positive integer n. We say that the sequence $\{a_n\}$ is *monotone* if either $\{a_n\}$ is increasing or $\{a_n\}$ is decreasing.

If $a_n < a_{n+1}$ ($a_n > a_{n+1}$) for every positive integer n, we say that $\{a_n\}$ is *strictly increasing (strictly decreasing)*. We say that $\{a_n\}$ is *strictly monotone* if either $\{a_n\}$ is strictly increasing or strictly decreasing.

If $\{a_n\}$ is an increasing sequence, then $\{a_n\}$ is bounded below by a_1 and hence $\{a_n\}$ is bounded if and only if $\{a_n\}$ is bounded above. Similarly, a decreasing sequence $\{a_n\}$ is bounded if and only if $\{a_n\}$ is bounded below.

Theorem 16.2   A monotone sequence $\{a_n\}$ is convergent if and only if $\{a_n\}$ is bounded.

Proof.   Suppose $\{a_n\}$ is increasing. (The proof is similar if $\{a_n\}$ is decreasing.)

If $\{a_n\}$ is convergent, then $\{a_n\}$ is bounded by Theorem 13.2.

Suppose $\{a_n\}$ is bounded. By the least-upper-bound axiom, the least upper bound L of the set

$$X = \{a_n \mid n \in \mathbf{P}\}$$

exists. Let $\varepsilon > 0$. Since $L - \varepsilon$ is not an upper bound of the set X, there exists a positive integer N such that $L - \varepsilon < a_N$. Since $\{a_n\}$ is increasing, if $n \geq N$ we have $a_N \leq a_n$. Thus if $n \geq N$, we have

$$L - \varepsilon < a_N \leq a_n \leq L < L + \varepsilon$$

Therefore $\lim_{n \to \infty} a_n = L$.   ■

One method of finding the limit of a sequence is to prove that the sequence is monotone and bounded, then derive an equation which can be solved for the limit. Theorems 16.3, 16.4, and 16.7 are examples of such a technique.

Theorem 16.3   If $|a| < 1$, then $\lim_{n \to \infty} a^n = 0$.

Proof.   First suppose that $0 < a < 1$. Then $\{a^n\}$ is a bounded monotone sequence, and thus $\{a^n\}$ converges by Theorem 16.2. Suppose $\lim_{n \to \infty} a^n = L$. By Theorems 11.2 and 12.3,

$$L = \lim_{n \to \infty} a^{n+1} = \lim_{n \to \infty} aa^n = a \lim_{n \to \infty} a^n = aL$$

If $L \neq 0$, then $a = 1$, which is impossible. Therefore, $\lim_{n \to \infty} a^n = 0$ if $0 < a < 1$.

Now suppose $-1 < a < 0$. Then $0 < -a < 1$ and thus $\lim_{n \to \infty} (-a)^n = 0$. By Theorem 13.3,

$$\lim_{n \to \infty} a^n = \lim_{n \to \infty} (-1)^n (-a)^n = 0$$

If $a = 0$, $\lim_{n \to \infty} a^n = 0$ by Theorem 12.1.   ■

Theorem 16.4 If $a > 0$, then $\lim_{n \to \infty} a^{1/n} = 1$.

Proof. First suppose that $a \geq 1$. Then $\{a^{1/n}\}$ is a decreasing sequence bounded below by 1, hence by Theorem 16.2, $\{a^{1/n}\}$ converges.

Suppose $\lim_{n \to \infty} a^{1/n} = L$. Since $a^{1/n} \geq 1$, by Theorem 14.2, $L \neq 0$. By Corollary 12.7,

$$\lim_{n \to \infty} a^{2/n} = L^2$$

By Theorem 11.2, the subsequence $\{a^{2/2n}\}$ of $\{a^{2/n}\}$ also converges to L^2, and since $a^{2/2n} = a^{1/n}$, $L = L^2$. Since $L \neq 0$, $L = 1$.

If $0 < a < 1$, then $1 < 1/a$, and by the special case above,

$$\lim_{n \to \infty} \frac{1}{a^{1/n}} = \lim_{n \to \infty} \left(\frac{1}{a}\right)^{1/n} = 1$$

By Theorem 12.9,

$$\lim_{n \to \infty} a^{1/n} = \lim_{n \to \infty} \frac{1}{(1/a)^{1/n}} = 1 \qquad \blacksquare$$

Theorem 16.2 will allow us to define the number e as the limit of the sequence $\{(1 + 1/n)^n\}$. We will use the inequality stated in Lemma 16.5 to prove that the sequence $\{(1 + 1/n)^n\}$ is increasing and bounded.

Lemma 16.5 Let a and b be numbers such that $0 \leq a < b$. Then

$$\frac{b^{n+1} - a^{n+1}}{b - a} < (n + 1)b^n$$

Proof. If $0 \leq a < b$, then

$$\begin{aligned}
\frac{b^{n+1} - a^{n+1}}{b - a} &= b^n + ab^{n-1} + a^2b^{n-2} + \cdots + a^{n-1}b + a^n \\
&< b^n + bb^{n-1} + b^2b^{n-2} + \cdots + b^{n-1}b + b^n \\
&= (n + 1)b^n \qquad \blacksquare
\end{aligned}$$

Theorem 16.6 The sequence $\{(1 + 1/n)^n\}$ is increasing and convergent. The limit is denoted e.

Proof. We first rewrite the inequality of Lemma 16.5 as

$$b^n[b - (n + 1)(b - a)] < a^{n+1}$$

If we set $a = 1 + 1/(n + 1)$ and $b = 1 + 1/n$ the term in brackets reduces to 1 and we have

$$\left(1 + \frac{1}{n}\right)^n < \left(1 + \frac{1}{n + 1}\right)^{n+1}$$

Next we set $a = 1$ and $b = 1 + 1/2n$. This time the term in brackets reduces to $\frac{1}{2}$, and we have

$$\left(1 + \frac{1}{2n}\right)^n < 2$$

Thus

$$\left(1 + \frac{1}{2n}\right)^{2n} < 4$$

Since $\{(1 + 1/n)^n\}$ is increasing,

$$\left(1 + \frac{1}{n}\right)^n < \left(1 + \frac{1}{2n}\right)^{2n} < 4$$

for every positive integer n. Therefore the sequence $\{(1 + 1/n)^n\}$ is increasing and bounded above by 4. By Theorem 16.2, $\{(1 + 1/n)^n\}$ converges. ∎

The proof of Theorem 16.6 shows that

$$\left(1 + \frac{1}{n}\right)^n < 4 \tag{16.1}$$

for every positive integer n. On the other hand, since $\{(1 + 1/n)^n\}$ is increasing and $(1 + 1/n)^n = 2$ if $n = 1$, we have

$$2 \leq \left(1 + \frac{1}{n}\right)^n \leq 4$$

for every $n \in \mathbf{P}$. By Theorem 14.2, $2 \leq e \leq 4$ (It can be shown that the decimal expansion of e begins $2.71828182 \cdots$).

The proof of Theorem 16.6 was recently recalled to attention by Johnsonbaugh (1974). It is apparently originally due to Fort in 1862 [see Chrystal (1964), vol. II, p. 77].

We conclude this section by combining inequality (16.1) with the method of the proofs of Theorems 16.3 and 16.4 to prove that $\lim_{n \to \infty} n^{1/n} = 1$.

Theorem 16.7   The sequence $\{n^{1/n}\}_{n=3}^{\infty}$ is decreasing and $\lim_{n \to \infty} n^{1/n} = 1$.

Proof.   We derive the following equivalent inequalities.

$$(n + 1)^{1/(n+1)} \leq n^{1/n} \tag{16.2}$$

$$(n + 1)^n \leq n^{n+1}$$

$$\frac{(n + 1)^n}{n^n} \leq n$$

$$\left(1 + \frac{1}{n}\right)^n \leq n \tag{16.3}$$

By inequality (16.1), $(1 + 1/n)^n \le n$ if $n \ge 4$. It may be verified that $(1 + \frac{1}{3})^3 \le 3$ and thus inequality (16.3) [and hence also (16.2)] holds for all $n \ge 3$.

Therefore $\{n^{1/n}\}_{n=3}^{\infty}$ is decreasing. Since $\{n^{1/n}\}_{n=3}^{\infty}$ is bounded below by 1, $\{n^{1/n}\}$ converges by Theorem 16.2.

Let $L = \lim_{n\to\infty} n^{1/n}$. Since $n^{1/n} \ge 1$ for every positive integer n, by Theorem 14.2, $L \ne 0$. By Corollary 12.7,

$$\lim_{n\to\infty} n^{2/n} = L^2$$

and by Theorem 16.4,

$$\lim_{n\to\infty} \left(\frac{1}{2}\right)^{2/n} = 1$$

so by Theorem 12.6,

$$\lim_{n\to\infty} \left(\frac{n}{2}\right)^{2/n} = L^2$$

By Theorem 11.2, the subsequence $\{(2n/2)^{2/2n}\}$ of $\{(n/2)^{2/n}\}$ also converges to L^2 and since $(2n/2)^{2/2n} = n^{1/n}$, $L = L^2$. Since $L \ne 0$, $L = 1$. ∎

Exercises

16.1 True or false? If the sequence $\{a_n\}$ is not increasing, then $\{a_n\}$ is decreasing.

16.2 Prove that if $\{a_n\}$ is a decreasing bounded sequence, then $\{a_n\}$ is convergent thus completing the proof of Theorem 16.2.

16.3 Prove that if $\{a_n\}$ is an increasing sequence which is not bounded, then $\lim_{n\to\infty} a_n = \infty$.

16.4 Prove that if $a > 1$, then $\lim_{n\to\infty} a^n = \infty$.

16.5 Find the following limits

(a) $\lim_{n\to\infty} \left(1 + \frac{1}{n^2}\right)^{n^2}$

(b) $\lim_{n\to\infty} \left(1 + \frac{1}{n+1}\right)^n$

(c) $\lim_{n\to\infty} \left(1 + \frac{1}{n}\right)^{n+1}$

(d) $\lim_{n\to\infty} \left(1 + \frac{1}{n^2}\right)^n$

(e) $\lim_{n\to\infty} \left(1 + \frac{1}{n}\right)^{n^2}$

(f) $\lim_{n\to\infty} \left(1 + \frac{1}{2n}\right)^n$

16.6 (a) Prove that if $0 \le a < b$, then
$$\frac{b^{n+1} - a^{n+1}}{b - a} > (n + 1)a^n \qquad \text{for } n = 1, 2, \ldots$$

(b) Take $a = 1 + 1/(n + 1)$ and $b = 1 + 1/n$ in part (a) and prove that
$$\left(1 + \frac{1}{n}\right)^{n+1} > \left(1 + \frac{1}{n+1}\right)^n \left[1 + \frac{1}{n+1} + \frac{1}{n}\right] \qquad \text{for } n = 1, 2, \ldots$$

(c) Prove that
$$\left(1 + \frac{1}{n+1}\right)^n \left[1 + \frac{1}{n+1} + \frac{1}{n}\right] > \left(1 + \frac{1}{n+1}\right)^{n+2} \qquad \text{for } n = 1, 2, \ldots$$

(d) Prove that $\{(1 + 1/n)^{n+1}\}$ is a decreasing sequence with limit e and that $e \le 3$.

16.7 Prove that every convergent sequence has a monotone subsequence.

16.8 Prove the nested interval theorem: If $\{[a_n, b_n]\}$ is a sequence of closed intervals such that
$$[a_n, b_n] \supset [a_{n+1}, b_{n+1}] \qquad \text{for } n = 1, 2, \ldots$$
then $\cap[a_n, b_n]$ is nonvoid.

Find a necessary and sufficient condition that $\cap_{n=1}^{\infty} [a_n, b_n]$ contains exactly one point.

16.9 Verify that
$$1^{\frac{1}{1}} \le 2^{\frac{1}{2}} \le 3^{\frac{1}{3}}$$

16.10 (a) Let x and y be positive numbers. Let $a_0 = y$, and let
$$a_n = \frac{(x/a_{n-1}) + a_{n-1}}{2} \qquad \text{for } n = 1, 2, \ldots$$

Prove that $\{a_n\}$ is a decreasing sequence with limit $\sqrt{x}$.

(b) Generalize (a) to nth roots.

16.11* Prove directly from Definition 10.2 that $\lim_{n \to \infty} n^{1/n} = 1$.

16.12 Let $a_n = \sqrt[n]{n}$ and $b_n = a_{n+1}/a_n$. Prove that $\{b_n\}_{n=5}^{\infty}$ is increasing and find the limit.

16.13 Let $\{a_n\}$ be any sequence of real numbers such that $\lim_{n \to \infty} na_n = 0$. Prove that
$$\lim_{n \to \infty} \left(1 + \frac{1}{n} + a_n\right)^n = e$$

16.14 (a) Prove that the positive sequence $\{a_n\}$ converges if the sequence $\{a_{n+1}/a_n\}$ is bounded above by 1.

(b) Prove that if $\lim_{n \to \infty} a_{n+1}/a_n$ exists and is less than 1, then $\lim_{n \to \infty} a_n = 0$.

16.15 Let $a_1 > 1$. Let $a_{n+1} = 2 - 1/a_n$ for $n = 1, 2, \ldots$. Prove that $\{a_n\}$ is a bounded monotone sequence and find the limit.

16.16 Let $\{a_n\}$ be a positive sequence which satisfies $a_{n+2} = a_{n+1} + a_n$ for $n = 1, 2, \ldots$.

(a)* Assuming that $\lim_{n \to \infty} (a_{n+1}/a_n)$ exists, prove that the value of this limit is $(1 + \sqrt{5})/2$.

(b) Prove that $\lim_{n \to \infty} (a_{n+1}/a_n)$ exists.

17. Real Exponents

In this section we make use of the convergence criterion for monotone sequences (Theorem 16.2) to define a^x, where $a > 0$ and x is a real (possibly irrational) number. This will complete the definition of exponents begun in Section 7.

It is fairly clear that $3^{\sqrt{2}}$ should be defined as the limit of a sequence such as

$$3^1, \quad 3^{1 \cdot 4}, \quad 3^{1 \cdot 41}, \quad 3^{1 \cdot 414}, \quad \ldots$$

We will define a^x, $a > 0$ and x real, as $\lim_{n \to \infty} a^{r_n}$, where $\{r_n\}$ is an increasing sequence of rational numbers with limit x. We will then verify that the laws of rational exponents carry over to real exponents. We first prove that a sequence such as $\{r_n\}$ exists.

Theorem 17.1 If x is a real number, there exists an increasing rational sequence $\{r_n\}$ with limit x.

Proof. Choose (Theorem 7.8) a rational number r_1 such that

$$x - 1 < r_1 < x.$$

Choose a rational number r_2 such that

$$\max \left\{ r_1, x - \frac{1}{2} \right\} < r_2 < x$$

In general, having chosen a rational number $r_n < x$, choose a rational number r_{n+1} such that

$$\max \left\{ r_n, x - \frac{1}{n + 1} \right\} < r_{n+1} < x$$

Clearly $\{r_n\}$ is an increasing rational sequence, and since

$$x - \frac{1}{n} < r_n < x$$

by the squeeze theorem (Theorem 14.3), $\lim_{n \to \infty} r_n = x$. ∎

If $a \geq 1$ and x is a real number, we choose an increasing rational sequence $\{r_n\}$ such that $\lim_{n \to \infty} r_n = x$. Then the sequence $\{a^{r_n}\}$ is increasing and if r is any rational number such that $r > x$, $\{a^{r_n}\}$ is bounded above by a^r. By Theorem 16.2, $\{a^{r_n}\}$ converges. We define $a^x = \lim_{n \to \infty} a^{r_n}$.

Definition 17.2 Let $a \geq 1$, and let x be a real number. We define

$$a^x = \lim_{n \to \infty} a^{r_n}$$

where $\{r_n\}$ is an increasing rational sequence with limit x.

If $0 < a < 1$ and x is a real number, we define

$$a^x = \left(\frac{1}{a}\right)^{-x}$$

We must show that a^x is well defined; that is, we must show that the value of a^x is independent of the choice of the sequence $\{r_n\}$.

Lemma 17.3   Let $a \geq 1$, and let x be a real number. Let $\{r_n\}$ and $\{s_n\}$ be increasing rational sequences such that $\lim_{n \to \infty} r_n = x = \lim_{n \to \infty} s_n$. Then

$$\lim_{n \to \infty} a^{r_n} = \lim_{n \to \infty} a^{s_n}$$

Proof.   Let

$$R_n = r_n - \frac{1}{n}, \qquad S_n = s_n - \frac{1}{n} \qquad n = 1, 2, \ldots$$

Then $\{R_n\}$ and $\{S_n\}$ are increasing rational sequences and

$$R_n < x \qquad S_n < x \qquad n = 1, 2, \ldots$$
$$\lim_{n \to \infty} R_n = x = \lim_{n \to \infty} S_n$$

By Theorems 12.6 and 16.4,

$$\lim_{n \to \infty} a^{r_n} = \lim_{n \to \infty} a^{R_n} \qquad \lim_{n \to \infty} a^{s_n} = \lim_{n \to \infty} a^{S_n}$$

Since $R_1 < x$, we may choose a positive integer m_1 such that

$$R_1 < S_{m_1}$$

Since $S_{m_1} < x$, we may choose a positive integer n_2 such that

$$S_{m_1} < R_{n_2}$$

Continuing in this way we choose an increasing sequence

$$R_{n_1} < S_{m_1} < R_{n_2} < S_{m_2} < \cdots$$

Let

$$b_{2k} = S_{m_k}, \qquad b_{2k-1} = R_{n_k} \qquad k = 1, 2, \ldots$$

Then $\{b_k\}$ is an increasing rational sequence which is bounded, and hence $\lim_{n \to \infty} a^{b_k}$ exists. Now

$$\lim_{n \to \infty} a^{r_n} = \lim_{n \to \infty} a^{R_n} = \lim_{k \to \infty} a^{R_{n_k}} = \lim_{k \to \infty} a^{b_{2k-1}}$$
$$= \lim_{k \to \infty} a^{b_{2k}} = \lim_{k \to \infty} a^{S_{m_k}} = \lim_{n \to \infty} a^{S_n} = \lim_{n \to \infty} a^{s_n}$$

where we have used Theorem 11.2. ∎

The technique used in the proof of Lemma 17.3 is known as the *interlacing method* and has other applications in mathematics. (See the proof of Theorem 84.1.)

If $a \geq 1$ and r is a rational number, we may choose the increasing sequence $\{r_n\}$, where $r_n = r$, $n = 1, 2, \ldots$, and thus we see that Definition 17.2 is consistent with our earlier definition (Definition 7.7) of a^r for r rational.

The laws of real exponents now follow from the laws of rational exponents by taking limits and using Definition 17.2.

Theorem 17.4 Let a and b be positive numbers. Then

(i) $a^{x+y} = a^x a^y$ $x, y \in \mathbf{R}$

(ii) $(a^x)^y = a^{xy}$ $x, y \in \mathbf{R}$

(iii) $(ab)^x = a^x b^x$ $x \in \mathbf{R}$

(iv) $a^{-x} = \dfrac{1}{a^x}$ $x \in \mathbf{R}$

(v) $\left(\dfrac{a}{b}\right)^x = \dfrac{a^x}{b^x}$ $x \in \mathbf{R}$

(vi) If $a > 1$ and $x < y$, then $a^x < a^y$.

(vii) If $0 < a < 1$ and $x < y$, then $a^x > a^y$.

(viii) If $x > 0$ and $a < b$, then $a^x < b^x$.

(ix) If $x < 0$ and $a < b$, then $a^x > b^x$.

Proof. We prove only (i) leaving the other parts as exercises. First suppose $a \geq 1$. Let $\{r_n\}$ and $\{s_n\}$ be increasing rational sequences with limits x and y, respectively. Then $\{r_n + s_n\}$ is an increasing rational sequence with limit $x + y$ and since

$$a^{r_n + s_n} = a^{r_n} a^{s_n}$$

we have

$$a^{x+y} = \lim_{n \to \infty} a^{r_n + s_n} = \lim_{n \to \infty} a^{r_n} a^{s_n} = \lim_{n \to \infty} a^{r_n} \lim_{n \to \infty} a^{s_n} = a^x a^y$$

If $0 < a < 1$, then $1/a > 1$, and therefore

$$a^{x+y} = \left(\frac{1}{a}\right)^{-x-y} = \left(\frac{1}{a}\right)^{-x}\left(\frac{1}{a}\right)^{-y} = a^x a^y \qquad \blacksquare$$

Exercises

17.1 Prove Theorem 17.4, parts (ii) to (ix).

17.2 Let $a > 0$, and let x be a real number. Prove that if $\{r_n\}$ is any decreasing rational sequence with limit x, then $a^x = \lim_{n \to \infty} a^{r_n}$.

17.3* Let $a > 0$, and let x be a real number. Let $\{b_n\}$ be any real sequence with limit x. Prove that $\lim_{n \to \infty} a^{b_n} = a^x$.

17.4 Let $\{a_n\}$ be a sequence with positive terms such that $\lim_{n \to \infty} a_n = L > 0$. Let x be a real number. Prove that $\lim_{n \to \infty} a_n^x = L^x$.

18. The Bolzano-Weierstrass Theorem

In Section 13 we proved that a convergent sequence is bounded (Theorem 13.2), and we noted that the converse of this theorem is false. The sequence $\{(-1)^n\}$ is bounded and divergent. Although $\{(-1)^n\}$ is divergent, it has a convergent subsequence, $\{(-1)^{2n}\}$. The Bolzano-Weierstrass theorem states that *every* bounded sequence has a convergent subsequence. This theorem states a fundamental property of the real numbers, and, in fact, it can be shown to be equivalent to the least-upper-bound axiom.

Theorem 18.1 (Bolzano-Weierstrass Theorem)   Every bounded real sequence has a convergent subsequence.

Proof.   Let $\{a_n\}$ be a bounded sequence. Then there is a closed interval $[c, d]$ such that $a_n \in [c, d]$ for every positive integer n.

Consider the two closed intervals

$$\left[c, \frac{c+d}{2}\right], \qquad \left[\frac{c+d}{2}, d\right]$$

One of these intervals must contain a_n for infinitely many positive integers n. We denote this interval by $[c_1, d_1]$.

We repeat this process with the interval $[c_1, d_1]$. One of the intervals

$$\left[c_1, \frac{c_1+d_1}{2}\right], \qquad \left[\frac{c_1+d_1}{2}, d_1\right]$$

must contain a_n for infinitely many positive integers n. We denote this interval by $[c_2, d_2]$. Continuing this process, we obtain a sequence

$$[c_1, d_1], [c_2, d_2], \ldots$$

of closed intervals such that

$$[c_1, d_1] \supset [c_2, d_2] \supset [c_3, d_3] \supset \cdots$$

$$d_k - c_k = \frac{d-c}{2^k}, \qquad k = 1, 2, \ldots \tag{18.1}$$

and each interval $[c_k, d_k]$ contains a_n for infinitely many positive integers n.

Choose a positive integer n_1 such that $a_{n_1} \in [c_1, d_1]$. Since $[c_2, d_2]$ contains a_n for infinitely many positive integers n, there exists a positive integer $n_2 > n_1$ such that $a_{n_2} \in [c_2, d_2]$. Continuing this process we obtain elements a_{n_1},

$a_{n_2}, \ldots$ satisfying

$$a_{n_k} \in [c_k, d_k], \qquad k = 1, 2, \ldots \qquad (18.2)$$

and
$$n_1 < n_2 < n_3 < \cdots$$

Therefore, $\{a_{n_k}\}_{k=1}^{\infty}$ is a subsequence of $\{a_n\}$. We will show that $\{a_{n_k}\}$ converges.

The sequence $\{c_k\}$ is monotone and bounded so by Theorem 16.2, $\{c_k\}$ is convergent. Let $L = \lim_{k \to \infty} c_k$. Similarly, the sequence $\{d_k\}$ converges to some number M. Using equation (18.1), we have

$$M - L = \lim_{k \to \infty} d_k - \lim_{k \to \infty} c_k = \lim_{k \to \infty} (d_k - c_k) = \lim_{k \to \infty} \frac{d - c}{2^k} = 0$$

Therefore $L = M$. Because of (18.2), we have

$$c_k \le a_{n_k} \le d_k$$

By the squeeze theorem (Theorem 14.3), $\{a_{n_k}\}$ converges to $L = M$. ∎

We will use the Bolzano-Weierstrass theorem to derive a general condition for convergence of a sequence (Section 19) and to define a kind of limit applicable to any sequence (Sections 20 and 21).

Exercises

18.1 Show (by example) that "bounded" cannot be omitted from the hypotheses of the Bolzano-Weierstrass theorem.

18.2* Prove that every sequence has a monotone subsequence.

18.3* Let $\{a_n\}$ be a bounded sequence. Prove that if every convergent subsequence of $\{a_n\}$ has a limit L, then $\lim_{n \to \infty} a_n = L$.

18.4 Let $\{a_n\}$ be a sequence. Prove that if every monotone subsequence of $\{a_n\}$ has limit L, then $\lim_{n \to \infty} a_n = L$.

18.5 Let $\{a_n\}$ be a sequence such that for some $\varepsilon > 0$,
$$|a_n - a_m| \ge \varepsilon \qquad \text{for all } n \ne m$$
Prove that $\{a_n\}$ has no convergent subsequence.

19. The Cauchy Condition

We begin by deriving a property of convergent sequences.

Theorem 19.1 Let $\{a_n\}$ be a convergent sequence. Then for every $\varepsilon > 0$, there exists a positive integer N such that if $m, n \ge N$, then

$$|a_m - a_n| < \varepsilon$$

Proof. Suppose $\lim_{n \to \infty} a_n = L$. Let $\varepsilon > 0$. There exists a positive integer N such that if $n \geq N$, then

$$|a_n - L| < \frac{\varepsilon}{2}$$

Therefore, if $m,n \geq N$, we have

$$|a_m - a_n| \leq |a_m - L| + |a_n - L| < \frac{\varepsilon}{2} + \frac{\varepsilon}{2} = \varepsilon \qquad \blacksquare$$

The condition stated in the conclusion of Theorem 19.1 is known as the *Cauchy condition*.

Definition 19.2   If $\{a_n\}$ is a sequence such that for every $\varepsilon > 0$, there exists a positive integer N such that if $m,n \geq N$, we have

$$|a_m - a_n| < \varepsilon$$

then we call $\{a_n\}$ a *Cauchy sequence*.

Theorem 19.1 states that a convergent sequence is a Cauchy sequence. The converse of Theorem 19.1 also holds; that is, if $\{a_n\}$ is a Cauchy sequence, then $\{a_n\}$ is convergent. The Cauchy condition does not involve the limit, and thus we have a convergence test for arbitrary sequences which does not assume that the value of the limit is known in advance.

Theorem 19.3   Let $\{a_n\}$ be a real sequence. Then $\{a_n\}$ is convergent if and only if $\{a_n\}$ is a Cauchy sequence.

Proof.   Theorem 19.1 states that if $\{a_n\}$ is convergent, then $\{a_n\}$ is a Cauchy sequence and thus we must prove the converse.

Let $\{a_n\}$ be a Cauchy sequence. We first show that $\{a_n\}$ is bounded, so that we may apply the Bolzano-Weierstrass theorem. If $\varepsilon = 1$, there exists a positive integer N such that if $m,n \geq N$, then

$$|a_m - a_n| < 1$$

Thus if $n \geq N$, we have

$$|a_n| \leq |a_n - a_N| + |a_N| < 1 + |a_N|$$

Therefore, $\{a_n\}$ is bounded by

$$\max \{|a_1|, |a_2|, \ldots, |a_{N-1}|, 1 + |a_N|\}$$

By Theorem 18.1, $\{a_n\}$ has a convergent subsequence $\{a_{n_k}\}$. Suppose

$$\lim_{k \to \infty} a_{n_k} = L$$

Let $\varepsilon > 0$. There exists a positive integer N such that if $m,n \geq N$, then

$$|a_m - a_n| < \frac{\varepsilon}{2}$$

There exists a positive integer N' such that if $k \geq N'$, then

$$|a_{n_k} - L| < \frac{\varepsilon}{2}$$

Choose a positive integer K such that $K \geq N'$ and $n_K \geq N$. Now if $n \geq N$, then

$$|a_n - L| \leq |a_n - a_{n_K}| + |a_{n_K} - L| < \frac{\varepsilon}{2} + \frac{\varepsilon}{2} = \varepsilon \qquad \blacksquare$$

Exercises

19.1 Let $\{a_n\}$ and $\{b_n\}$ be Cauchy sequences, and let c be a real number. Prove that $\{a_n + b_n\}$, $\{a_n - b_n\}$, $\{ca_n\}$, and $\{a_n b_n\}$ are Cauchy sequences.

19.2 Let $\{a_n\}$ be a sequence. Prove that $\{a_n\}$ is a Cauchy sequence if and only if for every $\varepsilon > 0$, there exists a positive integer N such that if $n \geq N$, then $|a_n - a_N| < \varepsilon$.

19.3 Let $0 \leq \alpha < 1$, and let f be a function from $\mathbf{R}$ into $\mathbf{R}$ which satisfies
$$|f(x) - f(y)| \leq \alpha |x - y| \qquad \text{for all } x, y \in \mathbf{R}$$
Let $a_1 \in \mathbf{R}$, and let $a_{n+1} = f(a_n)$ for $n = 1, 2, \ldots$ Prove that $\{a_n\}$ is a Cauchy sequence.

19.4 Let $\{a_n\}$ be the sequence defined by the relations $a_1 = 1$ and $a_{n+1} = a_n + (1/3^n)$ for $n = 1, 2, \ldots$ Prove that $\{a_n\}$ is a Cauchy sequence.

20. The lim sup and lim inf of Bounded Sequences

The "generalized limits" $\limsup_{n \to \infty} a_n$ and $\liminf_{n \to \infty} a_n$ are defined for arbitrary (not necessarily convergent) sequences $\{a_n\}$. In this section we will define $\limsup_{n \to \infty} a_n$ and $\liminf_{n \to \infty} a_n$ for bounded sequences $\{a_n\}$ and in Section 21 we will extend our definition to unbounded sequences.

If $\{a_n\}$ is a bounded sequence, the Bolzano-Weierstrass theorem assures us that $\{a_n\}$ has a convergent subsequence. The number $\limsup_{n \to \infty} a_n$ is the maximum value obtainable as the limit of a convergent subsequence of $\{a_n\}$ and $\liminf_{n \to \infty} a_n$ is the minimum value obtainable as the limit of a convergent subsequence of $\{a_n\}$.

Definition 20.1 Let $\{a_n\}$ be a bounded real sequence and let $\mathscr{L}_a$ denote the set of all L such that

$$L = \lim_{k \to \infty} a_{n_k}$$

where $\{a_{n_k}\}$ is a convergent subsequence of $\{a_n\}$. We define

$$\lim_{n \to \infty} \sup a_n = \text{lub } \mathscr{L}_a$$

and

$$\lim_{n \to \infty} \inf a_n = \text{glb } \mathscr{L}_a$$

The notations $\overline{\lim}_{n \to \infty} a_n$ and $\underline{\lim}_{n \to \infty} a_n$ are also used for $\lim \sup_{n \to \infty} a_n$ and $\lim \inf_{n \to \infty} a_n$, respectively.

Let $\{a_n\}$ be a bounded sequence. Then there exists a number M such that $|a_n| \leq M$ for every positive integer n. By the Bolzano-Weierstrass theorem, the set $\mathscr{L}_a$ is not empty. The set $\mathscr{L}_a$ is bounded above by M; so by the least-upper-bound axiom, lub $\mathscr{L}_a$ exists. Similarly, the set $\mathscr{L}_a$ is bounded below by $-M$; so glb $\mathscr{L}_a$ exists.

Our first theorem follows immediately from Definition 20.1.

Theorem 20.2   Let $\{a_n\}$ be a bounded sequence. Then

$$\lim_{n \to \infty} \inf a_n \leq \lim_{n \to \infty} \sup a_n.$$

Example.   Consider the sequence $\{a_n\}$ defined by

$$a_n = (-1)^n, \qquad n = 1, 2, \ldots$$

The subsequence $\{a_{2n}\}$ has limit 1; so

$$1 \leq \lim_{n \to \infty} \sup a_n$$

On the other hand, if $\{a_{n_k}\}$ is any convergent subsequence of $\{a_n\}$,

$$a_{n_k} \leq 1, \qquad k = 1, 2, \ldots$$

and thus

$$\lim_{k \to \infty} a_{n_k} \leq 1$$

Thus 1 is an upper bound for $\mathscr{L}_a$, and so

$$\lim_{k \to \infty} \sup a_n \leq 1$$

Therefore

$$\lim_{n \to \infty} \sup a_n = 1$$

Similar methods show that $\lim \inf_{n \to \infty} a_n = -1$.

The following theorem can be used to characterize $\lim \sup_{n \to \infty} a_n$ and $\lim \inf_{n \to \infty} a_n$. That is if L and M are numbers satisfying the *conclusions* of

Theorem 20.3, then $L = \lim\sup_{n\to\infty} a_n$ and $M = \lim\inf_{n\to\infty} a_n$ (see Exercise 20.3).

Theorem 20.3 Let $\{a_n\}$ be a bounded sequence, and let $L = \lim\sup_{n\to\infty} a_n$ and $M = \lim\inf_{n\to\infty} a_n$.

(i) If $\varepsilon > 0$, there exist infinitely many positive integers n such that $L - \varepsilon < a_n$ and there exists a positive integer N_1 such that if $n \geq N_1$, then $a_n < L + \varepsilon$.

(ii) If $\varepsilon > 0$, there exist infinitely many positive integers n such that $a_n < M + \varepsilon$ and there exists a positive integer N_2 such that if $n \geq N_2$, $M - \varepsilon < a_n$.

Proof. In this proof we use the notations a_{n_k} and $a_{n,k}$ interchangeably. We establish part (i) leaving part (ii) as an exercise. Suppose that it is false that $L - \varepsilon < a_n$ for infinitely many positive integers n. Then there exists a positive integer N_1 such that if $n \geq N_1$, then $a_n \leq L - \varepsilon$. Let $\{a_{n,k}\}$ be a convergent subsequence of $\{a_n\}$. If $k \geq N_1$, we have $a_{n,k} \leq L - \varepsilon$. By Theorem 14.2, $\lim_{k\to\infty} a_{n,k} \leq L - \varepsilon$. It follows that $L \leq L - \varepsilon$ which is a contradiction. Therefore $L - \varepsilon < a_n$ for infinitely many positive integers n.

We next show that $L + \varepsilon \leq a_n$ for at most finitely many positive integers n, from which it follows that there exists a positive integer N such that if $n \geq N$, then $a_n < L + \varepsilon$. Suppose that $L + \varepsilon \leq a_n$ for infinitely many positive integers n. Then there exists a subsequence $\{a_{n,k}\}$ of $\{a_n\}$ such that $L + \varepsilon \leq a_{n,k}$ for $k = 1, 2, \ldots$. By the Bolzano-Weierstrass theorem, $\{a_{n,k}\}$ has a convergent subsequence $\{a_{n,k,j}\}$. Thus $L + \varepsilon \leq a_{n,k,j}$ for $j = 1, 2, \ldots$. By Theorem 14.2, $L + \varepsilon \leq \lim_{j\to\infty} a_{n,k,j} \leq L$ which is a contradiction. Therefore $L + \varepsilon \leq a_n$ for at most finitely many positive integers n. ∎

If $\{a_n\}$ is a convergent sequence, then $\{a_n\}$ is bounded; so $\lim\sup_{n\to\infty} a_n$ and $\lim\inf_{n\to\infty} a_n$ are defined. It is easy to show that in this case, $\lim_{n\to\infty} a_n = \lim\sup_{n\to\infty} a_n = \lim\inf_{n\to\infty} a_n$.

Theorem 20.4

(i) Let $\{a_n\}$ be a sequence such that

$$\lim_{n\to\infty} a_n = L$$

Then
$$\lim\sup_{n\to\infty} a_n = L = \lim\inf_{n\to\infty} a_n$$

(ii) Let $\{a_n\}$ be a bounded sequence such that

$$\lim\sup_{n\to\infty} a_n = L = \lim\inf_{n\to\infty} a_n$$

Then
$$\lim_{n\to\infty} a_n = L$$

Proof. (i) By Theorem 11.2, every subsequence of $\{a_n\}$ has limit L. Thus $\mathscr{L}_a = \{L\}$ and the conclusion follows.

(ii) Let $\varepsilon > 0$. By Theorem 20.3(i), there exists a positive integer N_1 such that if $n \geq N_1$, then $a_n < L + \varepsilon$. By Theorem 20.3(ii), there exists a positive integer N_2 such that if $n \geq N_2$, then $L - \varepsilon < a_n$. Therefore, if $n \geq \max\{N_1, N_2\}$, then

$$L - \varepsilon < a_n < L + \varepsilon$$

Therefore $\lim_{n \to \infty} a_n = L$. ∎

Theorem 20.4 provides another test for convergence.

Certain theorems may be proved for lim sup and lim inf which resemble theorems for ordinary limits. In some cases the theorems are identical.

Theorem 20.5 Let $\{a_n\}$ and $\{b_n\}$ be bounded sequences such that $a_n \leq b_n$ for every positive integer n. Then

$$\limsup_{n \to \infty} a_n \leq \limsup_{n \to \infty} b_n$$

and

$$\liminf_{n \to \infty} a_n \leq \liminf_{n \to \infty} b_n$$

Proof. In this proof we use the notations a_{n_k} and $a_{n,k}$ interchangeably. We will establish the inequality involving lim sup and leave the other inequality as an exercise.

Let $\{a_{n,k}\}$ be a convergent subsequence of $\{a_n\}$ with limit L. Then $\{b_{n,k}\}$ is a subsequence (though not necessarily convergent) of $\{b_n\}$. Since $\{b_{n,k}\}$ is a bounded sequence, $\{b_{n,k}\}$ has a convergent subsequence $\{b_{n,k,j}\}$ by Theorem 18.1. Since $a_{n,k,j} \leq b_{n,k,j}$, we have

$$L = \lim_{k \to \infty} a_{n,k} = \lim_{j \to \infty} a_{n,k,j} \leq \lim_{j \to \infty} b_{n,k,j} \leq \limsup_{n \to \infty} b_n$$

(why?). Thus $\limsup_{n \to \infty} b_n$ is an upper bound for $\mathscr{L}_a$, and we conclude that

$$\limsup_{n \to \infty} a_n \leq \limsup_{n \to \infty} b_n \qquad ∎$$

Theorem 20.6 Let $\{a_n\}$ and $\{b_n\}$ be bounded sequences. Then

$$\limsup_{n \to \infty} (a_n + b_n) \leq \limsup_{n \to \infty} a_n + \limsup_{n \to \infty} b_n$$

and

$$\liminf_{n \to \infty} a_n + \liminf_{n \to \infty} b_n \leq \liminf_{n \to \infty} (a_n + b_n)$$

Proof. In this proof we use the notations a_{n_k} and $a_{n,k}$ interchangeably. We will establish the inequality involving lim inf and leave the other inequality as an exercise.

Let $\{a_{n,k} + b_{n,k}\}$ be a convergent subsequence of $\{a_n + b_n\}$ with limit L.

The sequence $\{a_{n,k}\}$ has a convergent subsequence $\{a_{n,k,j}\}$, and the sequence $\{b_{n,k,j}\}$ has a convergent subsequence $\{b_{n,k,j,i}\}$. Now

$$L = \lim_{k \to \infty} (a_{n,k} + b_{n,k}) = \lim_{i \to \infty} (a_{n,k,j,i} + b_{n,k,j,i})$$

$$= \lim_{i \to \infty} a_{n,k,j,i} + \lim_{i \to \infty} b_{n,k,j,i}$$

$$\geq \lim_{n \to \infty} \inf a_n + \lim_{n \to \infty} \inf b_n$$

Thus $\mathscr{L}_{a+b}$ is bounded below by $\lim \inf_{n \to \infty} a_n + \lim \inf_{n \to \infty} b_n$, and hence

$$\lim_{n \to \infty} \inf (a_n + b_n) \geq \lim_{n \to \infty} \inf a_n + \lim_{n \to \infty} \inf b_n \qquad \blacksquare$$

As an example of the use of the previous results, we prove the following theorem.

Theorem 20.7 Let $\{a_n\}$ be a sequence such that

$$\lim_{n \to \infty} a_n = L$$

Then

$$\lim_{n \to \infty} \frac{a_1 + a_2 + \cdots + a_n}{n} = L$$

Proof. Let $\varepsilon > 0$. There exists a positive integer N such that if $n \geq N$, then

$$L - \varepsilon < a_n < L + \varepsilon$$

Let

$$b_n = \frac{a_1 + a_2 + \cdots + a_n}{n} \qquad \text{for } n \geq N$$

Now

$$b_n = \frac{a_1 + a_2 + \cdots + a_N}{n} + \frac{a_{N+1} + \cdots + a_n}{n}$$

and since

$$\frac{(n - N)(L - \varepsilon)}{n} < \frac{a_{N+1} + \cdots + a_n}{n} < \frac{(n - N)(L + \varepsilon)}{n}$$

we have

$$\frac{C}{n} + \frac{(n - N)(L - \varepsilon)}{n} < b_n < \frac{C}{n} + \frac{(n - N)(L + \varepsilon)}{n} \qquad (20.1)$$

where

$$C = a_1 + a_2 \cdots + a_N$$

We apply Theorem 20.5 to inequality (20.1) to get

$$\lim_{n \to \infty} \sup \left(\frac{C}{n} + \frac{(n - N)(L - \varepsilon)}{n} \right) \leq \lim_{n \to \infty} \sup b_n$$

$$\leq \lim_{n \to \infty} \sup \left(\frac{C}{n} + \frac{(n - N)(L + \varepsilon)}{n} \right)$$

Since
$$\lim_{n \to \infty} \left(\frac{C}{n} + \frac{(n-N)(L-\varepsilon)}{n} \right) = 0 + (L - \varepsilon)$$

and
$$\lim_{n \to \infty} \left(\frac{C}{n} + \frac{(n-N)(L+\varepsilon)}{n} \right) = 0 + (L + \varepsilon)$$

by Theorem 20.4(i), we conclude that

$$L - \varepsilon \leq \limsup_{n \to \infty} b_n \leq L + \varepsilon, \qquad \text{for every } \varepsilon > 0$$

By Exercise 4.9,

$$L \leq \limsup_{n \to \infty} b_n \leq L$$

and so
$$L = \limsup_{n \to \infty} b_n$$

Similarly,
$$L = \liminf_{n \to \infty} b_n$$

By Theorem 20.4(ii),

$$L = \lim_{n \to \infty} b_n \qquad\blacksquare$$

We will later have occasion to use the following rather special result.

Theorem 20.8   Let $\{a_n\}$ and $\{b_n\}$ be sequences such that

$$\lim_{n \to \infty} a_n \geq 0$$

and $\{b_n\}$ is a bounded sequence. Then

$$\limsup_{n \to \infty} a_n b_n = \limsup_{n \to \infty} a_n \limsup_{n \to \infty} b_n = \lim_{n \to \infty} a_n \limsup_{n \to \infty} b_n$$

Proof.   The last equality follows from Theorem 20.4(i).

Let $L = \lim_{n \to \infty} a_n$. If $L = 0$, then $\lim_{n \to \infty} a_n b_n = 0$ by Theorem 13.3, and the conclusion follows.

Suppose $L > 0$. Let $\{b_{n_k}\}$ be a convergent subsequence of $\{b_n\}$. Then

$$\limsup_{n \to \infty} a_n b_n \geq \lim_{k \to \infty} a_{n_k} b_{n_k} = \lim_{k \to \infty} a_{n_k} \lim_{k \to \infty} b_{n_k} = L \lim_{k \to \infty} b_{n_k}$$

Thus
$$\frac{\displaystyle\limsup_{n \to \infty} a_n b_n}{L} \geq \lim_{n \to \infty} b_{n_k}$$

It follows that

$$\frac{\displaystyle\limsup_{n \to \infty} a_n b_n}{L} \geq \limsup_{n \to \infty} b_n$$

and hence
$$\limsup_{n \to \infty} a_n b_n \geq L \limsup_{n \to \infty} b_n$$

Let $\{a_{m_k} b_{m_k}\}$ be a convergent subsequence of $\{a_n b_n\}$. Since $\{b_{m_k}\}$ is the product of the convergent sequences $\{a_{m_k} b_{m_k}\}$ and $\{1/a_{m_k}\}$, it follows that $\{b_{m_k}\}$ converges.

Therefore, $\dfrac{1}{L} \lim_{k \to \infty} a_{m_k} b_{m_k} = \lim_{k \to \infty} \dfrac{1}{a_{m_k}} (a_{m_k} b_{m_k}) = \lim_{k \to \infty} b_{m_k} \leq \limsup_{n \to \infty} b_n$

Thus
$$\lim_{k \to \infty} a_{m_k} b_{m_k} \leq L \limsup_{n \to \infty} b_n$$

It follows that
$$\limsup_{n \to \infty} a_n b_n \leq L \limsup_{n \to \infty} b_n$$

and thus,
$$\limsup_{n \to \infty} a_n b_n = L \limsup_{n \to \infty} b_n = \lim_{n \to \infty} a_n \limsup_{n \to \infty} b_n$$
$$= \limsup_{n \to \infty} a_n \limsup_{n \to \infty} b_n \qquad \blacksquare$$

Exercises

20.1 Prove that $\liminf_{n \to \infty} (-1)^n = -1$.

20.2 Prove Theorem 20.3(ii).

20.3 Let L and M be numbers satisfying the conclusion of Theorem 20.3. Prove that $L = \limsup_{n \to \infty} a_n$ and $M = \liminf_{n \to \infty} a_n$.

20.4 Establish the inequality involving lim inf in Theorem 20.5.

20.5 Establish the inequality involving lim sup in Theorem 20.6.

20.6 Compute $\limsup_{n \to \infty} a_n$ and $\liminf_{n \to \infty} a_n$, where $a_n =$

   (a) $\dfrac{1}{n}$

   (b) $\left(1 + \dfrac{1}{n}\right)^n$

   (c) $(-1)^n \left(1 - \dfrac{1}{n}\right)$

20.7 Compute $\limsup_{n \to \infty} a_n$ and $\liminf_{n \to \infty} a_n$ and $\mathscr{L}_a$, where $a_1, a_2, \ldots$ is an enumeration of the rational numbers in the closed interval $[0, 1]$.

20.8 Let $\{a_n\}$ be a bounded sequence. Prove that there exist subsequences $\{a_{n_k}\}$ and $\{a_{m_k}\}$ of $\{a_n\}$ such that
$$\lim_{k \to \infty} a_{n_k} = \limsup_{n \to \infty} a_n \qquad \lim_{k \to \infty} a_{m_k} = \liminf_{n \to \infty} a_n$$

20.9 Let $\{a_n\}$ be a bounded sequence such that every convergent subsequence of $\{a_n\}$ has limit L. Prove that $\lim_{n \to \infty} a_n = L$.

20.10 Give an example of a sequence $\{a_n\}$ such that the sequence
$$\left\{\frac{a_1 + a_2 + \cdots + a_n}{n}\right\}$$
converges but $\{a_n\}$ diverges.

20.11 Prove that if $\{a_{n_k}\}$ is a subsequence of a bounded sequence $\{a_n\}$, then
$$\limsup_{k \to \infty} a_{n_k} \leq \limsup_{n \to \infty} a_n$$

20.12 Let $\{a_n\}$ be a bounded sequence. Prove that
$$\limsup_{n \to \infty} a_n = -\liminf_{n \to \infty} (-a_n)$$

20.13 Let $\{a_n\}$ and $\{b_n\}$ be sequences such that $\{a_n\}$ is convergent and $\{b_n\}$ is bounded. Prove that
$$\limsup_{n \to \infty} (a_n + b_n) = \limsup_{n \to \infty} a_n + \limsup_{n \to \infty} b_n$$
and
$$\liminf_{n \to \infty} (a_n + b_n) = \liminf_{n \to \infty} a_n + \liminf_{n \to \infty} b_n$$

20.14 Let $\{a_n\}$ be a bounded sequence. Suppose that for every bounded sequence $\{b_n\}$ we have
$$\limsup_{n \to \infty} (a_n + b_n) = \limsup_{n \to \infty} a_n + \limsup_{n \to \infty} b_n$$
Prove that $\{a_n\}$ is convergent.

20.15 Let $\{a_n\}$ and $\{b_n\}$ be bounded nonnegative sequences. Prove that
$$\liminf_{n \to \infty} a_n \liminf_{n \to \infty} b_n \leq \liminf_{n \to \infty} a_n b_n \leq \limsup_{n \to \infty} a_n b_n$$
$$\leq \limsup_{n \to \infty} a_n \limsup_{n \to \infty} b_n$$

20.16 Show (by giving examples) that any of the inequalities of Theorem 20.6 and Exercise 20.15 may be strict.

20.17 Show (by giving examples) that the first and third inequalities of Exercise 20.15 may fail if the nonnegative hypothesis is omitted.

20.18 State and prove a theorem analogous to Theorem 20.8 where it is assumed that $\lim_{n \to \infty} a_n \leq 0$.

20.19 Let $\{a_n\}$ and $\{b_n\}$ be sequences such that $\lim_{n \to \infty} a_n < 0$ and $\{b_n\}$ is a bounded sequence. Suppose that
$$\limsup_{n \to \infty} a_n b_n = \limsup_{n \to \infty} a_n \limsup_{n \to \infty} b_n$$
Prove that $\{b_n\}$ is convergent.

20.20 Let $\{a_n\}$ be a sequence of positive numbers such that $\lim_{n \to \infty} a_n = L$. Prove that
$$\lim_{n \to \infty} (a_1 a_2 \cdots a_n)^{1/n} = L$$

20.21 (Squeeze Theorem) Let $\{a_n\}$, $\{b_n\}$, and $\{c_n\}$ be bounded sequences such that
$$a_n \leq b_n \leq c_n$$
for every positive integer n and
$$\limsup_{n \to \infty} c_n \leq \liminf_{n \to \infty} a_n$$

Prove that

$$\lim_{n \to \infty} a_n = \lim_{n \to \infty} b_n = \lim_{n \to \infty} c_n$$

20.22 Let $\{a_n\}$ and $\{b_n\}$ be sequences such that

$$b_{n+1} = a_n + a_{n+1}$$

for every positive integer n, and suppose that $\{b_n\}$ is convergent. Prove that $\lim_{n \to \infty} (a_n/n) = 0$. Give an example to show that $\{a_n\}$ need not converge.

21. The lim sup and lim inf of Unbounded Sequences

In this section we first give another characterization of $\lim \sup_{n \to \infty} a_n$ and $\lim \inf_{n \to \infty} a_n$ for bounded sequences $\{a_n\}$ and then show how this characterization may be used to extend the definitions of $\lim \sup_{n \to \infty} a_n$ and $\lim \inf_{n \to \infty} a_n$ to unbounded sequences $\{a_n\}$.

If $\{a_n\}$ is a bounded sequence, then

$$A_n = \text{lub} \{a_n, a_{n+1}, \ldots\}$$

exists for every positive integer n. Since

$$\{a_{n+1}, a_{n+2}, \ldots\} \subset \{a_n, a_{n+1}, \ldots\}$$

it follows that

$$A_{n+1} \leq A_n \quad \text{for } n = 1, 2, \ldots$$

Because $\{a_n\}$ is bounded, the monotone sequence $\{A_n\}$ is also bounded and by Theorem 16.2, $\{A_n\}$ is convergent. The next theorem states that $\lim_{n \to \infty} A_n = \lim \sup_{n \to \infty} a_n$ and similarly,

$$\lim_{n \to \infty} \text{glb} \{a_n, a_{n+1}, \ldots\} = \lim_{n \to \infty} \inf a_n$$

Theorem 21.1 is often taken as the definition of lim sup and lim inf.

Theorem 21.1 Let $\{a_n\}$ be a bounded sequence. Then

$$\lim_{n \to \infty} \sup a_n = \lim_{n \to \infty} \text{lub} \{a_n, a_{n+1}, \ldots\}$$

and

$$\lim_{n \to \infty} \inf a_n = \lim_{n \to \infty} \text{glb} \{a_n, a_{n+1}, \ldots\}$$

Proof. We prove the first equation leaving the second as an exercise. Let $L = \lim \sup_{n \to \infty} a_n$ and let

$$A_n = \text{lub} \{a_n, a_{n+1}, \ldots\} \quad \text{for } n = 1, 2, \ldots$$

Since $a_n \leq A_n$ for every positive integer n, we have by Theorems 20.4 and

20.5 that

$$L = \lim_{n \to \infty} \sup a_n \le \lim_{n \to \infty} \sup A_n = \lim_{n \to \infty} A_n$$

Let $\varepsilon > 0$. By Theorem 20.3(i), there exists a positive integer N such that if $n \ge N$, then $a_n < L + \varepsilon$. It follows that $A_n \le L + \varepsilon$ for $n \ge N$. By Theorem 14.2, $\lim_{n \to \infty} A_n \le L + \varepsilon$. Since $\lim_{n \to \infty} A_n \le L + \varepsilon$ for every $\varepsilon > 0$, we have $\lim_{n \to \infty} A_n \le L$. Therefore $\lim_{n \to \infty} A_n = L$. ∎

Let $\{a_n\}$ be a sequence. If $\{a_n\}$ is bounded above, we let

$$A_n = \text{lub } \{a_n, a_{n+1}, \ldots\} \qquad \text{for } n = 1, 2, \ldots$$

Then $\{A_n\}$ is a decreasing sequence, and thus either $\{A_n\}$ converges or $\lim_{n \to \infty} A_n = -\infty$. Guided by Theorem 21.1, we would define $\lim \sup_{n \to \infty} a_n = \lim_{n \to \infty} A_n$. If $\{a_n\}$ is not bounded above, we could let $A_n = \infty$ for $n = 1, 2, \ldots$. If we want $\lim \sup_{n \to \infty} a_n = \lim_{n \to \infty} A_n$, we would have to define $\lim \sup_{n \to \infty} a_n = \infty$ in this case. This discussion motivates the next definition.

Definition 21.2   Let $\{a_n\}$ be a real sequence.
 (i) If $\{a_n\}$ is not bounded above, we define $\lim \sup_{n \to \infty} a_n = \infty$.
 (ii) If $\{a_n\}$ is bounded above, we define $\lim \sup_{n \to \infty} a_n = \lim_{n \to \infty} \text{lub } \{a_n, a_{n+1}, \ldots\}$.
 (iii) If $\{a_n\}$ is not bounded below, we define $\lim \inf_{n \to \infty} a_n = -\infty$.
 (iv) If $\{a_n\}$ is bounded below, we define $\lim \inf_{n \to \infty} a_n = \lim_{n \to \infty} \text{glb } \{a_n, a_{n+1}, \ldots\}$.

By Theorem 21.1, Definition 21.2 is consistent with Definition 20.1.

Examples.   (i) Let $\{a_n\}$ be the sequence defined by $a_n = n$ for $n = 1, 2, \ldots$. Since $\{a_n\}$ is not bounded above, $\lim \sup_{n \to \infty} a_n = \infty$. Now $\{a_n\}$ is bounded below and since

$$\text{glb } \{a_n, a_{n+1}, \ldots\} = n \qquad \text{for } n = 1, 2, \ldots$$

it follows that $\lim \inf_{n \to \infty} a_n = \infty$.
 (ii) Let $\{a_n\}$ be the sequence defined by

$$a_n = \begin{cases} -n & \text{if } n \text{ is even} \\ 0 & \text{if } n \text{ is odd} \end{cases}$$

Since $\{a_n\}$ is not bounded below, $\lim \inf_{n \to \infty} a_n = -\infty$. Since

$$\text{lub } \{a_n, a_{n+1}, \ldots\} = 0 \qquad \text{for } n = 1, 2, \ldots$$

it follows that $\lim \sup_{n \to \infty} a_n = 0$.

Many of the theorems of Section 20 remain valid for unbounded sequences if we accept the following conventions. For any x in $\mathbf{R}$, $-\infty < x$, $x < \infty$, and $-\infty < \infty$. The set $\mathbf{R} \cup \{-\infty\} \cup \{\infty\}$ with this ordering is called the *extended real number system*. The symbols "$-\infty$" and "∞" are *not* real numbers, but are introduced for convenience. We conclude by giving an example of an extension of an earlier result (Theorem 20.2) to arbitrary sequences.

Theorem 21.3 If $\{a_n\}$ is a sequence, then

$$\liminf_{n \to \infty} a_n \leq \limsup_{n \to \infty} a_n$$

Proof. If $\{a_n\}$ is not bounded above, then

$$\liminf_{n \to \infty} a_n \leq \infty = \limsup_{n \to \infty} a_n$$

If $\{a_n\}$ is not bounded below, then

$$\liminf_{n \to \infty} a_n = -\infty \leq \limsup_{n \to \infty} a_n$$

If $\{a_n\}$ is bounded above and below, the conclusion follows from Theorem 20.2. ∎

Exercises

21.1 Prove the second equation of Theorem 21.1.

21.2 Let $A_n = \text{lub } \{a_n, a_{n+1}, \ldots\}$ and $B_n = \text{glb } \{a_n, a_{n+1}, \ldots\}$ for $n = 1, 2, \ldots$. Compute A_n, B_n, $\lim_{n \to \infty} A_n$, and $\lim_{n \to \infty} B_n$, where $a_n =$

(a) $(-1)^n$

(b) $\dfrac{1}{n}$

(c) $\left(1 + \dfrac{1}{n}\right)^n$

(d) $\dfrac{(-1)^n}{n}$

(e) $(-1)^n \left(1 - \dfrac{1}{n}\right)$

21.3 Give another proof of the Bolzano-Weierstrass theorem by showing that if $\{a_n\}$ is a bounded sequence, there exists a subsequence $\{a_{n_k}\}$ of $\{a_n\}$ which converges to $\lim_{n \to \infty} \text{lub } \{a_n, a_{n+1}, \ldots\}$.

21.4 (a) Let $\{a_n\}$ be a sequence which is bounded above. Prove that either $\{a_n\}$ has a convergent subsequence in which case $\limsup_{n \to \infty} a_n = \text{lub } \mathscr{L}_a$ or $\{a_n\}$ diverges to $-\infty$ in which case $\limsup_{n \to \infty} a_n = -\infty$.

(b) State and prove the result corresponding to (a) for $\liminf_{n \to \infty} a_n$.

21.5 Compute lim sup and lim inf of the following sequences:
   (a) $0, 1, 0, 2, 0, 3, \ldots$
   (b) $1, -1, 2, -2, 3, -3, \ldots$
   (c) $-1, -2, -3, \ldots$

21.6 Let $\{a_n\}$ be a sequence. Prove that there exist subsequences $\{a_{n_k}\}$ and $\{a_{m_k}\}$ of $\{a_n\}$ such that

$$\lim_{k \to \infty} a_{n_k} = \limsup_{n \to \infty} a_n \quad \text{and} \quad \lim_{k \to \infty} a_{m_k} = \liminf_{n \to \infty} a_n$$

21.7 Prove Theorem 20.4 where L may assume the values $-\infty$ or ∞.

21.8 Prove Theorem 20.5 for arbitrary sequences $\{a_n\}$ and $\{b_n\}$.

21.9 State and prove a version of Theorem 20.3 valid for arbitrary sequences.

V

Infinite Series

The present chapter can be viewed as an extension of the ideas developed in Chapter IV on sequences. Infinite series play an important theoretical and practical role in analysis.

22. The Sum of an Infinite Series

In order to sum an infinite sequence of real numbers we must employ the notion of *limit*. We begin with the definition of an infinite series.

Definition 22.1   Let $\{a_n\}$ be a sequence. For each positive integer n, let

$$s_n = a_1 + a_2 + \cdots + a_n = \sum_{k=1}^{n} a_k$$

An *infinite series* is the ordered pair of sequences $(\{a_n\}, \{s_n\})$.

The number a_n is called the *nth term* of the infinite series, and the number s_n is called the *nth partial sum* of the infinite series. Instead of using the cumbersome ordered pair notation for an infinite series, we will use the notation

$$\sum_{n=1}^{\infty} a_n \qquad \text{or} \qquad a_1 + a_2 + a_3 + \cdots$$

to represent an infinite series. It should be understood that the symbolism $\sum_{n=1}^{\infty} a_n$ denotes *two sequences*, namely the sequence $\{a_n\}$ of terms of the infinite series and the sequence $\{s_n\}$ of partial sums of the infinite series.

It is important to distinguish between the sequence $\{a_n\}$ of terms and the sequence $\{s_n\}$ of partial sums of an infinite series $\sum_{n=1}^{\infty} a_n$. We illustrate this with an example. For the infinite series $\sum_{n=1}^{\infty} (-1)^n$, we have $a_n = (-1)^n$ and

$$s_n = \begin{cases} -1 & \text{if } n \text{ is odd} \\ 0 & \text{if } n \text{ is even} \end{cases} \qquad \text{for } n = 1, 2, \ldots$$

The sequence $\{(-1)^{2n}\}$ is a subsequence of $\{(-1)^n\}$. The nth term of the series $\sum_{n=1}^{\infty} (-1)^{2n}$ is 1, and the nth partial sum of this series is $t_n = n$ for $n = 1, 2, \ldots$. Notice that although $\{(-1)^{2n}\}$ is a subsequence of $\{(-1)^n\}$, $\{t_n\}$ bears little resemblance to $\{s_n\}$.

Sometimes it is convenient to begin the index of an infinite series with an integer other than 1. For example, we may consider the series $\sum_{n=0}^{\infty} a_n$ or, in general, $\sum_{n=p}^{\infty} a_n$, where p is an integer. The sequence of partial sums as defined in Definition 22.1 will be altered in the obvious way.

To sum an infinite series, we simply add on more and more "terms." This corresponds to taking the limit of the sequence of partial sums. We make this notion precise in the following definition.

Definition 22.2 Let $\sum_{n=1}^{\infty} a_n$ be an infinite series. If the sequence of partial sums $\{s_n\}$ ($s_n = a_1 + \cdots + a_n$) converges to L, we say that the *infinite series* $\sum_{n=1}^{\infty} a_n$ *converges to* L or that the *infinite series* $\sum_{n=1}^{\infty} a_n$ *has sum* L. If the sequence $\{s_n\}$ diverges, we say that the *infinite series* $\sum_{n=1}^{\infty} a_n$ *diverges*.

If the infinite series $\sum_{n=1}^{\infty} a_n$ converges we also use the symbolism $\sum_{n=1}^{\infty} a_n$ to denote its sum. We are using the notation $\sum_{n=1}^{\infty} a_n$ in two very different ways. For an arbitrary sequence $\{a_n\}$ we will speak of the infinite series $\sum_{n=1}^{\infty} a_n$. *Only* for a *convergent* infinite series will we write $\sum_{n=1}^{\infty} a_n = L$, where L is the sum of the series $\sum_{n=1}^{\infty} a_n$. The context will always make clear whether we are speaking of the infinite series or the sum of a convergent infinite series.

Our first theorem shows that if an infinite series is convergent, the terms of the series get small.

Theorem 22.3 If the infinite series $\sum_{n=1}^{\infty} a_n$ converges, then $\lim_{n \to \infty} a_n = 0$.

Proof. Suppose the infinite series $\sum_{n=1}^{\infty} a_n$ converges to L. Let $s_n = a_1 + \cdots + a_n$ be the nth partial sum. Then

$$s_{n+1} - s_n = a_{n+1}$$

and so

$$0 = L - L = \lim_{n \to \infty} s_{n+1} - \lim_{n \to \infty} s_n = \lim_{n \to \infty} (s_{n+1} - s_n) = \lim_{n \to \infty} a_{n+1}$$

Therefore

$$\lim_{n \to \infty} a_n = 0 \quad \blacksquare$$

By Theorem 22.3 the series $\sum_{n=1}^{\infty} n/(n + 1)$ and $\sum_{n=1}^{\infty} n^{1/n}$ diverge since

$$\lim_{n \to \infty} \frac{n}{n+1} = 1 = \lim_{n \to \infty} n^{1/n}$$

The converse of Theorem 22.3 is false. We will show later (see Section 24) that the series $\sum_{n=1}^{\infty} 1/n$ diverges even though $\lim_{n \to \infty} (1/n) = 0$.

Our first example of a convergent series is the geometric series.

Theorem 22.4 (The Geometric Series) Let a be a nonzero number. Then

(i) $\sum_{n=0}^{\infty} ar^n$ converges to $a/(1 - r)$ if $|r| < 1$.
(ii) $\sum_{n=0}^{\infty} ar^n$ diverges if $|r| \geq 1$.
    (r^0 is defined to be 1.)

Proof. First suppose $|r| < 1$. Then the nth partial sum of the series $\sum_{n=0}^{\infty} ar^n$ is

$$s_n = a + ar + \cdots + ar^{n-1} = \frac{a(1 - r^n)}{1 - r}$$

Since $|r| < 1$, $\lim_{n \to \infty} r^n = 0$, and hence $\lim_{n \to \infty} s_n = a/(1 - r)$.

If $|r| \geq 1$, then the sequence $\{ar^n\}$ does not converge to 0; so by Theorem 22.3, $\sum_{n=0}^{\infty} ar^n$ diverges. ∎

If $a = 0$, $\sum_{n=0}^{\infty} ar^n$ converges to 0 for any r.

Exercises

22.1 Prove that the series $\sum_{n=1}^{\infty} (-1)^n$ and $\sum_{n=1}^{\infty} n^2/(1 + n)$ diverge.

22.2 Prove that the arithmetic series $\sum_{n=1}^{\infty} (a + nb)$ converges if and only if $a = b = 0$.

22.3 Suppose $|x| < 1$. Prove that the series $\sum_{n=0}^{\infty} x^{2n}$ converges and find its sum.

22.4* Prove that the series $\sum_{n=1}^{\infty} 1/n(n + 1)$ converges and find its sum.

22.5 Prove that the series $\sum_{n=1}^{\infty} n/(n + 1)!$ converges and find its sum.

22.6 Suppose that the series $\sum_{n=1}^{\infty} a_n$ converges. Prove that the series $\sum_{n=p}^{\infty} a_n$ converges for every positive integer p and $\lim_{p \to \infty} \sum_{n=p}^{\infty} a_n = 0$.

22.7 Let $\{a_n\}$ and $\{b_n\}$ be sequences such that for some positive integer N,

$$a_n = b_n \quad \text{if } n \geq N$$

Prove that if $\sum_{n=1}^{\infty} a_n$ is convergent, then $\sum_{n=1}^{\infty} b_n$ is convergent. If the sum of the series $\sum_{n=1}^{\infty} a_n$ is L, find the sum of the series $\sum_{n=1}^{\infty} b_n$. (This exercise shows that altering a finite number of terms of an infinite series does not affect the convergence of the series, but that it may affect the sum of the series.)

22.8 Prove that the series $\sum_{n=1}^{\infty} a_n$ converges if and only if for every $\varepsilon > 0$, there exists a positive integer N such that if $n \geq m \geq N$, then

$$\left| \sum_{k=m}^{n} a_k \right| < \varepsilon$$

23. Algebraic Operations on Series

Because convergence of the series $\sum_{n=1}^{\infty} a_n$ is defined in terms of convergence of a sequence (namely, the sequence of partial sums), we may use many of the theorems about sequences in Chapter IV to prove theorems about series.

Theorem 23.1 Let $\sum_{n=1}^{\infty} a_n$ and $\sum_{n=1}^{\infty} b_n$ be series, and let c be a real number. If $\sum_{n=1}^{\infty} a_n$ converges to L and $\sum_{n=1}^{\infty} b_n$ converges to M, then $\sum_{n=1}^{\infty} (a_n + b_n)$ converges to $L + M$, $\sum_{n=1}^{\infty} (a_n - b_n)$ converges to $L - M$, and $\sum_{n=1}^{\infty} ca_n$ converges to cL.

Proof. Suppose $\sum_{n=1}^{\infty} a_n$ converges to L and $\sum_{n=1}^{\infty} b_n$ converges to M.

Let $$s_n = a_1 + \cdots + a_n$$

and $$t_n = b_1 + \cdots + b_n$$

be the nth partial sums of the series $\sum_{n=1}^{\infty} a_n$ and $\sum_{n=1}^{\infty} b_n$, respectively. Then $\{s_n + t_n\}$ is the sequence of partial sums of the series $\sum_{n=1}^{\infty} (a_n + b_n)$. Now

$$\lim_{n \to \infty} (s_n + t_n) = L + M$$

and hence $\sum_{n=1}^{\infty} (a_n + b_n)$ converges to $L + M$. Similarly, $\sum_{n=1}^{\infty} (a_n - b_n)$ converges to $L - M$ and $\sum_{n=1}^{\infty} ca_n$ converges to cL. ∎

There is no simple analogue of Theorem 23.1 for products. It is possible that the series $\sum_{n=1}^{\infty} a_n$ and $\sum_{n=1}^{\infty} b_n$ both converge yet the series $\sum_{n=1}^{\infty} a_n b_n$ diverges. We will show later (see Sections 24 and 25) that $\sum_{n=1}^{\infty} (-1)^n / \sqrt{n}$ converges, but that $\sum_{n=1}^{\infty} ((-1)^n / \sqrt{n})((-1)^n / \sqrt{n})) = \sum_{n=1}^{\infty} 1/n$ diverges. Even if $\sum_{n=1}^{\infty} a_n$ converges to L, $\sum_{n=1}^{\infty} b_n$ converges to M, and $\sum_{n=1}^{\infty} a_n b_n$ converges, the sum of $\sum_{n=1}^{\infty} a_n b_n$ is in general not remotely related to the product LM. The complication is that the product of the partial sums

$$\sum_{k=1}^{n} a_k \sum_{k=1}^{n} b_k = a_1 b_1 + a_1 b_2 + \cdots + a_n b_1 + \cdots + a_n b_n$$

involves cross terms $a_i b_j$, where $i \neq j$, whereas the partial sums

$$a_1 b_1 + \cdots + a_n b_n$$

of the series $\sum_{n=1}^{\infty} a_n b_n$ involve no terms of the form $a_i b_j$ ($i \neq j$).

In Section 29 (Theorem 29.9) we will prove a theorem which deals with the product of two infinite series.

Exercises

23.1 Complete the proof of Theorem 23.1 by showing that if $\sum_{n=1}^{\infty} a_n$ converges to L and $\sum_{n=1}^{\infty} b_n$ converges to M, then $\sum_{n=1}^{\infty} (a_n - b_n)$ converges to $L - M$ and $\sum_{n=1}^{\infty} ca_n$ converges to cL.

23.2* Find the sum of the series

$$\sum_{n=1}^{\infty} \left[\left(\frac{1}{2}\right)^n + \frac{2}{n(n+1)} \right]$$

23.3 Prove that the series $\sum_{n=1}^{\infty} a_n$ converges if and only if the series $\sum_{n=1}^{\infty} ca_n$ converges $(c \neq 0)$.

23.4 Prove that if $\sum_{n=1}^{\infty} a_n$ converges and $\sum_{n=1}^{\infty} b_n$ diverges, then $\sum_{n=1}^{\infty} (a_n + b_n)$ diverges.

23.5 Give an example of divergent series $\sum_{n=1}^{\infty} a_n$ and $\sum_{n=1}^{\infty} b_n$ such that $\sum_{n=1}^{\infty} (a_n + b_n)$ converges.

23.6 Let $a_n = \left(\frac{1}{2}\right)^n$ and $b_n = \left(\frac{1}{3}\right)^n$, $n = 1, 2, \dots$. Find the sums of the series

$$\sum_{n=1}^{\infty} a_n \qquad \sum_{n=1}^{\infty} b_n \qquad \text{and} \qquad \sum_{n=1}^{\infty} a_n b_n$$

23.7 Suppose that for constants c and d, where $c \neq d$, $\sum_{n=1}^{\infty} (a_{2n} + ca_{2n-1})$ and $\sum_{n=1}^{\infty} (a_{2n} + da_{2n-1})$ converge. Prove that $\sum_{n=1}^{\infty} a_n$ converges.

24. Series with Nonnegative Terms

One of the problems studied in the theory of infinite series is that of determining whether a given infinite series converges or diverges. In this section we will prove two tests for convergence of series whose terms are nonnegative. Another and often deeper problem is to determine the sum of a convergent infinite series. We will show (Corollary 24.3) that $\sum_{n=1}^{\infty} 1/n^2$ converges. However, to show that $\sum_{n=1}^{\infty} 1/n^2 = \pi^2/6$, we will have to employ more substantial methods; such as the theory of Fourier series (see the example following Corollary 78.7).

Theorem 24.1 which we will deduce from the test for convergence of a monotone *sequence* (Theorem 16.2) is the foundation of our subsequent results.

Theorem 24.1 Let $\sum_{n=1}^{\infty} a_n$ be a series with nonnegative terms. Then $\sum_{n=1}^{\infty} a_n$ converges if and only if the sequence of partial sums $\{s_n\}$ is bounded.

Proof. Let $\sum_{n=1}^{\infty} a_n$ be a series with nonnegative terms. Since $a_n \geq 0$ for every $n \in \mathbf{P}$, the sequence $\{s_n\}$ is increasing. By Theorem 16.2, $\{s_n\}$ converges if and only if $\{s_n\}$ is bounded. ∎

Like Theorem 16.2, Theorem 24.1 gives a test for convergence which does not involve a preknowledge of the limit.

If a series $\sum_{n=1}^{\infty} a_n$ with nonnegative terms converges to L and $\{s_n\}$ is the sequence of partial sums of $\sum_{n=1}^{\infty} a_n$, then $s_n \leq L$ for every positive integer n, for the proof of Theorem 16.2 shows that L is the least upper bound of the set $\{s_n \mid n \in \mathbf{P}\}$. The inequality $s_n \leq L$ may also be written $\sum_{k=1}^{n} a_k \leq \sum_{k=1}^{\infty} a_k$.

Example. $\sum_{n=1}^{\infty} 1/n$ diverges.

Proof. Let $s_n = 1 + 1/2 + \cdots + 1/n$. Then

$$s_1 = 1 \geq 1$$
$$s_2 = 1 + \tfrac{1}{2} \geq \tfrac{3}{2}$$
$$s_4 = s_2 + \tfrac{1}{3} + \tfrac{1}{4} \geq \tfrac{3}{2} + \tfrac{1}{4} + \tfrac{1}{4} = 2$$
$$s_8 = s_4 + \tfrac{1}{5} + \tfrac{1}{6} + \tfrac{1}{7} + \tfrac{1}{8} \geq 2 + \tfrac{1}{8} + \tfrac{1}{8} + \tfrac{1}{8} + \tfrac{1}{8} = \tfrac{5}{2}$$

In general one may show that $s_{2^n} \geq (n+2)/2$. Since the sequence $\{s_n\}$ is unbounded, $\sum_{n=1}^{\infty} 1/n$ diverges by Theorem 24.1.

The method of the preceding example may be used to prove the following theorem.

Theorem 24.2 (2^n Test) Let $\{a_n\}$ be a decreasing sequence of nonnegative numbers. Then $\sum_{n=1}^{\infty} a_n$ converges if and only if $\sum_{n=1}^{\infty} 2^n a_{2^n}$ converges.

Proof. First assume that $\sum_{n=1}^{\infty} 2^n a_{2^n}$ converges. Now

$$a_1 \leq a_1$$
$$a_2 + a_3 \leq a_2 + a_2 = 2a_2$$
$$a_4 + a_5 + a_6 + a_7 \leq 4a_4$$

and in general

$$a_{2^n} + a_{2^n+1} + \cdots + a_{2^{n+1}-1} \leq 2^n a_{2^n}$$

If we add these inequalities, we obtain

$$s_{2^{n+1}-1} = \sum_{k=1}^{2^{n+1}-1} a_k \leq \sum_{k=0}^{n} 2^k a_{2^k} \leq \sum_{k=0}^{\infty} 2^k a_{2^k}$$

Thus the subsequence $\{s_{2^{n+1}-1}\}$ of the sequence of partial sums of $\sum_{n=1}^{\infty} a_n$ is bounded by $\sum_{k=0}^{\infty} 2^k a_{2^k}$. Since the sequence $\{s_n\}$ is a monotone, it follows that the sequence of partial sums of $\sum_{n=1}^{\infty} a_n$ is bounded. By Theorem 24.1, $\sum_{n=1}^{\infty} a_n$ converges.

Now assume that $\sum_{n=1}^{\infty} a_n$ converges. Then

$$a_3 + a_4 \geq 2a_4$$
$$a_5 + a_6 + a_7 + a_8 \geq 4a_8$$

and in general

$$a_{2^n+1} + \cdots + a_{2^{n+1}} \geq 2^n a_{2^{n+1}}$$

so

$$\sum_{k=3}^{\infty} a_n \geq \frac{1}{2} \sum_{k=1}^{n} 2^{k+1} a_{2^{k+1}}$$

Thus the sequence of partial sums of the series $\sum_{n=1}^{\infty} 2^n a_{2^n}$ is bounded, and so $\sum_{n=1}^{\infty} 2^n a_{2^n}$ converges by Theorem 24.1. ∎

Corollary 24.3 The series $\sum_{n=1}^{\infty} 1/n^s$ diverges if $s \leq 1$ and converges if $s > 1$.

Proof. If $s < 0$, $\lim_{n \to \infty} (1/n^s) \neq 0$; so $\sum_{n=1}^{\infty} 1/n^s$ diverges by Theorem 22.3.

Suppose $s \geq 0$. By Theorem 24.2, $\sum_{n=1}^{\infty} 1/n^s$ converges if and only if

$$\sum_{n=1}^{\infty} 2^n \frac{1}{(2^n)^s} = \sum_{n=1}^{\infty} 2^{(1-s)n}$$

converges. By Theorem 22.4, the geometric series $\sum_{n=1}^{\infty} 2^{(1-s)n}$ converges if $2^{(1-s)} < 1$ and diverges if $2^{(1-s)} \geq 1$. Thus $\sum_{n=1}^{\infty} 1/n^s$ converges if $1 - s < 0$ and diverges if $1 - s \geq 0$. ∎

Exercises

24.1 Use induction to prove that if $s_n = 1 + 1/2 + \cdots + 1/n$, then $s_{2^n} \geq (n+2)/2$.

24.2 Assume that there exists an increasing function L from $[2, \infty)$ into $(0, \infty)$ which satisfies $L(x^n) = nL(x)$. (The natural logarithm is such a function.) Determine whether the following series converge or diverge.

(a) $\sum_{n=2}^{\infty} \frac{1}{nL(n)}$

(b) $\sum_{n=2}^{\infty} \frac{1}{L(n)}$

24.3* Prove that $\sum_{n=1}^{\infty} 1/n^2 \leq 2$.

24.4 Let $\sum_{n=1}^{\infty} a_n$ be a series with nonnegative terms which diverges, and let $\{s_n\}$ be the sequence of partial sums. Prove that $\lim_{n \to \infty} s_n = \infty$. (In this case we write $\sum_{n=1}^{\infty} a_n = \infty$.)

24.5 Let $\sum_{n=1}^{\infty} a_n$ be a series with nonnegative terms which converges. Let $\{a_{n_k}\}$ be a subsequence of $\{a_n\}$. Prove that $\sum_{k=1}^{\infty} a_{n_k}$ converges and $\sum_{k=1}^{\infty} a_{n_k} \leq \sum_{n=1}^{\infty} a_n$.

24.6 Let $\sum_{n=1}^{\infty} a_n$ be a series with nonnegative terms which converges. Let $\{b_n\}$ be a bounded nonnegative sequence. Prove that $\sum_{n=1}^{\infty} a_n b_n$ converges.

24.7 Let $\{a_n\}$ and $\{b_n\}$ be sequences such that $b_n > 0$ and $a_n \geq 0$ for every positive integer n. Prove that if $\sum_{n=1}^{\infty} b_n$ converges and the sequence $\{a_n/b_n\}$ is decreasing, then $\sum_{n=1}^{\infty} a_n$ converges.

24.8   State and prove an analogue of Exercise 24.7 where it is assumed that $\sum_{n=1}^{\infty}$ b_n diverges.

24.9*   Prove that if $\{a_n\}$ is a decreasing sequence of positive numbers and $\sum_{n=1}^{\infty} a_n$ converges, then $\lim_{n \to \infty} na_n = 0$. Deduce that $\sum_{n=1}^{\infty} 1/n^s$ diverges if $0 \leq s \leq 1$.

25. The Alternating Series Test

In the last section we proved two tests for series with nonnegative terms. In this section we will prove a very useful test for series whose terms are alternately positive and negative. In Section 29 we will prove several generalizations of the alternating series test.

Theorem 25.1 (Alternating Series Test)   Let $\{a_n\}$ be a decreasing sequence such that $\lim_{n \to \infty} a_n = 0$. Then $\sum_{n=1}^{\infty} (-1)^{n+1} a_n$ converges.

Proof.   If $\{a_n\}$ is a decreasing sequence with limit 0, it follows that $a_n \geq 0$ for every positive integer n. (Why?) We consider the subsequence $\{s_{2n}\}$ of the sequence $\{s_n\}$ of partial sums of $\sum_{n=1}^{\infty} (-1)^{n+1} a_n$. Since $\{a_n\}$ is decreasing, $a_{2n+1} \geq a_{2n+2}$, so that

$$s_{2n+2} - s_{2n} = -a_{2n+2} + a_{2n+1} \geq 0$$

and the sequence $\{s_{2n}\}$ is increasing.

Now

$$s_{2n} = a_1 - a_2 + \cdots + a_{2n-1} - a_{2n}$$
$$= a_1 - (a_2 - a_3) - \cdots - (a_{2n-2} - a_{2n-1}) - a_{2n} \leq a_1$$

Thus the monotone sequence $\{s_{2n}\}$ is bounded. By Theorem 16.2, $\{s_{2n}\}$ converges.

Since $s_{2n-1} = s_{2n} + a_{2n}$, it follows that $\{s_{2n-1}\}$ converges and

$$\lim_{n \to \infty} s_{2n-1} = \lim_{n \to \infty} s_{2n} + \lim_{n \to \infty} a_{2n} = \lim_{n \to \infty} s_{2n}$$

Since $\{s_{2n}\}$ and $\{s_{2n-1}\}$ converge to the same limit, $\{s_n\}$ converges. ∎

Corollary 25.2   If $s > 0$, then $\sum_{n=1}^{\infty} (-1)^{n+1}/n^s$ converges.

Proof.   Apply the alternating series test to the sequence $\{1/n^s\}$. ∎

Exercises

25.1   Prove that the series $\sum_{n=1}^{\infty} (-1)^{n+1}[e - (1 + 1/n)^n]$ and $\sum_{n=1}^{\infty} (-1)^{n+1} [n^{1/n} - 1]$ converge.

25.2 Let $\{a_n\}$ satisfy the hypotheses of the alternating series test. Let $\{s_n\}$ denote the sequence of partial sums of the series $\sum_{n=1}^{\infty} (-1)^{n+1}a_n$. Prove that the sequence $\{s_{2n-1}\}$ is decreasing and bounded below by 0.

25.3 Let $\{a_n\}$ be a decreasing sequence with limit 0. Let L and s_n denote, respectively, the sum and nth partial sum of the series $\sum_{n=1}^{\infty}(-1)^{n+1}a_n$. Prove that $|s_n - L| \le a_{n+1}$.

25.4 Give an example of a sequence $\{a_n\}$ of positive numbers such that $\lim_{n \to \infty} a_n = 0$, but the series $\sum_{n=1}^{\infty} (-1)^{n+1}a_n$ diverges.

25.5 In this exercise, let $\Delta a_n = a_n - a_{n+1}$.

(a) Verify the identity

$$\sum_{k=1}^{n} (-1)^{k+1}a_k = \frac{a_1}{2} + \frac{1}{2}\sum_{k=1}^{n-1} (-1)^{k+1} \Delta a_k + (-1)^{n+1}\frac{a_n}{2}$$

(b) Prove that if $\sum (-1)^{k+1}a_k$ converges, so does $\sum (-1)^{k+1}\Delta a_k$ and, moreover, $\sum_{k=1}^{\infty} (-1)^{k+1}a_k = a_1/2 + \frac{1}{2}\sum_{k=1}^{\infty} (-1)^{k+1} \Delta a_k$.

(c) Prove that if $\sum (-1)^{n+1}a_k$ converges, then

$$(-1)^{n+1}\frac{a_{n+1}}{2} + \frac{1}{2}\sum_{k=n+1}^{\infty} (-1)^{k+1} \Delta a_k \le \sum_{k=n+1}^{\infty} (-1)^{k+1}a_k$$

$$= (-1)^n\frac{a_n}{2} + \frac{1}{2}\sum_{k=n}^{\infty} (-1)^{k+1} \Delta a_k$$

(d)* Prove that if $\{a_n\}$ and $\{\Delta a_n\}$ strictly decrease to zero and $L = \sum_{k=1}^{\infty} (-1)^{k+1}a_k$, then $a_{n+1}/2 < |\sum_{k=1}^{n} (-1)^{k+1}a_k - L| < a_n/2$.
[Compare this with Exercise 25.3. The right-most inequality is due to Calabrese (1966). The left-most inequality is quoted by Boas (1978).]

(e) Use (d) above to compute exactly the number of terms one must sum in order to approximate $\sum_{k=1}^{\infty} (-1)^{k+1}/k$ to within $\frac{1}{2} \times 10^{-6}$ (six-place decimal accuracy).

26. Absolute Convergence

The tests of Section 24 apply to series with nonnegative terms. If we are given an arbitrary series $\sum_{n=1}^{\infty} a_n$, we can consider the related series $\sum_{n=1}^{\infty} |a_n|$ whose terms are nonnegative and often get useful information by applying the results of Section 24. This idea is captured in the following definition.

Definition 26.1 Let $\sum_{n=1}^{\infty} a_n$ be an infinite series. If $\sum_{n=1}^{\infty} |a_n|$ converges, we say $\sum_{n=1}^{\infty} a_n$ *converges absolutely*. If $\sum_{n=1}^{\infty} a_n$ converges and $\sum_{n=1}^{\infty} |a_n|$ diverges, we say $\sum_{n=1}^{\infty} a_n$ *converges conditionally*.

Any convergent series with nonnegative terms converges absolutely. The geometric series $\sum_{n=0}^{\infty} ar^n$ converges absolutely if $|r| < 1$. Since $\sum_{n=1}^{\infty} (-1)^n/n$ converges and $\sum_{n=1}^{\infty} 1/n$ diverges, $\sum_{n=1}^{\infty} (-1)^n/n$ converges conditionally. We will develop some tests for conditional convergence in Section 28.

Let $\sum_{n=1}^{\infty} a_n$ be an infinite series. We may relate the series $\sum_{n=1}^{\infty} |a_n|$ and $\sum_{n=1}^{\infty} a_n$ by separating $\{a_n\}$ into its positive and negative parts. We let

$$p_n = \begin{cases} a_n & \text{if } a_n \geq 0 \\ 0 & \text{if } a_n < 0 \end{cases}$$

and

$$q_n = \begin{cases} 0 & \text{if } a_n \geq 0 \\ -a_n & \text{if } a_n < 0 \end{cases}$$

Then $a_n = p_n - q_n$ and $|a_n| = p_n + q_n$ (verify).

We first justify the terminology of Definition 26.1 by proving the following theorem.

Theorem 26.2 If $\sum_{n=1}^{\infty} a_n$ converges absolutely, then $\sum_{n=1}^{\infty} a_n$ converges and $|\sum_{n=1}^{\infty} a_n| \leq \sum_{n=1}^{\infty} |a_n|$.

Proof. Suppose $\sum_{n=1}^{\infty} |a_n|$ converges. We let p_n and q_n be defined as above. Let $\{s_n\}$ be the sequence of partial sums of the series $\sum_{n=1}^{\infty} p_n$. Then

$$s_n = p_1 + \cdots + p_n \leq |a_1| + \cdots + |a_n| \leq \sum_{k=1}^{\infty} |a_k|$$

Thus the sequence of partial sums $\{s_n\}$ of the series $\sum_{n=1}^{\infty} p_n$ with nonnegative terms is bounded by $\sum_{k=1}^{\infty} |a_k|$. By Theorem 24.1, $\sum_{n=1}^{\infty} p_n$ converges. Similarly, $\sum_{n=1}^{\infty} q_n$ converges. By Theorem 23.1, $\sum_{n=1}^{\infty} a_n = \sum_{n=1}^{\infty} (p_n - q_n)$ converges. For any positive integer n, we have $|\sum_{k=1}^{n} a_k| \leq \sum_{k=1}^{n} |a_k|$. Taking limits we have $|\sum_{k=1}^{\infty} a_k| \leq \sum_{k=1}^{\infty} |a_k|$. ∎

We have already exploited the idea of comparing two series in Theorem 24.2. Now we formalize this idea.

Theorem 26.3 (Comparison Test) Let $\sum_{n=1}^{\infty} a_n$ and $\sum_{n=1}^{\infty} b_n$ be two series such that

$$|a_n| \leq |b_n|$$

for every positive integer n.

(i) If $\sum_{n=1}^{\infty} b_n$ converges absolutely, then $\sum_{n=1}^{\infty} a_n$ converges absolutely and

$$\sum_{n=1}^{\infty} |a_n| \leq \sum_{n=1}^{\infty} |b_n|$$

(ii) If $\sum_{n=1}^{\infty} |a_n|$ diverges, then $\sum_{n=1}^{\infty} |b_n|$ diverges.

Proof. If $\sum_{n=1}^{\infty} b_n$ converges absolutely, then

$$|a_1| + |a_2| + \cdots + |a_n| \leq |b_1| + \cdots + |b_n| \leq \sum_{k=1}^{\infty} |b_k|$$

Thus the sequence of partial sums of the series $\sum_{n=1}^{\infty} |a_n|$ is bounded by $\sum_{k=1}^{\infty} |b_k|$. By Theorem 24.1, $\sum_{n=1}^{\infty} a_n$ converges absolutely. Moreover, by Theorem 14.2,

$$\sum_{k=1}^{\infty} |a_k| = \lim_{n \to \infty} (|a_1| + \cdots + |a_n|) \leq \sum_{k=1}^{\infty} |b_k|$$

Part (ii) is a restatement of part (i) (really?). ∎

Example. Since

$$\left| \frac{(-1)^n}{n!} \right| = \frac{1}{n!} \leq \frac{1}{2^{n-1}}$$

for every position integer n and the geometric series $\sum_{n=1}^{\infty} 1/2^{n-1}$ converges, by the comparison test $\sum_{n=1}^{\infty} (-1)^n/n!$ converges. Also

$$\sum_{n=1}^{\infty} \frac{1}{n!} \leq \sum_{n=1}^{\infty} \frac{1}{2^{n-1}} = 2$$

The comparison test will be used to establish the next three tests.

Theorem 26.4 Let $\sum_{n=1}^{\infty} a_n$ be an infinite series and $\{b_n\}$ be a sequence.
(i) If the sequence $\{b_n\}$ is bounded and $\sum_{n=1}^{\infty} a_n$ converges absolutely, then $\sum_{n=1}^{\infty} a_n b_n$ converges absolutely.
(ii) If the sequence $\{1/b_n\}$ is bounded and $\sum_{n=1}^{\infty} |a_n|$ diverges, then $\sum_{n=1}^{\infty} |a_n b_n|$ diverges.

Proof. Suppose $\sum_{n=1}^{\infty} a_n$ converges absolutely and $\{b_n\}$ is bounded. Then there exists a number M such that $|b_n| < M$ for every positive integer n. Thus

$$|a_n b_n| \leq M |a_n|$$

for every positive integer n. Since $\sum_{n=1}^{\infty} M|a_n|$ converges, $\sum_{n=1}^{\infty} |a_n b_n|$ converges by the comparison test.

Suppose $\sum_{n=1}^{\infty} |a_n|$ diverges and $\{1/b_n\}$ is bounded. If $\sum_{n=1}^{\infty} |a_n b_n|$ converges, by part (i), $\sum_{n=1}^{\infty} |a_n b_n|/|b_n| = \sum_{n=1}^{\infty} |a_n|$ converges which is impossible. Thus $\sum_{n=1}^{\infty} |a_n b_n|$ diverges. ∎

Corollary 26.5 Let $\sum_{n=1}^{\infty} a_n$ be a series and $\{b_n\}$ be a convergent sequence.
(i) If $\sum_{n=1}^{\infty} a_n$ converges absolutely, then $\sum_{n=1}^{\infty} a_n b_n$ converges absolutely.
(ii) If $\sum_{n=1}^{\infty} |a_n|$ diverges and $\lim_{n \to \infty} b_n \neq 0$, then $\sum_{n=1}^{\infty} |a_n b_n|$ diverges.

Proof. Since a convergent sequence is bounded (Theorem 13.2), part (i) follows immediately from Theorem 26.4 (i).

If $\lim_{n \to \infty} b_n \neq 0$, then $\lim_{n \to \infty} (1/b_n)$ exists and hence $\{1/b_n\}$ is bounded, after a finite number of terms. Part (ii) now follows from Theorem 26.4 (ii). ∎

Examples. Consider the series $\sum_{n=1}^{\infty} n/(1 + n^2)$. Since $n/(1 + n^2) = 1/(1/n + n)$, for large n the series $\sum_{n=1}^{\infty} n/(1 + n^2)$ behaves like the series $\sum_{n=1}^{\infty} 1/n$. We expect that $\sum_{n=1}^{\infty} n/(1 + n^2)$ diverges. Now

$$\sum_{n=1}^{\infty} \frac{n}{1 + n^2} = \sum_{n=1}^{\infty} \frac{1}{n}\left[\frac{n^2}{1 + n^2}\right]$$

Thus if we take $a_n = 1/n$ and $b_n = n^2/(1 + n^2)$, we may use Corollary 26.5 (ii) to conclude that $\sum_{n=1}^{\infty} n/(1 + n^2)$ diverges.

Since

$$\frac{1}{n^2 + n} = \frac{1}{n^2}\left[\frac{n^2}{n^2 + n}\right]$$

we may take $a_n = 1/n^2$ and $b_n = n^2/(n^2 + n)$ in Corollary 26.5 (i) and conclude that $\sum_{n=1}^{\infty} 1/(n^2 + n)$ converges.

Theorem 26.6 (Ratio Test)   Let $\sum_{n=1}^{\infty} a_n$ be a series such that

$$L = \lim_{n \to \infty} \left|\frac{a_{n+1}}{a_n}\right| \qquad (L = \infty \text{ is allowed})$$

(i) If $L < 1$, then $\sum_{n=1}^{\infty} a_n$ converges absolutely.

(ii) If $L > 1$, then $\sum_{n=1}^{\infty} a_n$ diverges.

Proof. First suppose that $L < 1$. Choose a number M such that $L < M < 1$. Then there exists a positive integer N such that if $n \geq N$, then

$$\left|\frac{a_{n+1}}{a_n}\right| < M \qquad \text{(why?)}$$

In particular,

$$\left|\frac{a_{N+1}}{a_N}\right| < M$$

so

$$|a_{N+1}| < M|a_N|$$

Also

$$\left|\frac{a_{N+2}}{a_{N+1}}\right| < M$$

so

$$|a_{N+2}| < M|a_{N+1}| < M^2|a_N|$$

In general, if n is a positive integer,

$$|a_{N+n}| < M^n|a_N|$$

The geometric series $\sum_{n=1}^{\infty} M^n |a_N|$ converges since $0 < M < 1$. By the comparison test $\sum_{n=1}^{\infty} |a_{N+n}|$ converges, and hence $\sum_{n=1}^{\infty} a_n$ converges absolutely (verify).

Now suppose $L > 1$. There exists a positive integer N such that if $n \geq N$, then

$$\left| \frac{a_{n+1}}{a_n} \right| > 1$$

Thus $\qquad |a_N| < |a_{N+1}| < |a_{N+2}| < \cdots$

so that the sequence $\{a_n\}$ does not converge to 0 (why?). By Theorem 22.3. $\sum_{n=1}^{\infty} a_n$ diverges. ∎

The ratio test gives no information if $\lim_{n \to \infty} |a_{n+1}/a_n| = 1$. The series $\sum_{n=1}^{\infty} 1/n$ diverges and the series $\sum_{n=1}^{\infty} 1/n^2$ converges, but

$$\lim_{n \to \infty} \frac{1/(n+1)}{1/n} = 1 = \lim_{n \to \infty} \frac{1/(n+1)^2}{1/n^2}$$

Examples. The series $\sum_{n=1}^{\infty} 1/n!$ converges since the ratio

$$\frac{1/(n+1)!}{1/n!} = \frac{1}{n+1}$$

converges to 0.

The series $\sum_{n=1}^{\infty} n!/n^n$ converges since the ratio

$$\frac{(n+1)!/(n+1)^{n+1}}{n!/n^n} = \frac{n+1}{(n+1)^{n+1}} n^n = \frac{1}{(1 + 1/n)^n}$$

converges to $1/e \leq 1/2$. Similar arguments show that $\sum_{n=1}^{\infty} n^n/n!$ diverges.

Theorem 26.7 (Root Test) Let $\{a_n\}$ be a sequence and let $L = \limsup_{n \to \infty} |a_n|^{1/n}$. ($L = \infty$ is allowed.)

(i) If $L < 1$, then $\sum_{n=1}^{\infty} a_n$ converges absolutely.
(ii) If $L > 1$, then $\sum_{n=1}^{\infty} a_n$ diverges.

Proof. Suppose $L < 1$. Choose a number M such that $L < M < 1$. By Theorem 20.3, there exists a positive integer N such that if $n \geq N$, then

$$|a_n|^{1/n} < M$$

Thus if $n \geq N$, we have

$$|a_n| < M^n$$

Since the geometric series $\sum_{n=1}^{\infty} M^n$ converges, we have that $\sum_{n=1}^{\infty} a_n$ converges absolutely by the comparison test.

Next suppose $L > 1$. Then $|a_n|^{1/n} > 1$ for infinitely many positive integers n. Thus $|a_n| > 1$ for infinitely many positive integers n, and hence the sequence $\{a_n\}$ does not converge to 0. By Theorem 22.3, $\sum_{n=1}^{\infty} a_n$ diverges. ∎

As in the ratio test, if $\lim \sup_{n \to \infty} |a_n|^{1/n} = 1$, the root test gives no information. We have

$$\lim_{n \to \infty} \left(\frac{1}{n}\right)^{1/n} = 1 = \lim_{n \to \infty} \left(\frac{1}{n^2}\right)^{1/n}$$

but $\sum_{n=1}^{\infty} 1/n$ diverges and $\sum_{n=1}^{\infty} 1/n^2$ converges.

Examples. $\sum_{n=1}^{\infty} 2^n$ diverges since $\lim_{n \to \infty} (2^n)^{1/n} = 2$. $\sum_{n=1}^{\infty} (\frac{1}{2})^n$ converges since $\lim_{n \to \infty} [(\frac{1}{2})^n]^{1/n} = \frac{1}{2}$. $\sum_{n=1}^{\infty} n(\frac{1}{2})^n$ converges since $\lim_{n \to \infty} [n(\frac{1}{2})^n]^{1/n} = \lim_{n \to \infty} n^{1/n} \cdot \frac{1}{2} = \frac{1}{2}$.

Exercises

26.1 Determine whether each of the series converges absolutely, converges conditionally, or diverges.

(a) $\sum_{n=1}^{\infty} \dfrac{(-1)^{n+1}}{n^3 + n^2 + n}$ (b) $\sum_{n=1}^{\infty} \dfrac{(-1)^{n+1}}{n + \sqrt{n}}$

(c) $\sum_{n=1}^{\infty} \dfrac{(-1)^{n+1}(\sqrt{n+1} - \sqrt{n})}{n}$ (d) $\sum_{n=1}^{\infty} \dfrac{n(-1)^n}{2^n(n+1)}$

(e) $\sum_{n=1}^{\infty} \dfrac{n^2(-1)^n}{2^n}$ (f) $\sum_{n=1}^{\infty} \dfrac{(-1)^n}{(2n-1)!}$

(g) $\sum_{n=1}^{\infty} \dfrac{(-1)^n}{n^{1+1/n}}$ (h)* $\sum_{n=1}^{\infty} (-1)^n [n^{1/n} - 1]$

(i)* $\sum_{n=1}^{\infty} (-1)^n \left[e - \left(1 + \frac{1}{n}\right)^n \right]$ (j)* $\sum_{n=1}^{\infty} (-1)^n \dfrac{1 \cdot 3 \cdot 5 \cdots (2n-1)}{2 \cdot 4 \cdot 6 \cdots (2n)}$

26.2 Show (by example) that if $\{a_n\}$ is a sequence such that $\lim \inf_{n \to \infty} |a_n|^{1/n} < 1$, then $\sum_{n=1}^{\infty} a_n$ may either converge or diverge.
Prove that if $\lim \inf_{n \to \infty} |a_n|^{1/n} > 1$, then $\sum_{n=1}^{\infty} a_n$ diverges.

26.3 Use Theorem 19.3 to prove Theorem 26.2.

26.4 Prove that if $\sum_{n=1}^{\infty} a_n$ converges absolutely, then $\sum_{n=1}^{\infty} a_n^2$ converges.

26.5 Let $\sum_{n=1}^{\infty} a_n$ and $\sum_{n=1}^{\infty} b_n$ be absolutely convergent series. Prove that the series $\sum_{n=1}^{\infty} \sqrt{|a_n b_n|}$ converges.

26.6* Prove that the series $\sum_{n=1}^{\infty} a_n$ converges absolutely if and only if for every $\varepsilon > 0$, there exists a positive integer N such that

$$|a_{n_1} + a_{n_2} + \cdots + a_{n_k}| < \varepsilon$$

wherever $n_1, \ldots, n_k$ are distinct positive integers exceeding N.

26.7 Let $\{a_n\}$ be a sequence of nonzero numbers. Prove that if

$$\lim_{n \to \infty} \sup \left|\frac{a_{n+1}}{a_n}\right| < 1$$

then $\sum_{n=1}^{\infty} a_n$ converges absolutely. Prove that if

$$\liminf_{n \to \infty} \left| \frac{a_{n+1}}{a_n} \right| > 1$$

then $\sum_{n=1}^{\infty} a_n$ diverges.

26.8 Let $\{a_n\}$ be a sequence of positive numbers. Prove that if

$$\lim_{n \to \infty} \frac{a_{n+1}}{a_n} = L$$

then

$$\lim_{n \to \infty} a_n^{1/n} = L$$

Deduce

$$\lim_{n \to \infty} \frac{n}{(n!)^{1/n}} = e$$

26.9 Let $\{a_n\}$ be a sequence where each term is one of the integers 0, 1, 2, 3, 4, 5, 6, 7, 8, 9. Prove that the series $\sum_{n=1}^{\infty} a_n/10^n$ converges and the sum is in the closed interval [0,1]. The term $.a_1 a_2 \ldots$ is called the *decimal expansion* of the number $\sum_{n=1}^{\infty} a_n/10^n$.

26.10* Prove that if $x \in [0, 1]$, there exists a sequence $\{a_n\}$, where each term is one of $0, 1, 2, \ldots, 9$, such that $\sum_{n=1}^{\infty} a_n/10^n = x$.

26.11* Let $R_n = \sum_{k=n+1}^{\infty} 1/k!$ be the error in approximating $\sum_{k=1}^{\infty} 1/k!$ with $\sum_{k=1}^{n} 1/k!$. Prove that

$$\frac{1}{(n+1)!} < R_n < \frac{n+2}{n+1} \frac{1}{(n+1)!}$$

Find the least n necessary to have $R_n < \frac{1}{2} \times 10^{-10}$. (This gives 10-place decimal accuracy.)

27. Power Series

A power series is the infinite series generalization of a polynomial.

Definition 27.1 Let t be a fixed real number. A *power series* (*expanded about* t) is an infinite series of the form

$$\sum_{n=0}^{\infty} a_n (x - t)^n$$

where $\{a_n\}_{n=0}^{\infty}$ is a sequence and x is a real number. [$(x - t)^0$ is defined to be 1.]

A power series $\sum_{n=0}^{\infty} a_n (x - t)^n$ converges (absolutely) if $x = t$, for then the series becomes

$$a_0 + 0 + 0 + \cdots$$

We will show that there exists a value R, where R is either ∞ or a nonnegative number, such that $\sum_{n=0}^{\infty} a_n (x - t)^n$ converges absolutely if $|x - t| < R$ and diverges if $|x - t| > R$.

Theorem 27.2 Let $\sum_{n=0}^{\infty} a_n(x - t)^n$ be a power series.
Let $L = \limsup_{n\to\infty} |a_n|^{1/n}$. Let

$$R = \begin{cases} 0 & \text{if } L = \infty \\ \dfrac{1}{L} & \text{if } 0 < L < \infty \\ \infty & \text{if } L = 0 \end{cases}$$

Then the power series $\sum_{n=0}^{\infty} a_n(x - t)^n$ converges absolutely if $|x - t| < R$ and diverges if $|x - t| > R$.

Proof. Suppose $\{|a_n|^{1/n}\}$ is unbounded. Then $\limsup_{n\to\infty} |a_n|^{1/n} = \infty$. If $|x - t| > 0$, we have $|a_n|^{1/n} > 1/|x - t|$ for infinitely many positive integers n. Thus $|a_n(x - t)^n| > 1$ for infinitely many positive integers n, and it follows that $\sum_{n=0}^{\infty} a_n(x - t)^n$ diverges.

Now suppose that $\{|a_n|^{1/n}\}$ is bounded, then $\limsup_{n\to\infty} |a_n|^{1/n} = L$ is a real number. First, suppose $L > 0$. By the root test $\sum_{n=0}^{\infty} a_n(x - t)^n$ converges absolutely if $\limsup_{n\to\infty} |a_n(x - t)^n|^{1/n} < 1$ and diverges if $\limsup_{n\to\infty} |a_n(x - t)^n|^{1/n} > 1$. However,

$$\limsup_{n\to\infty} |a_n(x - t)^n|^{1/n} = \limsup_{n\to\infty} |a_n|^{1/n}|x - t|$$
$$= |x - t| \limsup_{n\to\infty} |a_n|^{1/n}$$
$$= |x - t|L$$

where we have used Theorem 20.8. Thus $\sum_{n=0}^{\infty} a_n(x - t)^n$ converges absolutely if $|x - t|L < 1$ and diverges if $|x - t|L > 1$.

Finally, suppose $L = 0$. Let x be any number. Again by Theorem 20.8,

$$\limsup_{n\to\infty} |a_n(x - t)^n|^{1/n} = |x - t|L = 0 < 1$$

and by the root test $\sum_{n=0}^{\infty} a_n(x - t)^n$ converges absolutely. ∎

Definition 27.3 The value R of Theorem 27.2 is called the *radius of convergence* of the power series $\sum_{n=0}^{\infty} a_n(x - t)^n$.

Examples. The radius of convergence of a power series is often easily obtained from the ratio test. Consider the power series $\sum_{n=0}^{\infty} n(x - 1)^n$. Here

$$\lim_{n\to\infty} \left| \frac{(n + 1)(x - 1)^{n+1}}{n(x - 1)^n} \right| = |x - 1|$$

so by the ratio test, the series converges absolutely if $|x - 1| < 1$ and diverges if $|x - 1| > 1$. Thus the radius of convergence of the power series $\sum_{n=0}^{\infty} n(x - 1)^n$ is $R = 1$. Similarly, the power series $\sum_{n=1}^{\infty} (x - 1)^n/n$ and $\sum_{n=1}^{\infty} (x - 1)^n/n^2$ both have radius of convergence $R = 1$. These examples also show that if $|x - t| = R$, no conclusions about convergence can be drawn. The series $\sum_{n=0}^{\infty} n(x - 1)^n$ diverges if $|x - 1| = 1$. The series $\sum_{n=1}^{\infty} (x - 1)^n/n$ diverges if $x - 1 = 1$ and converges if $x - 1 = -1$. The series $\sum_{n=1}^{\infty} (x - 1)^n/n^2$ converges absolutely if $|x - 1| = 1$. The ratio test is limited in its application to power series in that

$$\lim_{n \to \infty} \left| \frac{a_{n+1}(x - t)^{n+1}}{a_n(x - t)^n} \right|$$

may fail to exist.

Exercises

27.1 Find the radius of convergence R of each of the power series. Discuss the convergence of the power series at the points $|x - t| = R$.

(a) $\displaystyle\sum_{n=1}^{\infty} \frac{x^{2n-1}}{(2n-1)!}$ (b) $\displaystyle\sum_{n=1}^{\infty} \frac{n(x - 1)^n}{2^n}$

(c) $\displaystyle\sum_{n=1}^{\infty} \frac{n!(x + 2)^n}{2^n}$ (d) $\displaystyle\sum_{n=1}^{\infty} (n^{1/n} - 1)x^n$

(e) $\displaystyle\sum_{n=2}^{\infty} \frac{(x - \pi)^n}{n(n - 1)}$

27.2 Suppose that the power series $\sum_{n=0}^{\infty} a_n(x - t)^n$ has radius of convergence R. Let p be an integer. Prove that the power series $\sum_{n=0}^{\infty} n^p a_n(x - t)^n$ has radius of convergence R.

27.3 Let $\{a_n\}$ be a decreasing sequence such that $\lim_{n \to \infty} a_n = 0$. Prove that the power series $\sum_{n=0}^{\infty} a_n x^n$ converges if $|x| \leq 1$ and $x \neq 1$.

27.4 Let $\{a_n\}$ be a sequence such that $\lim_{n \to \infty} |a_{n+1}/a_n| = L$.
Let

$$R = \begin{cases} 0 & \text{if } L = \infty \\ \dfrac{1}{L} & \text{if } 0 < L < \infty \\ \infty & \text{if } L = 0 \end{cases}$$

Prove that the radius of convergence of the power series $\sum_{n=0}^{\infty} a_n(x - t)^n$ is R.

27.5 Let $\sum_{n=0}^{\infty} a_n(x - t)^n$ be a power series. Let $X = \{|x - t|: \sum_{n=0}^{\infty} |a_n||x - t|^n$ converges$\}$. Let $R = \text{lub } X$ if X is bounded and let $R = \infty$ if X is unbounded. Prove that R is the radius of convergence of the series $\sum_{n=0}^{\infty} a_n(x - t)^n$.

28. Conditional Convergence

We now develop some tests for convergent series which fail to converge absolutely. All of our conditional convergence tests depend upon the following theorem.

Theorem 28.1 (Summation by Parts) Let $\sum_{n=1}^{\infty} a_n$ be an infinite series and let $\{s_n\}$ be the sequence of partial sums of $\sum_{n=1}^{\infty} a_n$. Let $\{b_n\}$ be any sequence. Then for any positive integer n we have

$$\sum_{k=1}^{n} a_k b_k = \sum_{k=1}^{n} s_k(b_k - b_{k+1}) + s_n b_{n+1}.$$

Proof. The proof is a straightforward verification. Let $s_0 = 0$. Then

$$\begin{aligned}
\sum_{k=1}^{n} a_k b_k &= \sum_{k=1}^{n} (s_k - s_{k-1}) b_k \\
&= \sum_{k=1}^{n} s_k b_k - \sum_{k=1}^{n} s_{k-1} b_k \\
&= \sum_{k=1}^{n} s_k b_k - \sum_{k=1}^{n} s_k b_{k+1} + s_n b_{n+1} \\
&= \sum_{k=1}^{n} s_k(b_k - b_{k+1}) + s_n b_{n+1} \qquad\blacksquare
\end{aligned}$$

To prove that $\sum_{n=1}^{\infty} a_n b_n$ converges using Theorem 28.1, it is enough to prove that the series $\sum_{n=1}^{\infty} s_n(b_n - b_{n+1})$ and the sequence $\{s_n b_{n+1}\}$ converge.

Theorem 28.2 Let $\sum_{n=1}^{\infty} a_n$ be a series whose sequence of partial sums is bounded. If $\sum_{n=1}^{\infty} |b_n - b_{n+1}|$ converges and $\lim_{n \to \infty} b_n = 0$, then $\sum_{n=1}^{\infty} a_n b_n$ converges.

Proof. By Theorem 26.4 (i), $\sum_{n=1}^{\infty} s_n(b_n - b_{n+1})$ converges absolutely. Since the sequence $\{s_n\}$ is bounded and the sequence $\{b_{n+1}\}$ converges to 0, $\{s_n b_{n+1}\}$ converges to 0 by Theorem 13.3. By Theorem 28.1, $\sum_{n=1}^{\infty} a_n b_n$ converges. $\blacksquare$

Corollary 28.3 (Dirichlet's Test) If the sequence of partial sums of the series $\sum_{n=1}^{\infty} a_n$ is bounded and $\{b_n\}$ is a decreasing sequence with limit 0, then $\sum_{n=1}^{\infty} a_n b_n$ converges.

Proof. The sequence of partial sums of the series $\sum_{n=1}^{\infty} |b_n - b_{n+1}|$ is given by

$$s_n = |b_1 - b_2| + \cdots + |b_n - b_{n+1}|$$
$$= (b_1 - b_2) + \cdots + (b_n - b_{n+1})$$
$$= b_1 - b_{n+1}$$

which converges (to b_1). By Theorem 28.2, $\sum_{n=1}^{\infty} a_n b_n$ converges. ∎

Example.   Consider the series

$$1 + \tfrac{1}{2} - \tfrac{2}{3} + \tfrac{1}{4} + \tfrac{1}{5} - \tfrac{2}{6} + \cdots$$

Letting $\{a_n\}$ be the sequence

$$1, \quad 1, \quad -2, \quad 1, \quad 1, \quad -2, \ldots$$

and $b_n = 1/n$, we conclude the series converges by Dirichlet's test. Since the absolute value of the nth term of the series

$$1 + \tfrac{1}{2} - \tfrac{2}{3} + \tfrac{1}{4} + \tfrac{1}{5} - \tfrac{2}{6} + \cdots$$

is greater than or equal to $1/n$, this series converges conditionally.

Corollary 28.4   (Alternating Series Test Revisited)   If $\{b_n\}$ is a decreasing sequence with limit 0, then $\sum_{n=1}^{\infty} (-1)^{n+1} b_n$ converges.

Proof.   Let $a_n = (-1)^{n+1}$ and apply Dirichlet's test. ∎

We can deduce another theorem from the summation-by-parts formula by assuming less about $\sum_{n=1}^{\infty} |b_n - b_{n+1}|$ and more about $\sum_{n=1}^{\infty} a_n$.

Theorem 28.5   If $\sum_{n=1}^{\infty} a_n$ and $\sum_{n=1}^{\infty} |b_n - b_{n+1}|$ are convergent series, then $\sum_{n=1}^{\infty} a_n b_n$ converges.

Proof.   By Theorem 26.4 (i), $\sum_{n=1}^{\infty} s_n(b_n - b_{n+1})$ converges absolutely. Since $\sum_{n=1}^{\infty} (b_n - b_{n+1})$ converges,

$$\lim_{n \to \infty} [(b_1 - b_2) + (b_2 - b_3) + \cdots + (b_n - b_{n+1})] = \lim_{n \to \infty} (b_1 - b_{n+1})$$

exists. Thus the sequence $\{b_{n+1}\}$ converges, and hence $\{s_n b_{n+1}\}$ converges. By Theorem 28.1, $\sum_{n=1}^{\infty} a_n b_n$ converges. ∎

Corollary 28.6   (Abel's Test) If $\sum_{n=1}^{\infty} a_n$ converges and $\{b_n\}$ is a bounded monotone sequence, then $\sum_{n=1}^{\infty} a_n b_n$ converges.

Proof.   As in the proof of Corollary 28.3, one shows that $\sum_{n=1}^{\infty} |b_n - b_{n+1}|$ converges. The conclusion now follows from Theorem 28.5. ∎

Example.   The series

$$\sum_{n=1}^{\infty} \frac{(-1)^n(1 + 1/n)^n}{n}$$

converges since the series $\sum_{n=1}^{\infty} (-1)^n/n$ converges and $\{(1 + 1/n)^n\}$ is a bounded monotone sequence.

Exercises

28.1   Prove that the series

(a) $1 + \dfrac{1}{\sqrt{2}} - \dfrac{2}{\sqrt{3}} + \dfrac{1}{\sqrt{4}} + \dfrac{1}{\sqrt{5}} - \dfrac{2}{\sqrt{6}} + \cdots$

(b) $\displaystyle\sum_{n=1}^{\infty} (-1)^n n^{(1-n)/n}$

converge conditionally.

28.2   Suppose that $\sum_{n=1}^{\infty} \sqrt{n}a_n$ converges. Prove that $\sum_{n=1}^{\infty} a_n$ converges.

28.3   Give an example of a convergent series $\sum_{n=1}^{\infty} a_n$ and a positive sequence $\{b_n\}$ such that $\lim_{n\to\infty} b_n = 0$ and $\sum_{n=1}^{\infty} a_n b_n$ diverges.

28.4   Prove Abel's lemma: If $\{s_n\}$ is the sequence of partial sums of a series $\sum_{n=1}^{\infty} a_n$ and $\{b_n\}$ is a decreasing nonnegative sequence, then

$$\left|\sum_{k=1}^{n} a_k b_k\right| \leq b_1 \max\{|s_1|, \ldots, |s_n|\}$$

28.5   Prove the following result due to Shohat (1933). Let $\{a_n\}$ be a sequence and let $b_n = n(a_n - a_{n+1})$, for $n = 1, 2, \ldots$
  (a)   If $\lim_{n\to\infty} na_n$ exists and is nonzero, then $\sum_{n=1}^{\infty} a_n$ and $\sum_{n=1}^{\infty} b_n$ diverge.
  (b)   If $\sum_{n=1}^{\infty} a_n$ and $\sum_{n=1}^{\infty} b_n$ converge, then $\lim_{n\to\infty} na_n = 0$ and $\sum_{n=1}^{\infty} a_n = \sum_{n=1}^{\infty} b_n$.
Use (b) to prove the formula

$$\sum_{n=1}^{\infty} \frac{1}{n^2} = 2 - \sum_{n=1}^{\infty} \frac{1}{n(n+1)^2} = \frac{13}{8} + \frac{1}{2}\sum_{n=1}^{\infty} \frac{1}{(n+1)^2(n+2)^2}$$

Notice that the last two series converge more rapidly than the series $\sum_{n=1}^{\infty} 1/n^2$.

28.6   Let $\{a_n\}$ be a periodic sequence, $a_n = a_{n+p}$ for all n and for fixed p, for which $\sum_{n=1}^{p} a_n = 0$. Let $\{b_n\}$ be a decreasing sequence with limit 0. Prove that $\sum_{n=1}^{\infty} a_n b_n$ converges.

29.   Double Series and Applications

We begin by defining a double sequence. A double sequence is a function indexed by pairs of positive integers.

Definition 29.1 A (real) double sequence is a function from $\mathbf{P} \times \mathbf{P}$ into $\mathbf{R}$. The image of a double sequence a at a point (m, n) is denoted $a_{m,n}$. We will denote a double sequence as $\{a_{m,n}\}$.

A double sequence $\{a_{m,n}\}$ may be alternatively described as an infinite matrix

$$\begin{pmatrix} a_{1,1} & a_{1,2} & a_{1,3} & \cdots \\ a_{2,1} & a_{2,2} & a_{2,3} & \cdots \\ a_{3,1} & a_{3,2} & a_{3,3} & \cdots \\ \cdots\cdots\cdots\cdots\cdots\cdots \\ a_{m,1} & a_{m,2} & a_{m,3} & \cdots \\ \cdots\cdots\cdots\cdots\cdots\cdots \end{pmatrix} \qquad (29.1)$$

Definition 29.2 Let $\{a_{m,n}\}$ be a double sequence. We say that the *sum by rows of the series* $\sum_{m,n} a_{m,n}$ *exists* if for every positive integer m, the series $\sum_{n=1}^{\infty} a_{m,n}$ converges and the series $\sum_{m=1}^{\infty} (\sum_{n=1}^{\infty} a_{m,n})$ converges. In this case we say the sum by rows of the series $\sum a_{m,n}$ is the number $\sum_{m=1}^{\infty} (\sum_{n=1}^{\infty} a_{m,n})$. The sum by columns is similarly defined.

Notice that the sum $\sum_{n=1}^{\infty} a_{m,n}$ is just the sum of the entries in the mth row of the matrix (29.1).

We will be particularly interested in the situation in which the sum by rows and the sum by columns exist and are equal. Our first theorem which deals with this has many applications to ordinary infinite series. We will prove the main theorem of this section after giving an example.

Example. Let $|r| < 1$ and $|s| < 1$ and define $a_{m,n} = r^m s^n$ for $m,n = 0, 1, 2, \ldots$. The matrix of the double sequence $\{a_{m,n}\}$ is

$$\begin{pmatrix} 1 & s & s^2 & s^3 & \cdots \\ r & rs & rs^2 & rs^3 & \cdots \\ r^2 & r^2 s & r^2 s^2 & r^2 s^3 & \cdots \\ \cdots\cdots\cdots\cdots\cdots\cdots\cdots \end{pmatrix}$$

Now $\sum_{n=0}^{\infty} a_{m,n} = \sum_{n=0}^{\infty} r^m s^n = r^m \sum_{n=0}^{\infty} s^n = r^m/(1 - s)$. Thus the sum by rows is

$$\sum_{m=0}^{\infty} \left(\sum_{n=0}^{\infty} a_{m,n} \right) = \sum_{m=0}^{\infty} \frac{r^m}{1 - s} = \frac{1}{1 - s} \sum_{m=0}^{\infty} r^m = \frac{1}{1 - s} \frac{1}{1 - r}$$

Similarly, the sum by columns is $\sum_{n=0}^{\infty} (\sum_{m=0}^{\infty} a_{m,n}) = (1/(1 - r))(1/(1 - s))$. In this example, the sum by rows and the sum by columns both exist and they are equal.

In the proof of the next lemma, the following fact will be used. If

$\sum_{m=1}^{\infty} a_{m,n}$ converges for every positive integer n, then $\sum_{m=1}^{\infty} \left(\sum_{n=1}^{k} a_{m,n} \right)$ converges for every positive integer k and

$$\sum_{m=1}^{\infty} \left(\sum_{n=1}^{k} a_{m,n} \right) = \sum_{n=1}^{k} \left(\sum_{m=1}^{\infty} a_{m,n} \right)$$

This result follows immediately from Theorem 23.1 since

$$\sum_{n=1}^{k} \left(\sum_{m=1}^{\infty} a_{m,n} \right) = \sum_{m=1}^{\infty} a_{m,1} + \sum_{m=1}^{\infty} a_{m,2} + \cdots + \sum_{m=1}^{\infty} a_{m,k}$$

$$= \sum_{m=1}^{\infty} (a_{m,1} + a_{m,2} + \cdots + a_{m,k})$$

$$= \sum_{m=1}^{\infty} \left(\sum_{n=1}^{k} a_{m,n} \right)$$

Lemma 29.3 Let $\{a_{m,n}\}$ be a double sequence of nonnegative terms. If the sum by rows of the double series $\sum_{m,n} a_{m,n}$ exists, then the sum by columns also exists and they are equal.

Proof. We are given that the series $\sum_{n=1}^{\infty} a_{m,n}$ converges for every positive integer m and that the series $\sum_{m=1}^{\infty} \left(\sum_{n=1}^{\infty} a_{m,n} \right)$ converges.

Let j be a positive integer. For any positive integer m we have

$$a_{m,j} \le \sum_{n=1}^{\infty} a_{m,n}$$

Thus $\qquad \displaystyle \sum_{m=1}^{k} a_{m,j} \le \sum_{m=1}^{k} \left(\sum_{n=1}^{\infty} a_{m,n} \right) \le \sum_{m=1}^{\infty} \left(\sum_{n=1}^{\infty} a_{m,n} \right)$

By Theorem 24.1, $\sum_{m=1}^{\infty} a_{m,n}$ converges for every positive integer n.

Now $\qquad \displaystyle \sum_{n=1}^{k} \left(\sum_{m=1}^{\infty} a_{m,n} \right) = \sum_{m=1}^{\infty} \left(\sum_{n=1}^{k} a_{m,n} \right) \le \sum_{m=1}^{\infty} \left(\sum_{n=1}^{\infty} a_{m,n} \right)$

for every positive integer k. By Theorem 24.1, $\sum_{n=1}^{\infty} \left(\sum_{m=1}^{\infty} a_{m,n} \right)$ converges. Thus the sum by columns exists. Moreover, it follows that

$$\sum_{n=1}^{\infty} \left(\sum_{m=1}^{\infty} a_{m,n} \right) \le \left(\sum_{n=1}^{\infty} a_{m,n} \right)$$

Repeating the above argument with the roles of m and n interchanged, we have

$$\sum_{m=1}^{\infty} \left(\sum_{n=1}^{\infty} a_{m,n} \right) \le \sum_{n=1}^{\infty} \left(\sum_{m=1}^{\infty} a_{m,n} \right)$$

Thus the sum by rows and the sum by columns are equal. ∎

Theorem 29.4 Let $\{a_{m,n}\}$ be a double sequence. If the sum by rows of the series $\sum_{m,n} |a_{m,n}|$ exists, then the sum by rows and the sum by columns of the series $\sum_{m,n} a_{m,n}$ exist and they are equal. Furthermore, the series $\sum_{m=1}^{\infty} |\sum_{n=1}^{\infty} a_{m,n}|$ and $\sum_{n=1}^{\infty} |\sum_{m=1}^{\infty} a_{m,n}|$ converge.

Proof. As in Section 26, we let

$$p_{m,n} = \begin{cases} a_{m,n} & \text{if } a_{m,n} \geq 0 \\ 0 & \text{if } a_{m,n} < 0 \end{cases}$$

and

$$q_{m,n} = \begin{cases} 0 & \text{if } a_{m,n} \geq 0 \\ -a_{m,n} & \text{if } a_{m,n} < 0 \end{cases}$$

We first show that the sum by rows of the series $\sum_{m,n} p_{m,n}$ exists. For any positive integer i we have

$$\sum_{n=1}^{k} p_{i,n} \leq \sum_{n=1}^{k} |a_{i,n}| \leq \sum_{m=1}^{\infty} \left(\sum_{n=1}^{k} |a_{m,n}| \right) \leq \sum_{m=1}^{\infty} \left(\sum_{n=1}^{\infty} |a_{m,n}| \right)$$

for every positive integer k. By Theorem 24.1, $\sum_{n=1}^{\infty} p_{i,n}$ converges for every positive integer i. Also

$$\sum_{m=1}^{k} \left(\sum_{n=1}^{\infty} p_{m,n} \right) \leq \sum_{m=1}^{k} \left(\sum_{n=1}^{\infty} |a_{m,n}| \right) \leq \sum_{m=1}^{\infty} \left(\sum_{n=1}^{\infty} |a_{m,n}| \right)$$

By Theorem 24.1, the sum by rows of the series $\sum_{m,n} p_{m,n}$ exists. By Lemma 29.3, the sum by rows and the sum by columns of $\sum_{m,n} p_{m,n}$ exist and are equal. Similarly, the sum by rows and the sum by columns of the series $\sum_{m,n} q_{m,n}$ exist and are equal. Thus

$$\sum_{m=1}^{\infty} \left(\sum_{n=1}^{\infty} p_{m,n} \right) - \sum_{m=1}^{\infty} \left(\sum_{n=1}^{\infty} q_{m,n} \right) = \sum_{n=1}^{\infty} \left(\sum_{m=1}^{\infty} p_{m,n} \right) - \sum_{n=1}^{\infty} \left(\sum_{m=1}^{\infty} q_{m,n} \right)$$

By Theorem 23.1,

$$\sum_{m=1}^{\infty} \left(\sum_{n=1}^{\infty} p_{m,n} \right) - \sum_{m=1}^{\infty} \left(\sum_{n=1}^{\infty} q_{m,n} \right) = \sum_{m=1}^{\infty} \left(\sum_{n=1}^{\infty} p_{m,n} - \sum_{n=1}^{\infty} q_{m,n} \right)$$

$$= \sum_{m=1}^{\infty} \left[\sum_{n=1}^{\infty} (p_{m,n} - q_{m,n}) \right]$$

$$= \sum_{m=1}^{\infty} \left(\sum_{n=1}^{\infty} a_{m,n} \right)$$

Similarly,

$$\sum_{n=1}^{\infty} \left(\sum_{m=1}^{\infty} p_{m,n} \right) - \sum_{n=1}^{\infty} \left(\sum_{m=1}^{\infty} q_{m,n} \right) = \sum_{n=1}^{\infty} \left(\sum_{m=1}^{\infty} a_{m,n} \right)$$

and thus the sum by rows and the sum by columns are equal.

Finally, the series $\sum_{m=1}^{\infty} |\sum_{n=1}^{\infty} a_{m,n}|$ converges since for every positive integer k, we have

$$\sum_{m=1}^{k} \left| \sum_{n=1}^{\infty} a_{m,n} \right| \leq \sum_{m=1}^{k} \left(\sum_{n=1}^{\infty} |a_{m,n}| \right)$$

$$\leq \sum_{m=1}^{\infty} \left(\sum_{n=1}^{\infty} |a_{m,n}| \right)$$

By Lemma 29.3, $\sum_{n=1}^{\infty} \left(\sum_{m=1}^{\infty} |a_{m,n}| \right)$ converges so the above argument may be modified to show that $\sum_{n=1}^{\infty} \left| \sum_{m=1}^{\infty} a_{m,n} \right|$ converges. ∎

Theorem 29.4 is analogous to interchanging the order of integration in a multiple integral.

We next pursue some applications.

In a finite sum the order in which the terms occur is irrelevant. For example,

$$a_1 + a_2 + a_3 + a_4 = a_4 + a_1 + a_3 + a_2$$

We will show using Theorem 29.4, that this fact generalizes to *absolutely* convergent series. First we need a definition of a rearrangement of an infinite series. A rearrangement of $\sum_{n=1}^{\infty} a_n$ will be another infinite series whose terms are the same as those of $\sum_{n=1}^{\infty} a_n$, but occur in a different order.

Definition 29.5   A *permutation* of the set of positive integers is a one-to-one function from **P** onto **P**.

Definition 29.6   A *rearrangement* of the series $\sum_{n=1}^{\infty} a_n$ is a series $\sum_{n=1}^{\infty} a_{f(n)}$, where f is a permutation of the set of positive integers.

The function f of Definition 29.6 tells us the new order of terms in the rearrangement.

Theorem 29.7   Let $\sum_{n=1}^{\infty} a_n$ be an absolutely convergent series. Then the rearrangement $\sum_{n=1}^{\infty} a_{f(n)}$ of $\sum_{n=1}^{\infty} a_n$ converges absolutely, and

$$\sum_{n=1}^{\infty} a_{f(n)} = \sum_{n=1}^{\infty} a_n$$

Proof.   Let $\{b_{m,n}\}$ be the double sequence whose matrix is

$$
\begin{array}{c}
\quad\quad\quad 1 \quad\ f^{-1}(1) \quad\ n \\
\begin{array}{c} 1 \\ \\ f(1) \\ \\ f(n) \end{array}
\begin{pmatrix}
0 \cdots a_1 \cdots 0 \cdots \\
\cdots\cdots\cdots\cdots\cdots \\
a_{f(1)} \cdots\cdots\cdots\cdots \\
\cdots\cdots\cdots\cdots\cdots \\
0 \cdots\cdots\cdots a_{f(n)} \cdot \\
\cdots\cdots\cdots\cdots\cdots
\end{pmatrix}
\end{array}
$$

We place $a_{f(n)}$ in the $f(n)$th row and the nth column, and we place zeros everywhere else. The formula for $b_{m,n}$ is

$$b_{m,n} = \begin{cases} a_m & \text{if } m = f(n) \\ 0 & \text{otherwise} \end{cases}$$

Since $\sum_{n=1}^{\infty} |b_{m,n}| = |a_m|$, the series $\sum_{m=1}^{\infty} (\sum_{n=1}^{\infty} |b_{m,n}|) = \sum_{m=1}^{\infty} |a_m|$ converges. By Theorem 29.4, $\sum_{n=1}^{\infty} |\sum_{m=1}^{\infty} b_{m,n}|$ converges and

$$\sum_{n=1}^{\infty} \left(\sum_{m=1}^{\infty} b_{m,n} \right) = \sum_{m=1}^{\infty} \left(\sum_{n=1}^{\infty} b_{m,n} \right)$$

Since $|\sum_{m=1}^{\infty} b_{m,n}| = |a_{f(n)}|$ it follows that $\sum_{n=1}^{\infty} |a_{f(n)}|$ converges. Finally,

$$\sum_{n=1}^{\infty} a_{f(n)} = \sum_{n=1}^{\infty} \left(\sum_{m=1}^{\infty} b_{m,n} \right) = \sum_{m=1}^{\infty} \left(\sum_{n=1}^{\infty} b_{m,n} \right) = \sum_{m=1}^{\infty} a_m \quad \blacksquare$$

If one rearranges a *conditionally* convergent series the sum and even the convergence of the series may be affected. For example, consider the alternating (conditionally convergent) series

$$\sum_{n=1}^{\infty} \frac{(-1)^{n+1}}{n}$$

Denoting its sum by L, we have

$$\frac{L}{2} = \frac{1}{2} - \frac{1}{4} + \frac{1}{6} - \frac{1}{8} + \cdots$$

Inserting zeros does not affect the sum so that

$$\frac{L}{2} = 0 + \frac{1}{2} + 0 - \frac{1}{4} + 0 + \frac{1}{6} + 0 - \frac{1}{8} + \cdots$$

Also
$$L = 1 - \frac{1}{2} + \frac{1}{3} - \frac{1}{4} + \frac{1}{5} - \frac{1}{6} + \frac{1}{7} - \frac{1}{8} + \cdots$$

Adding these two convergent series, we have

$$\frac{3L}{2} = 1 + 0 + \frac{1}{3} - \frac{1}{2} + \frac{1}{5} + 0 + \frac{1}{7} - \frac{1}{4} + \cdots$$

$$= 1 + \frac{1}{3} - \frac{1}{2} + \frac{1}{5} + \frac{1}{7} - \frac{1}{4} + \cdots$$

which is a rearrangement of $\sum_{n=1}^{\infty} (-1)^{n+1}/n$ with a sum different from L. See Exercises 29.7 and 29.8 at the end of this section for further results concerning rearrangements of conditionally convergent series.

If we formally multiply the power series

$$a_0 + a_1 x + a_2 x^2 + \cdots \qquad \text{and} \qquad b_0 + b_1 x + b_2 x^2 + \cdots$$

and collect terms, we get the power series

$$a_0b_0 + (a_0b_1 + a_1b_0)x + (a_0b_2 + a_1b_1 + a_2b_0)x^2 + \cdots$$

This computation motivates the following definition.

Definition 29.8 The *Cauchy product* of the series $\sum_{n=0}^{\infty} a_n$ and $\sum_{n=0}^{\infty} b_n$ is the infinite series $\sum_{n=0}^{\infty} c_n$, where

$$c_n = a_0b_n + a_1b_{n-1} + \cdots + a_nb_0 = \sum_{k=0}^{n} a_k b_{n-k}$$

We will use Theorem 29.4 to prove the following theorem concerning products of absolutely convergent series.

Theorem 29.9 Let $\sum_{n=0}^{\infty} a_n$ and $\sum_{n=0}^{\infty} b_n$ be absolutely convergent series, and let $\sum_{n=0}^{\infty} c_n$ be their Cauchy product. Then $\sum_{n=0}^{\infty} c_n$ converges absolutely and

$$\sum_{n=0}^{\infty} c_n = \left(\sum_{n=0}^{\infty} a_n \right) \left(\sum_{n=0}^{\infty} b_n \right)$$

Proof. Let $\{d_{m,n}\}$ be the double sequence whose matrix is

$$\begin{pmatrix} a_0b_0 & a_0b_1 & a_0b_2 & a_0b_3 & \cdots \\ 0 & a_1b_0 & a_1b_1 & a_1b_2 & \cdots \\ 0 & 0 & a_2b_0 & a_2b_1 & \cdots \\ 0 & 0 & 0 & a_3b_0 & \cdots \\ 0 & 0 & 0 & 0 & \cdots \\ \cdots \cdots \cdots \cdots \cdots \cdots \cdots \cdots \end{pmatrix}$$

Then $\sum_{n=0}^{\infty} |d_{m,n}| = |a_m| \sum_{n=0}^{\infty} |b_n|$. Thus

$$\sum_{m=0}^{\infty} \left(\sum_{n=0}^{\infty} |d_{m,n}| \right) = \sum_{m=0}^{\infty} \left(|a_m| \sum_{n=0}^{\infty} |b_n| \right) = \left(\sum_{n=0}^{\infty} |b_n| \right) \left(\sum_{m=0}^{\infty} |a_m| \right) \quad (29.2)$$

By Theorem 29.4, $\sum_{n=0}^{\infty} |\sum_{m=0}^{\infty} d_{m,n}|$ converges and

$$\sum_{n=0}^{\infty} \left(\sum_{m=0}^{\infty} d_{m,n} \right) = \sum_{m=0}^{\infty} \left(\sum_{n=0}^{\infty} d_{m,n} \right)$$

Now $|\sum_{m=0}^{\infty} d_{m,n}| = |c_n|$ and thus $\sum_{m=0}^{\infty} |c_m|$ converges. Equation (29.2) also holds without absolute values and thus $\sum_{m=0}^{\infty} (\sum_{n=0}^{\infty} d_{m,n}) = (\sum_{n=0}^{\infty} b_n)(\sum_{m=0}^{\infty} a_m)$. Therefore

$$\sum_{n=0}^{\infty} c_n = \sum_{n=0}^{\infty} \left(\sum_{m=0}^{\infty} d_{m,n} \right) = \sum_{m=0}^{\infty} \left(\sum_{n=0}^{\infty} d_{m,n} \right) = \left(\sum_{n=0}^{\infty} b_n \right) \left(\sum_{n=0}^{\infty} a_n \right) \quad \blacksquare$$

Corollary 29.10 If $\sum_{n=0}^{\infty} a_n(x - t)^n$ and $\sum_{n=0}^{\infty} b_n(x - t)^n$ are power series with radii of convergence R_1 and R_2, respectively, then $\sum_{n=0}^{\infty} c_n(x - t)^n$ converges absolutely if $|x - t| < \min \{R_1, R_2\}$, and for such x we have

$$\sum_{n=0}^{\infty} c_n(x - t)^n = \left[\sum_{n=0}^{\infty} a_n(x - t)^n\right]\left[\sum_{n=0}^{\infty} b_n(x - t)^n\right]$$

where

$$c_n = \sum_{k=0}^{n} a_k b_{n-k}$$

Proof. The proof follows immediately from Theorem 29.9. ∎

We close this section with another application of Theorem 29.4 to power series. Suppose that $\sum_{n=0}^{\infty} a_n x^n$ is a power series with radius of convergence R. Is it possible to represent this power series as a power series $\sum_{n=0}^{\infty} b_n(x - t)^n$ with $t \neq 0$? Our next theorem answers this question affirmatively.

Theorem 29.11 Let $\sum_{n=0}^{\infty} a_n x^n$ be a power series with radius of convergence R. If $|t| < R$, there exists a sequence $\{b_n\}$ such that $\sum_{n=0}^{\infty} b_n(x - t)^n$ converges if $|x - t| < R - |t|$, and if $|x - t| < R - |t|$, then

$$\sum_{n=0}^{\infty} a_n x^n = \sum_{n=0}^{\infty} b_n(x - t)^n$$

Proof. Suppose $|x - t| < R - |t|$. Then $|x| \leq |x - t| + |t| < R$, and thus $\sum_{n=0}^{\infty} a_n x^n$ converges. Now

$$\sum_{n=0}^{\infty} a_n x^n = \sum_{n=0}^{\infty} a_n[(x - t) + t]^n = \sum_{n=0}^{\infty} a_n\left(\sum_{m=0}^{n} \binom{n}{m}(x - t)^m t^{n-m}\right)$$

This series is the sum by rows of the double sequence whose matrix is

$$\begin{pmatrix} \binom{0}{0}a_0 & 0 & 0 & 0 & \cdots \\ \binom{1}{0}a_1 t & \binom{1}{1}a_1(x - t) & 0 & 0 & \cdots \\ \binom{2}{0}a_2 t^2 & \binom{2}{1}a_2 t(x - t) & \binom{2}{2}a_2(x - t)^2 & 0 & \cdots \\ \binom{3}{0}a_3 t^3 & \binom{3}{1}a_3 t^2(x - t) & \binom{3}{2}a_3 t(x - t)^2 & \binom{3}{3}a_3(x - t)^3 & \cdots \\ \cdots\cdots\cdots\cdots\cdots\cdots\cdots\cdots\cdots\cdots\cdots\cdots \end{pmatrix} \quad (29.3)$$

Since $|x - t| + |t| < R$, the series $\sum_{n=0}^{\infty} |a_n|(|x - t| + |t|)^n$ converges. But $\sum_{n=0}^{\infty} |a_n|(|x - t| + |t|)^n$ is the sum by rows of the double sequence whose matrix is the same as (29.3) with each term T in (29.3) replaced by $|T|$ (check this). By Theorem 29.4, the sum by rows of (29.3) is equal to the sum by columns. The sum by rows of (29.3) is $\sum_{n=0}^{\infty} a_n x^n$, and the sum by columns of (29.3) is $\sum_{n=0}^{\infty} b_n(x - t)^n$, where $b_n = \sum_{m=n}^{\infty} \binom{m}{n}a_m t^{m-n}$.

Therefore $\qquad \sum\limits_{n=0}^{\infty} b_n(x-t)^n = \sum\limits_{n=0}^{\infty} a_n x^n$ ∎

for $|x-t| < R - |t|$.

Exercises

29.1 Let $\sum_{n=0}^{\infty} a_n(x-t)^n$ and $\sum_{n=0}^{\infty} b_n(x-t)^n$ be power series with radii of convergence R_1 and R_2, respectively. Prove that $\sum_{n=0}^{\infty}(a_n + b_n)(x-t)^n$ converges absolutely if $|x-t| < \min\{R_1, R_2\}$ and in this case

$$\sum_{n=0}^{\infty}(a_n + b_n)(x-t)^n = \sum_{n=0}^{\infty} a_n(x-t)^n + \sum_{n=0}^{\infty} b_n(x-t)^n$$

29.2 Prove that if $|x| < 1$, then

$$(1 + x + x^2 + \cdots)^2 = 1 + 2x + 3x^2 + \cdots$$

29.3 Prove that $\sum_{n=0}^{\infty} x^n/n!$ converges absolutely for every x and

$$\left(\sum_{n=0}^{\infty} \frac{x^n}{n!}\right)\left(\sum_{n=0}^{\infty} \frac{y^n}{n!}\right) = \sum_{n=0}^{\infty} \frac{(x+y)^n}{n!}$$

29.4 Let $\sum_{n=1}^{\infty} a_n$ and $\sum_{n=1}^{\infty} b_n$ be absolutely convergent series. Let f be a one-to-one function from $\mathbf{P}$ onto $\mathbf{P} \times \mathbf{P}$. Let α_n and β_n be defined by the equation $f(n) = (\alpha_n, \beta_n)$. Prove that the series $\sum_{n=1}^{\infty} a_{\alpha_n} b_{\beta_n}$ converges absolutely and $(\sum_{n=1}^{\infty} a_n)(\sum_{n=1}^{\infty} b_n) = \sum_{n=1}^{\infty} a_{\alpha_n} b_{\beta_n}$.

29.5 Give an example of convergent series $\sum_{n=0}^{\infty} a_n$ and $\sum_{n=0}^{\infty} b_n$ whose Cauchy product diverges.

29.6 Prove that if $\sum_{n=0}^{\infty} a_n$ converges absolutely and $\sum_{n=0}^{\infty} b_n$ converges, then their Cauchy product $\sum_{n=0}^{\infty} c_n$ converges and $\sum_{n=0}^{\infty} c_n = (\sum_{n=0}^{\infty} a_n)(\sum_{n=0}^{\infty} b_n)$. Give an example to show that in this case, $\sum_{n=0}^{\infty} c_n$ need not converge absolutely.

29.7 Let $\sum_{n=1}^{\infty} a_n$ be a conditionally convergent series, and let x be a real number. Prove that there exists a rearrangement of $\sum_{n=1}^{\infty} a_n$ which converges to x.

29.8 Let $\sum_{n=1}^{\infty} a_n$ be a conditionally convergent series. Let x and y be real numbers or $-\infty$ or ∞ satisfying $x \leq y$. Prove that there exists a rearrangement of $\sum_{n=1}^{\infty} a_n$ such that if $\{s_n\}$ denotes the sequence of partial sums of the rearrangement, then

$$\lim_{n \to \infty} \inf s_n = x \quad \text{and} \quad \lim_{n \to \infty} \sup s_n = y$$

29.9 Let $\sum_{n=0}^{\infty} a_n x^n$ be a power series with radius of convergence R. Let $\{b_{m,n}\}$ be a double sequence such that $|b_{m,n}| \leq |a_n|$ for all positive integers n and m. Suppose also that $\lim_{m \to \infty} b_{m,n} = b_n$ for every positive integer n. Prove that $\sum_{n=0}^{\infty} b_n x^n$ converges if $|x| < R$ and $\lim_{m \to \infty} \sum_{n=0}^{\infty} b_{m,n} x^n = \sum_{n=0}^{\infty} b_n x^n$ for $|x| < R$.

29.10 Consider the power series $\sum_{n=0}^{\infty} x^{2n}$ which has radius of convergence $R = 1$.
(a) Let $|t| \geq 1$. Disprove the following statement. There exists a sequence $\{b_n\}$ such that if A is the set of x for which $\sum_{n=0}^{\infty} b_n(x-t)^n$ converges,

then $A \cap (-1, 1)$ is nonvoid, and if $x \in A \cap (-1, 1)$, then $\sum_{n=0}^{\infty} b_n$ $(x - t)^n = \sum_{n=0}^{\infty} x^{2n}$. This exercise shows that in general in Theorem 29.11, one must have $|t| < R$.

(b) Let $|t| < 1$ and $\sum_{n=0}^{\infty} b_n(x - t)^n$ be a power series with radius of convergence $R^* > 0$. Suppose that if $|x| < 1$ and $|x - t| < R^*$, we have $\sum_{n=0}^{\infty} x^{2n} = \sum_{n=0}^{\infty} b_n(x - t)^n$. Prove that the radius of convergence of $\sum_{n=0}^{\infty} b_n(x - t)^n$ is $1 - |t|$. This exercise shows that, in general, the value $R - |t|$ of Theorem 29.11 is the best possible.

29.11 Let $\{a_{m,n}\}$ be a double sequence. We say that the double series $\sum_{m,n} a_{m,n}$ *converges to* L if for every $\varepsilon > 0$, there exists a positive integer N such that if $k, l \geq N$, then

$$\left| \sum_{m=1}^{k} \sum_{n=1}^{l} a_{m,n} - L \right| < \varepsilon$$

(a) Prove that under the hypotheses of Theorem 29.4, the double series converges.

(b) Prove that the double series $\sum_{m,n} a_{m,n}$ converges if and only if for every $\varepsilon > 0$, there exists a positive integer N such that if $k, l, p, q \geq N$, then

$$\left| \sum_{m=1}^{k} \sum_{n=1}^{l} a_{m,n} - \sum_{m=1}^{p} \sum_{n=1}^{q} a_{m,n} \right| < \varepsilon$$

(c)* Suppose the double series $\sum_{m,n} a_{m,n}$ converges to L and the series $\sum_{n=1}^{\infty} a_{m,n}$ converges for every positive integer m. Prove that the sum by rows of $\sum_{m=1}^{\infty} \left(\sum_{n=1}^{\infty} a_{m,n} \right)$ exists and equals L.

29.12 Prove that the set of permutations of the positive integers is uncountable.

29.13 Let X be a set of real numbers. Let $|X| = \{|x| : x \in X\}$. Let S be the set of all finite sums of members of $|X|$. Suppose that S is bounded.

(a)* Prove that X is countable.

(b) Define $\sum X$ and justify your definition.

29.14 Let $\sum_{n=0}^{\infty} a_n x^n$ be a power series with radius of convergence $R > 0$ and suppose that $a_0 \neq 0$. Prove that there exists a power series $\sum_{n=0}^{\infty} b_n x^n$ with radius of convergence $R^* > 0$ such that if $|x| < \min \{R, R^*\}$, we have

$$\sum_{n=0}^{\infty} b_n x^n = \frac{1}{\sum_{n=0}^{\infty} a_n x^n}.$$

29.15 Let $a_1 = 1$ and $a_n = -1/2^{n-1}$ for $n \geq 2$. Let $\{b_{m,n}\}$ be the double sequence whose matrix is

$$\begin{pmatrix} a_1 & a_1 & a_1 & a_1 & \cdots \\ 0 & a_2 & a_2 & a_2 & \cdots \\ 0 & 0 & a_3 & a_3 & \cdots \\ 0 & 0 & 0 & a_4 & \cdots \\ \cdots\cdots\cdots\cdots\cdots\cdots \end{pmatrix}$$

Show that the sum by columns of the series $\sum_{m,n} b_{m,n}$ exists, but that the sum by rows does not exist.

VI

Limits of Real-Valued Functions and Continuous Functions on the Real Line

In this chapter we will define the limit of a real-valued function whose domain is a subset of the real numbers. We will then derive several limit theorems which are analogous to those we established in Chapter IV for sequences. We will define continuity and prove the Heine-Borel theorem and then use this theorem to show that every continuous function on a closed interval has a maximum and minimum. In Chapter VII we will generalize the results of this chapter.

30. Definition of the Limit of a Function

We begin with two definitions.

Definition 30.1 Let X be a subset of real numbers and let $a \in \mathbf{R}$. We say that a is an *accumulation point* of X if for every $\delta > 0$, there exists a number $x \in X$ such that $0 < |x - a| < \delta$.

In other words, a is an accumulation point of X if there exist points in X different from a which are arbitrarily near a. The number a may or may not belong to X. In Section 32, we will need the concept of left and right accumulation points. The definition is obtained by modifying Definition 30.1 so that the inequality $0 < |x - a| < \delta$ is restricted in the obvious way. For example, a is a *left accumulation point* of X if for every $\delta > 0$, there exists a number $x \in X$ such that $0 < a - x < \delta$.

Definition 30.2 Let f be a function from a subset X of $\mathbf{R}$ into $\mathbf{R}$, and let a be an accumulation point of X. We say that the *limit of $f(x)$ as x approaches* *a is L* and write

$$\lim_{x \to a} f(x) = L$$

if for every $\varepsilon > 0$, there exists $\delta > 0$ such that if $0 < |x - a| < \delta$ and $x \in X$, then $|f(x) - L| < \varepsilon$.

The assumption that a is an accumulation point of X assures us that there are points $x \in X$ satisfying the inequality $0 < |x - a| < \delta$.

As in the case of sequences (see Theorem 10.3) we can prove (Exercise 30.7) that if f has a limit as x approaches a, the limit is unique. This justifies writing $\lim_{x \to a} f(x) = L$.

The inequality $|x - a| < \delta$ states that the distance between x and a is less than δ and the inequality $|f(x) - L| < \varepsilon$ states that the distance between $f(x)$ and L is less than ε. The inequality $0 < |x - a|$ is equivalent to the statement $x \neq a$. Thus Definition 30.1 says (roughly) that $\lim_{x \to a} f(x) = L$ provided that the distance between $f(x)$ and L is small (less than ε) whenever the distance between x and a is small (less than δ) and $x \neq a$. The value $f(a)$, if it is defined, is irrelevant as far as the limit at a is concerned.

The inequalities $|f(x) - L| < \varepsilon$ and $|x - a| < \delta$ are equivalent, respectively, to $L - \varepsilon < f(x) < L + \varepsilon$ and $a - \delta < x < a + \delta$. Thus $\lim_{x \to a} f(x) = L$ provided given any open interval $(L - \varepsilon, L + \varepsilon)$ there exists an open interval $(a - \delta, a + \delta)$ such that if $x \in (a - \delta, a + \delta)$ and $x \neq a$, then $f(x) \in (L - \varepsilon, L + \varepsilon)$. See Figure 30.1.

Examples. Let $f(x) = 3x - 1$. We will prove that $\lim_{x \to 1} f(x) = 2$ directly from Definition 30.2. We are required to show that if $\varepsilon > 0$, there exists $\delta > 0$ such that if $0 < |x - 1| < \delta$, then $|(3x - 1) - 2| < \varepsilon$. Now $|(3x - 1) - 2| < \varepsilon$ is equivalent to $|x - 1| < \varepsilon/3$, so if we take $\delta = \varepsilon/3$ the limit will be established. The formal proof would be given as follows.

Let $\varepsilon > 0$. Set $\delta = \varepsilon/3$ and suppose $0 < |x - 1| < \delta$. Then

$$|(3x - 1) - 2| = |3(x - 1)| < 3\delta = \varepsilon$$

Let $f(x) = x^2 + x$. We will prove that $\lim_{x \to 2} f(x) = 6$ directly from Definition 30.2. This time we are required to show that

$$|x + 3||x - 2| = |(x^2 + x) - 6| < \varepsilon$$

if $0 < |x - 2| < \delta$. Notice that the term $|x + 3|$ is bounded in any bounded interval. In particular if $|x - 2| < 1$, then $|x + 3| < 6$. Therefore, if $|x - 2| < \min \{\varepsilon/6, 1\}$, we have

$$|x + 3||x - 2| < 6\frac{\varepsilon}{6} = \varepsilon$$

The formal proof would be given as follows.

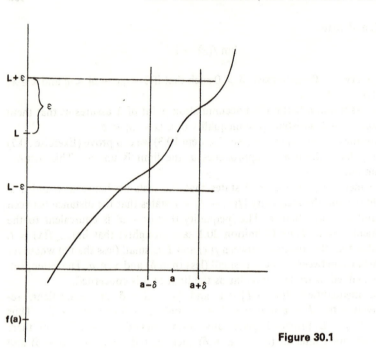

Figure 30.1

Let $\varepsilon > 0$. Set $\delta = \min\{\varepsilon/6, 1\}$ and suppose $0 < |x - 1| < \delta$. Then $|x + 3| < 6$; so

$$|(x^2 + x) - 6| = |x + 3||x - 2| < 6\frac{\varepsilon}{6} = \varepsilon$$

Exercises

In Exercises 30.1 to 30.5 prove the limit using Definition 30.2.

30.1 $\lim_{x \to a} f(x) = c$, where $f(x) = c$ for every real x.

30.2 $\lim_{x \to 2} (6x - 2) = 10$

30.3 $\lim_{x \to 2} x^2 = 4$

30.4 $\lim_{x \to 1} x^3 + x^2 + x + 1 = 4$

30.5 $\lim_{x \to 2} 2/x = 1$

30.6 Let

$$f(x) = \begin{cases} 0 & \text{if } x \text{ is rational} \\ 1 & \text{if } x \text{ is irrational} \end{cases}$$

Prove that $\lim_{x \to a} f(x)$ does not exist for any a.

30.7 Prove that if f has limit L and limit M as x approaches a, then $L = M$.

30.8 Suppose $\lim_{x \to a} f(x) = L > 0$. Prove that there exists $\delta > 0$ such that if $0 < |x - a| < \delta$, then $f(x) > 0$.

31. Limit Theorems for Functions

In this section we will prove some limit theorems for functions which are analogous to those we derived earlier for sequences (see Chapter IV, Section 12). We begin by defining algebraic methods of combining real-valued functions.

Definition 31.1 Let f and g be functions from a set X into $\mathbf{R}$, and let c be a real number. We define

 (i) $|f|(x) = |f(x)|, \qquad x \in X$

 (ii) $(cf)(x) = cf(x), \qquad x \in X$

 (iii) $(f + g)(x) = f(x) + g(x), \qquad x \in X$

 (iv) $(f - g)(x) = f(x) - g(x), \qquad x \in X$

 (v) $(f \cdot g)(x) = f(x)g(x), \qquad x \in X$

 (vi) $\left(\dfrac{f}{g}\right)(x) = \dfrac{f(x)}{g(x)}, \qquad x \in X$ and $g(x) \neq 0$

The operations imposed on the functions in Definition 31.1 are said to be *pointwise operations*. For example, in Definition 31.1, (iii), to compute the value of the function $f + g$ at the point x, we simply add the values of f and g computed at the point x.

We may prove directly from Definition 30.2 that if $\lim_{x \to a} f(x) = L$ and $\lim_{x \to a} g(x) = M$, then $\lim_{x \to a} f(x) + g(x) = L + M$. Let $\varepsilon > 0$. There exists $\delta_1 > 0$ such that if $0 < |x - a| < \delta_1$, then $|f(x) - L| < \varepsilon/2$. There exists $\delta_2 > 0$ such that if $0 < |x - a| < \delta_2$, then $|g(x) - M| < \varepsilon/2$. Let $\delta = \min\{\delta_1, \delta_2\}$. Now if $0 < |x - a| < \delta$, then

$$|f(x) + g(x) - (L + M)| \leq |f(x) - L| + |g(x) - M| < \frac{\varepsilon}{2} + \frac{\varepsilon}{2} = \varepsilon$$

and therefore $\lim_{x \to a} f(x) + g(x) = L + M$. The reader should compare this argument with the proof of Theorem 12.2 for sequences. The analogy is more than coincidental as is demonstrated by the following theorem which links the definition of the limit of a function with the definition of the limit of a sequence.

Theorem 31.2 Let $f: X \to \mathbf{R}$, $X \subset \mathbf{R}$ and suppose a is an accumulation point of X. Then $\lim_{x \to a} f(x) = L$ if and only if for every sequence $\{a_n\}$ in X

such that $\lim_{n \to \infty} a_n = a$ and $a_n \neq a$ for every positive integer n, we have $\lim_{n \to \infty} f(a_n) = L$.

Proof. First suppose that $\lim_{x \to a} f(x) = L$. Let $\{a_n\}$ be a sequence in X such that $\lim_{n \to \infty} a_n = a$ and $a_n \neq a$ for every positive integer n. Let $\varepsilon > 0$. There exists $\delta > 0$ such that if $0 < |x - a| < \delta$, then $|f(x) - L| < \varepsilon$. There exists a positive integer N such that if $n \geq N$, then $|a_n - a| < \delta$. Since $a_n \neq a$ for every positive integer n, we have $0 < |a_n - a|$. Therefore, if $n \geq N$, then

$$0 < |a_n - a| < \delta$$

so that

$$|f(a_n) - L| < \varepsilon$$

and consequently, $\lim_{n \to \infty} f(a_n) = L$.

Now suppose that $\lim_{n \to \infty} f(a_n) = L$ for every sequence $\{a_n\}$ in X such that $\lim_{n \to \infty} a_n = a$ and $a_n \neq a$ for every positive integer n. Assume by way of contradiction that $\lim_{x \to a} f(x) \neq L$. Then there exists $\varepsilon > 0$ such that for every $\delta > 0$, there exists an $x \in X$ with $0 < |x - a| < \delta$, such that $|f(x) - L| \geq \varepsilon$.

Let n be a positive integer. Taking $\delta = 1/n$, we find a number $a_n \in X$ such that

$$0 < |a_n - a| < \frac{1}{n} \quad \text{and} \quad |f(a_n) - L| \geq \varepsilon$$

Now $\{a_n\}$ is a sequence in X such that $\lim_{n \to \infty} a_n = a$ and $a_n \neq a$ for every positive integer n. Since $|f(a_n) - L| \geq \varepsilon$ for every positive integer n, $\lim_{n \to \infty} f(a_n) \neq L$, and we have the desired contradiction. ∎

We may use Theorem 31.2 to prove analogues of the theorems of Section 12 for sequences.

Theorem 31.3   Let f and g be two functions from X into $\mathbf{R}$ with $X \subset \mathbf{R}$. If $\lim_{x \to a} f(x) = L$ and $\lim_{x \to a} g(x) = M$, then

  (i) $\lim_{x \to a} |f(x)| = |L|$

  (ii) $\lim_{x \to a} cf(x) = cL \qquad c \in \mathbf{R}$

  (iii) $\lim_{x \to a} [f(x) + g(x)] = L + M$

  (iv) $\lim_{x \to a} [f(x) - g(x)] = L - M$

  (v) $\lim_{x \to a} f(x)g(x) = LM$

(vi) $\displaystyle\lim_{x \to a} \frac{f(x)}{g(x)} = \frac{L}{M}$

provided in (vi) that $M \neq 0$.

Proof. We prove only (iii). Let $\{a_n\}$ be a sequence such that $\lim_{n \to \infty} a_n = a$ and $a_n \neq a$ for every positive integer n. By Theorem 31.2, we have

$$\lim_{n \to \infty} f(a_n) = L \qquad \lim_{n \to \infty} g(a_n) = M$$

Hence by Theorem 12.2, $\lim_{n \to \infty} [f(a_n) + g(a_n)] = L + M$.

By Theorem 31.2, $\lim_{x \to a} [f(x) + g(x)] = L + M$. ∎

Exercises

31.1 Prove Theorem 31.3 (i), (ii), (iv), (v), and (vi) using Definition 30.2.

31.2 Prove Theorem 31.3 (i), (ii), (iv), (v), and (vi) using Theorem 31.2.

31.3 Let f and g be functions such that for some $\delta > 0$, $f(x) \le g(x)$ if $0 < |x - a| < \delta$. Suppose that $\lim_{x \to a} f(x) = L$ and $\lim_{x \to a} g(x) = M$. Prove that $L \le M$.

31.4 (Squeeze Theorem) Let f, g, and h be functions such that for some $\delta > 0$, $f(x) \le g(x) \le h(x)$ if $0 < |x - a| < \delta$. Suppose that $\lim_{x \to a} f(x) = L = \lim_{x \to a} h(x)$. Prove that $\lim_{x \to a} g(x) = L$.

31.5 (Cauchy Condition for Functions) Prove that f has a limit at a if and only if for every $\varepsilon > 0$, there exists $\delta > 0$ such that if $0 < |x - a| < \delta$ and $0 < |y - a| < \delta$, then $|f(x) - f(y)| < \varepsilon$.

31.6 Define $\limsup_{x \to a} f(x)$ and $\liminf_{x \to a} f(x)$ and derive some results analogous to those of Sections 20 and 21.

32. One-Sided and Infinite Limits

It is easy to show that the function

$$f(x) = \frac{|x|}{x}$$

does not have a limit at $x = 0$. However, for x close to 0 and $x > 0$, $f(x)$ is close to 1. The value 1 is called the *right-hand limit of f at* 0. Similarly, restricting our attention to $x < 0$, we see that f has a left-hand limit -1 at 0. See Figure 32.1. We now give the formal definition of one-sided limits.

Definition 32.1 Let f be a function from a subset X of **R** into **R**, and let a be a left (right) accumulation point of X. We say that the *limit of $f(x)$ as x*

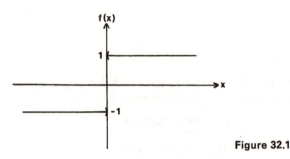

Figure 32.1

approaches a from the left (right) is L and write

$$\lim_{x \to a^-} f(x) = L \qquad (\lim_{x \to a^+} f(x) = L)$$

if for every $\varepsilon > 0$, there exists $\delta > 0$ such that if $0 < a - x < \delta$ $(0 < x - a < \delta)$, then $|f(x) - L| < \varepsilon$.

The following theorem connects Definitions 30.2 and 32.1.

Theorem 32.2 If a is a right and left accumulation point of X, then $\lim_{x \to a} f(x) = L$ if and only if $\lim_{x \to a^+} f(x) = L = \lim_{x \to a^-} f(x)$.

Proof. Suppose $\lim_{x \to a} f(x) = L$. Let $\varepsilon > 0$. There exists $\delta > 0$ such that if $0 < |x - a| < \delta$, then $|f(x) - L| < \varepsilon$. If $a < x < a + \delta$, then $0 < |x - a| < \delta$; so $|f(x) - L| < \varepsilon$. Therefore $\lim_{x \to a^+} f(x) = L$. Similarly, $\lim_{x \to a^-} f(x) = L$.

Now suppose $\lim_{x \to a^+} f(x) = L = \lim_{x \to a^-} f(x)$. Let $\varepsilon > 0$. There exists $\delta_1 > 0$ such that if $a < x < a + \delta_1$, then $|f(x) - L| < \varepsilon$. There exists $\delta_2 > 0$ such that if $a - \delta_2 < x < a$, then $|f(x) - L| < \varepsilon$. Let $\delta = \min\{\delta_1, \delta_2\}$. Then if $0 < |x - a| < \delta$ either $a < x < a + \delta_1$ or $a - \delta_2 < x < a$ so that $|f(x) - L| < \varepsilon$. ∎

The symbolism $\lim_{x \to \infty} f(x) = L$ should mean that $f(x)$ is close to L for large x. Just as "for large n" means $n \geq N$, "for large x" means $x > M$.

Definition 32.3 Let $f: X \to \mathbf{R}$ with $X \subset \mathbf{R}$. We say that *the limit of $f(x)$ as x approaches infinity (minus infinity) is L* and write $\lim_{x \to \infty} f(x) = L$ $[\lim_{x \to -\infty} f(x) = L]$ if for every $\varepsilon > 0$, there exists a number M such that if $x > M$ $(x < M)$, then $|f(x) - L| < \varepsilon$. However, if X is bounded above (below), we say $\lim_{x \to \infty} f(x)$ $[\lim_{x \to -\infty} f(x)]$ is undefined.

It is easy to show that $\lim_{x \to \infty} (1/x) = 0$ and $\lim_{x \to -\infty} (1/x) = 0$.

Infinite limits are linked to one-sided limits by the following fact, whose proof we leave as an exercise (Exercise 32.1). $\lim_{x \to \infty} f(x) = L$ if and only if $\lim_{x \to 0^+} f(1/x) = L$. A similar statement holds for minus infinity.

It is possible to generalize many of the results of Section 31 to one-sided and infinite limits. Some generalizations are given as exercises.

Exercises

32.1 Prove that $\lim_{x \to \infty} f(x) = L$ if and only if $\lim_{x \to 0^+} f(1/x) = L$. State and prove the analogue for minus infinity.

32.2 Prove analogues of Theorem 31.2 for one-sided and infinite limits.

32.3 Prove analogues of Theorem 31.3 for one-sided and infinite limits.

32.4 Prove that if $\lim_{x \to \infty} f(x) = L$, then $\lim_{n \to \infty} f(n) = L$.

32.5 Let f be an increasing function on $[a, b]$ (that is, if $x < y$ and $x, y \in [a, b]$, then $f(x) \le f(y)$). Prove that f has a right-hand limit at every point c in $[a, b)$ and that f has a left-hand limit at every point c in $(a, b]$.

33. Continuity

If $f(x) = x$, then $\lim_{x \to a} f(x) = a$ for every a in $\mathbf{R}$. For if $\varepsilon > 0$ and we set $\delta = \varepsilon$, then if $0 < |x - a| < \delta$, we have $|f(x) - a| < \varepsilon$. Using Theorem 31.3 (v), we have

$$\lim_{x \to a} x^n = a^n, \qquad n \in \mathbf{P} \tag{33.1}$$

Using Theorem 31.3 (ii) and equation (33.1), we may conclude that if c is in $\mathbf{R}$, then

$$\lim_{x \to a} cx^n = ca^n, \qquad n \in \mathbf{P} \tag{33.2}$$

If $p(x) = a_n x^n + \cdots + a_1 x + a_0$ is a polynomial, we may use Theorem 31.3 (iii) and equation (33.2) to conclude that

$$\lim_{x \to a} p(x) = \lim_{x \to a} (a_n x^n + \cdots + a_1 x + a_0)$$
$$= a_n a^n + \cdots + a_1 a + a_0$$
$$= p(a) \tag{33.3}$$

If $R(x) = p(x)/q(x)$ is a rational function [$p(x)$ and $q(x)$ are polynomials], we may use Theorem 31.3 (vi) and equation (33.3) to conclude that

$$\lim_{x \to a} R(x) = \lim_{x \to a} \frac{p(x)}{q(x)} = \frac{p(a)}{q(a)} = R(a) \tag{33.4}$$

provided that $q(a) \ne 0$.

In each of equations (33.1) through (33.4) we were able to calculate the limit of the function at a by simply substituting a for x. When this is possible we say that the function is continuous at a.

Definition 33.1 Let f be a function from a subset X of **R** into **R**. We say that f is *continuous at a* if either

(i) a is an accumulation point of X and $\lim_{x \to a} f(x) = f(a)$.
(ii) a is not an accumulation point of X.

Our next theorem follows immediately from our earlier work.

Theorem 33.2 If f and g are continuous at a, then each of the following functions is also continuous at a

  (i) $|f|$
 (ii) cf for each $c \in$ **R**
(iii) $f + g$
(iv) $f - g$
 (v) $f \cdot g$
(vi) $\dfrac{f}{g}$ provided that $g(a) \neq 0$

Proof. The proof follows immediately from Theorem 31.3. ∎

We next give an equivalent formulation of continuity.

Theorem 33.3 Let f be a function from a subset X of **R** into **R** and let $a \in X$. Then f is continuous at a if and only if for every $\varepsilon > 0$, there exists $\delta > 0$ such that if $|x - a| < \delta$ and $x \in X$, then $|f(x) - f(a)| < \varepsilon$.

Proof. Suppose that f is continuous at a. First suppose that a is not an accumulation point of X. Let $\varepsilon > 0$. There exists $\delta > 0$ such that

$$\{x \mid 0 < |x - a| < \delta\} \cap X = \varnothing$$

Therefore, if $|x - a| < \delta$ and $x \in X$, then $x = a$, and so $|f(x) - f(a)| = 0 < \varepsilon$.

Now suppose that a is an accumulation point of X. Then $\lim_{x \to a} f(x) = f(a)$. Let $\varepsilon > 0$. There exists $\delta > 0$ such that if $0 < |x - a| < \delta$ and $x \in X$, then $|f(x) - f(a)| < \varepsilon$. Suppose $|x - a| < \delta$ and $x \in X$. Either $|x - a| = 0$ or $0 < |x - a| < \delta$. If $|x - a| = 0$, then $x = a$, so that $|f(x) - f(a)| = 0 < \varepsilon$. If $0 < |x - a| < \delta$, then $|f(x) - f(a)| < \varepsilon$. In either case, $|f(x) - f(a)| < \varepsilon$.

Now assume that for every $\varepsilon > 0$, there exists $\delta > 0$ such that if $|x -$

$a| < \delta$ and $x \in X$, then $|f(x) - f(a)| < \varepsilon$. If a is not an accumulation point of X, then f is continuous at a; so suppose that a is an accumulation point of X. Then there exists $\delta > 0$ such that if $|x - a| < \delta$ and $x \in X$, then $|f(x) - f(a)| < \varepsilon$. Therefore, if $0 < |x - a| < \delta$ and $x \in X$, then $|f(x) - f(a)| < \varepsilon$. Thus $\lim_{x \to a} f(x) = f(a)$ so that f is continuous at a. ■

Our next theorem shows that the composition of continuous functions is continuous.

Theorem 33.4 If g is continuous at a and f is continuous at $g(a)$, then $f \circ g$ is continuous at a.

Proof. Let $\varepsilon > 0$. There exists $\delta_1 > 0$ such that if $|x - g(a)| < \delta_1$, then $|f(x) - f(g(a))| < \varepsilon$. There exists $\delta > 0$ such that if $|x - a| < \delta$, then $|g(x) - g(a)| < \delta_1$. Now suppose $|x - a| < \delta$. Then $|g(x) - g(a)| < \delta_1$; so $|f(g(x)) - f(g(a))| < \varepsilon$, and hence $f \circ g$ is continuous at a. ■

Example. $f(x) = |x|$ is continuous at every point of $\mathbf{R}$ and $g(x) = (3x^3 + 2x^2 + x + 1)/(x^2 + 1)$ is continuous at every point of $\mathbf{R}$; so by Theorem 33.4,

$$f(g(x)) = \left| \frac{3x^3 + 2x^2 + x + 1}{x^2 + 1} \right|$$

is continuous at every point of $\mathbf{R}$.

Exercises

33.1 Prove that f is continuous at a if and only if for every sequence $\{a_n\}$ such that $\lim_{n \to \infty} a_n = a$ we have $\lim_{n \to \infty} f(a_n) = f(a)$.

33.2 Let f be defined on $[0, 1]$ by the formula

$$f(x) = \begin{cases} 1 & \text{if } x \text{ is rational} \\ 0 & \text{if } x \text{ is irrational} \end{cases}$$

Prove that f is continuous at no point of $[0, 1]$.

33.3 Let f be defined on $[0, 1]$ by the formula

$$f(x) = \begin{cases} x & \text{if } x \text{ is rational} \\ 0 & \text{if } x \text{ is irrational} \end{cases}$$

Prove that f is continuous only at 0.

33.4 Let f be defined on $[0, 1]$ by the formula

$$f(x) = \begin{cases} \dfrac{1}{n} & \text{if } x = \dfrac{m}{n} \text{ is rational} \\ 0 & \text{if } x \text{ is irrational} \\ 1 & \text{if } x = 0 \end{cases}$$

($x = m/n$ is in lowest terms, $m, n \in P$). Prove that f is continuous only at the irrational points of $[0, 1]$.

33.5 Suppose that f is continuous at every point of $[a, b]$ and $f(x) = 0$ if x is rational. Prove that $f(x) = 0$ for every x in $[a, b]$.

33.6* Let f be an increasing function on $[a, b]$. Prove that the set of points at which f is not continuous is countable.

33.7 Prove that if f is continuous on **R** and $f(x + y) = f(x) + f(y)$ for $x, y \in R$, then $f(x) = ax$ for some $a \in R$.

34. The Heine-Borel Theorem and a Consequence for Continuous Functions

Let f be a function on $[a, b]$. We say that f is *continuous on* $[a, b]$ if f is continuous at every point of $[a, b]$ (at the end-points a and b we demand that $\lim_{x \to a+} f(x) = f(a)$ and $\lim_{x \to b-} f(x) = f(b)$). Our immediate goal is to prove that if f is continuous on $[a, b]$, then f is bounded on $[a, b]$.

In this section only, we will say that f is *bounded* on a set X if there exists a number M such that $|f(x)| < M$ for every $x \in X \cap$ (domain f). According to this definition, f can be bounded on a set X which is not a subset of the domain of f.

There is a certain finiteness property associated with being bounded. If f is bounded on X and Y, then f is bounded on $X \cup Y$. For if $|f(x)| < M_1$ for $x \in X \cap$ (domain f) and $|f(x)| < M_2$ for $x \in Y \cap$ (domain f), then $|f(x)| < \max\{M_1, M_2\}$ for $x \in (X \cup Y) \cap$ (domain f). By induction we see that if f is bounded on $X_1, X_2, \ldots, X_n$, then f is bounded on $X_1 \cup X_2 \cup \cdots \cup X_n$. It is easy to see that this result does *not* extend to infinite unions (verify).

We now show that if f is continuous at c, then f is bounded on an open interval containing c.

Lemma 34.1 If f is continuous at c, there exists $\delta > 0$ such that f is bounded on $(c - \delta, c + \delta)$.

Proof. Taking $\varepsilon = 1$, there exists $\delta > 0$ such that if $|x - c| < \delta$ and x is in the domain of f, then $|f(x) - f(c)| < 1$. Therefore, if $x \in (c - \delta, c + \delta) \cap$ (domain f), we have

$$|f(x)| \leq |f(x) - f(c)| + |f(c)| < 1 + |f(c)| \qquad \blacksquare$$

If f is continuous on $[a, b]$, then for each c in $[a, b]$ there exists an open interval I_c containing c such that f is bounded on I_c (Lemma 34.1). Of course the collection $\{I_c \mid c \in [a, b]\}$ is infinite, and so we cannot conclude that f is bounded on the *infinite* union $\cup \{I_c \mid c \in [a, b]\}$. However, the Heine-Borel theorem (Theorem 34.2) shows that some *finite* subcollection $\{I_{c_1}, \ldots, I_{c_n}\}$ has the property that

$$[a, b] \subset \bigcup_{i=1}^{n} I_{c_i}$$

Since f is bounded on I_{c_i} for $i = 1, \ldots, n$, it will follow that f is bounded on $\bigcup_{i=1}^{n} I_{c_i}$ and hence also on $[a, b]$.

Theorem 34.2 (Heine-Borel Theorem) Let $\mathscr{J}$ be a collection of open intervals such that

$$\cup \mathscr{J} \supset [a, b]$$

Then there exists a finite subset $\{I_1, \ldots, I_n\}$ of $\mathscr{J}$ such that

$$\bigcup_{i=1}^{n} I_i \supset [a, b]$$

Proof. Let

$$X = \left\{ x \in (a, b] \,|\, [a, x] \subset \bigcup_{i=1}^{n} I_i \text{ for some } I_1, \ldots, I_n \, \varepsilon \, \mathscr{J} \right\}$$

Now $a \in I = (s_0, t_0)$ for some $I \in \mathscr{J}$, and hence if we choose x_0, $a < x_0 < t_0$, it follows that $x_0 \in X$. Thus X is a nonvoid subset of **R** which is bounded above; so by the least-upper-bound axiom, X has a least upper bound c. Since $a < x_0 \le c$, we have $a < c$.

Now $c \in I = (s_1, t_1)$ for some $I \in \mathscr{J}$. Since $s_1 < c$, s_1 is not an upper bound for X, and hence there exists $x > a$, $x \in X$, such that $s_1 < x \le c$. By the definition of X, there exists $I_1, \ldots, I_n \in \mathscr{J}$ such that

$$[a, x] \subset \bigcup_{i=1}^{n} I_i$$

Therefore
$$[a, c] \subset \left[\bigcup_{i=1}^{n} I_i \right] \cup I$$

We conclude that $c \in X$ and the proof will be complete when we show that $c = b$.

Suppose $c < b$. Since $c \in X$, there exists $I_1, \ldots, I_n \in \mathscr{J}$ such that

$$[a, c] \subset \bigcup_{i=1}^{n} I_i$$

In particular, $c \in I_j = (s_2, t_2)$ for some j, $1 \le j \le n$. Choose d such that $c < d < b$ and $c < d < t_2$. Then

$$[a, d] \subset \bigcup_{i=1}^{n} I_i$$

and thus $d \in X$. Since $d > c$, we have a contradiction, and therefore $c = b$. ∎

The property of the closed interval $[a, b]$ given in Theorem 34.2 is called *compactness*, and this property will be thoroughly investigated in Chapter VII, Sections 42 to 44.

Theorem 34.3 If f is continuous on $[a, b]$, then f is bounded on $[a, b]$.

Proof. By Lemma 34.1, for each $c \in [a, b]$, there exists an open interval I_c containing c such that f is bounded on I_c. Since

$$[a, b] \subset \cup \{I_c \mid c \in [a, b]\}$$

by Theorem 34.2, there exist $c_1, \ldots, c_n$ in $[a, b]$ such that

$$[a, b] \subset \bigcup_{i=1}^{n} I_{c_i}$$

Now f is bounded on $\bigcup_{i=1}^{n} I_{c_i}$, and therefore f is bounded on $[a, b]$. ∎

Theorem 34.4 If f is continuous on $[a, b]$, there exist points c and d in $[a, b]$ such that

$$f(c) \leq f(x) \leq f(d)$$

for all x in $[a, b]$. That is, if f is continuous on $[a, b]$, then f attains a maximum and a minimum on $[a, b]$.

Proof. By Theorem 34.3, f is bounded on $[a, b]$. Let

$$M = \text{lub } \{f(x) \mid x \in [a, b]\}$$

We must show that $f(d) = M$ for some d in $[a, b]$. Suppose this fails. Then $f(x) < M$ for every x in $[a, b]$. Let

$$g(x) = \frac{1}{M - f(x)} \qquad \text{for } x \in [a, b]$$

By Theorem 33.2, g is continuous on $[a, b]$. By Theorem 34.3, g is bounded on $[a, b]$, and thus there exists a positive number N such that $g(x) < N$ for every x in $[a, b]$. We have

$$\frac{1}{M - f(x)} < N$$

for every x in $[a, b]$ which reduces to

$$f(x) < M - \frac{1}{N} \tag{34.1}$$

for every x in $[a, b]$. Since $M - 1/N < M$, inequality (34.1) contradicts the fact that M is the least upper bound of the set $\{f(x) \mid x \in [a, b]\}$. Therefore $f(d) = M$ for some d in $[a, b]$.

The existence of a minimum is proved in a similar way. ∎

In Section 42, we will generalize Theorems 34.3 and 34.4 (see Theorem 42.6 and Corollary 42.7).

Exercises

34.1 Prove that if f is continuous on $[a, b]$, there exists a point c in $[a, b]$ such that
$$f(c) = \text{glb } \{f(x) \mid x \in [a, b]\}$$
thus completing the proof of Theorem 34.4.

34.2 (a) Use Exercise 30.8 and the Heine-Borel theorem to prove that if f is continuous on $[a, b]$ and $f(x) > 0$ for every x in $[a, b]$, then there exists $\varepsilon > 0$ such that $f(x) \geq \varepsilon$ for every x in $[a, b]$.

    (b) Prove the statement in part (a) by considering the function $g(x) = 1/f(x)$

34.3 Give another proof of Theorem 34.3 by completing the following argument. If f is not bounded on $[a, b]$, there exists a sequence $\{a_n\}$ in $[a, b]$ such that $|f(a_n)| > n$. By the Bolzano-Weierstrass theorem there exists a convergent subsequence $\{a_{n_k}\}$ of $\{a_n\}$. Let $c = \lim_{k \to \infty} a_{n_k}$. Then $f(c) = \lim_{k \to \infty} f(a_{n_k})$, which is a contradiction.

34.4 Give an example of a collection $\mathscr{I}$ of open intervals such that $\cup \mathscr{I} \supset (a, b)$ and such that for no finite subcollection $\mathscr{J}$ of $\mathscr{I}$ do we have $\cup \mathscr{J} \supset (a, b)$.

34.5 Deduce the Bolzano-Weierstrass theorem (Theorem 18.1) from the Heine-Borel theorem (Theorem 34.2).

34.6 Prove that if f is continuous on $[a, b]$ and $\varepsilon > 0$, there exists $\delta > 0$ such that if $|x - y| < \delta$ and $x, y \in [a, b]$, then $|f(x) - f(y)| < \varepsilon$.

VII

Metric Spaces

In this chapter, we present an introduction to the theory of metric spaces. Our results generalize much of the material in Chapters IV, V, and VI, which were concerned with real numbers. The theory of metric spaces generalizes those parts of our earlier work which relied on the notion of distance.

35. The Distance Function

If x and y are real numbers, we may interpret $|x - y|$ geometrically as the distance between x and y. (See Figure 35.1.) Using the notion of distance, we may rephrase the definition of the limit of a sequence as follows: $\lim_{n \to \infty} a_n = L$ if and only if for every $\varepsilon > 0$, there exists a positive integer N such that the distance between a_n and L is less than ε if $n \geq N$. The definition of the limit of a function also involves the notion of distance. Letting $d(x, y) = |x - y|$ denote the distance between x and y, we may say that $\lim_{x \to a} f(x) = L$ if and only if for every $\varepsilon > 0$, there exists $\delta > 0$ such that if $d(x, y) < \delta$, then $d(f(x), L) < \varepsilon$. Thus both of our fundamental definitions of limit involve the notion of distance, and we would expect to be able to define limits in arbitrary sets provided an adequate notion of distance were defined.

The distance function $d(x, y) = |x - y|$ is a function from $\mathbf{R} \times \mathbf{R}$ into $[0, \infty)$ which satisfies

(i) $d(x, y) = 0$ if and only if $x = y$

(ii) $d(x, y) = d(y, x)$ for all $x, y \in \mathbf{R}$

(iii) $d(x, z) \leq d(x, y) + d(y, z)$ for all $x, y, z \in \mathbf{R}$

It turns out that properties (i), (ii), and (iii) above are enough to determine

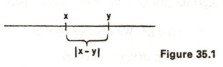

Figure 35.1

a useful notion of distance in an arbitrary set and we are led to the following definition.

Definition 35.1 Let M be a set. A *metric* on M is a function d from $M \times M$ into $[0, \infty)$ which satisfies

(i) $d(x, y) = 0$ if and only if $x = y$
(ii) $d(x, y) = d(y, x)$ for all $x, y \in M$
(iii) $d(x, z) \leq d(x, y) + d(y, z)$ for all $x, y, z \in M$

Definition 35.2 A *metric space* is an ordered pair (M, d), where M is a set and d is a metric for M.

A metric space is (roughly) a set M together with a distance function (metric) d on M.

Property (iii) of Definition 35.1 is called the *triangle inequality* because of its geometric interpretation in the plane. (See Example 35.4.)

Example 35.3 The pair $(\mathbf{R}, d)$ is a metric space, where d is defined by $d(x, y) = |x - y|$. The metric d is called the *usual* (or *absolute value* or *Euclidean*) *metric* for $\mathbf{R}$. The space $\mathbf{R}$ is understood to have the usual metric unless otherwise specified.

Let $\mathbf{R}^2 = \mathbf{R} \times \mathbf{R}$. Geometrically, $\mathbf{R}^2$ is the plane.

Example 35.4 We will show that

$$d(x, y) = \sqrt{(x_1 - y_1)^2 + (x_2 - y_2)^2}$$

where $x = (x_1, x_2)$, $y = (y_1, y_2) \in \mathbf{R}^2$ defines a metric on $\mathbf{R}^2$.

Now

$$d(x, y) = \sqrt{(x_1 - y_1)^2 + (x_2 - y_2)^2} = 0$$

if and only if $x_1 = y_1$ and $x_2 = y_2$. Therefore, $d(x, y) = 0$ if and only if $x = y$, and hence part (i) of Definition 35.1 holds.

Part (ii) of Definition 35.1 holds since

$$d(x, y) = \sqrt{(x_1 - y_1)^2 + (x_2 - y_2)^2}$$
$$= \sqrt{(y_1 - x_1)^2 + (y_2 - x_2)^2} = d(y, x)$$

Finally, we must establish the triangle inequality [part (iii) of Definition

35.1]. As is often the case, the triangle inequality is the most difficult part of Definition 35.1 to verify.

Let $x = (x_1, x_2)$, $y = (y_1, y_2)$, $z = (z_1, z_2) \in \mathbf{R}^2$. We must prove that

$$\sqrt{(x_1 - z_1)^2 + (x_2 - z_2)^2} = d(x, z) \leq d(x, y) + d(y, z)$$
$$= \sqrt{(x_1 - y_1)^2 + (x_2 - y_2)^2}$$
$$+ \sqrt{(y_1 - z_1)^2 + (y_2 - z_2)^2} \quad (35.1)$$

Letting $a_i = (x_i - y_i)$, $b_i = (y_i - z_i)$, and writing $x_i - z_i = (x_i - y_i) + (y_i - z_i)$, $i = 1, 2$, inequality (35.1) becomes

$$\sqrt{(a_1 + b_1)^2 + (a_2 + b_2)^2} \leq \sqrt{a_1^2 + a_2^2} + \sqrt{b_1^2 + b_2^2}$$

Squaring both sides and simplifying, we obtain the equivalent inequality

$$a_1 b_1 + a_2 b_2 \leq \sqrt{(a_1^2 + a_2^2)(b_1^2 + b_2^2)}$$

We show that

$$|a_1 b_1 + a_2 b_2| \leq \sqrt{(a_1^2 + a_2^2)(b_1^2 + b_2^2)} \quad (35.2)$$

for any real numbers a_1, a_2, b_1, b_2 thus establishing the triangle inequality.

Squaring both sides of inequality (35.2) and simplifying, we obtain the equivalent inequality

$$0 \leq a_1^2 b_2^2 - 2a_1 b_1 a_2 b_2 + a_2^2 b_1^2$$

which in turn may be written

$$0 \leq (a_1 b_2 - a_2 b_1)^2 \quad (35.3)$$

Since (35.3) holds for all real numbers a_1, a_2, b_1, b_2, (35.2) and hence also (35.1) are established. Thus the triangle inequality holds for d, and therefore d is a metric for $\mathbf{R}^2$.

The metric d of Example 35.4 is called the *usual* (or *Euclidean*) *metric* for $\mathbf{R}^2$. This metric gives the usual notion of distance in the plane. The triangle inequality is illustrated in Figure 35.2.

We will generalize Example 35.4 to n-space in the next section.

Figure 35.2

Example 35.5 We let l^1 denote the set of all real sequences $\{a_n\}$ such that $\sum_{n=1}^{\infty} a_n$ converges absolutely.

We will show that

$$d(\{a_n\}, \{b_n\}) = \sum_{n=1}^{\infty} |a_n - b_n| \qquad (35.4)$$

defines a metric on l^1.

The series in equation (35.4) converges by the comparison test since $|a_n - b_n| \leq |a_n| + |b_n|$ for every positive integer n.

It is easy to verify that parts (i) and (ii) of Definition 35.1 hold for d. We will verify the triangle inequality. Let $\{a_n\}$, $\{b_n\}$, and $\{c_n\}$ be elements of l^1. Then

$$|a_n - c_n| \leq |a_n - b_n| + |b_n - c_n|$$

for every positive integer n, and hence

$$\sum_{n=1}^{k} |a_n - c_n| \leq \sum_{n=1}^{k} |a_n - b_n| + \sum_{n=1}^{k} |b_n - c_n|$$

$$d(\{a_n\}, \{c_n\}) = \sum_{n=1}^{\infty} |a_n - c_n|$$

$$\leq \sum_{n=1}^{\infty} |a_n - b_n| + \sum_{n=1}^{\infty} |b_n - c_n|$$

$$= d(\{a_n\}, \{b_n\}) + d(\{b_n\}, \{c_n\})$$

Therefore d is a metric on l^1.

Example 35.6 Let X be a set. It is easily verified that

$$d(x, y) = \begin{cases} 0 & \text{if } x = y \\ 1 & \text{if } x \neq y \end{cases}$$

defines a metric on X. The metric d is called the *discrete metric* for X, and we will refer to the metric space (X, d) as X *with the discrete metric*.

Exercises

35.1 Verify that the function d of Example 35.5 satisfies Definition 35.1 (i) and (ii).

35.2 Prove that the function d of Example 35.6 is a metric.

35.3 Let d be a metric on a set M. Prove that
$$|d(x, z) - d(y, z)| \leq d(x, y)$$
for all $x, y, z \in M$.

35.4 Let M be a set and let d be a function from $M \times M$ into $\mathbf{R}$ which satisfies the three properties of Definition 35.1. Prove that $d: M \times M \to [0, \infty)$.

35.5 Let (M, d) be a metric space and let X be a subset of M. Prove that $(X, d | X \times X)$ is a metric space.

35.6 Let l^∞ denote the set of all bounded real sequences, and let c_0 denote the set of all real sequences which converge to 0.

(a) Prove that $l^1 \subset c_0 \subset l^\infty$.

(b) Prove that the containments in (a) are proper.

(c) Prove that
$$d'(\{a_n\}, \{b_n\}) = \text{lub } \{|a_n - b_n| \mid n \in \mathbf{P}\}$$
defines a metric on l^∞ [and thus by (a) on c_0 and l^1 also].

(d) Let d be the metric on l^1 defined in Example 35.5. Prove that $d'(x, y) \le d(x, y)$ for $x, y \in l^1$.

35.7 Let (M, d) be a metric space. Prove that
$$d'(x, y) = \frac{d(x, y)}{1 + d(x, y)} \qquad d''(x, y) = \min \{d(x, y), 1\}$$
define metrics on M. Prove that d' and d'' are bounded by 1.

35.8 Let (M_1, d_1) and (M_2, d_2) be metric spaces. Prove that $(M_1 \times M_2, d)$ is a metric space, where d is defined by the formula
$$d[(x_1, x_2), (y_1, y_2)] = d_1(x_1, y_1) + d_2(x_2, y_2)$$
The space $(M_1 \times M_2, d)$ is called the *product metric space*.

35.9 Let H^∞ denote the set of all real sequences $\{a_n\}$ such that $|a_n| \le 1$ for every positive integer n. H^∞ is called the *Hilbert cube*.

(a) Let $\{a_n\}, \{b_n\} \in H^\infty$. Prove that the series
$$\sum_{n=1}^{\infty} \frac{|a_n - b_n|}{2^n}$$
converges.

(b) Prove that
$$d(\{a_n\}, \{b_n\}) = \sum_{n=1}^{\infty} \frac{|a_n - b_n|}{2^n}$$
defines a metric on H^∞.

36. $\mathbf{R}^n$, l^2, and the Cauchy-Schwarz Inequality

We first generalize Example 35.4 to n-space. We let
$$\mathbf{R}^n = \{(x_1, x_2, \ldots, x_n) \mid x_i \in \mathbf{R} \text{ for } i = 1, \ldots, n\}$$
The space $\mathbf{R}^n$ is called *real Euclidean n-space*. In case $n = 1$, $\mathbf{R}^1$ is identified with the real line; in case $n = 2$, $\mathbf{R}^2$ is identified with the plane; and in case $n = 3$, $\mathbf{R}^3$ is identified with ordinary 3-space.

We will prove that

$$d(x, y) = \sqrt{\sum_{k=1}^{n} (x_k - y_k)^2} \tag{36.1}$$

where $x = (x_1, \ldots, x_n)$, $y = (y_1, \ldots, y_n) \in \mathbf{R}^n$ defines a metric on $\mathbf{R}^n$. For $n = 1$, d is the absolute value metric for the line; for $n = 2$, d is the usual metric of Example 35.4 for $\mathbf{R}^2$; and for $n = 3$, d gives the usual notion of distance in 3-space. The function d of equation (36.1) is called the *usual* (or *Euclidean*) *metric* for $\mathbf{R}^n$.

To prove that the triangle inequality holds in $\mathbf{R}^n$, we must show that if $(x_1, \ldots, x_n)$, $(y_1, \ldots, y_n)$, and $(z_1, \ldots, z_n) \in \mathbf{R}^n$, then

$$\sqrt{\sum_{k=1}^{n} (x_k - z_k)^2} \leq \sqrt{\sum_{k=1}^{n} (x_k - y_k)^2} + \sqrt{\sum_{k=1}^{n} (y_k - z_k)^2}$$

Letting $a_k = x_k - y_k$ and $b_k = y_k - z_k$ for $k = 1, \ldots, n$ and arguing as in Example 35.4, we can show that the triangle inequality will be established if we can prove that

$$\left| \sum_{k=1}^{n} a_k b_k \right| \leq \sqrt{\left(\sum_{k=1}^{n} a_k^2 \right) \left(\sum_{k=1}^{n} b_k^2 \right)}$$

This famous inequality is known as the *Cauchy-Schwarz inequality*.

Theorem 36.1 (Cauchy-Schwarz Inequality for R^n) Let $a_1, \ldots, a_n$ and $b_1, \ldots, b_n$ be real numbers. Then

$$\left| \sum_{k=1}^{n} a_k b_k \right| \leq \sqrt{\left(\sum_{k=1}^{n} a_k^2 \right) \left(\sum_{k=1}^{n} b_k^2 \right)}$$

Proof. If $b_k = 0$ for $1 \leq k \leq n$, the inequality follows immediately (both sides are zero). Thus assume that $b_k \neq 0$ for some k, $1 \leq k \leq n$. Then

$$\sum_{k=1}^{n} b_k^2 > 0$$

If x is any real number, we have

$$0 \leq \sum_{k=1}^{n} (a_k - x b_k)^2$$

so that

$$0 \leq \sum_{k=1}^{n} a_k^2 - 2x \sum_{k=1}^{n} a_k b_k + x^2 \sum_{k=1}^{n} b_k^2$$

for any x. If we let

$$A = \sum_{k=1}^{n} a_k^2, \qquad B = \sum_{k=1}^{n} a_k b_k, \qquad C = \sum_{k=1}^{n} b_k^2$$

and take $x = B/C$, then

$$0 \le A - 2\left(\frac{B}{C}\right)B + \left(\frac{B}{C}\right)^2 C$$

which reduces to

$$0 \le AC - B^2$$

and the inequality now follows. ∎

Theorem 36.2 The equation

$$d(x, y) = \sqrt{\sum_{k=1}^{n} (x_k - y_k)^2}$$

where $x = (x_1, \ldots, x_n)$, $y = (y_1, \ldots, y_n) \in \mathbf{R}^n$ defines a metric on $\mathbf{R}^n$.

Proof. Parts (i) and (ii) of Definition 35.1 are easily seen to hold for d, and part (iii) follows from Theorem 36.1. ∎

We next generalize $\mathbf{R}^n$ to "infinite-tuples" which are sequences and generalize the metric of Theorem 36.2 to the function

$$d(\{a_n\}, \{b_n\}) = \sqrt{\sum_{k=1}^{\infty} (a_k - b_k)^2} \qquad (36.2)$$

If equation (36.2) is to be meaningful, the series $\sum_{k=1}^{\infty} (a_k - b_k)^2$ must converge, and thus we must restrict our attention to those sequences $\{a_k\}$ for which the series $\sum_{k=1}^{\infty} a_k^2$ converges.

We let l^2 denote the set of all real sequences $\{a_k\}$ for which the series $\sum_{k=1}^{\infty} a_k^2$ converges. We next show that if $\{a_k\}, \{b_k\} \in l^2$, then the series $\sum_{k=1}^{\infty} (a_k - b_k)^2$ converges, and thus equation (36.2) defines a function from $l^2 \times l^2$ into $[0, \infty)$. Actually we will deduce this result from another theorem which is important in its own right.

Theorem 36.3 If $\{a_k\}, \{b_k\} \in l^2$, then the series $\sum_{k=1}^{\infty} a_k b_k$ converges absolutely.

Proof. By the Cauchy-Schwarz inequality (Theorem 36.1),

$$\sum_{k=1}^{n} |a_k||b_k| \le \sqrt{\left(\sum_{k=1}^{n} |a_k|^2\right)\left(\sum_{k=1}^{n} |b_k|^2\right)}$$

for every positive integer n. Therefore

$$\sum_{k=1}^{n} |a_k b_k| \le \sqrt{\left(\sum_{k=1}^{n} a_k^2\right)\left(\sum_{k=1}^{n} b_k^2\right)}$$

$$\le \sqrt{\left(\sum_{k=1}^{\infty} a_k^2\right)\left(\sum_{k=1}^{\infty} b_k^2\right)}$$

for every positive integer n. By Theorem 24.1, $\sum_{k=1}^{\infty} a_k b_k$ converges absolutely. ∎

Corollary 36.4 If $\{a_k\},\{b_k\} \in l^2$, then the series $\sum_{k=1}^{\infty} (a_k - b_k)^2$ converges.

Proof. The series $\sum_{k=1}^{\infty} (a_k - b_k)^2$ is the sum of the three convergent series

$$\sum_{k=1}^{\infty} a_k^2, \qquad -2\sum_{k=1}^{\infty} a_k b_k, \qquad \sum_{k=1}^{\infty} b_k^2$$

and therefore converges by Theorem 23.1. ∎

Theorem 36.5 The equation

$$d(\{a_k\}, \{b_k\}) = \sqrt{\sum_{k=1}^{\infty} (a_k - b_k)^2}$$

defines a metric on l^2.

Proof. The series $\sum_{k=1}^{\infty} (a_k - b_k)^2$ converges by Corollary 36.4. Parts (i) and (ii) of Definition 35.1 are easily seen to hold for d, and thus we verify only the triangle inequality.

Let $\{a_k\},\{b_k\},\{c_k\} \in l^2$. From the triangle inequality for $\mathbf{R}^n$ we have

$$\sqrt{\sum_{k=1}^{n} (a_k - c_k)^2} \leq \sqrt{\sum_{k=1}^{n} (a_k - b_k)^2} + \sqrt{\sum_{k=1}^{n} (b_k - c_k)^2}$$

for every positive integer n. Taking limits we have

$$d(\{a_k\}, \{c_k\}) = \sqrt{\sum_{k=1}^{\infty} (a_k - c_k)^2}$$
$$\leq \sqrt{\sum_{k=1}^{\infty} (a_k - b_k)^2} + \sqrt{\sum_{k=1}^{\infty} (b_k - c_k)^2}$$
$$= d(\{a_k\}, \{b_k\}) + d(\{b_k\}, \{c_k\}) \qquad ∎$$

By taking limits in Theorem 36.1, we can derive the Cauchy-Schwarz inequality for l^2.

Theorem 36.6 (Cauchy-Schwarz Inequality for l^2) If $\{a_k\},\{b_k\} \in l^2$, then the series $\sum_{k=1}^{\infty} a_k b_k$ converges absolutely and

$$\left| \sum_{k=1}^{\infty} a_k b_k \right| \leq \sqrt{\left(\sum_{k=1}^{\infty} a_k^2\right)\left(\sum_{k=1}^{\infty} b_k^2\right)}$$

Proof. The series $\sum_{k=1}^{\infty} a_k b_k$ converges absolutely by Theorem 36.3. The conclusion now follows by taking limits in Theorem 36.1. ∎

Exercises

36.1 Give the details of the proof of Theorem 36.2.

36.2 Prove that parts (i) and (ii) of Definition 35.1 hold for the metric of Theorem 36.5.

36.3 Prove that $l^1 \subset l^2 \subset c_0$ and that the containments are proper.

36.4 Prove that if $\sum_{n=1}^{\infty} a_n^2$ converges, then $\sum_{n=1}^{\infty} a_n/n$ converges absolutely.

36.5 Let $a_1, \ldots, a_n$ be real numbers. Prove that

$$\left| \frac{a_1 + \cdots + a_n}{n} \right| \leq \sqrt{\frac{a_1^2 + \cdots + a_n^2}{n}}$$

This exercise shows that the absolute value of the arithmetic average does not exceed the root mean square average.

36.6* Show that

$$\left| \sum_{k=1}^{n} a_k b_k \right| = \sqrt{ \left(\sum_{k=1}^{n} a_k^2 \right) \left(\sum_{k=1}^{n} b_k^2 \right) }$$

if and only if there exist numbers s and t such that $s a_k = t b_k$ for $1 \leq k \leq n$. Interpret this result geometrically for $n = 2$.

36.7* Show that the function $f(x) = A - 2Bx + Cx^2$, $C > 0$, has a minimum at $x = B/C$.

36.8 Let $\{a_n\} \in l^1$ and $\{b_n\} \in l^{\infty}$. Prove that $\{a_n b_n\} \in l^1$.

36.9 Let $\{a_n\} \in c_0$ and $\{b_n\} \in l^{\infty}$. Prove that $\{a_n b_n\} \in c_0$. Give an example of sequences $\{a_n\}$ and $\{b_n\}$ such that $\{a_n\} \in c_0$ and $\{b_n\} \in l^{\infty}$, but $\{a_n b_n\} \notin l^2$.

36.10 Let $\{a_n\}, \{b_n\} \in l^{\infty}$. Prove that $\{a_n b_n\} \in l^{\infty}$. Give an example of sequences $\{a_n\}$ and $\{b_n\}$ such that $\{a_n\}, \{b_n\} \in l^{\infty}$, but $\{a_n b_n\} \notin c_0$.

36.11 Let $\{a_n\}$ be a sequence such that $\{a_n b_n\} \in l^1$ for every sequence $\{b_n\} \in l^1$. Prove that $\{a_n\} \in l^{\infty}$. Show (by example) that the above statement is false if l^{∞} is replaced by c_0.

36.12 State and prove other results obtained by replacing either or both l^1's in Exercise 36.11 by any of l^1, l^2, l^{∞}, or c_0.

36.13 This alternative proof of the Cauchy-Schwarz inequality for $\mathbf{R}^n$ was published by Tolsted (1964).

(a) Prove that for any real numbers a and b

$$ab \leq \frac{a^2 + b^2}{2}$$

(b) Let $a_1, \ldots, a_n$ and $b_1, \ldots, b_n$ be real numbers. Set $A = (\sum_1^n a_k^2)^{1/2}$ and $B = (\sum_1^n b_k^2)^{1/2}$. Take $a = a_k/A$ and $b = b_k/B$ in (a) to derive

$$\frac{a_k b_k}{AB} \leq \frac{1}{2} \left(\frac{a_k^2}{A^2} + \frac{b_k^2}{B^2} \right)$$

(c) Sum the inequalities in (b) to derive the Cauchy-Schwarz inequality (Theorem 36.1).

37. Sequences in Metric Spaces

If (M, d) is a metric space, we will often abbreviate (M, d) to M. When we refer to a metric space M, it should be understood that there is a metric defined on M, but that we are not mentioning it explicitly.

A sequence in a metric space M is simply a function from $\mathbf{P}$ into M, and we denote such a sequence as $\{a_n\}_{n=1}^{\infty}$, or more simply $\{a_n\}$. (We will sometimes find it convenient to use the notation $\{a^{(k)}\}_{k=1}^{\infty}$ or $\{a^{(k)}\}$.) The definition of the limit of a sequence in a metric space is a direct generalization of the definition of the limit of a real sequence (Definition 10.2).

Definition 37.1 Let $\{a_n\}$ be a sequence in a metric space (M, d). We say that $\{a_n\}$ *converges to* (or *has limit*) L, where $L \in M$, and write

$$\lim_{n \to \infty} a_n = L$$

if for every $\varepsilon > 0$, there exists a positive integer N such that if $n \geq N$, then $d(a_n, L) < \varepsilon$. If $\{a_n\}$ has a limit, we call $\{a_n\}$ a *convergent sequence*. If $\{a_n\}$ has no limit, we call $\{a_n\}$ a *divergent sequence*.

As in the case for real sequences, we can show that the limit of a sequence in a metric space, if it exists, is unique, and so we are justified in writing $\lim_{n \to \infty} a_n = L$. One can also prove that if $\{a_n\}$ is a constant sequence ($a_n = L$ for every n), then $\lim_{n \to \infty} a_n = L$. Another important fact is that if a sequence $\{a_n\}$ in a metric space converges to L, then every subsequence of $\{a_n\}$ converges to L. In each case the proof is an imitation of the proof of the corresponding result for real sequences. In an arbitrary metric space we will not be able to prove generalizations of the algebraic theorems of Chapter IV, Section 12, for in an arbitrary metric space, there may be no algebraic operations defined.

We now examine the meaning of sequential convergence in some specific metric spaces. In the metric space $(\mathbf{R}, d)$, where $d(x, y) = |x - y|$ is the usual metric, Definition 37.1 is identical to Definition 10.2, and thus a real sequence $\{a_n\}$ converges to L in the metric space $(\mathbf{R}, d)$ if and only if $\{a_n\}$ converges to L according to Definition 10.2.

Let X be a set with the discrete metric d. Suppose that $\{a_n\}$ converges to L in (X, d). Taking $\varepsilon = \frac{1}{2}$, there exists a positive integer N such that if $n \geq N$, then $d(a_n, L) < \frac{1}{2}$. But this implies that $d(a_n, L) = 0$ if $n \geq N$. Therefore, $a_n = L$ if $n \geq N$. We will say that a sequence $\{a_n\}$ is *eventually constant* if there exists a positive integer N such that $a_n = a_N$ if $n \geq N$. We have just shown that if $\{a_n\}$ converges in (X, d), then $\{a_n\}$ is eventually constant. It is easy to verify that the converse of this statement also holds.

Let $\{a^{(k)}\}_{k=1}^\infty$ be a sequence in $\mathbf{R}^n$. Then each element $a^{(k)}$ is an n-tuple which we denote $(a_1^{(k)}, a_2^{(k)}, \ldots, a_n^{(k)})$. A listing of the points $\{a^{(k)}\}_{k=1}^\infty$ is given below.

$$a^{(1)} = (a_1^{(1)}, a_2^{(1)}, \ldots, a_j^{(1)}, \ldots, a_n^{(1)})$$
$$a^{(2)} = (a_1^{(2)}, a_2^{(2)}, \ldots, a_j^{(2)}, \ldots, a_n^{(2)})$$
$$\cdots\cdots\cdots\cdots\cdots\cdots\cdots\cdots\cdots\cdots\cdots$$
$$a^{(k)} = (a_1^{(k)}, a_2^{(k)}, \ldots, a_j^{(k)}, \ldots, a_n^{(k)})$$
$$\cdots\cdots\cdots\cdots\cdots\cdots\cdots\cdots\cdots\cdots\cdots$$

In the jth column we find a *real* sequence $\{a_j^{(k)}\}_{k=1}^\infty$. The link between convergence of the real sequences $\{a_j^{(k)}\}_{k=1}^\infty$ for $1 \le j \le n$ and convergence of the original sequence $\{a^{(k)}\}_{k=1}^\infty$ in $\mathbf{R}^n$ is given in the next theorem.

Theorem 37.2 Let $\{a^{(k)}\}_{k=1}^\infty$ be a sequence of points in $\mathbf{R}^n$, and let $a = (a_1, a_2, \ldots, a_n) \in \mathbf{R}^n$. Let $a^{(k)} = (a_1^{(k)}, a_2^{(k)}, \ldots, a_n^{(k)})$, for $k = 1, 2, \ldots$. Then $\{a^{(k)}\}_{k=1}^\infty$ converges to a if and only if $\lim_{k \to \infty} a_j^{(k)} = a_j$ for $j = 1, \ldots, n$.

Proof. First suppose that $\{a^{(k)}\}_{k=1}^\infty$ converges to a. Let $\varepsilon > 0$. There exists a positive integer N such that if $k \ge N$, then $d(a^{(k)}, a) < \varepsilon$. Let j be a positive integer such that $1 \le j \le n$. If $k \ge N$, then

$$|a_j^{(k)} - a_j| \le \sqrt{\sum_{i=1}^n (a_i^{(k)} - a_i)^2} = d(a^{(k)}, a) < \varepsilon$$

and thus $\lim_{k \to \infty} a_j^{(k)} = a_j$ for $j = 1, \ldots, n$.

Now suppose that $\lim_{k \to \infty} a_j^{(k)} = a_j$ for $j = 1, \ldots, n$. Let $\varepsilon > 0$. For each $j, 1 \le j \le n$, there exists a positive integer N_j such that if $k \ge N_j$, then

$$|a_j^{(k)} - a_j| < \frac{\varepsilon}{\sqrt{n}}$$

If $k \ge \max\{N_1, \ldots, N_n\}$, then

$$d(a^{(k)}, a) = \sqrt{\sum_{j=1}^n (a_j^{(k)} - a_j)^2}$$
$$< \sqrt{\frac{\varepsilon^2}{n} + \frac{\varepsilon^2}{n} + \cdots + \frac{\varepsilon^2}{n}} = \sqrt{\frac{n\varepsilon^2}{n}} = \varepsilon \qquad \blacksquare$$

For example, the sequence $\{((1 + 1/k)^k, 1/k)\}_{k=1}^\infty$ of points in $\mathbf{R}^2$ converges to $(e, 0)$, since $\lim_{k \to \infty} (1 + 1/k)^k = e$ and $\lim_{k \to \infty} (1/k) = 0$.

The statement, "Let $\{a_n\}$ be a sequence in l^1," is ambiguous. Do we mean that a_n is a real number for every positive integer n and that $\sum_{n=1}^\infty a_n$ converges absolutely, or do we mean that $a_n \in l^1$ for every positive integer n and that $\{a_n\}$ is a sequence of points in the metric space l^1? To avoid such ambiguity, we agree that when we say that $\{a_n\}$ is a sequence in l^1, we mean that a_n is a real number for every positive integer n and that $\sum_{n=1}^{\infty} a_n$ converges absolutely, that is, $\{a_n\} \in l^1$. If we wish to speak of a sequence of points in the metric space l^1, we will say that $\{a^{(k)}\}_{k=1}^{\infty}$ is a *sequence of points* in l^1. In this case $a^{(k)} \in l^1$ for every positive integer k. We will use a notation similar to that employed for $\mathbf{R}^n$. We let $a^{(k)} = \{a_n^{(k)}\}_{n=1}^{\infty}$. Then $a_n^{(k)}$ is the nth term of the kth sequence. A listing of the points $\{a^{(k)}\}_{k=1}^{\infty}$ is given below.

$$a^{(1)} = (a_1^{(1)}, a_2^{(1)}, \ldots)$$
$$a^{(2)} = (a_1^{(2)}, a_2^{(2)}, \ldots)$$
$$\cdots\cdots\cdots\cdots\cdots\cdots$$
$$a^{(k)} = (a_1^{(k)}, a_2^{(k)}, \ldots)$$
$$\cdots\cdots\cdots\cdots\cdots$$

Theorem 37.2 does not hold for l^1 (nor for l^2, c_0, l^∞). For let $\{\delta^{(k)}\}_{k=1}^{\infty}$ be the sequence of points in l^1 defined by

$$\delta_n^{(k)} = \begin{cases} 1 & \text{if } n = k \\ 0 & \text{if } n \neq k \end{cases}$$

This sequence is listed below.

$$\delta^{(1)} = (1, 0, 0, 0, \ldots)$$
$$\delta^{(2)} = (0, 1, 0, 0, \ldots)$$
$$\delta^{(3)} = (0, 0, 1, 0, \ldots)$$
$$\cdots\cdots\cdots\cdots\cdots$$

Let $a_n = 0$ for every positive integer n, and let $a = \{a_n\}$. It is clear that $\lim_{k \to \infty} \delta_n^{(k)} = a_n$ for every positive integer n. However, $\{\delta^{(k)}\}$ does not converge to a in l^1, since $d(\delta^{(k)}, a) = 1$ for every positive integer k. One-half of Theorem 37.2 does generalize to l^1 (and to l^2, c_0, l^∞) as the next theorem demonstrates.

Theorem 37.3 Let $\{a^{(k)}\}$ be a sequence of points in l^1, and let $a = \{a_n\} \in l^1$. If $\{a^{(k)}\}$ converges to a, then $\lim_{k \to \infty} a_n^{(k)} = a_n$ for every positive integer n.

Proof. Let $\varepsilon > 0$. There exists a positive integer N such that if $k \geq N$, then $d(a^{(k)}, a) < \varepsilon$. Let n be a positive integer. If $k \geq N$, then

$$|a_n^{(k)} - a_n| \leq \sum_{i=1}^{\infty} |a_i^{(k)} - a_i| = d(a^{(k)}, a) < \varepsilon \qquad \blacksquare$$

Exercises

37.1 Prove that the limit of a sequence in a metric space is unique.

37.2 Let M be a metric space. Prove that if $\{a_n\}$ is a sequence in M such that $a_n = L$ for every positive integer n, then $\lim_{n \to \infty} a_n = L$.

37.3 Let M be a metric space. Let $\{a_n\}$ be a sequence in M such that $\lim_{n \to \infty} a_n = L$. Prove that every subsequence of $\{a_n\}$ has limit L.

37.4 Show (by examples) that Theorem 37.2 does not generalize to l^2, c_0, or l^∞.

37.5 Prove Theorem 37.3 with l^1 replaced by l^2, c_0, l^∞.

37.6 Let $\{(x_n, y_n)\}$ be a sequence of points in $\mathbf{R}^2$ such that the sequences $\{x_n\}$ and $\{y_n\}$ are bounded. Prove that $\{(x_n, y_n)\}$ has a convergent subsequence.

37.7 Let $\{a^{(k)}\}$ be a convergent sequence of points in l^1. Prove that $\{a^{(k)}\}$ converges in l^∞.

37.8 Let M_1 and M_2 be metric spaces. Let $\{(x_n, y_n)\}$ be a sequence of points in the product metric space $M_1 \times M_2$. (See Exercise 35.8.) Let $(x, y) \in M_1 \times M_2$. Prove that $\lim_{n \to \infty} (x_n, y_n) = (x, y)$ if and only if $\lim_{n \to \infty} x_n = x$ and $\lim_{n \to \infty} y_n = y$.

37.9 (a) Prove that the equation
$$d'(x, y) = \sum_{i=1}^{n} |x_i - y_i|$$
where $x = (x_1, \ldots, x_n)$, $y = (y_1, \ldots, y_n) \in \mathbf{R}^n$ defines a metric on $\mathbf{R}^n$.

   (b) Let $\{a^{(k)}\}$ be a sequence of points in $\mathbf{R}^n$, and let $a \in \mathbf{R}^n$. Prove that $\{a^{(k)}\}$ converges to a in $(\mathbf{R}^n, d)$ if and only if $\{a^{(k)}\}$ converges to a in $(\mathbf{R}^n, d')$, where d is the usual metric for $\mathbf{R}^n$.

37.10 Let (M, d) be a metric space, and let d' and d'' be defined as in Exercise 35.7. Let $\{a_n\}$ be a sequence in M and let $a \in M$. Prove that the following statements are equivalent.

   (a) $\{a_n\}$ converges to a in (M, d).

   (b) $\{a_n\}$ converges to a in (M, d').

   (c) $\{a_n\}$ converges to a in (M, d'').

37.11* Let $\{a^{(k)}\}_{k=1}^{\infty}$ be a sequence of points in H^∞. (H^∞ is defined in Exercise 35.9.) Let $a = \{a_n\}_{n=1}^{\infty} \in H^\infty$. Prove that $\{a^{(k)}\}_{k=1}^{\infty}$ converges to a in H^∞ if and only if $\lim_{k \to \infty} a_n^{(k)} = a_n$ for every positive integer n.

38. Closed Sets

A closed interval $[a, b]$ of the real line has the following property. If $\{x_n\}$ is a convergent sequence of points in $[a, b]$ and $\lim_{n \to \infty} x_n = x$, then x also belongs to the interval $[a, b]$. For if $a \le x_n \le b$ for every positive integer n, then $a \le \lim_{n \to \infty} x_n \le b$ (Theorem 14.2). We generalize this property to arbitrary metric spaces.

Definition 38.1 Let M be a metric space, and let X be a subset of M. We say that a point x in M is a *limit point* of X if there is a sequence $\{x_n\}$ such that $x_n \in X$ for every positive integer n and $\lim_{n \to \infty} x_n = x$.

On the real line 1 is a limit point of either of the sets $[0, 1]$ or $(0, 1)$ (Verify). Thus a limit point of a set X may or may not belong to X. If *every* limit point of a set X belongs to X, we say that X is closed.

Definition 38.2 Let M be a metric space, and let X be a subset of M. If every limit point of X belongs to X, we say that X is *closed* (in M).

The discussion preceding Definition 38.1 shows that a closed interval of the real line is a closed subset of **R**. An open interval (a, b) of **R** is not a closed subset of **R** since a is a limit point of (a, b), but $a \notin (a, b)$.

A closed rectangle $R = [a, b] \times [c, d]$ is a closed subset of $\mathbf{R}^2$. For let (x, y) be a limit point of R. Then there exists a sequence $\{(x_n, y_n)\}$ in R which converges to (x, y). By Theorem 37.2, $\lim_{n \to \infty} x_n = x$ and $\lim_{n \to \infty} y_n = y$. Since $a \le x_n \le b$ and $c \le y_n \le d$ for every positive integer n, $a \le x \le b$ and $c \le y \le d$ by Theorem 14.2. Therefore, $(x, y) \in R$, and R is closed.

Let X be a set with the discrete metric. We show that every subset of X is closed. Let $Y \subset X$, and let y be a limit point of Y. There exists a sequence $\{y_n\}$ in Y such that $\lim_{n \to \infty} y_n = y$. Now $\{y_n\}$ is a convergent sequence in X, and thus $\{y_n\}$ is eventually constant (Section 37). Therefore, $y_n = y$ for some positive integer n, and thus $y \in Y$ and Y is closed.

We return now to arbitrary metric spaces.

Definition 38.3 Let M be a metric space, and let X be a subset of M. We let $\overline{X}$ denote the set of limit points of X.

It is easy to verify that $X \subset \overline{X}$ for if $x \in X$, we may let $x_n = x$ for every positive integer n and then $\{x_n\}$ is a sequence of points in X such that $\lim_{n \to \infty} x_n = x$; therefore $x \in \overline{X}$. We may combine Definitions 38.2 and 38.3 into a single statement. A subset X of a metric space M is closed if and only if $X = \overline{X}$.

Theorem 38.4 Let M be a metric space and let $x \in M$. Then M, $\varnothing$, and $\{x\}$ are closed subsets of M.

Proof. If x is a limit point of M, then $x \in M$ by definition, and therefore M is closed.

Suppose $\varnothing$ is not closed. Then there exists $x \in M$ such that x is a limit point of $\varnothing$, but $x \notin \varnothing$. Thus there exists a sequence $\{x_n\}$ in $\varnothing$ such that $\lim_{n \to \infty} x_n = x$. But now $x_n \in \varnothing$ for every positive integer n, which is impossible; therefore $\varnothing$ is closed.

Let y be a limit point of $\{x\}$. Then there exists a sequence $\{x_n\}$ in $\{x\}$ such that $\lim_{n \to \infty} x_n = y$. Since $x_n \in \{x\}$ for every positive integer n, it follows that $x_n = x$ for every positive integer n. Thus $y = \lim_{n \to \infty} x_n = x$, and so $y \in \{x\}$. Therefore $\{x\}$ is closed. ∎

Theorem 38.5 Let X and Y be closed subsets of a metric space. Then $X \cup Y$ is closed.

Proof. Let x be a limit point of $X \cup Y$. Then there exists a sequence $\{x_n\}$ in $X \cup Y$ such that $\lim_{n \to \infty} x_n = x$. Either $x_n \in X$ for infinitely many positive integers n or $x_n \in Y$ for infinitely many positive integers n. We may assume $x_n \in X$ for infinitely many positive integers n. Suppose $x_{n_k} \in X$, $n_1 < n_2 < n_3 < \cdots$. Now $\lim_{k \to \infty} x_{n_k} = x$; so x is a limit point of X. Since X is closed, $x \in X$. Therefore $x \in X \cup Y$; so $X \cup Y$ is closed. ∎

Corollary 38.6 If $X_1, X_2, \ldots, X_n$ are closed subsets of a metric space, then $X_1 \cup X_2 \cup \cdots \cup X_n$ is closed.

Proof. The proof follows by induction from Theorem 38.5. ∎

An infinite union of closed subsets of a metric space need not be closed. For example,

$$(0, 1) = \bigcup_{n=2}^{\infty} \left[\frac{1}{n}, \frac{n}{n + 1} \right]$$

(verify), and $[1/n, n/(n + 1)]$ is closed in $\mathbf{R}$ for each positive integer n, but $(0, 1)$ is not a closed subset of $\mathbf{R}$. An arbitrary intersection of closed sets is closed, as we will show in Theorem 38.8 after deducing one more corollary of Theorem 38.5.

Corollary 38.7 Any finite subset of a metric space is closed.

Proof. If $\{x_1, \ldots, x_n\}$ is a finite subset of a metric space, then $\{x_1, \ldots, x_n\} = \{x_1\} \cup \cdots \cup \{x_n\}$ is closed by Theorem 38.3 and Corollary 38.6. ∎

Theorem 38.8 Let $\mathscr{G}$ be a collection of closed subsets of a metric space. Then $\cap \, \mathscr{G}$ is closed.

Proof. Let x be a limit point of $\cap \, \mathscr{G}$. Then there exists a sequence $\{x_n\}$

in $\cap \, \mathscr{G}$ such that $\lim_{n \to \infty} x_n = x$. Let $C \in \mathscr{G}$. Then $x_n \in C$ for every positive integer n. Since $\lim_{n \to \infty} x_n = x$, x is a limit point of C. Since C is closed, $x \in C$. Thus $x \in C$ for every $C \in \mathscr{G}$. Therefore, $x \in \cap \, \mathscr{G}$ and hence $\cap \, \mathscr{G}$ is closed. ∎

Exercises

38.1   Give the inductive proof of Corollary 38.6.

38.2   Let (a, b) be an open interval in **R**. Describe sequences $\{a_n\}$ and $\{b_n\}$ in (a, b) such that $\lim_{n \to \infty} a_n = a$ and $\lim_{n \to \infty} b_n = b$.

38.3   Prove that a half-open interval I is not a closed subset of **R** unless $I = (-\infty, b]$ or $I = [a, \infty)$.

38.4   Let M be a metric space such that M is a finite set. Prove that every subset of M is closed.

38.5   Let X be a subset of a metric space M. We say that a point x in M is an *accumulation point* of X if there exists a sequence $\{x_n\}$ in X such that $\lim_{n \to \infty} x_n = x$ and $x_n \neq x$ for every positive integer n. We let X^a denote the set of accumulation points of X.
  (a)   Prove that X is closed if and only if $X^a \subset X$.
  (b)   Prove the following form of the Bolzano-Weierstrass theorem: If X is a bounded infinite subset of **R**, then $X^a \neq \varnothing$.
  (c)   Prove that if X is an uncountable subset of **R**, then $X^a \neq \varnothing$.

38.6   Let f be a continuous function from **R** into **R**. Prove that $\{x \,|\, f(x) = 0\}$ is a closed subset of **R**.

38.7   Let $\{\delta^{(k)}\}_{k=1}^{\infty}$ be as in Section 37. Prove that $\{\delta^{(k)} \,|\, k \in \mathbf{P}\}$ is a closed subset of l^1, l^2, c^0, and l^{∞}.

38.8   Let X and Y be closed subsets of **R**. Prove that $X \times Y$ is a closed subset of $\mathbf{R}^2$. State and prove a generalization to $\mathbf{R}^n$.

38.9   Let (M, d) be a metric space. Let $\varepsilon > 0$ and let $y \in M$. Prove that the closed ball $\{x \,|\, d(x, y) \leq \varepsilon\}$ is a closed subset of M.

38.10   Let (M, d) be a metric space. Let d' and d'' be defined as in Exercise 35.7. Let $X \subset M$. Prove that the following are equivalent:
  (a) X is closed in (M, d).
  (b) X is closed in (M, d').
  (c) X is closed in (M, d'').

38.11   Let $\{x_n\}$ be a sequence in a metric space M which converges to x. Prove that $\{x_n \,|\, n \in \mathbf{P}\} \cup \{x\}$ is a closed subset of M.

38.12   Let $\{[a_n \, b_n]\}_{n=1}^{\infty}$ be a sequence of closed intervals such that $|a_n| \leq 1$ and $|b_n| \leq 1$ for every positive integer n. Prove that $\{\{x_n\}_{n=1}^{\infty} \,|\, x_n \in [a_n, b_n]\}$ is a closed subset of H^{∞}.

38.13   Let M be a metric space. Prove the following:
  (a) $\bar{X} = \bar{\bar{X}}$ for $X \subset M$.

    (b) $\bar{X}$ is closed for all $X \subset M$.

    (c) If $X \subset Y \subset M$, then $\bar{X} \subset \bar{Y}$.

    (d) $\overline{X \cup Y} = \bar{X} \cup \bar{Y}$ for $X, Y \subset M$.

    (e) If Y is a closed subset of M such that $\bar{X} \subset Y$, then $X \subset Y$.

    (f) If $X \subset M$, then $\bar{X} = \cap \{Y \mid Y$ is a closed subset of M containing $X\}$.

38.14 Let $\{x_n\}$ be a sequence in a metric space M with no convergent subsequence. Prove that $\{x_n \mid n \in P\}$ is a closed subset of M.

39. Open Sets

Let (a, b) be an open interval where a and b are in $\mathbf{R}$. If we let $c = (a + b)/2$ and $\varepsilon = (b - a)/2$, we may write

$$(a, b) = \{x \in \mathbf{R} \mid d(x, c) = |x - c| < \varepsilon\}$$

that is, (a, b) is the set of real numbers x such that the distance between x and c is less than ε. The generalized open interval in a metric space is called an *open ball*.

Definition 39.1 Let (M, d) be a metric space. Let $\varepsilon > 0$ and let $x \in M$. We let

$$B_\varepsilon(x) = \{y \in M \mid d(x, y) < \varepsilon\}$$

$B_\varepsilon(x)$ is called *the open ball of radius ε centered at x.*

The open ball in $\mathbf{R}$ of radius ε centered at x is the open interval $(x - \varepsilon, x + \varepsilon)$. An open ball in $\mathbf{R}^2$ is the interior of a disk and an open ball in $\mathbf{R}^3$ is the interior of a sphere. See Figure 39.1.

Definition 39.2 Let M be a metric space and let X be a subset of M. We

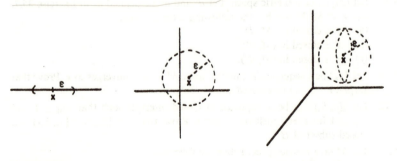

Figure 39.1

say that X is *open* if for every $x \in X$, there exists an open ball $B_\varepsilon(x)$ (centered at x) such that $B_\varepsilon(x) \subset X$.

An open interval (a, b) in $\mathbf{R}$ is open in $\mathbf{R}$, for if $x \in (a, b)$ and we let $\varepsilon = \min \{b - x, x - a\}$, then $(x - \varepsilon, x + \varepsilon) \subset (a, b)$ (verify). (We will show in Theorem 39.4 that an open ball in a metric space M is always an open subset of M.)

Let X be a set with the discrete metric. Then every subset of X is open. For if $Y \subset X$ and $x \in Y$, we have $B_{\frac{1}{2}}(x) \subset Y$.

We can prove a theorem analogous to Theorem 38.4 with the exception that in general $\{x\}$ is *not* open.

Theorem 39.3 Let M be a metric space. Then M and $\varnothing$ are open subsets of M.

Proof. Let $x \in M$. Then $B_\varepsilon(x) \subset M$ (for any $\varepsilon > 0$), and thus M is open. If $\varnothing$ is not open, there exists $x \in \varnothing$ such that $B_\varepsilon(x) \not\subset \varnothing$ for any $\varepsilon > 0$. However $x \notin \varnothing$, and thus we have a contradiction. ∎

We next show that open balls are open sets. The proof of this theorem (Theorem 39.4) for $\mathbf{R}^2$ is illustrated in Figure 39.2.

Theorem 39.4 Let M be a metric space. Let $x \in M$ and let $\varepsilon > 0$. Then the open ball $B_\varepsilon(x)$ is an open subset of M.

Proof. Let (M, d) be a metric space. We must show that if $y \in B_\varepsilon(x)$, there exists $\delta > 0$ such that $B_\delta(y) \subset B_\varepsilon(x)$.

Let $y \in B_\varepsilon(x)$. Let $\delta = \varepsilon - d(x, y)$. Then $\delta > 0$. (Why?) We will show that $B_\delta(y) \subset B_\varepsilon(x)$. Let $z \in B_\delta(y)$. Then $d(y, z) < \delta$. Now

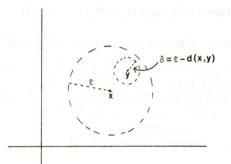

Figure 39.2

$$d(x, z) \leq d(x, y) + d(y, z)$$
$$< d(x, y) + \delta$$
$$= \varepsilon$$

and hence $z \in B_\varepsilon(x)$. ∎

The next theorem gives the relationship between closed and open sets.

Theorem 39.5 Let M be a metric space and let $X \subset M$. Then X is open if and only if X' is closed.

Proof. Let (M, d) be a metric space. Suppose that X is an open subset of M. Let x be a limit point of X'. We must show that $x \in X'$. If $x \notin X'$, then $x \in X$. Since X is open there exists an open ball $B_\varepsilon(x)$ such that $B_\varepsilon(x) \subset X$. Since x is a limit point of X', there exists a sequence $\{x_n\}$ of points in X' such that $\lim_{n \to \infty} x_n = x$. Therefore, there exists a positive integer N such that $d(x_N, x) < \varepsilon$. Now $x_N \in B_\varepsilon(x)$, and so $x_N \in X \cap X'$, which is impossible. Thus $x \in X'$, and therefore X' is closed.

Suppose X' is closed. If X is not open, there exists $x \in X$ such that for any $\varepsilon > 0$, $B_\varepsilon(x) \not\subset X$. Equivalently, we have $B_\varepsilon(x) \cap X' \neq \varnothing$ for every $\varepsilon > 0$. For each positive integer n, there exists a point x_n in $B_{1/n}(x) \cap X'$. Now $\{x_n\}$ is a sequence of points in X', and since $d(x_n, x) < 1/n$, we have $\lim_{n \to \infty} x_n = x$. Thus x is a limit point of X' and since X' is closed, $x \in X'$. This contradiction yields the theorem. ∎

Let X be a set with the discrete metric. In Section 38 we proved that every subset of X is closed. It follows from Theorem 39.5 that every subset of X is open.

The next theorem is analogous to Corollary 38.6 and Theorem 38.8.

Theorem 39.6 Let M be a metric space.
 (i) If $U_1, \ldots, U_n$ are open subsets of M, then $U_1 \cap U_2 \cap \cdots \cap U_n$ is an open subset of M.
(ii) If $\mathscr{U}$ is a collection of open subsets of M, then $\cup \, \mathscr{U}$ is an open subset of M.

Proof. Let $U_1, U_2, \ldots, U_n$ be open subsets of M. By Theorem 39.5 it is enough to show that $(U_1 \cap \cdots \cap U_n)'$ is closed. By Theorem 39.5, each of $U_1', U_2', \ldots, U_n'$ is closed and by Corollary 38.6, $U_1' \cup U_2' \cup \cdots \cup U_n'$ is closed. By Theorem 1.7, $(U_1 \cap \cdots \cap U_n)' = U_1' \cup \cdots \cup U_n'$. Thus $(U_1 \cap \cdots \cap U_n)'$ is closed, and hence $U_1 \cap \cdots \cap U_n$ is open.

Part (ii) is proved similarly using Theorems 38.8 and 1.7. ∎

An arbitrary intersection of open sets need not be open. For example,

$${0} = \bigcap_{n=1}^{\infty} \left(\frac{-1}{n}, \frac{1}{n} \right)$$

is not open in **R**.

If a subset of a metric space is not open, this does not mean in general that it is closed. Indeed, there are subsets of **R** which are neither open nor closed. For example, the half-open interval [0, 1) is neither open nor closed in **R**.

If X is a set and $\mathscr{T}$ is a collection of subsets of X satisfying

(i) $X, \varnothing \in \mathscr{T}$

(ii) The union of a subcollection of $\mathscr{T}$ is a member of $\mathscr{T}$

(iii) The intersection of a finite subcollection of $\mathscr{T}$ is a member of $\mathscr{T}$

then $\mathscr{T}$ is called a *topology* for X. By Theorems 39.3 and 39.6, the collection of open subsets of a metric space M is a topology for M.

Exercises

39.1 Prove that a half-open interval not of the form $(-\infty, b]$ or $[a, \infty)$ is neither closed nor open in **R**.

39.2 Prove Theorem 39.6 (ii).

39.3 Deduce Theorem 39.6 from Definition 39.2.

39.4 Let M be a metric space such that M is a finite set. Prove that every subset of M is open.

39.5 Prove that the interior of a rectangle in $\mathbf{R}^2$

$${(x, y) | a < x < b, c < y < d}$$

is an open subset of $\mathbf{R}^2$.

39.6 Prove that if X and Y are open subsets of **R**, then $X \times Y$ is an open subset of $\mathbf{R}^2$. State and prove a generalization to $\mathbf{R}^n$.

39.7 Let f be a continuous function from **R** into **R**. Prove that ${x | f(x) > 0}$ is an open subset of **R**.

39.8 Let X be an open, nonempty subset of **R**. Prove that there exists a unique countable set of open intervals ${(a_n, b_n)}_{n=1}^{\infty}$ such that

(a) $\bigcup_{n=1}^{\infty} (a_n, b_n) = X$

(b) $(a_m, b_m) \cap (a_n, b_n) = \varnothing$ if $n \neq m$ ($a_n = -\infty$ and $b_n = \infty$ may occur)

39.9 Let X be a subset of a metric space M. Prove that X is an open subset of M if and only if X is the union of open balls.

39.10 Let X be a subset of a metric space M. Prove that X is a closed subset of M if and only if whenever x is a point in M such that $B_\varepsilon(x) \cap X \neq \varnothing$ for every $\varepsilon > 0$, then $x \in X$.

39.11 If X is a subset of a metric space M, we say that a point x in X is an *interior point* of X if $B_\varepsilon(x) \subset X$ for some $\varepsilon > 0$, and we let X^0 denote the set of interior points of X.

Let M be a metric space. Prove the following:

(a) $X^0 \subset X$ for $X \subset M$.

(b) X is open if and only if $X^0 = X$, $X \subset M$.

(c) $(X^0)^0 = X^0$ for $X \subset M$.

(d) X^0 is open for all X, $X \subset M$.

(e) If $X \subset Y \subset M$, then $X^0 \subset Y^0$.

(f) $(X \cap Y)^0 = X^0 \cap Y^0$ for $X, Y \subset M$.

(g) If Y is an open subset of M such that $Y \subset X \subset M$, then $Y \subset X^0$.

(h) If $X \subset M$, then $X^0 = \cup \{Y \mid Y \subset X$ and Y is open$\}$.

(i) $((X')^-)' = X^0$, for all $X \subset M$.

39.12 If X is a subset of a metric space M, we define the *boundary* of X to be the set $\partial X = \bar{X} \cap (X')^-$.

Let M be a metric space. Prove the following:

(a) ∂X is closed for all X, $X \subset M$.

(b) $X \cup \partial X = \bar{X}$ for all X, $X \subset M$.

(c) $X \setminus \partial X = X^0$ for all X, $X \subset M$.

(d) If X is a proper nonempty subset of $\mathbf{R}$, then $\partial X \neq \varnothing$.

40. Continuous Functions on Metric Spaces

We will define continuity of a function in an arbitrary metric space analogously to the situation for the real line. Guided by Theorem 33.3 we make the following definition.

Definition 40.1 Let (M_1, d_1) and (M_2, d_2) be metric spaces, let $a \in M_1$, and let f be a function from M_1 into M_2. We say that *f is continuous at a* if for every $\varepsilon > 0$, there exists $\delta > 0$ such that if $d_1(x, a) < \delta$, then $d_2(f(x), f(a)) < \varepsilon$. We say that *$f$ is continuous on M_1* if f is continuous at every point of M_1.

In case $M_1 = M_2 = \mathbf{R}$ and $d_1 = d_2$ is the usual metric for $\mathbf{R}$, Definition 40.1 gives exactly the definition of continuity for real-valued functions defined on a subset of $\mathbf{R}$ (see Theorem 33.3).

Examples. Let $f: l^1 \to \mathbf{R}$ be defined by $f(\{a_n\}_{n=1}^\infty) = a_1$. Let $\{a_n\}_{n=1}^\infty \in l^1$. We will prove that f is continuous at every point of l^1. Let $\varepsilon > 0$ and take $\delta = \varepsilon$. If $0 < d(\{x_n\}, \{a_n\}) < \delta$, then $\sum_{n=1}^\infty |x_n - a_n| < \delta$, so that $|x_1 - a_1| \le \sum_{n=1}^\infty |x_n - a_n| < \delta = \varepsilon$, and thus $|f(\{x_n\}) - f(\{a_n\})| < \varepsilon$, and we have established continuity.

Let $I: l^1 \to l^2$ be defined by $I(x) = x$, for $x \in l^1$. Let $\{a_n\}_{n=1}^\infty \in l^1$. We will prove that I is continuous at every point of l^1. Let d^1 and d^2 denote the metrics on l^1 and l^2, respectively. It is easy to verify that $d^2(x, y) \le d^1(x, y)$ for $x, y \in l^1$. Let $\varepsilon > 0$. Set $\delta = \varepsilon$. If $0 < d^1(x, \{a_n\}) < \delta$, then $d^2(I(x), \{a_n\}) = d^2(x, \{a_n\}) \le d^1(x, \{a_n\} < \delta = \varepsilon$, and we have established continuity.

Theorem 40.2 Let f be a function from a metric space M_1 into a metric space M_2. Let $a \in M_1$. Then f is continuous at a if and only if whenever $\{x_n\}$ is a sequence in M_1 such that $\lim_{n \to \infty} x_n = a$, then $\lim_{n \to \infty} f(x_n) = f(a)$.

Proof. Let (M_1, d_1) and (M_2, d_2) be metric spaces, let f be a function from M_1 into M_2, and let $a \in M_1$.

First, suppose that f is continuous at a. Let $\{x_n\}$ be a sequence in M_1 such that $\lim_{n \to \infty} x_n = a$. Let $\varepsilon > 0$. There exists $\delta > 0$ such that if $d_1(x, a) < \delta$, then $d_2(f(x), f(a)) < \varepsilon$. Since $\lim_{n \to \infty} x_n = a$, there exists a positive integer N such that if $n \ge N$, then $d_1(x_n, a) < \delta$. If $n \ge N$, then $d_1(x_n, a) < \delta$, and hence $d_2(f(x_n), f(a)) < \varepsilon$.

Now suppose that whenever $\{x_n\}$ is a sequence in M_1 such that $\lim_{n \to \infty} x_n = a$, then $\lim_{n \to \infty} f(x_n) = f(a)$. If f is not continuous at a, there exists $\varepsilon > 0$ such that for any $\delta > 0$, $d_1(x, a) < \delta$ for some $x \in M_1$, but $d_2(f(x), f(a)) \ge \varepsilon$. Thus for every positive integer n, there exists $x_n \in M_1$ such that

$$d_1(x_n, a) < \frac{1}{n} \quad \text{and} \quad d_2(f(x_n), f(a)) \ge \varepsilon$$

Now, $\lim_{n \to \infty} x_n = a$, but $\lim_{n \to \infty} f(x_n) \ne f(a)$, and we have a contradiction. ∎

The equation $\lim_{n \to \infty} f(x_n) = f(a)$ in Theorem 40.2 can also be written $\lim_{n \to \infty} f(x_n) = f(\lim_{n \to \infty} x_n)$ (since $\lim_{n \to \infty} x_n = a$). Thus f is continuous at a if and only if "lim" and "f" can be interchanged. Many problems in analysis reduce to showing that "lim" may be interchanged with some other symbol. Inevitably there is a question of continuity lurking in the background.

Of particular importance are continuous real-valued functions on metric spaces. [Of course, when we say that f is a continuous real-valued function on a metric space M, we mean that f is a continuous function from M into $\mathbf{R}$, where the metric on $\mathbf{R}$ is the usual (absolute value) metric.] Clearly any constant function is continuous on a metric space. The next theorem shows that there are nonconstant continuous real-valued functions on any metric space with more than one point. (See Exercise 40.14 for a related result.)

Theorem 40.3 Let (M, d) be a metric space and let $a \in M$. The function f defined by the equation

$$f(x) = d(x, a)$$

is a continuous real-valued function on M. Moreover, $f(x) = 0$ if and only if $x = a$.

   Proof. Let $\varepsilon > 0$. Let $\delta = \varepsilon$. Suppose $d(x, y) < \delta$. Then

$$d(x, a) \leq d(x, y) + d(y, a)$$

so
$$f(x) - f(y) = d(x, a) - d(y, a) \leq d(x, y)$$

Similarly
$$f(y) - f(x) \leq d(x, y)$$

Therefore
$$|f(x) - f(y)| \leq d(x, y) < \delta = \varepsilon$$

Since f is continuous at every point of M, f is continuous on M.

   $f(x) = 0$ if and only if $d(x, a) = 0$, which by Definition 35.1 occurs if and only if $x = a$. ∎

   Theorem 40.3 is often summarized by stating that the distance function (or metric) is a continuous function.

   Our next theorem is analogous to a theorem (Theorem 33.2) about continuous real-valued functions on **R**.

Theorem 40.4 Let f and g be continuous real-valued functions on a metric space M. Then

   (i) $|f|$ is continuous on M.
   (ii) $f + g$ is continuous on M.
   (iii) cf is continuous on M for each $c \in \mathbf{R}$.
   (iv) $f - g$ is continuous on M.
   (v) $f \cdot g$ is continuous on M.
   (vi) f/g is continuous on M if $g(x) \neq 0$ for every $x \in M$.

   Proof. We prove only part (ii). The proofs of the other parts are similar and are left as exercises.

   Let $a \in M$. Let $\{x_n\}$ be a sequence in M such that $\lim_{n \to \infty} x_n = a$. Since f and g are continuous at a, by Theorem 40.2, $\lim_{n \to \infty} f(x_n) = f(a)$ and $\lim_{n \to \infty} g(x_n) = g(a)$. Now $\{f(x_n)\}$ and $\{g(x_n)\}$ are real sequences; so by Theorem 12.2, $\lim_{n \to \infty} [f(x_n) + g(x_n)] = f(a) + g(a)$. By Theorem 40.2, $f + g$ is continuous at a for every $a \in M$. ∎

   We return now to arbitrary continuous functions on metric spaces. We first characterize continuity in terms of open and closed sets.

Theorem 40.5 Let f be a function from a metric space M_1 into a metric space M_2. The following are equivalent:

   (i) f is continuous on M_1.

(ii) $f^{-1}(C)$ is closed whenever C is a closed subset of M_2.

(iii) $f^{-1}(U)$ is open whenever U is an open subset of M_2.

Proof. (i) implies (ii). Suppose that f is continuous on M_1, and let C be a closed subset of M_2. To show that $f^{-1}(C)$ is closed, we must show that $f^{-1}(C)$ contains its limit points.

Let a be a limit point of $f^{-1}(C)$. Then there exists a sequence $\{x_n\}$ in $f^{-1}(C)$ such that $\lim_{n \to \infty} x_n = a$. Since f is continuous at a, $\lim_{n \to \infty} f(x_n) = f(a)$. Now $x_n \in f^{-1}(C)$ for every positive integer n, and hence $f(x_n) \in C$ for every positive integer n. Therefore, $f(a)$ is a limit point of C, and since C is closed, $f(a) \in C$. But this means that $a \in f^{-1}(C)$, and thus $f^{-1}(C)$ is closed.

(ii) implies (iii). Let U be an open subset of M_2. By Theorem 39.5, U' is closed. By hypothesis, $f^{-1}(U')$ is closed. Since $f^{-1}(U') = [f^{-1}(U)]'$, $[f^{-1}(U)]'$ is closed. By Theorem 39.5, $[f^{-1}(U)]'' = f^{-1}(U)$ is open.

(iii) implies (i). Suppose now that $f^{-1}(U)$ is open whenever U is an open subset of M_2. We will prove that f is continuous on M_1 directly from the definition of continuity (Definition 40.1).

Let d_1 and d_2 be the metrics for M_1 and M_2, respectively. Let $a \in M_1$ and let $\varepsilon > 0$. By Theorem 39.4, $B_\varepsilon(f(a))$ is an open subset of M_2, and so $f^{-1}(B_\varepsilon(f(a)))$ is open in M_1. Since $a \in f^{-1}(B_\varepsilon(f(a)))$, there exists $\delta > 0$ such that $B_\delta(a) \subset f^{-1}(B_\varepsilon(f(a)))$. If $d_1(x, a) < \delta$, then $x \in B_\delta(a)$, and hence $x \in f^{-1}(B_\varepsilon(f(a)))$. Therefore, $f(x) \in B_\varepsilon(f(a))$, but this means that $d_2(f(x), f(a)) < \varepsilon$. Thus f is continuous at a for every $a \in M_1$. ∎

Let X be a set with the discrete metric and let M be a metric space. We will show that every function f from X into M is continuous. For if U is open in M, $f^{-1}(U)$ is open in X since every subset of X is open (see Section 39).

Corollary 40.6 Let M_1, M_2, and M_3 be metric spaces, and suppose that g is a continuous function from M_1 into M_2 and that f is a continuous function from M_2 into M_3. Then $f \circ g$ is a continuous function from M_1 into M_3.

Proof. By Theorem 40.5, it is enough to show that $(f \circ g)^{-1}(U)$ is open whenever U is an open subset of M_3. Let U be open in M_3. By Theorem 40.5, $f^{-1}(U)$ is open in M_2. Again by Theorem 40.5, $g^{-1}(f^{-1}(U))$ is open in M_1. Since $(f \circ g)^{-1}(U) = g^{-1}(f^{-1}(U))$, we have the desired result. ∎

Corollary 40.6 can be summarized by stating that the composition of continuous functions is continuous.

Exercises

40.1 Prove Theorem 40.4(i), (iii), (iv), (v), and (vi).

40.2 Prove Theorem 40.4 using Definition 40.1.

40.3 Prove that Theorem 40.5(i) implies Theorem 40.5(iii) using Definition 40.1.

40.4 Prove Corollary 40.6 using Theorem 40.2.

40.5 Prove Corollary 40.6 using Definition 40.1.

40.6 Let M_1 and M_2 be metric spaces and let $c \in M_2$. Let $f(x) = c$ for all $x \in M_1$. Prove that f is continuous on M_1.

40.7 Let f be a function from a metric space (M_1, d_1) into a metric space (M_2, d_2). Let $a \in M_1$. Prove that the following are equivalent.
   (a) f is continuous at a.
   (b) If U is an open subset of M_2 which contains $f(a)$, there exists an open subset V of M_1 which contains a such that $V \subset f^{-1}(U)$.

40.8 Let f and g be continuous functions from $\mathbf{R}$ into $\mathbf{R}$. Prove that $h(x) = (f(x), g(x))$ defines a continuous function from $\mathbf{R}$ into $\mathbf{R}^2$. State and prove generalizations involving continuous functions from $\mathbf{R}^m$ into $\mathbf{R}^n$.

40.9 Let k be a positive integer. If $\{a_n\}$ is a real sequence, let $p_k(\{a_n\}) = a_k$. Prove that p_k is a continuous real-valued function on any of the spaces l^1, l^2, c_0, l^∞, or H^∞.

40.10 Let $\{a_n\} \in l^\infty$. Prove that f defined by

$$f(\{b_n\}) = \sum_{n=1}^{\infty} a_n b_n$$

is a continuous real-valued function on l^1.

40.11 Let $\{a_n\} \in l^2$. Prove that f defined by

$$f(\{b_n\}) = \sum_{n=1}^{\infty} a_n b_n$$

is a continuous real-valued function on l^2.

40.12 Let (M, d) be a metric space. Let $M \times M$ be the product metric space (see Exercise 35.8). Prove that d is a continuous real-valued function on $M \times M$.

40.13 Let a and b be real numbers. Prove that

$$\max \{a, b\} = \frac{a + b + |a - b|}{2}$$

and

$$\min \{a, b\} = \frac{a + b - |a - b|}{2}$$

If f and g are real-valued functions on a set X, we define

$$[\max \{f, g\}](x) = \max \{f(x), g(x)\} \qquad \text{for } x \in X$$

and

$$[\min \{f, g\}](x) = \min \{f(x), g(x)\} \qquad \text{for } x \in X$$

Let f and g be continuous real-valued functions on a metric space M. Prove that $\max \{f, g\}$ and $\min \{f, g\}$ are continuous real-valued functions on M.

40.14 Let (M, d) be a metric space and let X be a subset of M. If $x \in M$, we define

$$d(x, X) = \text{glb } \{d(x, y) | y \in X\}$$

(a) Prove that $f(x) = d(x, X)$ defines a continuous real-valued function on M.

(b) Prove that $f(x) = 0$ if and only if $x \in \bar{X}$.

(c) Let A and B be closed subsets of M such that $A \cap B = \varnothing$. Let

$$g(x) = \frac{d(x, A)}{d(x, A) + d(x, B)}$$

Prove that g is a continuous function from M into $[0, 1]$ such that

$$g(x) = \begin{cases} 0 & \text{if } x \in A \\ 1 & \text{if } x \in B \end{cases}$$

(d)* Let A and B be closed subsets of M such that $A \cap B = \varnothing$. Prove that there exist open subsets U and V of M such that $A \subset U$, $B \subset V$, and $U \cap V = \varnothing$.

40.15 Let f be a real-valued function on a metric space M. Prove that f is continuous on M if and only if the sets

$$\{x | f(x) < c\} \qquad \{x | f(x) > c\}$$

are open in M for every c in $\mathbf{R}$.

40.16 Let (M, d) be a metric space such that $d(x, y) \leq 1$ for all $x, y \in M$, and let $\{a_n\}$ be a sequence of points in M. Set $f(x) = \{d(x, a_n)\}_{n=1}^{\infty}$ for $x \in M$.

(a)* Prove that f is a continuous function from M into H^{∞}.

(b)* Prove that if $\{a_n | n \in \mathbf{P}\}^- = M$, then f is one-to-one.

40.17 Let M be a set and let d and d' be metrics for M. We say that d and d' are *equivalent metrics* for M if the collection of open subsets of (M, d) is identical with the collection of open subsets of (M, d').

(a) Prove that the following are equivalent.

    (i) d and d' are equivalent metrics.

    (ii) The collection of closed subsets of (M, d) is identical with the collection of closed subsets of (M, d').

    (iii) The sequence $\{x_n\}$ converges in (M, d) if and only if $\{x_n\}$ converges in (M, d').

(b) Prove that the metrics d, d', and d'' of Exercise 35.7 are equivalent.

(c) Prove that the metric d' of Exercise 37.9 is equivalent to the usual (Euclidean) metric on $\mathbf{R}^n$.

41. The Relative Metric

Let $\mathbf{R}$ have the usual metric, $d(x, y) = |x - y|$. Since $[0, 1]$ is a subset of $\mathbf{R}$, we can regard $[0, 1]$ as a metric space by restricting the metric d to $[0, 1]$. Thus the metric d' for $[0, 1]$ is given by

$$d'(x, y) = |x - y| \quad \text{for} \quad x, y \in [0, 1]$$

We will use the notation $B_\varepsilon^{[0, 1]}(x)$ to denote the open ball of radius ε centered at x in the metric space $([0, 1], d')$. For example,

$$B_{1/4}^{[0,1]}\left(\tfrac{1}{2}\right) = \left(\tfrac{1}{4}, \tfrac{3}{4}\right) \qquad B_{1/2}^{[0,1]}(0) = [0, \tfrac{1}{2})$$

By Theorem 39.4, an open ball is an open set, and therefore $(\tfrac{1}{4}, \tfrac{3}{4})$ and $[0, \tfrac{1}{2})$ are open sets *in the metric space* $([0, 1], d')$. We must be careful to specify the metric space in which we are working, for $[0, \tfrac{1}{2})$ is *not* open *in the metric space* $(\mathbf{R}, d)$. Notice that the open ball $B_{1/2}^{[0,1]}(0) = [0, \tfrac{1}{2})$ (in $[0, 1]$) is simply the open ball $B_{1/2}^{\mathbf{R}}(0) = (-\tfrac{1}{2}, \tfrac{1}{2})$ (in $\mathbf{R}$) "cut down" to $[0, 1]$. More precisely, we have

$$B_{1/2}^{[0,1]}(0) = B_{1/2}^{\mathbf{R}}(0) \cap [0, 1]$$

It is easy to see that this will always be the case. The open ball $B_{\varepsilon}^{[0,1]}(x)$ (in $([0, 1])$ is simply $B_{\varepsilon}^{\mathbf{R}}(x) \cap [0, 1]$. We know that open balls in $\mathbf{R}$ are open intervals; hence the open balls in $[0, 1]$ are simply open intervals intersected with $[0, 1]$. We now analyze this situation in an arbitrary metric space.

Definition 41.1 Let (M, d) be a metric space and let X be a subset of M. The function d' defined by

$$d'(x, y) = d(x, y) \qquad \text{for} \quad x, y \in X$$

is called the *metric for X relative to M* or more simply, the *relative metric* for X.

If X is a subset of a metric space M and we wish to regard X as a metric space, we will always use the relative metric of Definition 41.1 for X. We will refer to X as a *relative metric space*.

Keeping the notation of Definition 41.1, an open ball in X of radius ε centered at a is the set

$$\{x \in X \mid d'(x, a) < \varepsilon\}$$

which we denote $B_{\varepsilon}^{X}(a)$. Letting

$$B_{\varepsilon}(a) = \{x \in M \mid d(x, a) < \varepsilon\}$$

be the open ball in M of radius ε centered at a, we see that

$$B_{\varepsilon}^{X}(a) = B_{\varepsilon}(a) \cap X \tag{41.1}$$

(Verify.) A relationship like that given in equation (41.1) also holds for open and closed sets as the next theorem shows.

Theorem 41.2 Let M be a metric space and let X be a subset of M with the relative metric. Let Y be a subset of X.

(i) Y is open in X if and only if $Y = X \cap U$, where U is open in M.

(ii) Y is closed in X if and only if $Y = X \cap C$, where C is closed in M.

Proof. Suppose Y is open in X. Then for every $x \in Y$, there exists $\varepsilon_x > 0$ such that

$$B_{\varepsilon_x}^X(x) \subset Y$$

Let

$$U = \bigcup_{x \in Y} B_{\varepsilon_x}(x)$$

Then U is open in M and

$$
\begin{aligned}
X \cap U &= X \cap (\bigcup_{x \in Y} B_{\varepsilon_x}(x)) \\
&= \bigcup_{x \in Y} (X \cap B_{\varepsilon_x}(x)) \\
&= \bigcup_{x \in Y} B_{\varepsilon_x}^X(x) \\
&= Y
\end{aligned}
$$

Next suppose that Y is a subset of X such that $Y = X \cap U$, where U is open in M. Let $x \in Y$. Then $x \in U$ and since U is open in M, there exists $\varepsilon > 0$ such that $B_\varepsilon(x) \subset U$. Now

$$B_\varepsilon^X(x) = B_\varepsilon(x) \cap X \subset U \cap X = Y$$

and thus Y is open in X. We have established part (i) of the theorem.

Suppose Y is closed in X. Let $C = \overline{Y}$, where the closure is taken in M. Then C is closed in M. Clearly, $Y \subset X \cap C$. Let $x \in X \cap C$. Since $x \in \overline{Y}$, there exists a sequence $\{x_n\}$ in Y such that $\lim_{n \to \infty} x_n = x$. Since $x \in X$, x is a limit point (in X) of Y. Since Y is closed in X, $x \in Y$. Thus $X \cap C = Y$.

Finally, suppose that Y is a subset of X such that $Y = X \cap C$, where C is closed in M. Let x be a limit point (in X) of Y. Then there exists a sequence $\{x_n\}$ in Y such that $\lim_{n \to \infty} x_n = x$. Now $\{x_n\}$ is also in C, and hence x is a limit point of C, and since C is closed, $x \in C$. Thus $x \in X \cap C = Y$, and so Y is closed in X. ∎

Corollary 41.3 (i) Let X be an open subset of a metric space M and let $Y \subset X$. Then Y is open in X if and only if Y is open in M.

(ii) Let X be a closed subset of a metric space M and let $Y \subset X$. Then Y is closed in X if and only if Y is closed in M.

Proof. Let X be an open subset of M and let $Y \subset X$. If Y is open in M, then Y is open in X by Theorem 41.2 (i) since $Y = X \cap Y$. Now suppose that Y is open in X. By Theorem 41.2 (i), there exists a set U open in M such that $Y = X \cap U$. Since X and U are open in M, $X \cap U$ is open in M [Theorem 39.6 (i)].

Part (ii) is proved similarly. ∎

Exercises

41.1 Deduce Theorem 41.2(ii) from Theorem 41.2(i) using Theorem 39.5.

41.2 Prove Corollary 41.3(ii).

41.3 Let $A = [0, 1]$, $B = (\frac{1}{2}, 1]$, $C = (\frac{1}{4}, \frac{3}{4})$.
   (a) Is B open (closed) in A?
   (b) Is C open (closed) in A?
   (c) Is A open (closed) in $\mathbf{R}$?
   (d) Is C open (closed) in $\mathbf{R}$?
   (e) Is A open (closed) in $\mathbf{R}^2$?
   (f) Is C open (closed) in $\mathbf{R}^2$?
   (g) Is $\mathbf{R}$ open (closed) in $\mathbf{R}^2$?

41.4 Let M be a metric space and let X be a subset of M with the relative metric. Prove that if f is a continuous function on M, then $f|X$ is a continuous function on X.

41.5 Let M be a metric space and let X be a subset of M with the relative metric. If Y is a subset of X, let $Y^{-(X)}$ denote the closure of Y in the metric space X. Prove that $Y^{-(X)} = Y^{-} \cap X$. State and prove a corresponding result for Y°.

42. Compact Metric Spaces

We recall (Theorem 34.2) that a closed interval $[a, b]$ of the real line has the following property. If $\mathcal{U}$ is a collection of open intervals such that $[a, b] \subset \cup \mathcal{U}$, then there exists a finite subcollection $\{U_1, U_2, \ldots, U_n\}$ of $\mathcal{U}$ such that $[a, b] \subset U_1 \cup U_2 \cup \cdots \cup U_n$. We now generalize this property to arbitrary metric spaces. "Open interval" is replaced by "open set."

Definition 42.1 An *open cover* of a metric space M is a collection $\mathcal{U}$ of open subsets of M such that $M = \cup \mathcal{U}$. A *subcover of $\mathcal{U}$* is a subcollection $\mathcal{U}^*$ of $\mathcal{U}$ such that $M = \cup \mathcal{U}^*$.

Definition 42.2 A metric space M is said to be *compact* if every open cover of M has a finite subcover.

In other words a metric space M is compact if whenever $\mathcal{U}$ is a collection of open subsets of M such that $M = \cup \mathcal{U}$, there exist $U_1, U_2, \ldots, U_n$ in $\mathcal{U}$ such that $M = U_1 \cup U_2 \cup \cdots \cup U_n$. The crucial part of the definition is that we must be able to reduce *every* open cover to a finite subcover.

Examples. Let M be a finite metric space. Then every open cover $\mathcal{U}$ is finite in the first place, and therefore M is compact.

Let X be a set with the discrete metric. If X is finite, then X is compact.

However, if X is infinite X is not compact, since the open cover

$$\mathcal{U} = \{\{x\} \mid x \in X\}$$

has no finite subcover.

The open interval $(0, 1)$ (with the relative metric) is not compact since the open cover

$$\mathcal{U} = \left\{ \left(\frac{1}{n}, 1 \right) \,\middle|\, n \in \mathbf{P}, n \geq 2 \right\}$$

has no finite subcover (verify). [The sets $(1/n, 1)$ are open in $(0, 1)$ by Corollary 41.3 (i).]

The space $\mathbf{R}$ is not compact since the open cover

$$\{(-n, n) \mid n \in \mathbf{P}\}$$

has no finite subcover. In a similar way, one can show that $\mathbf{R}^n$, l^1, l^2, c_0, and l^∞ are not compact.

The Heine-Borel theorem (Theorem 34.2) is virtually the statement that a closed interval $[a, b]$ is compact. We must only sort out the relative metric.

Theorem 42.3 A closed interval $[a, b]$ is compact.

Proof. Let $\mathcal{U}$ be an open cover of $[a, b]$. Then for every x in $[a, b]$, there exists an open set U_x in $\mathcal{U}$ such that $x \in U_x$. Since U_x is open, there exists an open ball (in $[a, b]$) $B_{\varepsilon_x}^{[a,b]}(x) \subset U_x$. Now the collection

$$\mathcal{B} = \{B_{\varepsilon_x}^{\mathbf{R}}(x) \mid x \in [a, b]\}$$

is a family of open intervals such that $[a, b] \subset \cup \mathcal{B}$. By Theorem 34.2, there exist $x_1, \ldots, x_n \in [a, b]$ such that

$$[a, b] \subset \bigcup_{i=1}^{n} B_{\varepsilon_{x_i}}^{\mathbf{R}}(x_i)$$

Now
$$[a, b] = \bigcup_{i=1}^{n} ([a, b] \cap B_{\varepsilon_{x_i}}^{\mathbf{R}}(x_i)) = \bigcup_{i=1}^{n} B_{\varepsilon_{x_i}}^{[a,b]}(x_i) \subset \bigcup_{i=1}^{n} U_{x_i}$$

Therefore $[a, b] = \bigcup_{i=1}^{n} U_{x_i}$, and hence $[a, b]$ is compact. ∎

We now imitate the proof of Theorem 34.3, which is concerned with the compact space $[a, b]$, to prove that a continuous function on a compact metric space attains a maximum and a minimum.

Definition 42.4 Let f be a real-valued function on a set X. We say that f is *bounded on* X if there exists a number M such that $|f(x)| \leq M$ for every $x \in X$.

Lemma 42.5 Let f be a real-valued function on a metric space M. If f is continuous at a, there exists an open set U containing a such that f is bounded on U.

Proof. Let (M, d) be a metric space.
Taking $\varepsilon = 1$, there exists $\delta > 0$ such that if $d(x, a) < \delta$, then $|f(x) - f(a)|$ < 1. Thus if x is in the open set $B_\delta(a)$, we have

$$|f(x)| \leq |f(x) - f(a)| + |f(a)| < 1 + |f(a)| \qquad \blacksquare$$

Theorem 42.6 If f is a continuous real-valued function on a compact metric space M, then f is bounded on M.

Proof. By Lemma 42.5, for each $a \in M$ there exists an open set U_a containing a such that f is bounded on U_a. Now $\{U_a \mid a \in M\}$ is an open cover of M, and since M is compact, there exists $a_1, \ldots, a_n \in M$ such that

$$M = U_{a_1} \cup U_{a_2} \cup \cdots \cup U_{a_n}$$

Since f is bounded on U_{a_i} for $1 \leq i \leq n$, f is bounded on $U_{a_1} \cup \cdots \cup U_{a_n} = M$. $\blacksquare$

Corollary 42.7 If f is a continuous real-valued function on a compact metric space M, there exist $c, d \in M$ such that $f(c) \leq f(x) \leq f(d)$ for all $x \in M$. That is, f attains a maximum and a minimum on M.

Proof. By Theorem 42.6, f is bounded, and hence the least upper bound T of the set

$$X = \{f(x) \mid x \in M\}$$

exists. We must show that $f(d) = T$ for some $d \in M$. If this fails, $f(x) < T$ for all $x \in M$. By Theorem 40.4, $g(x) = 1/(T - f(x))$ is a continuous function on M. By Theorem 42.6, g is bounded. Suppose $g(x) < S$ for all x in M. Then $f(x) < T - 1/S < T$ for all x in M. This contradicts the assumption that $T = \text{lub } X$.

Similarly, f attains a minimum on M. $\blacksquare$

Exercises

42.1 Prove that none of the spaces $\mathbf{R}^n$, l^1, l^2, c_0, or l^∞ is compact.

42.2* Let X be a compact subset of a metric space M. Prove that X is closed.

42.3 Let $X_1, \ldots, X_n$ be a finite collection of compact subsets of a metric space M. Prove that $X_1 \cup X_2 \cup \cdots \cup X_n$ is a compact metric space. Show (by example) that this result does not generalize to infinite unions.

42.4 Let $\mathscr{C}$ be a collection of compact subsets of a metric space. Prove that $\cap \mathscr{C}$ is compact.

42.5 A collection $\mathscr{C}$ of subsets of a set X is said to have the *finite intersection property* if whenever $\{C_1, \ldots, C_n\}$ is a finite subcollection of $\mathscr{C}$, we have $C_1 \cap C_2 \cap \cdots \cap C_n \neq \varnothing$. Prove that a metric space M is compact if and only if whenever $\mathscr{C}$ is a collection of closed subsets of M having the finite intersection property, we have $\cap \mathscr{C} \neq \varnothing$.

42.6 Let f be a continuous real-valued function on a compact metric space M. Suppose that $f(x) > 0$ for all $x \in M$. Prove that there exists $T > 0$ such that $f(x) > T$ for all $x \in M$.

42.7 Let X be a compact subset of $\mathbf{R}$ and let $y \in \mathbf{R}$. Prove that the set $\{x + y \mid x \in X\}$ is compact.

42.8 Let f be a continuous, real-valued function on a metric space M which is never zero. Prove that the collection of open sets U for which either $f(x) > 0$ for $x \in U$ or $f(x) < 0$ for $x \in U$ is an open cover of M.

42.9 Call an open cover $\mathscr{U}$ of a metric space M an *additive cover* if whenever $U, V \in \mathscr{U}$, we have $U \cup V \in \mathscr{U}$. Prove that M is compact if and only if every additive open cover of M contains M (Johnsonbaugh, 1977).

42.10 Let $\{X_n\}$ be a sequence of compact subsets of a metric space M with $X_1 \supset X_2 \supset X_3 \supset \cdots$. Prove that if U is an open set containing $\cap X_n$, then there exists $X_n \subset U$.

42.11 Let f be a function on $[a, b]$. Let K be a compact subset of $[a, b]$ on which f is continuous. Suppose there exists $c > 0$ such that for each $x \in K$, there exists $h_x > 0$ with

$$\left| \frac{f(x + h_x) - f(x)}{h_x} \right| < c$$

Prove that there exists a finite subset $\{x_1, \ldots, x_n\} \subset K$ and positive numbers $h_1, \ldots, h_n$ such that

(a) $x_1 < x_1 + h_1 \leq x_2 < x_2 + h_2 \leq x_3 < \cdots$

(b) $\left| \dfrac{f(x_i + h_i) - f(x_i)}{h_i} \right| < c \qquad$ for $i = 1, \ldots, n$

(c) $K \subset \bigcup\limits_{i=1}^{n} [x_i, x_i + h_i]$

42.12* A *contractive mapping* on M is a function f from the metric space (M, d) into itself satisfying

$$d(f(x), f(y)) < d(x, y)$$

whenever $x, y \in M$ with $x \neq y$. Prove that if f is a contractive mapping on a compact metric space M, there exists a unique point $x \in M$ with $f(x) = x$. (Such a point is called a *fixed point*.)

43. The Bolzano-Weierstrass Characterization of a Compact Metric Space

The Bolzano-Weierstrass theorem (Theorem 18.1) states that every bounded real sequence has a convergent subsequence. We will prove (Theorem 43.5) that a metric space M is compact if and only if *every* sequence in M has a convergent subsequence. The proof is divided into four lemmas.

Lemma 43.1 If M is a compact metric space, then every sequence in M has a convergent subsequence.

Proof. Suppose there exists a sequence $\{x_n\}$ in M which has no convergent subsequence. We claim that for every $x \in M$, there exists an open ball $B_\varepsilon(x)$ such that the set

$$\{n \in \mathbf{P} \mid x_n \in B_\varepsilon(x)\}$$

is finite. For if some $x \in M$ has the property that every open ball $B_\varepsilon(x)$ contains x_n for infinitely many positive integers n, we may produce a convergent subsequence of $\{x_n\}$ as follows. Choose $x_{n_1} \in B_1(x)$. Choose $x_{n_2} \in B_{\frac{1}{2}}(x)$ such that $n_2 > n_1$. Having chosen $x_{n_k} \in B_{1/k}(x)$, choose $x_{n_{k+1}} \in B_{1/(k+1)}(x)$ such that $n_{k+1} > n_k$. Then $\lim_{k \to \infty} x_{n_k} = x$, and hence $\{x_{n_k}\}_{k=1}^{\infty}$ is a convergent subsequence of $\{x_n\}$. This is a contradiction, and hence for every $x \in M$, there exists an open ball $B_{\varepsilon_x}(x)$ such that the set

$$\{n \in \mathbf{P} \mid x_n \in B_{\varepsilon_x}(x)\}$$

is finite.

Now $\mathcal{U} = \{B_{\varepsilon_x}(x) \mid x \in M\}$ is an open cover of M. No finite subcollection of $\mathcal{U}$ covers M, for otherwise the set

$$\{n \in \mathbf{P} \mid x_n \in M\}$$

is finite, which is impossible. This contradicts the definition of compactness and we have the desired conclusion. ∎

Let M be a metric space in which every sequence has a convergent subsequence. To prove the converse of Lemma 43.1, we must show that every open cover of M has a finite subcover. Let $\varepsilon > 0$. We first show that the special open cover

$$\mathcal{U} = \{B_\varepsilon(x) \mid x \in M\}$$

has a finite subcover.

Lemma 43.2 Let M be a metric space in which every sequence has a convergent subsequence. Let $\varepsilon > 0$. Then there exist $x_1, \ldots, x_n \in M$ such that

$$M = B_\varepsilon(x_1) \cup \cdots \cup B_\varepsilon(x_n)$$

Proof. Let d denote the metric on M. Let $x_1 \in M$. If $B_\varepsilon(x_1) = M$, we stop. Otherwise, there exists $x_2 \in M \backslash B_\varepsilon(x_1)$, and thus $d(x_2, x_1) \geq \varepsilon$. If $M = B_\varepsilon(x_1) \cup B_\varepsilon(x_2)$, we stop. Otherwise, there exists $x_3 \in M \backslash (B_\varepsilon(x_1) \cup B_\varepsilon(x_2))$, and thus $d(x_3, x_1) \geq \varepsilon$ and $d(x_3, x_2) \geq \varepsilon$. This process must stop after a finite number of steps, for otherwise we obtain a sequence $\{x_n\}$ in M such that

$$d(x_n, x_m) \geq \varepsilon \qquad \text{if } n \neq m$$

Such a sequence cannot have a convergent subsequence (verify). ∎

Let M be a metric space in which every sequence has a convergent subsequence, and let $\mathcal{U}$ be an open cover of M. If for some $\varepsilon > 0$, each open ball $B_\varepsilon(x)$ were contained in some $U \in \mathcal{U}$, then M would be compact. For

$$M = B_\varepsilon(x_1) \cup \cdots \cup B_\varepsilon(x_n)$$

by Lemma 43.2 and since $B_\varepsilon(x_i)$ is contained in U_i for some $U_i \in \mathcal{U}$, $i = 1, \ldots, n$, we would have

$$M = B_\varepsilon(x_1) \cup \cdots \cup B_\varepsilon(x_n) \subset U_1 \cup \cdots \cup U_n$$

in which case M would be compact. Our next lemma shows that such an ε exists. The proof of Lemma 43.3 is illustrated in Figure 43.1.

Lemma 43.3 Let M be a metric space in which every sequence has a con-

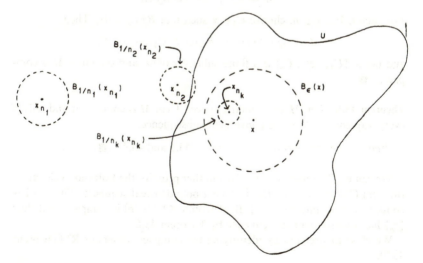

Figure 43.1

vergent subsequence. If $\mathcal{U}$ is an open cover of M, there exists $\varepsilon > 0$ such that if $x \in M$, $B_\varepsilon(x) \subset U$ for some $U \in \mathcal{U}$.

Proof. Suppose the conclusion is false. Then for every $\varepsilon > 0$, there exists $x \in M$ such that $B_\varepsilon(x) \not\subset U$ for any $U \in \mathcal{U}$. Taking $\varepsilon = 1/n$, there exists $x_n \in M$ such that $B_{1/n}(x_n) \not\subset U$ for any $U \in \mathcal{U}$. By our hypothesis, $\{x_n\}$ has a convergent subsequence $\{x_{n_k}\}$, and we suppose $\lim_{k \to \infty} x_{n_k} = x$. Now $x \in U$ for some $U \in \mathcal{U}$, and since U is open, there exists $\varepsilon > 0$ such that $B_\varepsilon(x) \subset U$. Choose a positive integer k such that

$$d(x_{n_k}, x) < \varepsilon/2 \qquad \text{and} \qquad \frac{1}{n_k} < \varepsilon/2$$

where d denotes the metric for M. We will show that $B_{1/n_k}(x_{n_k}) \subset U$ which will be a contradiction. Suppose $y \in B_{1/n_k}(x_{n_k})$. Then $d(y, x_{n_k}) < 1/n_k$. Now

$$d(y, x) \leq d(y, x_{n_k}) + d(x_{n_k}, x) < \frac{1}{n_k} + d(x_{n_k}, x) < \frac{\varepsilon}{2} + \frac{\varepsilon}{2} = \varepsilon$$

Thus $y \in B_\varepsilon(x)$, and hence $y \in U$. ∎

Lemma 43.4 If M is a metric space in which every sequence has a convergent subsequence, then M is compact.

Proof. Let $\mathcal{U}$ be an open cover of M. By Lemma 43.3, there exists $\varepsilon > 0$ such that if $x \in M$, $B_\varepsilon(x) \subset U$ for some $U \in \mathcal{U}$. By Lemma 43.2, there exists $x_1, \ldots, x_n \in M$ such that

$$M = B_\varepsilon(x_1) \cup \cdots \cup B_\varepsilon(x_n)$$

For each i, $1 \leq i \leq n$, choose $U_i \in \mathcal{U}$ such that $B_\varepsilon(x_i) \subset U_i$. Then

$$M = B_\varepsilon(x_1) \cup \cdots \cup B_\varepsilon(x_n) \subset U_1 \cup \cdots \cup U_n$$

and hence $\{U_1, \ldots, U_n\}$ is a finite subcover of $\mathcal{U}$, and therefore M is compact. ∎

Theorem 43.5 Let M be a metric space. Then M is compact if and only if every sequence in M has a convergent subsequence.

Proof. The theorem restates Lemmas 43.1 and 43.4. ∎

We can use Theorem 43.5 to give another proof of the Bolzano-Weierstrass theorem (Theorem 18.1). Let $\{a_n\}$ be a bounded real sequence. Then $\{a_n\}$ is contained in an interval $[a, b]$. By Theorem 42.3 $[a, b]$ is compact, and thus $\{a_n\}$ has a convergent subsequence by Theorem 43.5.

We close this section by identifying the compact subsets of $\mathbf{R}^n$ (Theorem 43.8).

Definition 43.6 A subset S of a metric space (M, d) is *bounded* if there exists a number A such that $d(x, y) \leq A$ for all $x, y \in S$.

Theorem 43.7 If C is a compact subset of a metric space M, then C is closed and bounded (in M).

Proof. Let d denote the metric on M. Let $x \in M$. Then

$$\mathscr{U} = \{B_n^C(x) \mid n \in \mathbf{P}\}$$

is an open cover of C and hence some finite subcollection $\{B_{n_1}^C(x), \ldots, B_{n_k}^C(x)\}$ of $\mathscr{U}$ covers C. Let $N = \max\{n_1, \ldots, n_k\}$. If $y, z \in C$, then $d(y, z) \leq 2N$, and therefore C is bounded.

Let x be a limit point of C. Then there exists a sequence $\{x_n\}$ in C such that $\lim_{n \to \infty} x_n = x$. Since C is compact, the sequence $\{x_n\}$ has a subsequence $\{x_{n_k}\}$ which converges to a point y in C (Theorem 43.5). But now $x = \lim_{n \to \infty} x_n = \lim_{k \to \infty} x_{n_k} = y \in C$, and therefore C is closed. ∎

The converse of Theorem 43.7 holds for $\mathbf{R}^n$, as we will show next, but the converse does not hold in general. Let 0 denote the sequence $\{a_n\}$, where $a_n = 0$ for every positive integer n. Let

$$X = \{x \in l^1 \mid d(x, 0) = 1\}$$

By Theorem 40.4, $f(x) = d(x, 0)$ is continuous and by Theorem 40.6 $f^{-1}(1) = X$ is closed. Obviously X is bounded. However, X is not compact. The sequence $\{\delta^{(k)}\}$ (defined in Section 37) has no convergent subsequence for if $\{\delta^{(k)}\}$ had a convergent subsequence $\{\delta^{(k_j)}\}$, by Theorem 37.3 $\{\delta^{(k_j)}\}$ would converge to 0, and we have already noted (Section 37) that this is not the case.

Theorem 43.8 Let C be a closed subset of a compact metric space M. Then C is compact.

Proof. If $\{x_n\}$ is a sequence in C, $\{x_n\}$ has a subsequence $\{x_{n_k}\}$ which converges to a point x in M, since M is compact. But x is a limit point of C, and since C is closed, $x \in C$. By Theorem 43.5, C is compact. ∎

Theorem 43.9 A subset X of $\mathbf{R}^n$ is compact if and only if X is closed and bounded.

Proof. Let X be a closed and bounded subset of $\mathbf{R}^n$, and let $\{a^{(k)}\}_{k=1}^{\infty}$ be a sequence in X. Since X is bounded, the real sequence $\{a_1^{(k)}\}_{k=1}^{\infty}$ is bounded and thus has a convergent subsequence $\{a_1^{(k_j)}\}_{j=1}^{\infty}$. The real sequence $\{a_2^{(k_j)}\}_{j=1}^{\infty}$ is bounded and thus has a convergent subsequence $\{a_2^{(k_{j_i})}\}_{i=1}^{\infty}$. Now $\{a_1^{(k_{j_i})}\}_{i=1}^{\infty}$ is a subsequence of the convergent sequence $\{a_1^{(k_j)}\}_{j=1}^{\infty}$ and thus converges.

We continue in this way producing a subsequence $\{b^{(k)}\}_{k=1}^{\infty}$ of $\{a^{(k)}\}_{k=1}^{\infty}$ such that $\{b_j^{(k)}\}_{k=1}^{\infty}$ converges for $j = 1, \ldots, n$. By Theorem 37.2, $\{b^{(k)}\}_{k=1}^{\infty}$ converges to a point a of $\mathbf{R}^n$. Since X is closed, $a \in X$, and by Theorem 43.5, X is compact.

The converse follows from Theorem 43.7. ∎

Corollary 43.10 If f is a continuous real-valued function on a closed and bounded subset X of $\mathbf{R}^n$, then f attains a maximum and a minimum on X.

Proof. Combine Corollary 42.7 and Theorem 43.9. ∎

Exercises

43.1 Prove that the set $\{x \in M \mid d(x, 0) = 1\}$ is closed and bounded in M, but not compact if M is l^2, c_0, or l^∞.

43.2 Let M_1 and M_2 be compact metric spaces. Prove that the product metric space $M_1 \times M_2$ is compact.

43.3* Let M be a metric space. If there exists a countable subset X of M such that $\bar{X} = M$, M is said to be *separable*.
Prove that a compact metric space is separable.

43.4 If (M, d) is a bounded metric space, we let diam $M = \text{lub} \{d(x, y) \mid x, y \in M\}$. Prove that if (M, d) is a compact metric space, there exist $x, y \in M$ such that $d(x, y) = \text{diam } M$.

43.5 If X is a compact subset of $\mathbf{R}$, describe X'.

43.6 Prove that H^∞ is a compact metric space.

43.7 Let X be a compact subset of a metric space M. If $y \in X'$, prove that there exists a point $a \in X$ such that
$$d(a, y) \leq d(x, y) \qquad \text{for all } x \in X$$
Give an example to show that the conclusion may fail if "compact" is replaced by "closed."

44. Continuous Functions on Compact Metric Spaces

Continuous functions on compact metric spaces have many nice properties. We have already proved (Corollary 42.7) that a continuous real-valued function on a compact metric space attains a maximum and a minimum. In this section we will derive further results.

Theorem 44.1 If f is a continuous function from a compact metric space M_1 into a metric space M_2, then $f(M_1)$ is compact.

Proof. Let $\{f(x_n)\}$ be a sequence in $f(M_1)$. Then $\{x_n\}$ is a sequence in M_1, and since M_1 is compact, $\{x_n\}$ has a subsequence $\{x_{n_k}\}$ which converges to a in M_1 (Theorem 43.5). Since f is continuous at a, $\lim_{k\to\infty} f(x_{n_k}) = f(a)$. Therefore $\{f(x_{n_k})\}$ is a convergent subsequence of $\{f(x_n)\}$ in M_2. By Theorem 43.5, $f(M_1)$ is compact. ∎

Corollary 44.2 If f is a continuous function from a compact metric space M_1 into a metric space M_2, then $f(M_1)$ is closed and bounded.

Proof. The proof follows immediately from Theorems 43.7 and 44.1. ∎

The next theorem shows that a one-to-one continuous function on a compact metric space has a continuous inverse.

Theorem 44.3 If f is a one-to-one continuous function from a compact metric space M_1 onto a metric space M_2, then f^{-1} is continuous on M_2.

Proof. By Theorem 40.6, we must show that if C is a closed subset of M_1, then $(f^{-1})^{-1}(C) = f(C)$ is a closed subset of M_2.

Let C be a closed subset of M_1. Then C is compact by Theorem 43.8. By Corollary 44.2, $f(C)$ is closed in M_2. ∎

Example. Let n be a positive integer and let $f(x) = x^n$, $x \geq 0$. In Section 33, we showed that f is continuous. Let $a > 0$. Then f is a continuous function from $[0, a]$ onto $[0, a^n]$. Now $f^{-1}(x) = x^{1/n}$, and by Theorem 44.3, f^{-1} is a continuous function from $[0, a^n]$ onto $[0, a]$ for every $a > 0$. Therefore $f^{-1}(x) = x^{1/n}$ is continuous on $[0, \infty)$. If n is an odd positive integer, the argument can be modified to show that $f^{-1}(x) = x^{1/n}$ is continuous on **R**.

Let f be a continuous function from a metric space (M_1, d_1) into a metric space (M_2, d_2). Let $\varepsilon > 0$. For each $y \in M_1$, there exists $\delta > 0$ such that if $d_1(x, y) < \delta$, then $d_2(f(x), f(y)) < \varepsilon$. In general, δ will depend upon y. However if there exists $\delta > 0$ which is independent of y, we say that f is uniformly continuous on M.

Definition 44.4 Let (M_1, d_1) and (M_2, d_2) be metric spaces, and let f be a function from M_1 into M_2. We say that f is *uniformly continuous* on M_1 if for every $\varepsilon > 0$, there exists $\delta > 0$ such that if $d_1(x, y) < \delta$, then $d_2(f(x), f(y)) < \varepsilon$.

Examples. The function $f(x) = x$ is uniformly continuous on **R** for if $\varepsilon > 0$ we may take $\delta = \varepsilon$. If $|x - y| < \delta$, then $|f(x) - f(y)| = |x - y| < \delta = \varepsilon$.

The function $f(x) = x^2$ is not uniformly continuous on **R**. We will show that for $\varepsilon = 1$, there is no δ such that if $|x - y| < \delta$, then $|f(x) - f(y)| < 1$.

For suppose such a δ exists. Choose a positive integer n such that

$$n \geq \frac{2/\delta - \delta/2}{2}$$

Let $x = n + \delta/2$, $y = n$. Then $|x - y| = \delta/2 < \delta$. But $|x^2 - y^2| = (x - y)(x + y) = (\delta/2)(2n + \delta/2) \geq 1$. Therefore f is not uniformly continuous on $\mathbf{R}$.

We will show that a continuous function on a compact metric space is uniformly continuous.

Theorem 44.5 If f is a continuous function from a compact metric space M_1 into a metric space M_2, then f is uniformly continuous on M_1.

Proof. Let f be a continuous function from a compact metric space (M_1, d_1) into a metric space (M_2, d_2). Let $\varepsilon > 0$. For each $z \in M_1$, there exists $\delta_z > 0$ such that if $d_1(x, z) < \delta_z$, then $d_2(f(x), f(z)) < \varepsilon/2$. The collection

$$\{B_{\delta_z}(z) \mid z \in M_1\}$$

is an open cover of M_1. By Lemma 43.3, there exists $\delta > 0$ such that for any $x \in M_1$, $B_\delta(x) \subset B_{\delta_z}(z)$ for some $z \in M_1$.

Suppose $d_1(x, y) < \delta$. Now $B_\delta(x) \subset B_{\delta_z}(z)$ for some $z \in M_1$. Since $x, y \in B_\delta(x)$, we have $x, y \in B_{\delta_z}(z)$. Thus $d_1(x, z) < \delta_z$ and $d_1(y, z) < \delta_z$, and therefore $d_2(f(x), f(z)) < \varepsilon/2$ and $d_2(f(y), f(z)) < \varepsilon/2$. Therefore, if $d_1(x, y) < \delta$, then

$$d_2(f(x), f(y)) \leq d_2(f(x), f(z)) + d_2(f(z), f(y)) < \frac{\varepsilon}{2} + \frac{\varepsilon}{2} = \varepsilon. \quad \blacksquare$$

Corollary 44.6 If f is a continuous real-valued function on a closed and bounded subset X of $\mathbf{R}^n$, then f is uniformly continuous on X.

Proof. Use Theorems 43.9 and 44.5. $\blacksquare$

Exercises

44.1 Give an example of metric spaces M_1 and M_2 and a continuous function f from M_1 onto M_2 such that M_2 is compact, but M_1 is not compact.

44.2 (a) Prove that $f(x) = \sqrt{x}$ is uniformly continuous on $[0, \infty)$.
(b) Prove that $f(x) = x^3$ is not uniformly continuous on $\mathbf{R}$.

44.3 Let f and g be uniformly continuous real-valued functions on a metric space, M. Let $c \in \mathbf{R}$. Prove that cf and $f + g$ are uniformly continuous on M.

44.4 Let f be a function from $\mathbf{R}$ into a set X. We say that f is *periodic* if there exists $p > 0$ such that $f(x + p) = f(x)$ for all $x \in \mathbf{R}$. Prove that if f is a continuous periodic function from **R** into a metric space M, then f is uniformly continuous on **R**.

44.5 Let M_1, M_2, and M_3 be metric spaces. Let g be a uniformly continuous function from M_1 into M_2, and let f be a uniformly continuous function from M_2 into M_3. Prove that $f \circ g$ is uniformly continuous on M_1.

44.6 Let f be a one-to-one function from a metric space M_1 onto a metric space M_2. If f and f^{-1} are continuous, we say that f is a *homeomorphism* and that M_1 and M_2 are *homeomorphic* metric spaces.

(a) Prove that any two closed intervals of **R** are homeomorphic.

(b) Prove (a) with "closed" replaced by "open"; with "closed" replaced by "half-open."

(c) Prove that a closed interval is not homeomorphic to either an open interval or a half-open interval.

(d) Let M be a metric space, and let $G(M)$ denote the set of homeomorphisms of M onto M.

  (i) Prove that $G(M)$ is a group under composition.

  (ii) Identify the group $G(M)$ in case M is finite.

  (iii) Prove that if M_1 and M_2 are homeomorphic metric spaces, then $G(M_1)$ is isomorphic to $G(M_2)$.

  (iv) Show, by example, that the converse of (iii) does not hold.

(e)* Prove that any metric space M is homeomorphic to a metric space (M^*, d) where d is bounded by 1.

(f)* Let M be a separable metric space. Prove that there is a one-to-one, continuous function f from M into H^∞. (*Separable* is defined in Exercise 43.3.)

(g)* Prove the following theorem. A metric space M is compact if and only if M is homeomorphic to a closed subset of H^∞.

44.7* A *contraction mapping* on M is a function f from the metric space (M, d) into itself satisfying

$$d(f(x), f(y)) \le c\, d(x, y)$$

for some c, $0 \le c < 1$ and all x and y in M.

(a) Prove that a contraction mapping on M is uniformly continuous on M.

(b) Give an example of a contraction mapping from **R** onto **R**.

(c) Prove that there is no contraction mapping from a compact metric space (with more than one point) onto itself.

44.8 Let X be a compact subset of **R**, and let f be a real-valued function on X. Prove that f is continuous if and only if $\{(x, f(x)) \mid x \in X\}$ is a compact subset of $\mathbf{R}^2$.

45. Connected Metric Spaces

Under any reasonable definition of connectedness, the metric space $X = [0, 1] \cup [2, 3]$ should *not* be connected. Notice that $[0, 1]$ is open and closed in X (verify) and $[0, 1]$ is neither X nor $\emptyset$. By Theorems 38.4 and 39.3, if M

is a metric space, then M and $\emptyset$ are both open and closed. If these are the only subsets of M which are both open and closed, we say that M is connected.

Definition 45.1 Let M be a metric space. If the only subsets of M which are both open and closed are M and $\emptyset$, then we say that M is *connected*.

To prove that a metric space is connected, we often argue by contradiction, and so it is convenient to have a characterization of "not connected."

Theorem 45.2 Let M be a metric space. The following are equivalent.

(i) M is not connected.
(ii) There exist open nonempty subsets U and V of M such that $M = U \cup V$ and $U \cap V = \emptyset$.
(iii) There exist closed nonempty subsets C and D of M such that $M = C \cup D$ and $C \cap D = \emptyset$.

Proof. Suppose M is not connected. Then there exists an open and closed subset U of M such that U is neither M nor $\emptyset$. If we let $V = U'$, then V is open and $V \neq \emptyset$. Therefore $M = U \cup V$, $U \cap V = \emptyset$, and U and V are open and nonempty.

Now suppose that there exist open nonempty subsets U and V of M such that $M = U \cup V$ and $U \cap V = \emptyset$. Since $U' = V$, U is closed. Thus U is an open and closed subset of M which is neither M nor $\emptyset$, and hence M is not connected. Therefore parts (i) and (ii) are equivalent.

The equivalence of parts (i) and (iii) is proved in a similar way. ∎

We may characterize the connected subsets of the real line using the next theorem.

Theorem 45.3 A subset X of $\mathbf{R}$ is connected if and only if whenever $a,b \in X$ with $a < b$, we have $[a, b] \subset X$.

Proof. Let X be a connected subset of $\mathbf{R}$. Suppose there exist $a,b \in X$ with $a < b$, but $c \notin X$ for some $c \in [a, b]$. Then $U = \{x \in X \mid x < c\}$ and $V = \{x \in X \mid c < x\}$ are open nonempty subsets of X such that $X = U \cup V$ and $U \cap V = \emptyset$. By Theorem 45.2, X is not connected, and we have a contradiction. Thus whenever $a,b \in X$ with $a < b$, we have $[a, b] \subset X$.

Now suppose that whenever $a,b \in X$ with $a < b$, we have $[a, b] \subset X$. If X is not connected, by Theorem 45.2, there exist closed nonempty subsets C and D of X such that $X = C \cup D$ and $C \cap D = \emptyset$. Let $a \in C$ and $b \in D$. We may assume that $a < b$. Let

$$c = \text{lub } \{x \in C \mid x < b\}$$

For every positive integer n, there exists $x_n \in C$ such that $c - 1/n < x_n \leq c$. Thus $\lim_{n \to \infty} x_n = c$, and c is a limit point of C. Since C is closed, $c \in C$. Now $c < b$ and by our hypothesis, $[c, b] \subset X$.

We next show that $(c, b) \subset D$. If there exists $y \in (c, b)$ such that $y \notin D$, then $y \in C$. Now $y < b$ and $y \in C$; so $y \in \{x \in C \mid x < b\}$. Thus

$$y \leq c = \text{lub } \{x \in C \mid x < b\}$$

Since $y \in (c, b)$, $y > c$ which is impossible, and thus $(c, b) \subset D$.

Because $(c, b) \subset D$, c is a limit point of D, and since D is closed, $c \in D$. But now $c \in C \cap D = \varnothing$, which is a contradiction, and therefore X is connected. ∎

Corollary 45.4 A nonempty subset X of **R** is connected if and only if X is either a point or an interval. In particular, **R** is connected.

Proof. Points and intervals are connected by Theorem 45.3.

Now suppose that X is a connected subset of **R**. Let $a = \text{glb } X$ and $b = \text{lub } X$. (We let $a = -\infty$ if X is not bounded below, and we let $b = \infty$ if X is not bounded above). If $a = b$, then $X = \{a\}$; so suppose $a < b$. Let $c \in (a, b)$. There exist x and y in X such that $a < x < c < y < b$. By Theorem 45.3, $c \in X$. Thus $(a, b) \subset X$. If $x < a$ or if $x > b$, then $x \notin X$. Therefore X is one of (a, b), $[a, b)$, $(a, b]$, or $[a, b]$. ∎

The next theorem is identical to Theorem 44.1 with *compact* replaced by *connected*.

Theorem 45.5 If f is a continuous function from a connected metric space M_1 into a metric space M_2, then $f(M_1)$ is connected.

Proof. If $M = f(M_1)$ is not connected, by Theorem 45.2 there exist open nonempty subsets U and V of M such that $U \cup V = M$ and $U \cap V = \varnothing$. Since f is continuous, $f^{-1}(U)$ and $f^{-1}(V)$ are open subsets of M_1. Now $M_1 = f^{-1}(U) \cup f^{-1}(V)$, where $f^{-1}(U)$ and $f^{-1}(V)$ are open nonempty subsets of M_1 such that $f^{-1}(U) \cap f^{-1}(V) = \varnothing$. Therefore M_1 is not connected and we have a contradiction. ∎

Corollary 45.6 (Intermediate-Value Theorem) Suppose f is a continuous real-valued function on $[a, b]$ such that $f(a) < f(b)$. If y is a real number satisfying $f(a) < y < f(b)$, then $f(x) = y$ for some x in (a, b).

Proof. By Theorem 45.5, $f([a, b])$ is connected. By Theorem 45.3, $[f(a), f(b)] \subset f([a, b])$. ∎

We will use Theorem 45.5, to show that $\mathbf{R}^n$ is connected.

Theorem 45.7   The metric space $\mathbf{R}^n$ is connected.

Proof.   Suppose $\mathbf{R}^n$ is not connected. Then there exist open nonempty subsets U and V of $\mathbf{R}^n$ such that $\mathbf{R}^n = U \cup V$ and $U \cap V = \emptyset$. Let $x = (x_1, \ldots, x_n) \in U$ and $y = (y_1, \ldots, y_n) \in V$. Let

$$f(t) = (tx_1 + (1 - t)y_1, tx_2 + (1 - t)y_2, \ldots, tx_n + (1 - t)y_n)$$

for $0 \leq t \leq 1$. We show that f is a continuous function from $[0, 1]$ into $\mathbf{R}^n$. $[f([0, 1])$ is a line joining x and y].

Let $\varepsilon > 0$. Let $\delta = \varepsilon \Big/ \sqrt{\sum_{k=1}^{n} (x_k - y_k)^2}$. If $|t_1 - t_2| < \delta$, then

$$\begin{aligned}
d(f(t_1), f(t_2)) &= \sqrt{\sum_{k=1}^{n} \{[(t_1 x_k + (1 - t_1)y_k] - [t_2 x_k + (1 - t_2)y_k]\}^2} \\
&= |t_1 - t_2| \sqrt{\sum_{k=1}^{n} (x_k - y_k)^2} \\
&< \varepsilon
\end{aligned}$$

and thus f is continuous.

By Corollary 45.4 and Theorem 45.5, $X = f([0, 1])$ is connected. Let $U_1 = X \cap U$ and $V_1 = X \cap V$. Then U_1 and V_1 are open subsets of X such that $X = U_1 \cup V_1$ and $U_1 \cap V_1 = \emptyset$. Now $x = f(1) \in U_1$ and $y = f(0) \in V_1$ and thus U_1 and V_1 are nonempty. This contradicts Theorem 45.2 and therefore $\mathbf{R}^n$ is connected. ■

The method of proof of Theorem 45.7 may also be used to prove that l^1 and l^2 are connected.

Exercises

45.1   Prove that l^1, l^2, c_0, l^∞, and H^∞ are connected metric spaces.

45.2   (a)   Give an example of a subset of $\mathbf{R}$ which is connected but not compact.
(b)   Give an example of a subset of $\mathbf{R}$ which is compact but not connected.
(c)   Characterize the compact, connected subsets of $\mathbf{R}$.

45.3   Let f be a continuous function from a compact, connected metric space M into $\mathbf{R}$. Prove that $f(M)$ is a closed interval.

45.4   Let M be a metric space. Prove that the following are equivalent.
(a) M is not connected.
(b)   There exist nonempty subsets X and Y of M such that $M = X \cup Y$, $\bar{X} \cap Y = \emptyset = X \cap \bar{Y}$.

45.5 Let X be a connected subset of metric space M. Prove that $\bar{X}$ is connected. Is X^0 necessarily connected?

45.6 Let M_1 and M_2 be connected metric spaces. Prove that the product metric space $M_1 \times M_2$ is connected.

45.7 (a) Show, by examples, that unions and intersections of connected sets are not necessarily connected.

(b) Prove that if X and Y are connected subsets of **R**, then $X \cap Y$ is connected.

(c) Let $\mathscr{F}$ be a collection of connected subsets of a metric space M such that $\cap \mathscr{F} \neq \varnothing$. Prove that $\cup \mathscr{F}$ is connected.

45.8 Prove that an open interval of **R** is not homeomorphic to a half-open interval of **R**. (*Homeomorphic* is defined in Exercise 44.6.)

45.9 (a) Call an open cover $\mathscr{U}$ of a metric space M *infinitely additive* if whenever $\{U_\alpha\}$ is a subcollection of $\mathscr{U}$ such that $\cap U_\alpha \neq \varnothing$, then $\cup U_\alpha \in \mathscr{U}$. Prove that M is connected if and only if every infinitely additive open cover of M contains M.

(b) Call an open cover $\mathscr{U}$ of a metric space M *strongly additive* if whenever $U, V \in \mathscr{U}$ and $U \cap V \neq \varnothing$, then $U \cup V \in \mathscr{U}$. Prove that M is compact and connected if and only if every strongly additive open cover of M contains M (Johnsonbaugh, 1977).

46. Complete Metric Spaces

We define a Cauchy sequence in a metric space analogously to Definition 19.2 which applies to real sequences.

Definition 46.1 Let (M, d) be a metric space. A sequence $\{x_n\}$ in M is a *Cauchy sequence* if for every $\varepsilon > 0$, there exists a positive integer N such that if $m, n \geq N$, then $d(x_m, x_n) < \varepsilon$.

As in the case of the real line, every convergent sequence in a metric space is a Cauchy sequence.

Theorem 46.2 If $\{x_n\}$ is a convergent sequence in a metric space, then $\{x_n\}$ is a Cauchy sequence.

Proof. Let $\{x_n\}$ be a convergent sequence in a metric space (M, d). Suppose $\lim_{n \to \infty} x_n = x$. Let $\varepsilon > 0$. There exists a positive integer N such that if $n \geq N$, then $d(x_n, x) < \varepsilon/2$. If $m, n \geq N$, then

$$d(x_m, x_n) \leq d(x_n, x) + d(x_m, x) < \frac{\varepsilon}{2} + \frac{\varepsilon}{2} = \varepsilon \qquad \blacksquare$$

The converse of Theorem 46.2 holds for the real line (Theorem 19.3); however, the converse of Theorem 46.2 does *not* hold for an arbitrary metric space. For example, the sequence $\{1/n\}_{n=1}^{\infty}$ is a Cauchy sequence in $(0, 2)$, but $\{1/n\}_{n=1}^{\infty}$ does not converge in $(0, 2)$. If the converse of Theorem 46.2 holds for a metric space M, M is said to be complete.

Definition 46.3   Let M be a metric space. If every Cauchy sequence in M is convergent, we say that M is a *complete* metric space.

Thus $\mathbf{R}$ is a complete metric space, but $(0, 2)$ is not a complete metric space.

If X is a set with the discrete metric d, X is complete. For let $\{x_n\}$ be a Cauchy sequence in X. There exists a positive integer N such that if $m,n \geq N$, then $d(x_m, x_n) < \frac{1}{2}$. Therefore, if $n \geq N$, $x_n = x_N$, and thus $\{x_n\}$ converges to x_N.

Theorem 46.4   The metric space $\mathbf{R}^n$ is complete.

Proof. Let $\{a^{(k)}\}$ be a Cauchy sequence in $\mathbf{R}^n$. Let $\varepsilon > 0$. There exists a positive integer N such that if $k,m \geq N$, then

$$\sqrt{\sum_{i=1}^{n} (a_i^{(k)} - a_i^{(m)})^2} = d(a^{(k)}, a^{(m)}) < \varepsilon$$

If j is a positive integer with $1 \leq j \leq n$, we have

$$|a_j^{(k)} - a_j^{(m)}| \leq \sqrt{\sum_{i=1}^{n} (a_i^{(k)} - a_i^{(m)})^2} < \varepsilon$$

for $k,m \geq N$. Thus for $1 \leq j \leq n$, the real sequence $\{a_j^{(k)}\}_{k=1}^{\infty}$ is a Cauchy sequence. By Theorem 19.3, $\{a_j^{(k)}\}_{k=1}^{\infty}$ converges for $1 \leq j \leq n$. By Theorem 37.2, $\{a^{(k)}\}$ converges, and therefore $\mathbf{R}^n$ is complete. ∎

A variation of the method used to prove that $\mathbf{R}^n$ is complete may be used to prove that l^1 (as well as l^2, c_0, and l^{∞}) is complete.

Theorem 46.5   The metric space l^1 is complete.

Proof. Let $\{a^{(n)}\}$ be a Cauchy sequence of points in l^1. Let $\varepsilon > 0$. There exists a positive integer N such that if $m,n \geq N$, then

$$|a_i^{(m)} - a_i^{(n)}| \leq \sum_{k=1}^{\infty} |a_k^{(m)} - a_k^{(n)}| = d(\{a^{(m)}\}, \{a^{(n)}\}) < \varepsilon \qquad (46.1)$$

for any positive integer i. Thus for any positive integer i, $\{a_i^{(n)}\}_{n=1}^{\infty}$ is a

Cauchy sequence in $\mathbf{R}$. By Theorem 19.3, $\{a_i^{(n)}\}_{n=1}^{\infty}$ is convergent. We let $a_i = \lim_{n \to \infty} a_i^{(n)}$.

From equation (46.1), we have

$$\sum_{k=1}^{\infty} |a_k^{(n)}| \leq \sum_{k=1}^{\infty} |a_k^{(n)} - a_k^{(N)}| + \sum_{k=1}^{\infty} |a_k^{(N)}| < \varepsilon + \sum_{k=1}^{\infty} |a_k^{(N)}|$$

if $n \geq N$. Thus for any positive integer p, if $n \geq N$,

$$\sum_{k=1}^{p} |a_k^{(n)}| < T$$

where $T = \varepsilon + \sum_{k=1}^{\infty} |a_k^{(N)}|$. Taking the limit on n, we have

$$\sum_{k=1}^{p} |a_k| \leq T$$

for every positive integer p. By Theorem 24.1, $\{a_k\} \in l^1$.

Again using equation (46.1), we have for any positive integer p,

$$\sum_{k=1}^{p} |a_k^{(m)} - a_k^{(n)}| < \varepsilon$$

if $m, n \geq N$. Taking the limit on m, we have

$$\sum_{k=1}^{p} |a_k - a_k^{(n)}| \leq \varepsilon \qquad \text{for} \quad n \geq N$$

Taking the limit on p, we have

$$d(\{a_k\}, \{a_k^{(n)}\}) = \sum_{k=1}^{\infty} |a_k - a_k^{(n)}| \leq \varepsilon$$

if $n \geq N$, and hence $\{a^{(n)}\}_{n=1}^{\infty}$ converges to $\{a_k\}_{k=1}^{\infty}$ in l^1. ∎

Definition 46.6 Let (M, d) be a metric space. We say that a metric space (M_1, d_1) is a *completion* of (M, d) if

1. (M_1, d_1) is a complete metric space
2. $M \subset M_1$.
3. $d(x, y) = d_1(x, y)$ for all x and y in M.

Theorem 46.7 Every metric space (M, d) has a completion. Moreover there exists a completion (M_1, d_1) of (M, d) such that $\overline{M} = M_1$, i.e., M is dense in M_1 (see Definition 47.1).

Proof. Let (M, d) be any metric space. Define

$$\mathscr{L} = \{\{x_n\}_{n=1}^{\infty} \mid \{x_n\}_{n=1}^{\infty} \text{ is a Cauchy sequence in } (M, d)\}$$

We say that two Cauchy sequences $\{x_n\}$ and $\{y_n\}$ in (M, d) are equivalent and write $\{x_n\} \sim \{y_n\}$ if $\lim_{n \to \infty} d(x_n, y_n) = 0$.

$\sim$ is an equivalence relation on $\mathscr{L}$, that is, $\sim$ satisfies

1. $\{x_n\} \sim \{x_n\}$ for all $\{x_n\} \in \mathscr{L}$.
2. If $\{x_n\} \sim \{y_n\}$, then $\{y_n\} \sim \{x_n\}$ for $\{x_n\}$ and $\{y_n\}$ in $\mathscr{L}$.
3. If $\{x_n\} \sim \{y_n\}$ and $\{y_n\} \sim \{z_n\}$, then $\{x_n\} \sim \{z_n\}$ for $\{x_n\}$, $\{y_n\}$, and $\{z_n\}$ in $\mathscr{L}$.

(1) to (3) are easily verified.

The relation $\sim$ partitions $\mathscr{L}$ into disjoint equivalence classes X, i.e., $\{x_n\}$ and $\{y_n\}$ are in X if and only if $\{x_n\} \sim \{y_n\}$.

We define

$$M_1 = \{X \mid X \text{ is an equivalence class of the set } \mathscr{L} \text{ under the equivalence relation } \sim\}$$

We now make M_1 into a metric space by defining $d_1 : X \times X \to [0, \infty)$ by

$$d_1(X, Y) = \lim_{n \to \infty} d(x_n, y_n)$$

where $\{x_n\}_{n=1}^{\infty} \in X$ and $\{y_n\}_{n=1}^{\infty} \in Y$.

CLAIM 1. (M_1, d_1) is a metric space.

4. $d_1(X, Y) \in [0, \infty)$ and is independent of the elements $\{x_n\}$ and $\{y_n\}$ chosen from X and Y, since first if $\{x_n\} \in X$ and $\{y_n\} \in Y$, we have for each $\varepsilon > 0$ there exists a positive integer N such that $d(x_n, x_m) < \varepsilon/2$ and $d(y_n, y_m) < \varepsilon/2$ for all $n, m \geq N$. Therefore using the triangle inequality we get

$$d(x_n, y_n) \leq d(x_n, x_m) + d(x_m, y_m) + d(y_m, y_n)$$

and

$$d(x_m, y_m) \leq d(x_m, x_n) + d(x_n, y_n) + d(y_n, y_m)$$

which implies

$$d(x_n, y_n) - d(x_m, y_m) \leq d(x_n, x_m) + d(y_m, y_n)$$
$$d(x_m, y_m) - d(x_n, y_n) \leq d(x_n, x_m) + d(y_m, y_n)$$

and this implies

$$|d(x_n, y_n) - d(x_m, y_m)| \leq d(x_n, x_m) + d(y_m, y_n) < \frac{\varepsilon}{2} + \frac{\varepsilon}{2} = \varepsilon$$

This implies that the real sequence $\{d(x_n, y_n)\}_{n=1}^{\infty}$ is a Cauchy sequence; therefore, since $\mathbf{R}$ is complete, $\lim_{n \to \infty} d(x_n, y_n)$ exists, and since each $d(x_n, y_n) \in [0, \infty)$, $\lim_{n \to \infty} d(x_n, y_n) \in [0, \infty)$. Now if $\{x_n'\} \in X$ and $\{y_n'\} \in Y$ are any other elements of X and Y, we have $\lim_{n \to \infty} d(x_n, x_n') = 0$ and $\lim_{n \to \infty} d(y_n, y_n') = 0$; therefore,

$$|d(x_n, y_n) - d(x_n', y_n')| \le d(x_n, x_n') + d(y_n, y_n')$$

Thus
$$\lim_{n \to \infty} d(x_n, y_n) = \lim_{n \to \infty} d(x_n', y_n')$$

so $d_1(X, Y)$ is well defined.

5. $d_1(X, Y) = d_1(Y, X)$ for all X and Y in M_1 is obviously true.

6. $d_1(X, Y) \le d_1(X, Z) + d_1(Z, Y)$ for all X, Y, and Z in M_1 since

$$d_1(X, Y) = \lim_{n \to \infty} d(x_n, y_n) \le \lim_{n \to \infty} [d(x_n, Z_n) + d(Z_n, y_n)]$$

$$= \lim_{n \to \infty} d(x_n, z_n) + \lim_{n \to \infty} d(z_n, y_n) = d_1(X, Z) + d_1(Z, Y)$$

using the triangle inequality for (M, d) and 4.

By (4)–(6) we have that (M_1, d_1) is a metric space.

CLAIM 2. (M_1, d_1) is a complete metric space.

7. Let $\{X_n\}_{n=1}^{\infty}$ be a Cauchy sequence in (M_1, d_1). We must find $X \in M_1$ such that $\lim_{n \to \infty} X_n = X$.

First we prove a useful result. If S is a subset of M, we define $\rho(S) = $ lub $\{d(x, y) \mid x$ and y are elements of $S\}$. ρ is called the *diameter of S* in (M, d).

Take $X \in M_1$. We want to show that for every $\varepsilon > 0$, there exists a Cauchy sequence $\{x_n\} \in X$ such that the diameter of $\{x_n\}$ in (M, d) is less than ε, i.e., $\rho(\{x_n\}) < \varepsilon$.

Choose any $\{y_n\}_{n=1}^{\infty} \in X$. Since $\{y_n\}$ is a Cauchy sequence, there exists a positive integer N such that for $m, n \ge N$, $d(y_n, y_m) < \varepsilon$. Now the subsequence $\{y_{N+i}\}_{i=1}^{\infty}$ has diameter less than ε in (M, d). Define $\{x_n\}_{n=1}^{\infty}$ by $x_i = y_{N+i}$ for $i = 1, 2, \dots$. Now $\{x_n\} \in X$ since $\lim_{n \to \infty} d(x_n, y_n) = \lim_{n \to \infty} d(y_{N+n}, y_n) = 0$. For each X_n in our given sequence choose $\{x_i^{(n)}\}_{i=1}^{\infty} \in X_n$ such that $\rho(\{x_i^{(n)}\}_{i=1}^{\infty}) < 1/n$. Recall $d_1(X_n, X_m) = \lim_{i \to \infty} d(x_i^{(n)}, x_i^{(m)})$. Consider the sequence $\{x_1^{(n)}\}_{n=1}^{\infty}$. Let $\varepsilon > 0$. Since $\{X_n\}_{n=1}^{\infty}$ is a Cauchy sequence in M_1, there exists a positive integer N such that if $m, n \ge N$, $d_1(X_n, X_m) < \varepsilon/3$. Choose a positive integer L, such that $L > N$ and $1/L < \varepsilon/3$. Now for each $m, n \ge L$ we have that $d_1(X_n, X_m) < \varepsilon/3$, i.e., $\lim_{i \to \infty} d(x_i^{(n)}, x_i^{(m)}) < \varepsilon/3$. Thus there exists a positive integer k, such that $d(x_k^{(n)}, x_k^{(m)}) < \varepsilon/3$. Now

$$d(x_1^{(m)}, x_1^{(n)}) \le d(x_1^{(m)}, x_k^{(m)}) + d(x_k^{(m)}, x_k^{(n)}) + d(x_k^{(n)}, x_1^{(n)})$$

$$< \frac{1}{m} + \frac{\varepsilon}{3} + \frac{1}{n} < \frac{\varepsilon}{3} + \frac{\varepsilon}{3} + \frac{\varepsilon}{3} = \varepsilon$$

which implies $\{x_1^{(n)}\}_{n=1}^{\infty}$ is a Cauchy sequence in (M, d). Therefore by our definition of M_1, there exists $X \in M_1$ such that $\{x_1^{(n)}\}_{n=1}^{\infty} \in X$.

We now prove that $\lim_{n \to \infty} d_1(X_n, X) = 0$. A picture is handy at this point. We have

$$(x_1^{(1)}, x_2^{(1)}, x_3^{(1)}, \ldots) \in X_1$$
$$(x_1^{(2)}, x_2^{(2)}, x_3^{(2)}, \ldots) \in X_2$$
$$\ldots\ldots\ldots\ldots\ldots\ldots\ldots\ldots\ldots$$
$$(x_1^{(n)}, x_2^{(n)}, x_3^{(n)}, \ldots) \in X_n$$
$$\ldots\ldots\ldots\ldots\ldots\ldots\ldots\ldots\ldots$$
$$(x_1^{(1)}, x_1^{(2)}, x_1^{(3)}, \ldots) \in X$$

and we are showing that $X_n \to X$ in (M_1, d_1).

Now $\lim_{n \to \infty} d_1(X_n, X) = \lim_{n \to \infty} \lim_{i \to \infty} d(x_i^{(n)}, x_1^{(i)})$. Using L from the previous argument, we have that if $n, i \geq L$, then there exists J such that $d(x_J^{(n)}, x_J^{(i)}) < \varepsilon/3$. Now if $n, i \geq L$, then

$$d(x_i^{(n)}, x_1^{(i)}) \leq d(x_i^{(n)}, x_J^{(n)}) + d(x_J^{(n)}, x_J^{(i)}) + d(x_J^{(i)}, x_1^{(i)})$$

$$< \frac{1}{n} + \frac{\varepsilon}{3} + \frac{1}{i} < \frac{1}{L} + \frac{\varepsilon}{3} + \frac{1}{L} < \frac{\varepsilon}{3} + \frac{\varepsilon}{3} + \frac{\varepsilon}{3} = \varepsilon$$

which implies $\lim_{n \to \infty} d_1(X_n, X) = 0$. Therefore, (M_1, d_1) is complete.

8. We now have two metric spaces (M, d) and (M_1, d_1). We now imbed M into M_1 in such a way as to preserve its metric.

For each $x \in M$, the sequence $\{x, x, \ldots\}$ is a Cauchy sequence in M; so there exists an $X \in M_1$ such that $\{x, x, \ldots\} \in X$. We define $\Pi(x) = X$. $\Pi: M \to M_1$ and for each pair $x, y \in M$ we have $\Pi(x) = X$ and $\Pi(y) = Y$, but also $d_1(\Pi(x), \Pi(y)) = \lim_{n \to \infty} d(x, y) = d(x, y)$ by definition of d_1 and the fact that $\{x, x, \ldots\} \in X$ and $\{y, y, \ldots\} \in Y$. Therefore M and $\Pi(M)$ are *identical* as metric spaces, since Π is a one-to-one, onto, metric preserving mapping from M onto $\Pi(M)$. (We say M and $\Pi(M)$ are *isomorphic metric spaces*, and Π is a metric space *isomorphism*.) But $\Pi(M) \subset M_1$ and M_1 is complete; so M_1 is a completion of M, i.e., M_1 is a completion of M, using the identifying function Π.

Our last step is to show that $\overline{\Pi(M)} = M_1$. Let $X \in M_1$. We must show that there exists a sequence $\{X_n\}_{n=1}^{\infty}$ in $\Pi(M)$ such that $X_n \to X$ in M_1. Take any $\{x_n\}_{n=1}^{\infty} \in X$. Define $X_i = \Pi(x_i)$. Recall that $(x_i, x_i, x_i, \ldots) \in \Pi(X_i)$. Now $X_i \to X$, since $\lim_{i \to \infty} d_1(X_i, X) = \lim_{i \to \infty} \lim_{n \to \infty} d(x_i, x_n) = 0$ since $\{x_i\}_{i=1}^{\infty}$ is a Cauchy sequence in M. Therefore $\overline{\Pi(M)} = M_1$. We call (M_1, d_1) *the* completion of the metric space (M, d). Recalling that it is not true *directly* that $M \subset M_1$, but under the metric space isomorphism Π, it is true that $\Pi(M) \subset M_1$ and $\overline{\Pi(M)} = M_1$. ∎

Example 46.8 Let us apply our results from Theorem 46.7 to the metric space $(M, d) = (Q, d)$, where Q is the set of rational numbers and $d(x, y) =$

$|x - y|$ for all $x,y \in Q$ is the usual Euclidean distance. Now M_1 will be the collection of all equivalence classes of Cauchy sequences of rational numbers. If X and $Y \in M_1$, then by taking any $\{r_n\} \in X$ and $\{q_n\} \in Y$, we have $d_1(X, Y) = \lim_{n \to \infty} |r_n - q_n|$.

The mapping $\Pi : Q \to M_1$ is, for example, $\Pi(\frac{1}{2}) = X$, where X is the set of all Cauchy sequences of rational numbers which converge to $\frac{1}{2}$. Note that $(\frac{1}{2}, \frac{1}{2}, \frac{1}{2}, \ldots)$ is one such sequence.

(M_1, d_1) can be shown to satisfy Definition 3.2. That is, $M_1 = \mathbf{R}$. Thus the proof of Theorem 46.7 can be used to construct the real numbers from the rational numbers.

Exercises

46.1 Prove that every finite subset of a metric space is complete.

46.2 Give an example of a complete metric space which is not compact.

46.3 Give an example of a connected metric space which is not complete.

46.4 Give an example of a complete metric space which is not connected.

46.5 Let M be a metric space.
    (a)   Prove that if C is a complete subset of M, then C is closed.
    (b)   Prove that if M is complete, then every closed subset of M is complete.

46.6 Let $\{a_n\}$ be a Cauchy sequence in a metric space M. Prove that $\{a_n \mid n \in \mathbf{P}\}$ is bounded.

46.7 Prove that a compact metric space is complete.

46.8 Prove that l^2, c_0, and l^∞ are complete metric spaces.

46.9 Let M_1 and M_2 be complete metric spaces. Prove that the product metric space $M_1 \times M_2$ is complete.

46.10 Let $\{X_n\}$ be a sequence of closed and bounded subsets of a complete metric space such that $X_n \supset X_{n+1}$ for every positive integer n and $\lim_{n \to \infty}$ (diam $X_n) = 0$. (See Exercise 43.4.) Prove that $\bigcap_{n=1}^{\infty} X_n$ contains exactly one point. Exercise 46.10 is a generalization of the nested interval theorem for $\mathbf{R}$ (see Exercise 16.8).

46.11 (Banach's theorem) Let f be a contraction mapping on a complete metric space M. (See Exercise 44.7 for the definition of a contraction mapping.)
    (a)   Let $x_1 \in M$. Let $x_{n+1} = f(x_n)$ for $n = 1, 2, \ldots$. Prove that $\{x_n\}$ is a Cauchy sequence in M.
    (b)   Prove that if $x = \lim_{n \to \infty} x_n$, where $\{x_n\}$ is the sequence defined in part (a), then $f(x) = x$.
    (c)   Prove that there is only one point $x \in M$ such that $f(x) = x$.
    Banach's theorem provides another example of a fixed point theorem. (See also Exercise 42.12.)

46.12 Let M_1 and M_2 be metric spaces such that M_2 is complete. Let f be a uniformly continuous function from a subset X of M_1 into M_2. Suppose that $\bar{X} = M_1$. Prove that f has a unique **uniformly** continuous extension from M_1 into M_2 (that is, prove that there exists a unique uniformly continuous function g from M_1 into M_2 such that $g \mid X = f$.)

46.13　We say that a metric space M is *totally bounded* if for every $\varepsilon > 0$, there exist $x_1, \ldots, x_n \in M$ such that $M = B_\varepsilon(x_1) \cup \cdots \cup B_\varepsilon(x_n)$.

(a)　Prove that if M is a totally bounded metric space, then M is bounded. Give an example to show that the converse is false.

(b)　A metric space in which every sequence has a Cauchy subsequence is said to be *conditionally compact*. Prove that a metric space M is conditionally compact if and only if M is totally bounded.

(c)　Let M be a metric space. Prove that M is compact if and only if M is complete and totally bounded.

46.14　Find the completion of each of the following spaces with the given metric and justify your answer.

(a)　$M = \{(x, y) \in \mathbf{R}^2 \mid x, y \in Q\}$ with the usual metric for $\mathbf{R}^2$.

(b)　M as in (a) with the discrete metric for $\mathbf{R}^2$.

(c)　$M = \{(x, \sin(1/x) \in \mathbf{R}^2 \mid 0 < x < 1\}$ with the usual metric for $\mathbf{R}^2$.

(d)　$M = \{\{a_n\} \in l^1 \mid$ there exists $N \in \mathbf{P}$ such that $a_n = 0$ for all $n \geq N\}$ with the usual metric for l^1.

47.　Baire Category Theorem

We begin with a definition.

Definition 47.1　Let M be a metric space and let X be a subset of M. We say that X is *dense* in M if $\bar{X} = M$.

Thus X is dense in (M, d) if whenever $x \in M$ and $\varepsilon > 0$, there exists $y \in X$ such that $d(x, y) < \varepsilon$. Alternatively, X is dense in M if whenever $x \in M$ and $\varepsilon > 0$, we have $B_\varepsilon(x) \cap X \neq \varnothing$. For example, the set of rational numbers (or the set of irrational numbers) is dense in $\mathbf{R}$.

We come now to our main theorem from which we will deduce the Baire category theorem (Theorem 47.5). The proof is illustrated in Figure 47.1.

Theorem 47.2　If $\{U_n\}$ is a sequence of dense open sets in a complete metric space M, then $\bigcap_{n=1}^\infty U_n$ is dense in M.

Proof.　Let $x \in M$ and let $\varepsilon > 0$. We must show that there exists $y \in \bigcap_{n=1}^\infty U_n$ such that $d(x, y) < \varepsilon$. Since U_1 is dense in M, there exists $y_1 \in U_1 \cap B_\varepsilon(x)$. Since $U_1 \cap B_\varepsilon(x)$ is open, there exists an open ball $B_{\varepsilon_1}(y_1)$ such that $B_{\varepsilon_1}(y_1) \subset U_1 \cap B_\varepsilon(x)$. Let $\delta_1 = \min\{\varepsilon_1/2, 1\}$. Since U_2 is dense

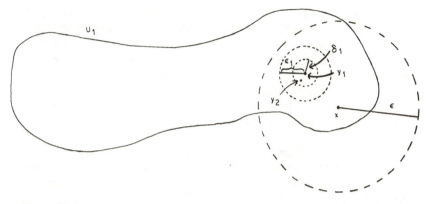

Figure 47.1

in M, there exists $y_2 \in U_2 \cap B_{\delta_1}(y_1)$. Since $U_2 \cap B_{\delta_1}(y_1)$ is open, there exists an open ball $B_{\varepsilon_2}(y_2)$ such that $B_{\varepsilon_2}(y_2) \subset U_2 \cap B_{\delta_1}(y_1)$. Let $\delta_2 = \min \{\varepsilon_2/2, 1/2\}$. We continue this process inductively. At the next step we choose $y_3 \in U_3 \cap B_{\delta_2}(y_2)$ and an open ball $B_{\varepsilon_3}(y_3) \subset U_3 \cap B_{\delta_2}(y_2)$ and we let $\delta_3 = \min \{\varepsilon_3/2, 1/3\}$.

We now have a sequence $\{y_n\}$ and open balls $\{B_{\varepsilon_n}(y_n)\}$ and $\{B_{\delta_n}(y_n)\}$ satisfying

$$B_{\varepsilon_{n+1}}(y_{n+1}) \subset U_{n+1} \cap B_{\delta_n}(y_n)$$

where
$$\delta_n = \min \left\{\frac{\varepsilon_n}{2}, \frac{1}{n}\right\}$$

for $n = 1, 2, \ldots$. Now $B_{\delta_{n+1}}(y_{n+1}) \subset B_{\varepsilon_{n+1}}(y_{n+1}) \subset B_{\delta_n}(y_n)$, thus

$$y_m \in B_{\delta_n}(y_n) \qquad \text{if } m > n \tag{47.1}$$

Since $\delta_n \leq 1/n$, we have $d(y_m, y_n) < 1/n$ if $m > n$, and therefore $\{y_n\}$ is a Cauchy sequence. Since M is a complete metric space, $\lim_{n \to \infty} y_n = y$ for some y in M. It follows from (47.1) that $y \in B_{\delta_n}(y_n)^-$ for $n = 1, 2, \ldots$. Since $\delta_n \leq \varepsilon_n/2$, we have

$$y \in B_{\delta_n}(y_n)^- \subset B_{\varepsilon_n/2}(y_n)^- \subset B_{\varepsilon_n}(y_n) \subset U_n$$

for $n = 1, 2, \ldots$. Therefore $y \in \bigcap_{n=1}^{\infty} U_n$. Finally, $y \in B_{\varepsilon_1}(y_1) \subset B_{\varepsilon}(x)$, and thus $d(x, y) < \varepsilon$. ∎

Definition 47.3 Let M be a metric space and let X be a subset of M. We say that X is *nowhere dense* (in M) if $X^{-\prime -} = M$.

For example, a point of **R** is nowhere dense in **R**. However, a point of **P** is not nowhere dense in **P**. These examples show that the concept of a set X being nowhere dense is defined only relative to a metric space of which X is a subset. This situation is different from such concepts as compactness or connectedness. A set X is compact or connected in and of itself. It is immaterial in which metric space it resides.

Before giving the next definition we note that a closed set X is nowhere dense in M if its complement is dense in M.

Definition 47.4 Let M be a metric space and let X be a subset of M. We say that X is of *first category* (in M) if X is a countable union of nowhere dense sets. We say that X is of *second category* (in M) if X is not of first category (in M).

Just as the definition of X being nowhere dense is relative to the metric space M of which X is a subset, so too are the concepts of first and second category relative to the containing metric space M.

It is easy to see that a countable union of sets of first category (in M) is of first category (in M). We pointed out before that a point is nowhere dense in **R**, and thus any countable subset of **R** is of first category in **R**. In particular, the set of rational numbers is of first category in **R**. If the set of irrational numbers were of first category in **R**, then $\mathbf{R} = \mathbf{Q} \cup \mathbf{Q}'$ would be of first category. This is not, however, the case as the Baire category theorem shows.

Theorem 47.5 (Baire Category Theorem) If M is a nonempty complete metric space, then M is of second category (in M).

Proof. Suppose that M is of first category. Then $M = \bigcup_{n=1}^{\infty} A_n$, where each A_n is nowhere dense in M. Now $M = \bigcup_{n=1}^{\infty} A_n^-$ and thus $\varnothing = \bigcap_{n=1}^{\infty} (A_n^{-\prime})$. Now each set $A_n^{-\prime}$ is an open dense subset of M; so by Theorem 47.2, $\varnothing = \bigcap_{n=1}^{\infty} (A_n^{-\prime})$ is dense in M. This contradiction establishes the theorem. ∎

The Baire category theorem has many applications in analysis. The usual application is to show that a point x of a complete metric space exists with a particular property P. A typical argument runs as follows. Let $X = \{x \in M \mid x$ does not have property $P\}$. By some argument, we show that X is of first category. Since M is of second category, there exists $x \in M \cap X'$. Thus x has property P.

We give one interesting application of the Baire category theorem.

Definition 47.6 Let M be a metric space and let $x \in M$. We say that x is an *isolated point* of M if for some $\varepsilon > 0$, $B_\varepsilon(x) \cap M = \{x\}$.

Theorem 47.7   If M is a complete metric space with no isolated points, then M is uncountable.

Proof.   Suppose M is countable. Then $M = \bigcup_{n=1}^{\infty} \{a_n\}$, and each set is nowhere dense, since M has no isolated points. Thus M is of first category, and this contradicts Theorem 47.5.   ■

Corollary 47.8   If M is a compact metric space with no isolated points, then M is uncountable.

Proof.   A compact metric space is complete. Now apply Theorem 47.7.   ■

Exercises

47.1   Use the Baire category theorem to prove that irrational numbers exist.

47.2   Do Exercise 9.10 using the Baire category theorem.

47.3   Prove that $\{0\}$ is of first category in **R**, but that $\{0\}$ is of second category in itself.

47.4   Assuming Theorem 47.5 (Baire category theorem), deduce Theorem 47.2.

47.5   If M is a metric space and X is a subset of M, we say that X is a G_δ set if X is a countable intersection of open sets. Prove that the set of rational numbers is not a G_δ set in **R**.

47.6   If f is a real-valued function which is bounded on an open interval I containing a, we let

$$\omega_f(I) = \text{lub } \{f(x) \mid x \in I\} - \text{glb } \{f(x) \mid x \in I\}$$

and $\omega_f(a) = \text{glb } \{\omega_f(J) \mid J \text{ is an open interval containing } a\}$

(a)   Prove that f is continuous at a if and only if $\omega_f(a) = 0$.

(b)   Prove that if $\varepsilon > 0$, then $\{x \mid \omega_f(x) < \varepsilon\}$ is open.

(c)*   Prove that there is no function f on **R** such that f is continuous at every rational point and discontinuous at every irrational point.

In Section 33 (Exercise 33.4), we defined a function which is continuous at every irrational point and discontinuous at every rational point.

47.7   (a)   Let M_1 be a compact metric space and let (M_2, d) be a complete metric space. Let $\mathscr{C}(M_1, M_2)$ denote the set of all continuous functions from M_1 into M_2. Prove that

$$d^1(f, g) = \text{lub } \{d(f(x), g(x)) \mid x \in M_1\}$$

defines a metric on $\mathscr{C}(M_1, M_2)$ and that $\mathscr{C}(M_1, M_2)$ is a complete metric space.

(b)*   Let M_2 and (M_1, d) be metric spaces and let f be a continuous function from M_1 into M_2. We say that f is an ε-*mapping* if whenever $x, x' \in M_1$ and $f(x) = f(x')$, we have $d(x, x') < \varepsilon$. ("f is within ε of being one to one.") Prove the following result due to Hurewicz (see Hurewicz and Wallman, 1941). Let M_1 be a compact metric space and

let M_2 be a complete metric space. If for every $\varepsilon > 0$, the set G_ε of
ε-mappings of M_1 into M_2 is dense in $\mathscr{C}(M_1, M_2)$, then M_1 is homeomorphic to a subset of M_2.

47.8* Prove that the closed interval $[a, b]$ cannot be written as a disjoint union of
closed intervals of length less than $b - a$.

47.9* Let $f(x, y)$ be a function from $\mathbf{R}^2$ into $\mathbf{R}$ such that for each $x \in \mathbf{R}$, $f_x(y) = f(x, y)$ is a polynomial of finite degree in y (the degree dependent only on x).
Also for each $y \in \mathbf{R}$, $f_y(x) = f(x, y)$ is a polynomial of finite degree in x
(the degree dependent only on y). Prove that $f(x, y)$ is a finite polynomial in
x and y, that is,

$$f(x, y) = \sum_{i=0}^{n} \sum_{j=0}^{m} a_{ij} x^i y^j \qquad \text{for all } x \text{ and } y$$

VIII

Differential Calculus of the Real Line

In this chapter we review many of the topics which are covered in introductory calculus texts.

48. Basic Definitions and Theorems

Definition 48.1 Let f be a real-valued function defined at all points of an open interval (a, b) in $\mathbf{R}$.

For each x in (a, b), we define

$$f'(x) = \lim_{y \to x} \frac{f(y) - f(x)}{y - x}$$

if this limit exists. If $f'(x)$ is defined, we say f is *differentiable at x*, and we call the number $f'(x)$ the *derivative of f at x*. If $f'(x)$ is not defined, we say f is *not differentiable* at x. If f is differentiable at each point of (a, b), we say that f is *differentiable on (a, b)*.

In dealing with functions which are defined only on closed intervals $[a, b]$ it is sometimes useful to use the concept of right and left derivatives at the endpoints b and a; so we have the following definition.

Definition 48.2 Let f be a real-valued function defined at all points of a closed interval $[a, b]$ in $\mathbf{R}$.

For each x in $[a, b]$, we define

$$f'_r(x) = \lim_{y \to x^+} \frac{f(y) - f(x)}{y - x} \quad \text{and} \quad f'_l(x) = \lim_{y \to x^-} \frac{f(y) - f(x)}{y - x}$$

if these limits exist. $f'_r(x)$ and $f'_l(x)$ are called the *right* and *left derivatives* of f at x (when they are defined).

By Theorem 32.2 we see that $f'(x)$ exists if and only if $f'_r(x)$ and $f'_l(x)$ exist and $f'_r(x) = f'_l(x) \ [=f'(x)]$.

We say a real-valued function f is differentiable on the closed interval $[a, b]$ with $a < b$ if f is differentiable on (a, b) and $f'_r(a)$ and $f'_l(b)$ both exist.

Examples 48.3   (a) Let $f(x) = |x|$ on $[-1, 1]$. Then $f'_r(0) = 1$ and $f'_l(0) = -1$ both exist, but $f'(0)$ does not exist, since $f'_r(0) \neq f'_l(0)$. Note that

$$f'(x) = -1 \qquad \text{for all } -1 < x < 0$$
$$f'(x) = 1 \qquad \text{for all } 0 < x < 1$$
$$f'_r(-1) = -1 \qquad \text{and} \qquad f'_l(1) = 1$$

(Verify.) (See Figure 48.1.)

  (b) Let

$$f(x) = \begin{cases} 0 & \text{for} \quad -2 \le x \le -1 \\ +\sqrt{1 - x^2} & \text{for} \quad -1 < x \le 1 \end{cases}$$

(See Figure 48.2.)

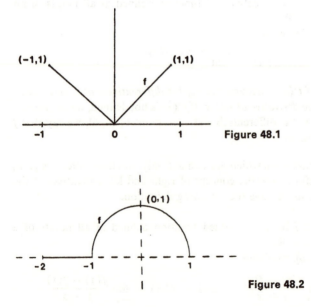

Figure 48.1

Figure 48.2

Verify that $f'_l(-1) = 0$ and $f'_r(-1)$ does not exist and that therefore $f'(-1)$ does not exist. Verify that $f'(x) = 0$ for all $-2 < x < -1$ and $f'(x) = -x/\sqrt{1-x^2}$ for all $-1 < x < 1$. Also show that $f'_l(1)$ does not exist.

Theorem 48.4 If f is differentiable at a point c, then f is continuous at c.

Proof. We are given that

$$\lim_{y \to c} \frac{f(y) - f(c)}{y - c} = f'(c)$$

exists.

But
$$\lim_{y \to c} f(y) - f(c) = \lim_{y \to c} \frac{f(y) - f(c)}{y - c} \cdot (y - c)$$

$$= \lim_{y \to c} \frac{f(y) - f(c)}{y - c} \cdot \lim_{y \to c} (y - c) = f'(c) \cdot 0 = 0$$

So $\lim_{y \to c} f(y) = f(c)$, which says f is continuous at c. ∎

The converse of Theorem 48.4 is not true. For example, Example 48.3(a) is continuous at 0, but not differentiable at 0.

Theorem 48.5 Suppose f and g are defined on an open interval (a, b) and are differentiable at a point c in (a, b). Then $f + g$, $f \cdot g$, and f/g $[g(c) \neq 0]$ are differentiable at c and

(a) $(f + g)'(c) = f'(c) + g'(c)$

(b) $(f \cdot g)'(c) = f'(c) \cdot g(c) + f(c) \cdot g'(c)$

(c) $\left(\dfrac{f}{g}\right)'(c) = \dfrac{g(c) \cdot f'(c) - g'(c) \cdot f(c)}{[g(c)]^2}$ $\qquad [g(c) \neq 0]$

Proof. (a) is left as an exercise.

(b) $(f \cdot g)'(c) = \lim_{y \to c} \dfrac{(f \cdot g)(y) - (f \cdot g)(c)}{y - c} = \lim_{y \to c} \dfrac{f(y) \cdot g(y) - f(c) \cdot g(c)}{y - c}$

$\qquad = \lim_{y \to c} \left(f(y) \dfrac{[g(y) - g(c)]}{y - c} + g(c) \dfrac{[f(y) - f(c)]}{y - c} \right)$

$\qquad = f(c) \cdot g'(c) + g(c) \cdot f'(c)$

The last equality holds by Theorem 48.4 and the algebraic properties of limits.

(c) We have

$$\left(\frac{f}{g}\right)'(c) = \lim_{y \to c} \frac{(f/g)(y) - (f/g)(c)}{y - c} = \lim_{y \to c} \frac{f(y)/g(y) - f(c)/g(c)}{y - c}$$

$$= \lim_{y \to c} \left[\frac{1}{g(y) \cdot g(c)} \cdot \left(g(c) \cdot \frac{f(y) - f(c)}{y - c} - f(c) \cdot \frac{g(y) - g(c)}{y - c}\right)\right]$$

$$= \frac{g(c) \cdot f'(c) - f(c) \cdot g'(c)}{[g(c)]^2} \qquad \blacksquare$$

Examples. (a) If $f(x) = c$ for all $x \in \mathbf{R}$ (where c is a fixed constant), then $f'(x) = 0$ for all $x \in \mathbf{R}$, since

$$f'(x) = \lim_{y \to x} \frac{f(y) - f(x)}{y - x} = \lim_{y \to x} \frac{c - c}{y - x} = 0$$

(b) If $f(x) = x$ for all $x \in \mathbf{R}$. Then $f'(x) = 1$ for all $x \in \mathbf{R}$, since

$$f'(x) = \lim_{y \to x} \frac{f(y) - f(x)}{y - x} = \lim_{y \to x} \frac{y - x}{y - x} = 1$$

(c) If $f(x) = x^2 = x \cdot x$ for all $x \in \mathbf{R}$. Then by Theorem 48.5(b) $f'(x) = 1 \cdot x + x \cdot 1 = 2x$ for all $x \in \mathbf{R}$. Using induction on n and Theorem 48.5(b), we may prove the following.

(d) If $f(x) = x^n$ for n a positive integer and for all $x \in \mathbf{R}$, then

$$f'(x) = nx^{n-1} \qquad \text{for all } x \in \mathbf{R}$$

Theorem 48.6 (The Chain Rule) Suppose f is a function on (a, b) and $f'(c)$ exists at some $c \in (a, b)$. Suppose g is defined on an interval I, which contains the range of f, and suppose g is differentiable at the point $f(c)$ in I. If we define $K(x) = (g \circ f)(x) = g(f(x))$ for all x in (a, b), then K is differentiable at c and

$$K'(c) = g'(f(c)) \cdot f'(c)$$

Proof. $$K'(c) = \lim_{y \to c} \frac{K(y) - K(c)}{y - c} = \lim_{y \to c} \frac{g(f(y)) - g(f(c))}{y - c}$$

CASE 1. If there exists an $\varepsilon > 0$ such that for all y with $0 < |y - c| < \varepsilon$, we have $f(y) \neq f(c)$, then since f is differentiable at c, f is continuous at c (Theorem 48.4); so

$$\lim_{y \to c} f(y) = f(c)$$

Therefore

$$K'(c) = \lim_{y \to c} \frac{g(f(y)) - g(f(c))}{y - c} = \lim_{\substack{y \to c \\ 0 < |y - c| < \varepsilon}} \frac{g(f(y)) - g(f(c))}{y - c}$$

$$= \lim_{\substack{y \to c \\ 0 < |y - c| < \varepsilon}} \frac{g(f(y)) - g(f(c))}{f(y) - f(c)} \cdot \frac{f(y) - f(c)}{y - c} = g'(f(c)) \cdot f'(c)$$

since $g'(f(c))$ and $f'(c)$ both exist.

CASE 2. If there exists no $\varepsilon > 0$ such that for all y with $0 < |y - c| < \varepsilon$, we have $f(y) \neq f(c)$, then this implies there exists a sequence $\{a_n\}_{n=1}^{\infty}$ such that for each n, we have $0 < |a_n - c| < 1/n$ and $f(a_n) = f(c)$. In this case we have

$$f'(c) = \lim_{y \to c} \frac{f(y) - f(c)}{y - c} = \lim_{n \to \infty} \frac{f(a_n) - f(c)}{a_n - c} = \lim_{n \to \infty} \frac{0}{a_n - c} = 0$$

Let $\{a_n\}_{n=1}^{\infty}$ be any sequence such that $\lim_{n \to \infty} a_n = c$, and $a_n \neq c$ for each n. Let $M = \{y \mid y \neq c \text{ and } f(y) = f(c)\}$. Then

$$\lim_{\substack{n \to \infty \\ a_n \in M}} \frac{g(f(a_n)) - g(f(c))}{a_n - c} = \lim_{\substack{n \to \infty \\ a_n \in M}} \frac{g(f(c)) - g(f(c))}{a_n - c} = 0$$

and $$\lim_{\substack{n \to \infty \\ a_n \notin M}} \frac{g(f(a_n)) - g(f(c))}{a_n - c} = \lim_{\substack{n \to \infty \\ a_n \notin M}} \frac{g(f(a_n)) - g(f(c))}{f(a_n) - f(c)} \cdot \frac{f(a_n) - f(c)}{a_n - c}$$

$$= g'(f(c)) \cdot f'(c) = g'(f(c)) \cdot 0 = 0$$

Therefore, we conclude that

$$\lim_{n \to \infty} \frac{g(f(a_n)) - g(f(c))}{a_n - c} = 0$$

for each sequence $\{a_n\}_{n=1}^{\infty}$ with $\lim_{n \to \infty} a_n = c$ and $a_n \neq c$ for each n. This implies, by Theorem 31.2, that

$$\lim_{y \to c} \frac{g(f(y)) - g(f(c))}{y - c} = 0$$

Therefore, $K'(c)$ exists and equals 0. But $0 = K'(c) = g'(f(c)) \cdot 0 = g'(f(c)) \cdot f'(c)$. ∎

Examples. (a) Let $K(x) = (x^2 + 3)^{1/2}$ on $(-1, 1)$. Then $K(x) = g(f(x))$, where $f(x) = x^2 + 3$ on $(-1, 1)$ and $g(y) = y^{1/2}$ on the interval $I = (3, 4)$, which is the range of $f \cdot g$ is differentiable at each point of I and

$$g'(y) = \frac{1}{2y^{\frac{1}{2}}} \qquad \text{(Verify.)}$$

f is differentiable on $(-1, 1)$ and $f'(x) = 2x$. Theorem 48.6 gives us that

$$K'(x) = g'(f(x)) \cdot f'(x)$$

for all x in $(-1, 1)$. So

$$K'(x) = \frac{1}{2(f(x))^{\frac{1}{2}}} \cdot f'(x) = \frac{1}{2(x^2 + 3)^{\frac{1}{2}}} \cdot 2x = \frac{x}{(x^2 + 3)^{\frac{1}{2}}}$$

for all x in $(-1, 1)$.

(b) Let $K(x) = (x^3 + 2)^{10}$ on **R**. Using Theorem 48.6, we see that $K'(x) = 30(x^3 + 2)^9 x^2$ for all x in **R**.

Exercises

48.1 Suppose f is a real-valued function on (a, b) and $f_l'(c)$ and $f_r'(c)$ exist for some $c \in (a, b)$. Can f be discontinuous at c? Prove your answer.

48.2 Suppose $f'(x) = 0$ for all $x \in (a, b)$. Prove that f is constant on (a, b).

48.3 Suppose f' is continuous on $[a, b]$ and $\varepsilon > 0$. Prove that there exists $\delta > 0$ such that

$$\left| \frac{f(t) - f(x)}{t - x} - f'(x) \right| < \varepsilon$$

whenever $0 < |t - x| < \delta$, with x and t in $[a, b]$.

48.4 (a) Prove that f is differentiable at a if and only if there exists a function $E(h)$ defined on an open interval $(-\delta, \delta)$ and a constant c such that

$$f(a + h) = f(a) + ch + E(h)$$

where $\lim_{h \to 0} (E(h)/h) = 0$.

(b) Prove that in case f is differentiable at a, the constant c of part (a) is equal to $f'(a)$.

(c) Use part (a) above to give another proof of the chain rule.

48.5 Suppose f is continuous on $[a, b]$ and $f_r'(x) = 0$ for all $x \in [a, b)$. Prove that f is constant on $[a, b]$.

49. Mean-Value Theorems and L'Hospital's Rule

Definition 49.1 Let f be a real-valued function defined on **R**. We say f has a *local maximum* at a point $c \in \mathbf{R}$ if there exists a $\delta > 0$ such that

$$f(x) \leq f(c) \qquad \text{for all } x \in \mathbf{R} \text{ such that } |c - x| < \delta$$

In other words in some open interval about c, $f(c)$ is the largest value which f takes on in this interval.

We say f has a *local minimum* at a point $d \in \mathbf{R}$ if there exists a $\delta > 0$ such that

$$f(x) \geq f(d) \qquad \text{for all } x \in \mathbf{R} \text{ such that } |d - x| < \delta$$

Examples 49.2 (See Figure 49.1.) (a) Let $f(x) = -x^2 + 1$ on $\mathbf{R}$. Then $c = 0$ is the only local maximum of f on $\mathbf{R}$. f has no local minima on $\mathbf{R}$.

(b) Let $g(x) = x(x^2 - 3)$ on $\mathbf{R}$. Then $c = -1$ is the only local maximum and $d = 1$ is the only local minimum for g on $\mathbf{R}$.

(c) Let $h(x) = x^5/5 - \frac{5}{3}x^3 + 4x$ on $\mathbf{R}$. Then $c_1 = -2$ and $c_2 = 1$ are the only local maxima, and $d_1 = -1$ and $d_2 = 2$ are the only local minima for h on $\mathbf{R}$.

(d) Let $k(x) = |x|$ for x in $\mathbf{R}$. Then $d = 0$ is the only local minimum for k in $\mathbf{R}$.

The following theorem is the basis of many of the applications of differentiation.

Theorem 49.3 Let f be defined on $[a, b]$. If f has a local maximum at a point $c \in (a, b)$ and $f'(c)$ exists, then $f'(c) = 0$.

Proof. Suppose $f(c)$ is a local maximum and $f'(c)$ exists. Then

$$f'(c) = f'_l(c) = \lim_{y \to c^-} \frac{f(y) - f(c)}{y - c}$$

and since $f(c) \geq f(y)$ for all y sufficiently close to c, we must have $f(y) - f(c) \leq 0$ and therefore

$$\frac{f(y) - f(c)}{y - c} \geq 0$$

for all y sufficiently close to c and with $y - c < 0$. Therefore, taking the limit,

$$f'(c) = f'_l(c) \geq 0 \qquad\qquad (49.1)$$

Now $$f'(c) = f'_r(c) = \lim_{y \to c^+} \frac{f(y) - f(c)}{y - c}$$

and again $f(y) - f(c) \leq 0$ for all y, sufficiently close to c; so $[f(y) - f(c)]/(y - c) \leq 0$ for all y sufficiently close to c and with $y - c > 0$. This implies

$$f'(c) = f'_r(c) \leq 0 \qquad\qquad (49.2)$$

Inequalities (49.1) and (49.2) give that

$$0 \leq f'_l(c) = f'(c) = f'_r(c) \leq 0; \text{ so } f'(c) = 0 \qquad \blacksquare$$

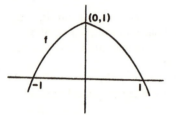

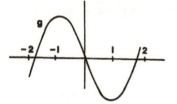

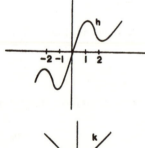

Figure 49.1

Corollary 49.4 Let f be defined on $[a, b]$. If f has a local minimum at a point d in (a, b) and if $f'(d)$ exists, then $f'(d) = 0$.

Proof. $-f$ has a local maximum at d and $(-f)'(d) = -(f'(d))$ exists; so Theorem 49.3 says $(-f)'(d) = 0$. Therefore $-(f'(d)) = 0$; so $f'(d) = 0$. ∎

If f is differentiable for all points in (a, b), then by investigating the behavior of f only at the points where $f'(x) = 0$, we are able to find all the local maxima and minima of f on (a, b). However, $f'(x) = 0$ does not always imply that $f(x)$ is either a local maximum or local minimum. For example $f(x) = x^3$ on $(-1, 1)$ has no local maximum or local minimum, however, $f'(0) = 0$. Also note that in Examples 49.2(a), (b), and (c) we could find all the local maxima and minima by finding where f', g', and h' were 0. However, in (d) the local minimum occurs at a point where k' does not exist.

Theorem 49.5 If f and g are continuous real functions on $[a, b]$ which are differentiable in (a, b), then there is a point $c \in (a, b)$ at which

$$[f(b) - f(a)] \cdot g'(c) = [g(b) - g(a)] \cdot f'(c) \qquad (49.3)$$

Proof. Define

$$h(x) = [f(b) - f(a)]g(x) - [g(b) - g(a)]f(x) \qquad \text{for all } x \in [a, b].$$

Then h is continuous on $[a, b]$ and differentiable in (a, b) and

$$h(a) = f(b)g(a) - f(a)g(b) = h(b) \qquad (49.4)$$

All we need is that there exists $c \in (a, b)$ such that $h'(c) = 0$, for then

$$0 = h'(c) = [f(b) - f(a)]g'(c) - [g(b) - g(a)]f'(c)$$

If h is a constant function, then $h'(c) = 0$ for all c in (a, b), which completes the proof.

Suppose $h(t) > h(a)$ for some t in (a, b). By Theorem 34.4, h attains a maximum at some c in $[a, b]$. But by (49.4), since $h(c) \geq h(t) > h(a)$, we must have $c \in (a, b)$; so by Theorem 49.3, $h'(c) = 0$.

If $h(t) < h(a)$ for some t in (a, b) a similar argument gives us a $c \in (a, b)$ with $h'(c) = 0$. ∎

The above theorem is often called the *generalized mean-value theorem*.

Theorem 49.6 (The Mean-Value Theorem) If f is real-valued and continuous on $[a, b]$ and f is differentiable in (a, b), then there exists a point $c \in (a, b)$ such that

$$f(b) - f(a) = (b - a)f'(c)$$

Proof. Take $g(x) = x$ in Theorem 49.5. ∎

Corollary 49.7 (Rolle's Theorem) Let f be a real-valued continuous function on $[a, b]$ which is differentiable in (a, b) and suppose $f(a) = f(b)$. Then there exists $c \in (a, b)$ such that $f'(c) = 0$. (See Figure 49.2.)

Proof. By Theorem 49.6 there exists a point $c \in (a, b)$ such that $0 = f(b) - f(a) = (b - a)f'(c)$. This implies $f'(c) = 0$. ∎

Definition 49.8 Let f be a real-valued function defined on an interval I. We say f is *increasing* (*decreasing*) *on* I if for each x and y in I with $x < y$, we have $f(x) \leq f(y)$ [$f(x) \geq f(y)$]. We say f is *strictly increasing* (*strictly decreasing*) on I if $f(x) < f(y)$ [$f(x) > f(y)$] for all x and y in I with $x < y$.

Theorem 49.9 Suppose f is real-valued and differentiable in (a, b).

(a) If $f'(x) \geq 0$ for all $x \in (a, b)$, then f is increasing on (a, b).
(b) If $f'(x) = 0$ for all $x \in (a, b)$, then f is constant on (a, b).
(c) If $f'(x) \leq 0$ for all $x \in (a, b)$, then f is decreasing on (a, b).

Proof. All conclusions can be read from the equation $f(x_2) - f(x_1) = (x_2 - x_1) \cdot f'(c)$, which is valid for each pair of points $x_1 < x_2$ in (a, b) and for some c with $x_1 < c < x_2$ (Theorem 49.6). ∎

Theorem 49.10 If $f'(x) > 0$ for all x in (a, b), then f is strictly increasing in (a, b).

Let g be the inverse of f. Then g is differentiable at each point in the range

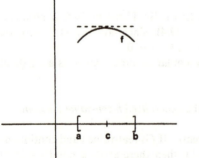

Figure 49.2

of f and

$$g'(f(x)) = \frac{1}{f'(x)} \qquad \text{for all } x \in (a, b)$$

Proof. First, for each $x_1 < x_2$ in (a, b), there exists, by Theorem 49.6, $c \in (a, b)$ such that

$$f(x_2) - f(x_1) = (x_2 - x_1)f'(c)$$

So
$$\frac{f(x_2) - f(x_1)}{x_2 - x_1} = f'(c) > 0$$

thus $f(x_2) - f(x_1) > 0$. Therefore, f is strictly increasing on (a, b). Furthermore, f has an inverse, call it g, with domain $g = (f(a), f(b))$ (verify). Recall $g(y) = x$ if and only if $f(x) = y$. f is differentiable in (a, b); so f is continuous on (a, b) by Theorem 48.4.

Let $y \in (f(a), f(b))$. Then $y = f(x)$ for a unique x in (a, b). Therefore,

$$g'(y) = g'(f(x)) = \lim_{z \to f(x)} \frac{g(z) - g(f(x))}{z - f(x)} = \lim_{z \to f(x)} \frac{g(z) - x}{z - f(x)}$$

For all z sufficiently close to $f(x)$, we have $z \in (f(a), f(b))$; so since f is one-to-one, there exists a unique x_z in (a, b) such that $f(x_z) = z$. Since f is continuous and strictly increasing,

$$\lim_{z \to f(x)} x_z = x \tag{49.5}$$

We now have

$$g'(y) = g'(f(x)) = \lim_{z \to f(x)} \frac{g(z) - x}{z - f(x)} = \lim_{z \to f(x)} \frac{g(f(x_z)) - x}{f(x_z) - f(x)}$$

$$= \lim_{z \to f(x)} \frac{x_z - x}{f(x_z) - f(x)} = \lim_{z \to f(x)} \frac{1}{[f(x_z) - f(x)]/(x_z - x)} = \frac{1}{f'(x)}$$

The last equality follows by (49.5) and the fact that $f'(x)$ exists. ∎

If we had known that g was differentiable at $f(x)$, since

$$g(f(x)) = x \qquad \text{for all } x \text{ in } (a, b) \tag{49.6}$$

we could have applied Theorem 48.6 to the left side of (49.6) and usual differentiation rule on the right to get

$$g'(f(x)) \cdot f'(x) = 1$$

or
$$g'(f(x)) = \frac{1}{f'(x)} \tag{49.7}$$

Example.   Let $f(x) = x^2$ on $(0, 2)$. Then $f'(x) = 2x > 0$ for all x in $(0, 2)$.
$g(y) = \sqrt{y}$ is the inverse of f on $(f(0), f(2)) = (0, 4)$. By Theorem 49.10,

$$g'(y) = g'(f(x)) = \frac{1}{f'(x)} = \frac{1}{2x} = \frac{1}{2\sqrt{y}}$$

since $\sqrt{y} = g(y) = x$. For example, if $y = 3$, then $f(\sqrt{3}) = 3$; so $g'(3) = g'(f(\sqrt{3})) = 1/2\sqrt{3}$.

The following theorem, which arises from Theorem 49.5, gives a simple procedure for the evaluation of limiting values of quotient functions.

Theorem 49.11 (L'Hospital's Rule)   Let f and g be differentiable on the interval $[a, b)$, with $g'(x) \neq 0$ on $[a, b)$.

(i)   If $\displaystyle\lim_{x \to b^-} f(x) = 0$ and $\displaystyle\lim_{x \to b^-} g(x) = 0$ or

(ii)   If $\displaystyle\lim_{x \to b^-} f(x) = \infty$ and $\displaystyle\lim_{x \to b^-} g(x) = \infty$

   and if $\displaystyle\lim_{x \to b^-} \frac{f'(x)}{g'(x)} = L$

   then $\displaystyle\lim_{x \to b^-} \frac{f(x)}{g(x)} = L$

The end "point" b may be finite or "∞," and L may be finite or "∞."

Proof.   We will prove the theorem under hypothesis (i), and we shall assume that b is finite; therefore for convenience we take $b = 1$.

Since $\lim_{x \to 1^-} f(x) = \lim_{x \to 1^-} g(x) = 0$, we define $f(1) = g(1) = 0$ and therefore have f and g continuous on $[a, 1]$. Applying Theorem 49.5, we have that for each $a < x < 1$, there is a point x^* with $x < x^* < 1$ such that

$$\frac{f'(x^*)}{g'(x^*)} = \frac{f(1) - f(x)}{g(1) - g(x)} = \frac{f(x)}{g(x)}$$

Since $\displaystyle\lim_{x^* \to 1^-} \frac{f'(x^*)}{g'(x^*)} = L$

we may choose $\delta_n > 0$ so that

$$\left| \frac{f(x)}{g(x)} - L \right| = \left| \frac{f'(x^*)}{g'(x^*)} - L \right| < \frac{1}{n}   \text{whenever}   1 - \delta_n < x < 1$$

This proves that $\lim_{x \to 1^-} [f(x)/g(x)] = L$ as desired. Under hypothesis (ii), we again assume $b = 1$. Given $\varepsilon > 0$, we choose δ so that whenever $1 - \delta < \bar{x} < 1$,

$$\left| \frac{f'(\bar{x})}{g'(\bar{x})} - L \right| < \varepsilon   \text{and}   f(\bar{x}) \neq 0$$

Set $x_0 = 1 - \delta$ and take any point x with $x_0 < x < 1$. By Theorem 49.5 there exists $\bar{x}$, with $x_0 < \bar{x} < 1$, such that

$$\frac{f(x) - f(x_0)}{g(x) - g(x_0)} = \frac{f'(\bar{x})}{g'(\bar{x})}$$

Therefore

$$\left| \frac{f(x) - f(x_0)}{g(x) - g(x_0)} - L \right| < \varepsilon$$

Define $h(x) = (1 - f(x_0)/f(x))/(1 - g(x_0)/g(x))$ for all x where this expression is well-defined. Then

$$\left| \frac{f(x)}{g(x)} \cdot h(x) - L \right| < \varepsilon$$

is valid for all x with $x_0 < x < 1$. Since $\lim_{x \to 1^-} f(x) = \lim_{x \to 1^-} g(x) = \infty$, it follows that $\lim_{x \to 1^-} h(x) = 1$. Choose $x_1 \in (x_0, 1)$ such that $|h(x) - 1| < \varepsilon$ and $h(x) > \frac{1}{2}$ whenever $x \in (x_1, 1)$. For such values of x, we have

$$\left| \left(\frac{f(x)}{g(x)} - L \right) \cdot h(x) \right| = \left| \frac{f(x)}{g(x)} h(x) - L \cdot h(x) \right|$$

$$\leq \left| \frac{f(x)}{g(x)} \cdot h(x) - L \right| + |L \cdot [1 - h(x)]| < \varepsilon + |L| \cdot \varepsilon$$

and

$$\left| \frac{f(x)}{g(x)} - L \right| < \frac{(1 + |L|)\varepsilon}{h(x)} < 2(1 + |L|)\varepsilon$$

This proves that $\lim_{x \to 1^-} [f(x)/g(x)] = L$. Proofs of the other modifications of the theorem are straightforward. For example, in the case $b = \infty$, we consider

$$F(x) = f\left(\frac{1}{x}\right) \quad \text{and} \quad G(x) = g\left(\frac{1}{x}\right)$$

If f and g obey (i) or (ii) for $x \to \infty$, then F and G obey them for $x \to 0^+$. Since

$$\frac{F'(x)}{G'(x)} = \frac{f'(1/x)}{g'(1/x)}$$

we may apply the previously proved form and conclude that if

$$\lim_{x \to \infty} \frac{f'(x)}{g'(x)} = L$$

then

$$\lim_{x \to \infty} \frac{f(x)}{g(x)} = L \qquad \blacksquare$$

Examples. (If you are unfamiliar with the functions sin (x) and cos (x), see Chapter XI.)
   (a) Evaluate

$$\lim_{x \to 0^-} \frac{\sin x^2}{x^4}$$

By L'Hospital's rule,

$$\lim_{x \to 0^-} \frac{\sin x^2}{x^4} = \lim_{x \to 0^-} \frac{2x \cos x^2}{4x^3} = \lim_{x \to 0^-} \frac{\cos x^2}{2x^2} = \infty$$

   (b) Evaluate

$$\lim_{x \to 0^+} \frac{1 - \cos x^2}{x^4}$$

It is clear that L'Hospital's rule may be modified to hold for right-hand limits; so applying L'Hospital's rule twice, we have

$$\lim_{x \to 0^+} \frac{1 - \cos x^2}{x^4} = \lim_{x \to 0^+} \frac{2x \sin x^2}{4x^3} = \lim_{x \to 0^+} \frac{\sin x^2}{2x^2}$$

$$= \lim_{x \to 0^+} \frac{2x \cos x^2}{4x} = \lim_{x \to 0^+} \frac{\cos x^2}{2} = \frac{1}{2}$$

Exercises

49.1 Let f be continuous on (a, b) and let $x \in (a, b)$. Suppose $f'(y)$ exists for all y in $(a, b)\backslash\{x\}$ and $\lim_{y \to x} f'(y)$ exists. Prove that $f'(x)$ exists.

49.2 Suppose $f'(x)$ exists and is bounded on all of **R**. Prove that f is uniformly continuous on **R**.

49.3 If f, g, h are continuous functions on $[a, b]$ which are differentiable on (a, b), then prove that there exists $c \in (a, b)$ such that

$$f(a)[g(b)h'(c) - h(b)g'(c)] + h(a)[f(b)g'(c) - g(b)f'(c)]$$
$$= g(a)[f(b)h'(c) - h(b)f'(c)]$$

49.4 Suppose
   (a) f is continuous on $[0, \infty)$
   (b) $f'(x)$ exists on $(0, \infty)$
   (c) $f(0) = 0$
   (d) $f'(x)$ is increasing on $(0, \infty)$.
   Define $g(x) = f(x)/x$ for all x in $(0, \infty)$ and prove that g is increasing on $(0, \infty)$.

49.5 Evaluate
   (a) $\lim_{x \to 0} \dfrac{1 - \cos x^2}{x^3 \sin x}$

(b) $\lim\limits_{x \to 0+} x^x$

(The functions $\sin x$, $\cos x$, $\log x$ and e^x are defined in Chapter XI.)

49.6 Suppose f is defined on $[a, b]$ and $c \in (a, b)$ and suppose f' exists in an open interval containing c and that $(f')'(c)$ exists. Show that

$$\lim_{h \to 0} \frac{f(c + 2h) - 2f(c + h) + f(c)}{h^2} = (f')'(c)$$

Give an example to show that the limit above may exist even though $f''(c)$ does not exist.

49.7 Let f be a function on (a, b) such that f' is bounded on (a, b). Prove that there exists $c \neq 0$ such that $g(x) = x + cf(x)$ is one to one.

49.8 Suppose that f is nonnegative and f''' exists on $(0, 1)$. Suppose also that there exist points $c \neq d$ in $(0, 1)$ such that $f(c) = 0 = f(d)$. Prove that $f'''(x) = 0$ for some $x \in (0, 1)$.

50. Taylor's Theorem

In this section we look at the problem of approximating a real-valued function f by a polynomial function P. That is, given f on (a, b), find a polynomial P such that $|f(x) - P(x)|$ is small for all x in (a, b). Of course, there are functions f where it is impossible to find P as above, but we shall develop a method applying to certain types of functions f, which produces a polynomial P, where we have some information about the accuracy of the approximation.

Definition 50.1 Let f be a real-valued function on (a, b). $f^{(1)}$ is the function on a subset of (a, b) defined by

$$f^{(1)}(x) = f'(x) \qquad \text{with domain } f^{(1)} = \{x \in (a, b) \,|\, f'(x) \text{ exists}\}$$

$f^{(2)}$ is the function

$$f^{(2)}(x) = (f^{(1)})'(x) \qquad \text{with domain } f^{(2)} = \{x \in (a, b) \,|\, (f^{(1)})'(x) \text{ exists}\}$$

$f^{(3)}, f^{(4)}, f^{(5)}, \ldots$ are defined similarly as the successive derivatives of f on (a, b). For convenience we define $f^{(0)}(x) = f(x)$ for all x in (a, b).

We always have $(a, b) \supset$ domain $f^{(1)} \supset$ domain $f^{(2)} \supset \cdots \supset$ domain $f^{(n)} \supset \cdots$.

Definition 50.2 A real-valued function f is said to be of *class* C^n on (a, b) if $f^{(n)}(x)$ exists and is continuous for all x in (a, b). Clearly, if f is of class C^n, then f must be of classes $C^{n-1}, C^{n-2}, \ldots, C^1$. If f is of class C^n for all positive integers n, then f is said to be of *class* C^∞.

Let $f \in C^n$ on (a, b) and let $c \in (a, b)$. Among all the polynomials of degree

n, there is exactly one polynomial P_n such that

$$P_n^{(k)}(c) = f^{(k)}(c) \qquad \text{for} \quad k = 0, 1, \ldots, n \tag{50.1}$$

We call this polynomial the nth degree *Taylor polynomial* for f at c. Writing $P_n(x)$ as

$$P_n(x) = A_0 + A_1(x - c) + A_2(x - c)^2 + \cdots + A_n(x - c)^n \tag{50.2}$$

and using (50.1), we find that

$$A_0 = f(c), \quad A_1 = f^{(1)}(c), \ldots, A_n = \frac{f^{(n)}(c)}{n!}$$

Therefore the nth degree Taylor polynomial for f at c is

$$P_n(x) = f(c) + f^{(1)}(c)\cdot(x - c) + \frac{f^{(2)}(c)}{2!}(x - c)^2 + \cdots + \frac{f^{(n)}(c)}{n!}(x - c)^n \tag{50.3}$$

We see that $P_n(c) = f(c)$. We now ask the question: How good an approximation is $P_n(x)$ to $f(x)$ for all x in (a, b)? Since we are concerned with the size of $f(x) - P_n(x)$ on (a, b), we estimate this difference in the following theorem.

Theorem 50.3 (Taylor's Theorem) Let $f \in C^{n+1}$ on (a, b), and let c and d be any points in (a, b). Then there exists a point t between c and d such that

$$f(d) = P_n(d) + \frac{f^{(n+1)}(t)}{(n + 1)!}(d - c)^{n+1} \tag{50.4}$$

where P_n is the nth degree Taylor polynomial for f at c.

The number t in (50.4) depends on both c and d, but the importance of (50.4) is that if

$$\left|\frac{f^{(n+1)}(t)}{(n + 1)!}\right|$$

is small for *all* t in some interval $I \subset (a, b)$ with $c \in I$, then we see that since

$$|f(d) - P_n(d)| = \left|\frac{f^{(n+1)}(t)}{(n + 1)!}\right|\cdot|d - c|^{n+1} \tag{50.5}$$

we can say that $|f(d) - P_n(d)|$ is small for all d in the interval I.

Proof. Let M be the number defined by the formula

$$f(d) = P_n(d) + M(d - c)^{n+1}$$

and define $g(x) = f(x) - P_n(x) - M(x - c)^{n+1}$ for all x in (a, b)

$$\tag{50.6}$$

We must show that there exists a number t between c and d such that $f^{(n+1)}(t) = (n+1)!M$.

We know $P_n^{(n+1)}(x) = 0$ for all x in (a, b) by definition; so (50.6) gives

$$g^{(n+1)}(x) = f^{(n+1)}(x) - (n+1)!M \qquad \text{for all } x \text{ in } (a, b) \qquad (50.7)$$

Therefore, our proof will be complete if we can find a number t between c and d such that $g^{(n+1)}(t) = 0$. Since $P_n^{(k)}(c) = f^{(k)}(c)$ for $k = 0, 1, 2, \dots, n$, we have

$$g(c) = g^{(1)}(c) = \cdots = g^{(n)}(c) = 0$$

Our choice of M shows that $g(d) = 0$. But since $g(d) = 0$ and $g(c) = 0$, the mean-value theorem states there is a number t_1 between c and d such that $g^{(1)}(t_1) = 0$. Now $g^{(1)}(t_1) = 0$ and $g^{(1)}(c) = 0$ implies by the mean-value theorem applied to $g^{(1)}$ that there exists a number t_2 between t_1 and c with $g^{(2)}(t_2) = 0$. After $n + 1$ such steps, we arrive at the conclusion that there is a number t_{n+1} between t_n and c such that $g^{(n+1)}(t_{n+1}) = 0$. By construction t_{n+1} is between c and d; so let $t = t_{n+1}$, and our proof is complete. ∎

We now give a simple example to demonstrate the way Theorem 50.3 may be used to determine a polynomial approximation of a given function.

Example. Find a polynomial P such that

$$\left| P(x) - \frac{1}{x} \right| < \frac{1}{10{,}000} \qquad \text{for all } x \text{ in } (100, 102)$$

Let us use the Taylor polynomials expanded about $c = 101$, to find the desired P. By our note above we look at $f(x) = 1/x$ and investigate

$$\left| \frac{f^{(n+1)}(t)}{(n+1)!} \right| \qquad \text{for all } t \text{ in } (100, 102)$$

But $\qquad \dfrac{f^{(n+1)}(t)}{(n+1)!} = \dfrac{(-1)^{n+1}(n+1)!\, t^{-(n+2)}}{(n+1)!} = (-1)^{n+1} t^{-(n+2)}$

so $\qquad \left| \dfrac{f^{(n+1)}(t)}{(n+1)!} \right| = |t|^{-(n+2)} < \dfrac{1}{(100)^{n+2}} \qquad \text{for all } t \text{ in } (100, 102)$

Now $|d - c|^{n+1} = |d - 101|^{n+1} \le 1^{n+1} = 1$ for all d in $(100, 102)$. By Theorem 50.3 for $n = 1$, we have for *each* d in $(100, 102)$, there is a number t_d in $(100, 102)$ such that

$$|f(d) - P_1(d)| = \left| \frac{f^{(1+1)}(t_d)}{(1+1)!} \right| |d - 101|^{1+1} < \frac{1}{(100)^3} \cdot 1 = \frac{1}{1{,}000{,}000}$$

$$P_1(x) = f(101) + \frac{f^{(1)}(101)}{1!}(x - 101) = \frac{1}{101} - \frac{1}{(101)^2}(x - 101)$$

$$= \frac{2}{101} - \frac{1}{(101)^2}x$$

and by the above we know that $|1/x - P_1(x)| < 10^{-6}$ for all x in $(100, 102)$, which is an even better approximation than we had set out for.

Exercises

50.1 Find a polynomial P which approximates the function $f(x) = \sqrt{x}$ within 10^{-8} on the interval $(4, 5)$.

50.2 Sketch the graphs of the functions $f(x) = 1/(1 + x^2)$ and $P_1(x)$, $P_2(x)$, and $P_3(x)$, the Taylor polynomials for f expanded at $c = 0$.

50.3 Let $f \in C^{n+1}$ on (a, b), let $[c, d] \subset (a, b)$, and let p be an integer, $1 \le p \le n + 1$. Let

$$F(x) = f(d) - \sum_{k=0}^{n} \frac{f^{(k)}(x)(d - x)^k}{k!}$$

and

$$G(x) = F(x) - \left(\frac{d - x}{d - c}\right)^p F(c)$$

Apply Rolle's theorem to G on $[c, d]$ to obtain

$$f(d) = \sum_{k=1}^{n} \frac{f^{(k)}(c)}{k!}(d - c)^k + R_n$$

where

$$R_n = \frac{f^{(n+1)}(t)(d - t)^{n+1-p}(d - c)^p}{p(n!)} \qquad \text{for some } t \in (c, d)$$

This formula for R_n is called *Schömilch's form of the remainder* for Taylor's series.

Show that if $p = n + 1$, we obtain Theorem 50.3. This formula for R_n is called *Lagrange's form of the remainder*. For $p = 1$, we obtain *Cauchy's form of the remainder*. Show that if $\{(d - x)/(d - c)\}^p$ is replaced by any function G where $G(c) = 1$ and $G(d) = 0$ additional forms of the remainder result. Obtain additional formulas for the remainder.

50.4 Show that if $n = 0$ in Taylor's theorem, we obtain the mean-value theorem.

50.5 Derive Leibniz's formula: If $h = fg$ and f and g are n-times differentiable, then

$$D^n(fg) = \sum_{k=0}^{n} \binom{n}{k} f^{(k)} g^{(n-k)}$$

where D^n denotes the nth derivative.

50.6 Suppose that f'' exists on $[0,1]$ and that $f(0) = 0 = f(1)$. Suppose also that $|f''(x)| \le K$ for $x \in (0,1)$. Prove that $|f'(\frac{1}{2})| \le K/4$ and that $|f'(x)| \le K/2$ for $x \in [0, 1]$.

IX

The Riemann-Stieltjes Integral

The Riemann integral solves the problem of defining the area under the graph of a continuous function f defined on a closed interval $[a, b]$.

Consider the function $f(x) = x^2$ on $[1, 2]$. (See Figure IX.1.) We may approximate the area under f from below by choosing a finite set of numbers $1 = x_0 < x_1 < \cdots < x_n = 2$ and computing the sum

$$\sum_{i=1}^{n} m_i(x_i - x_{i-1})$$

where m_i is the minimum of f on the interval $[x_{i-1}, x_i]$, for $i = 1, \ldots, n$. In this case, $m_i = f(x_{i-1})$. For example, if we choose the points $1 < \frac{4}{3} < \frac{5}{3} < 2$, we obtain the sum

$$1 \cdot \tfrac{1}{3} + \tfrac{16}{9} \cdot \tfrac{1}{3} + \tfrac{25}{9} \cdot \tfrac{1}{3} = \tfrac{50}{27} \simeq 1.85$$

This sum is the area of the rectangles in Figure IX.2.

In a similar way, we may approximate the area under f from above by choosing a finite set of numbers $1 = x_0 < x_1 < \cdots < x_n = 2$ and computing the sum

$$\sum_{i=1}^{n} M_i(x_i - x_{i-1})$$

where M_i is the maximum of f on the interval $[x_{i-1}, x_i]$, for $i = 1, \ldots, n$. One can show that the number $\frac{7}{3}$ has the property that

$$\sum_{i=1}^{n} m_i(x_i - x_{i-1}) \leq \tfrac{7}{3} \leq \sum_{i=1}^{n} M_i(x_i - x_{i-1})$$

for every set of numbers $1 = x_0 < x_1 < \cdots < x_n = 2$. Moreover, by choosing the points $1 = x_0 < x_1 < \cdots < x_n = 2$ close enough together, the upper

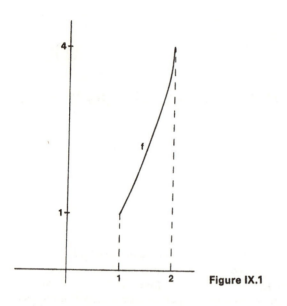

Figure IX.1

and lower approximations can be made as close to $\frac{7}{3}$ as we like. Thus, it is reasonable to define the area under the curve to be $\frac{7}{3}$.

What we have done is average the function f over an interval $[a, b]$ weighted by the length of the subintervals $[x_{i-1}, x_i]$. The Riemann-Stieltjes integral generalizes this process by allowing weights other than length. This more general integral, which includes the Riemann integral as a special case, has many applications as, for example, in the theory of probability.

51. Riemann-Stieltjes Integration with Respect to an Increasing Integrator

We begin with some definitions.

Definition 51.1 Let $[a, b]$ be a closed interval. A *partition* P of $[a, b]$ is a finite set $\{x_0, x_1, \ldots, x_n\}$ of real numbers such that

$$a = x_0 < x_1 < \cdots < x_n = b$$

The closed intervals $[x_{i-1}, x_i]$ for $i = 1, \ldots, n$ are called the *component intervals* of the partition P.

Definition 51.2 Let f be a bounded function on $[a, b]$, let α be an increasing function on $[a, b]$, and let $P = \{x_0, x_1, \ldots, x_n\}$ be a partition of $[a, b]$.

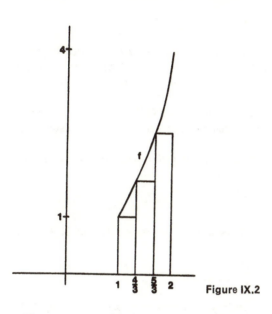

Figure IX.2

The *upper sum* of f with respect to α for the partition P is the sum

$$U(f, P) = \sum_{i=1}^{n} M_i[\alpha(x_i) - \alpha(x_{i-1})]$$

where

$$M_i = \text{lub } \{f(x) \mid x \in [x_{i-1}, x_i]\}$$

for $i = 1, \ldots, n$. The *lower sum* of f with respect to α for the partition P is the sum

$$L(f, P) = \sum_{i=1}^{n} m_i[\alpha(x_i) - \alpha(x_{i-1})]$$

where

$$m_i = \text{glb } \{f(x) \mid x \in [x_{i-1}, x_i]\}$$

for $i = 1, \ldots, n$.

We will often abbreviate $\alpha(x_i) - \alpha(x_{i-1})$ to $\Delta\alpha_i$. The number $\Delta\alpha_i$ is the "weight" assigned to the component interval $[x_{i-1}, x_i]$. The assumption that α is increasing is equivalent to assuming that the weight assigned to each component interval is nonnegative. In Section 53 we will consider arbitrary (not necessarily nonnegative) weights. If $\alpha(x) = x$, the weight assigned to a component interval $[x_{i-1}, x_i]$ is $\alpha(x_i) - \alpha(x_{i-1}) = x_i - x_{i-1}$, which is the length of the interval $[x_{i-1}, x_i]$. Thus our definitions will include the Riemann integral as a special case.

Let f and α be as in Definition 51.2. Suppose that $m \leq f(x) \leq M$ for all x in $[a, b]$. Let $P = \{x_0, x_1, \ldots, x_n\}$ be a partition of $[a, b]$. Then

$$m[\alpha(b) - \alpha(a)] = \sum_{i=1}^{n} m \, \Delta\alpha_i \leq \sum_{i=1}^{n} M_i \Delta\alpha_i = U(f, P)$$

Thus the set of upper sums for f with respect to α is bounded below by $m[\alpha(b) - \alpha(a)]$. Similarly, the set of lower sums for f with respect to α is bounded above by $M[\alpha(b) - \alpha(a)]$. This justifies our next definition.

Definition 51.3 Let f be a bounded function on $[a, b]$ and let α be an increasing function on $[a, b]$. Let $\mathcal{U}(\mathcal{L})$ denote the set of upper (lower) sums for f with respect to α.

The *upper Riemann-Stieltjes integral of f with respect to α on $[a, b]$* is the number

$$\overline{\int_a^b} f \, d\alpha = \text{glb } \mathcal{U}$$

The *lower Riemann-Stieltjes integral of f with respect to α on $[a, b]$* is the number

$$\underline{\int_a^b} f \, d\alpha = \text{lub } \mathcal{L}$$

We say that *f is Riemann-Stieltjes integrable with respect to α on $[a, b]$*, if

$$\overline{\int_a^b} f \, d\alpha = \underline{\int_a^b} f \, d\alpha$$

and in this case we define the *Riemann-Stieltjes integral of f with respect to α on $[a, b]$*, denoted

$$\int_a^b f \, d\alpha \quad \text{or} \quad \int_a^b f(x) \, d\alpha(x)$$

to be the common value $\overline{\int}_a^b f \, d\alpha = \underline{\int}_a^b f \, d\alpha$. We let $\mathcal{R}_\alpha[a, b]$ denote the set of functions which are Riemann-Stieltjes integrable with respect to α on $[a, b]$.

For the special case, $\beta(x) = x$, we let $\mathcal{R}_\beta[a, b] = \mathcal{R}[a, b]$. If $f \in \mathcal{R}[a, b]$, we say that f is *Riemann integrable on $[a, b]$*. The integral of a function f in $\mathcal{R}[a, b]$ is denoted

$$\int_a^b f \quad \text{or} \quad \int_a^b f(x) \, dx$$

Examples. (a) If $\alpha(x) = k$ for all x in $[a, b]$ and f is any bounded function on $[a, b]$, then every upper and lower sum of f is 0, and thus

$$\underline{\int_a^b} f \, d\alpha = 0 = \overline{\int_a^b} f \, d\alpha$$

(verify). Therefore $f \in \mathcal{R}_\alpha[a, b]$ and $\int_a^b f \, d\alpha = 0$.

(b) If α is an increasing function on $[a, b]$ and $f(x) = k$ for all x in $[a, b]$, then every upper and lower sum for f is equal to $k[\alpha(b) - \alpha(a)]$. Thus $f \in \mathscr{R}_{\alpha}[a, b]$ and

$$\int_a^b f \, d\alpha = k[\alpha(b) - \alpha(a)]$$

(c) Let

$$f(x) = \begin{cases} 1 & \text{if } x \text{ is rational} \\ 0 & \text{if } x \text{ is irrational} \end{cases}$$

for x in $[0, 1]$. Let $\alpha(x) = x$ on $[0, 1]$. Then every upper sum for f is 1 and every lower sum for f is 0. Therefore

$$\underline{\int_0^1} f = 0 \qquad \overline{\int_0^1} f = 1$$

and thus $f \notin \mathscr{R}[0, 1]$.

We next derive Riemann's condition (Theorem 51.8) which gives a necessary and sufficient condition that $f \in \mathscr{R}_{\alpha}[a, b]$. In 51.5 to 51.7, f is a bounded function on $[a, b]$ and α is an increasing function on $[a, b]$.

Definition 51.4 If S and P are partitions of $[a, b]$ and $P \subset S$, we say that S is a *refinement* of P.

Lemma 51.5 If P and S are partitions of $[a, b]$ and S is a refinement of P,

$$U(f, S) \leq U(f, P) \qquad \text{and} \qquad L(f, S) \geq L(f, P)$$

Proof. First suppose that S contains one more point than P. Let $P = \{x_0, x_1, \ldots, x_n\}$. Then $S = P \cup \{x^*\}$. Suppose $x^* \in [x_{j-1}, x_j]$. Let

$$M_j^* = \text{lub } \{f(x) \mid x \in [x_{j-1}, x^*]\}$$
$$M_j^{**} = \text{lub } \{f(x) \mid x \in [x^*, \dot{x}_j]\}$$

Then $M_j^* \leq M_j$ and $M_j^{**} \leq M_j$, and we have

$$\begin{aligned} U(f, P) &= \sum_{i=1}^n M_i \, \Delta\alpha_i \\ &= \sum_{i=1}^{j-1} M_i \, \Delta\alpha_i + M_j[\alpha(x^*) - \alpha(x_{j-1})] \\ &\quad + M_j[\alpha(x_j) - \alpha(x^*)] + \sum_{i=j+1}^n M_i \, \Delta\alpha_i \\ &\geq \sum_{i=1}^{j-1} M_i \, \Delta\alpha_i + M_j^*[\alpha(x^*) - \alpha(x_{j-1})] \\ &\quad + M_j^{**}[\alpha(x_j) - \alpha(x^*)] + \sum_{i=j+1}^n M_i \, \Delta\alpha_i \\ &= U(f, S) \end{aligned}$$

If S contains k more points than P, we repeat the above argument k times to conclude that $U(f, S) \leq U(f, P)$.

Similarly, $L(f, S) \geq L(f, P)$. ■

Corollary 51.6 If P and S are partitions of $[a, b]$, then

$$L(f, S) \leq U(f, P)$$

Proof. Let $T = P \cup S$. Then T is a refinement of both P and S; so by Lemma 51.5, we have

$$L(f, S) \leq L(f, T) \leq U(f, T) \leq U(f, P) \qquad\qquad ■$$

Corollary 51.7

$$\underline{\int_a^b} f \, d\alpha \leq \overline{\int_a^b} f \, d\alpha$$

Proof. Let P be a partition of $[a, b]$. Then for any partition S of $[a, b]$, we have (by Corollary 51.6)

$$L(f, S) \leq U(f, P)$$

Therefore, $U(f, P)$ is an upper bound for the set of lower sums, and thus

$$\underline{\int_a^b} f \, d\alpha \leq U(f, P)$$

for any partition P of $[a, b]$. Thus $\underline{\int_a^b} f \, d\alpha$ is a lower bound for the set of upper sums and hence

$$\underline{\int_a^b} f \, d\alpha \leq \overline{\int_a^b} f \, d\alpha \qquad\qquad ■$$

Theorem 51.8 (Riemann's Condition) Let f be a bounded function on $[a, b]$ and let α be an increasing function on $[a, b]$. Then $f \in \mathcal{R}_\alpha[a, b]$ if and only if for every $\varepsilon > 0$, there exists a partition P of $[a, b]$ such that

$$U(f, P) - L(f, P) < \varepsilon$$

Proof. Suppose $f \in \mathcal{R}_\alpha[a, b]$. Let $\varepsilon > 0$. There exist partitions S and T of $[a, b]$ such that

$$\underline{\int_a^b} f \, d\alpha - \frac{\varepsilon}{2} < L(f, T) \qquad U(f, S) < \overline{\int_a^b} f \, d\alpha + \frac{\varepsilon}{2}$$

Let $P = S \cup T$. By Lemma 51.5,

$$U(f, P) - L(f, P) \leq U(f, S) - L(f, T)$$

$$< \left(\overline{\int_a^b} f \, d\alpha + \frac{\varepsilon}{2} \right) + \left(-\underline{\int_a^b} f \, d\alpha + \frac{\varepsilon}{2} \right) = \varepsilon$$

since

$$\overline{\int_a^b} f \, d\alpha = \underline{\int_a^b} f \, d\alpha.$$

Now suppose the condition holds. Let $\varepsilon > 0$. There exists a partition P of $[a, b]$ such that

$$U(f, P) - L(f, P) < \varepsilon$$

Thus

$$\overline{\int_a^b} f \, d\alpha - \underline{\int_a^b} f \, d\alpha \leq U(f, P) - L(f, P) < \varepsilon$$

Therefore, using Corollary 51.7, we have

$$0 \leq \overline{\int_a^b} f \, d\alpha - \underline{\int_a^b} f \, d\alpha < \varepsilon$$

for every $\varepsilon > 0$. It follows that $\overline{\int_a^b} f \, d\alpha = \underline{\int_a^b} f \, d\alpha$, and hence $f \in \mathscr{R}_\alpha[a, b]$. ∎

Example 51.9 Let $[a, b]$ be a closed interval and let $c \in (a, b)$. Let $k_1 < k_2$ and define

$$\alpha(x) = \begin{cases} k_1 & a \leq x < c \\ k_2 & c < x \leq b \end{cases}$$

(The value of $\alpha(c) \in [k_1, k_2]$ is immaterial.) (See Figure 51.1.) Let f be a bounded function on $[a, b]$ which is continuous at c. We use our preceding results to show that $f \in \mathscr{R}_\alpha[a, b]$ and

$$\int_a^b f \, d\alpha = f(c) \cdot (k_2 - k_1)$$

Let $\varepsilon > 0$. Since f is continuous at c, there exist points x_1 and x_2 with

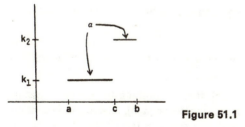

Figure 51.1

$a < x_1 < c < x_2 < b$ such that if $x \in [x_1, x_2]$, then

$$f(c) - \frac{\varepsilon}{3(k_2 - k_1)} < f(x) < f(c) + \frac{\varepsilon}{3(k_2 - k_1)}$$

Let $P = \{a, x_1, x_2, b\}$. The numbers m_2 and M_2 denote the glb and lub, respectively, of f on the interval $[x_1, x_2]$; thus

$$f(c) - \frac{\varepsilon}{3(k_2 - k_1)} \leq m_2 \leq M_2 \leq f(c) + \frac{\varepsilon}{3(k_2 - k_1)}$$

Now $U(f, P) = M_2[\alpha(x_2) - \alpha(x_1)] = M_2(k_2 - k_1) \leq f(c)(k_2 - k_1) + \frac{\varepsilon}{3}$

Similarly, $f(c)(k_2 - k_1) - \frac{\varepsilon}{3} \leq L(f, P)$

Therefore, $U(f, P) - L(f, P) \leq \frac{2\varepsilon}{3} < \varepsilon$

and by Theorem 51.8, $f \in \mathcal{R}_\alpha[a, b]$.
  Now

$$\int_a^b f \, d\alpha \leq U(f, P) \leq f(c)(k_2 - k_1) + \frac{\varepsilon}{3}$$

hence $$\int_a^b f \, d\alpha \leq f(c)(k_2 - k_1) + \frac{\varepsilon}{3}$$

for every $\varepsilon > 0$. Therefore,

$$\int_a^b f \, d\alpha \leq f(c)(k_2 - k_1)$$

Similarly, $$\int_a^b f \, d\alpha \geq f(c)(k_2 - k_1)$$

and thus $$\int_a^b f \, d\alpha = f(c)(k_2 - k_1)$$

  We are now in a position to prove that a continuous function is Riemann-Stieltjes integrable.

Theorem 51.10   If f is continuous on $[a, b]$ and α is increasing on $[a, b]$, then $f \in \mathcal{R}_\alpha[a, b]$.

  Proof. If α is constant on $[a, b]$, there is nothing to prove; so assume $\alpha(a) < \alpha(b)$. Let $\varepsilon > 0$. Since f is continuous on $[a, b]$, f is uniformly continuous on $[a, b]$ (Corollary 44.6); so there exists $\delta > 0$ such that if $|s - t| < \delta$,

then

$$|f(s) - f(t)| < \frac{\varepsilon}{2[\alpha(b) - \alpha(a)]}$$

Let $P = \{x_0, x_1, \ldots, x_n\}$ be a partition such that $x_i - x_{i-1} < \delta$ for $i = 1, \ldots, n$. Then

$$M_i - m_i \le \frac{\varepsilon}{2[\alpha(b) - \alpha(a)]}$$

for $i = 1, \ldots, n$ (verify). Thus

$$U(f, P) - L(f, P) = \sum_{i=1}^{n} (M_i - m_i)\Delta\alpha_i$$

$$\le \frac{\varepsilon}{2[\alpha(b) - \alpha(a)]} \sum_{i=1}^{n} \Delta\alpha_i$$

$$= \frac{\varepsilon}{2} < \varepsilon$$

By Theorem 51.8, $f \in \mathcal{R}_\alpha[a, b]$. ∎

The remainder of this section is devoted to deriving properties of the set $\mathcal{R}_\alpha[a, b]$ where α is an increasing function on $[a, b]$.

Theorem 51.11 Let α be an increasing function on $[a, b]$. If $f \in \mathcal{R}_\alpha[a, b]$, then $f \in \mathcal{R}_\alpha[c, d]$ for every closed subinterval $[c, d]$ of $[a, b]$ and

$$\int_a^b f \, d\alpha = \int_a^c f \, d\alpha + \int_c^b f \, d\alpha$$

for all c in (a, b).

Proof. Let $c \in (a, b)$. Let P be a partition of $[a, b]$ and let $P^* = P \cup \{c\}$. Then $P_1 = P^* \cap [a, c]$ and $P_2 = P^* \cap [c, b]$ are partitions of $[a, c]$ and $[c, b]$, respectively. Now

$$U(f, P) \ge U(f, P^*) = U(f, P_1) + U(f, P_2) \ge \overline{\int_a^c} f \, d\alpha + \overline{\int_c^b} f \, d\alpha$$

It follows that

$$\overline{\int_a^b} f \, d\alpha \ge \overline{\int_a^c} f \, d\alpha + \overline{\int_c^b} f \, d\alpha$$

Similarly,

$$\underline{\int_a^b} f \, d\alpha \le \underline{\int_a^c} f \, d\alpha + \underline{\int_c^b} f \, d\alpha$$

Therefore,
$$\int_a^b f \, d\alpha \geq \overline{\int_a^c} f \, d\alpha + \overline{\int_c^b} f \, d\alpha \geq \overline{\int_a^c} f \, d\alpha$$

$$+ \overline{\int_c^b} f \, d\alpha \geq \underline{\int_a^c} f \, d\alpha + \underline{\int_c^b} f \, d\alpha \geq \int_a^b f \, d\alpha$$

Thus
$$\overline{\int_a^c} f \, d\alpha = \underline{\int_a^c} f \, d\alpha \quad \text{and} \quad \overline{\int_c^b} f \, d\alpha = \underline{\int_c^b} f \, d\alpha$$

and hence $f \in \mathscr{R}_\alpha[a, c] \cap \mathscr{R}_\alpha[c, b]$. Furthermore,

$$\int_a^b f \, d\alpha = \int_a^c f \, d\alpha + \int_c^b f \, d\alpha$$

Let $[c, d]$ be a closed subinterval of $[a, b]$. By the above result $f \in \mathscr{R}_\alpha[a, d]$ and another application of the above result yields $f \in \mathscr{R}_\alpha[c, d]$. ∎

Theorem 51.12 Let f be a bounded function on $[a, b]$ and let α be an increasing function on $[a, b]$. Suppose that $m \leq f(x) \leq M$ for all x in $[a, b]$. If $f \in \mathscr{R}_\alpha[a, b]$ and g is continuous on $[m, M]$, then $g \circ f \in \mathscr{R}_\alpha[a, b]$.

Proof. Let $h = g \circ f$ and let k satisfy $k > |g(t)|$ for all t in $[m, M]$. Let $\varepsilon > 0$. Since g is uniformly continuous on $[m, M]$, there exists δ, $0 < \delta < \varepsilon/(2k + \alpha(b) - \alpha(a))$ such that if $|s - t| \leq \delta$, then

$$|g(s) - g(t)| < \frac{\varepsilon}{2k + \alpha(b) - \alpha(a)} \tag{51.1}$$

Since $f \in \mathscr{R}_\alpha[a, b]$, by Theorem 51.8, there exists a partition $P = \{x_0, x_1, \ldots, x_n\}$ of $[a, b]$ such that

$$U(f, P) - L(f, P) < \delta^2$$

Let
$$M_i = \text{lub } \{f(x) \mid x \in [x_{i-1}, x_i]\}$$

$$m_i = \text{glb } \{f(x) \mid x \in [x_{i-1}, x_i]\}$$

$$M_i^* = \text{lub } \{h(x) \mid x \in [x_{i-1}, x_i]\}$$

$$m_i^* = \text{glb } \{h(x) \mid x \in [x_{i-1}, x_i]\}$$

for $i = 1, \ldots, n$. Let

$$A = \{i \mid 1 \leq i \leq n \text{ and } M_i - m_i < \delta\}$$

$$B = \{i \mid 1 \leq i \leq n \text{ and } M_i - m_i \geq \delta\}$$

If $i \in A$, we have by inequality (51.1) that $M_i^* - m_i^* \leq \varepsilon/(2k + \alpha(b) - \alpha(a))$. If $i \in B$, we have $M_i^* - m_i^* \leq 2k$. Now

$$\delta \sum_{i \in B} \Delta\alpha_i \leq \sum_{i \in B} (M_i - m_i)\Delta\alpha_i \leq \sum_{i=1}^{n} (M_i - m_i)\Delta\alpha_i$$
$$= U(f, P) - L(f, P) < \delta^2$$

and it follows that

$$\sum_{i \in B} \Delta\alpha_i < \delta$$

Therefore,

$$U(h, P) - L(h, P) = \sum_{i=1}^{n} (M_i^* - m_i^*)\Delta\alpha_i$$
$$= \sum_{i \in A} (M_i^* - m_i^*)\Delta\alpha_i + \sum_{i \in B} (M_i^* - m_i^*)\Delta\alpha_i$$
$$\leq \frac{\varepsilon}{2k + \alpha(b) - \alpha(a)} \sum_{i \in A} \Delta\alpha_i + 2k \sum_{i \in B} \Delta\alpha_i$$
$$\leq \frac{\varepsilon}{2k + \alpha(b) - \alpha(a)} \sum_{i=1}^{n} \Delta\alpha_i + 2k\delta$$
$$< \frac{\varepsilon}{2k + \alpha(b) - \alpha(a)} (\alpha(b) - \alpha(a))$$
$$+ 2k \frac{\varepsilon}{2k + \alpha(b) - \alpha(a)} = \varepsilon$$

By Theorem 51.8, $h \in \mathcal{R}_\alpha[a, b]$. ∎

Theorem 51.13   Let α be an increasing function on $[a, b]$. If f and g are in $\mathcal{R}_\alpha[a, b]$ and $c \in \mathbf{R}$, then

(i) $cf \in \mathbf{R}_\alpha[a, b]$ and $\displaystyle\int_a^b cf \, d\alpha = c \int_a^b f \, d\alpha$.

(ii) $f + g \in \mathbf{R}_\alpha[a, b]$ and $\displaystyle\int_a^b (f + g) \, d\alpha = \int_a^b f \, d\alpha + \int_a^b g \, d\alpha$.

(iii) $f \cdot g \in \mathbf{R}_\alpha[a, b]$.

(iv) If $f(x) \leq g(x)$ for all x in $[a, b]$, then $\displaystyle\int_a^b f \, d\alpha \leq \int_a^b g \, d\alpha$.

(v) $|f| \in \mathbf{R}_\alpha[a, b]$ and $\displaystyle\left| \int_a^b f \, d\alpha \right| \leq \int_a^b |f| \, d\alpha$.

Proof.   (i) Let $g(x) = cx$. By Theorem 51.12, $g \circ f(x) = cf(x)$ is in $\mathcal{R}_\alpha[a, b]$. Suppose $c \geq 0$. Let P be a partition of $[a, b]$. Then

$$U(cf, P) = cU(f, P) \geq c \int_a^b f \, d\alpha$$

and it follows that

$$\int_a^b cf \, d\alpha \geq c \int_a^b f \, d\alpha$$

Similarly,

$$\int_a^b cf \, d\alpha \leq c \int_a^b f \, d\alpha$$

Therefore,

$$\int_a^b cf \, d\alpha = c \int_a^b f \, d\alpha$$

The case $c < 0$ is treated in a similar way.

(ii) For any partition P of $[a, b]$, we have

$$L(f, P) + L(g, P) \leq L(f + g, P) \leq U(f + g, P) \leq U(f, P) + U(g, P)$$

(verify).

Let $\varepsilon > 0$. By Theorem 51.8, there exist partitions S and T of $[a, b]$ such that

$$U(f, S) - L(f, S) < \frac{\varepsilon}{2} \quad \text{and} \quad U(g, T) - L(g, T) < \frac{\varepsilon}{2}$$

Let $P = S \cup T$. Then we have

$$U(f + g, P) - L(f + g, P) < \frac{\varepsilon}{2} + \frac{\varepsilon}{2} = \varepsilon$$

By Theorem 51.8, $f + g \in \mathcal{R}_\alpha[a, b]$. Also

$$U(f, P) \leq U(f, S) < L(f, S) + \frac{\varepsilon}{2} \leq \int_a^b f \, d\alpha + \frac{\varepsilon}{2}$$

Similarly,

$$U(g, P) < \int_a^b g \, d\alpha + \frac{\varepsilon}{2}$$

Therefore,

$$\int_a^b (f + g) \, d\alpha \leq U(f + g, P) \leq U(f, P) + U(g, P)$$

$$< \int_a^b f \, d\alpha + \int_a^b g \, d\alpha + \varepsilon$$

Since $\varepsilon > 0$ was arbitrary, we have

$$\int_a^b (f + g) \, d\alpha \leq \int_a^b f \, d\alpha + \int_a^b g \, d\alpha$$

Similarly,

$$\int_a^b (f + g) \, d\alpha \geq \int_a^b f \, d\alpha + \int_a^b g \, d\alpha$$

and therefore

$$\int_a^b (f + g)\, d\alpha = \int_a^b f\, d\alpha + \int_a^b g\, d\alpha$$

(iii) It follows from Theorem 51.12 that if $f \in \mathscr{R}_\alpha[a, b]$, then $f^2 \in \mathscr{R}_\alpha[a, b]$ for if $f \in \mathscr{R}_\alpha[a, b]$, since $g(x) = x^2$ is continuous, we have $g \circ f = f^2 \in \mathscr{R}_\alpha[a, b]$.

Now let $f, g \in \mathscr{R}_\alpha[a, b]$. By parts (i) and (ii), $f + g, f - g \in \mathscr{R}_\alpha[a, b]$. By the result above $(f + g)^2, (f - g)^2 \in \mathscr{R}_\alpha[a, b]$. By parts (i) and (ii),

$$f \cdot g = \tfrac{1}{4}[(f + g)^2 - (f - g)^2]$$

is in $\mathscr{R}_\alpha[a, b]$.

(iv) Suppose $f \in \mathscr{R}_\alpha[a, b]$ and $f(x) \geq 0$ for x in $[a, b]$. Let P be a partition of $[a, b]$. Then

$$\int_a^b f\, d\alpha \geq L(f, P) \geq 0$$

If $f, g \in \mathscr{R}_\alpha[a, b]$ and $g(x) \geq f(x)$ for x in $[a, b]$, then $g(x) - f(x) \geq 0$ for x in $[a, b]$ and by our result above

$$\int_a^b (g - f)\, d\alpha \geq 0$$

By parts (i) and (ii)

$$\int_a^b (g - f)\, d\alpha = \int_a^b g\, d\alpha - \int_a^b f\, d\alpha$$

and therefore

$$\int_a^b g\, d\alpha \geq \int_a^b f\, d\alpha$$

(v) Since $g(x) = |x|$ is continuous, $g \circ f = |f| \in \mathscr{R}_\alpha[a, b]$. Define c by the equation

$$c \int_a^b f\, d\alpha = \left| \int_a^b f\, d\alpha \right|$$

(The number c is either 1 or -1.) Now $cf(x) \leq |f(x)|$ for all x in $[a, b]$; so by part (iv) we have

$$\int_a^b cf\, d\alpha \leq \int_a^b |f|\, d\alpha$$

Using part (i) we have

$$\left| \int_a^b f\, d\alpha \right| = c \int_a^b f\, d\alpha = \int_a^b cf\, d\alpha \leq \int_a^b |f|\, d\alpha \qquad \blacksquare$$

Theorem 51.14 (Mean-Value Theorem of the Integral) If f is continuous on

$[a, b]$ and α is increasing on $[a, b]$, there exists $c \in [a, b]$ such that

$$\int_a^b f \, d\alpha = f(c)[\alpha(b) - \alpha(a)]$$

Proof. If $\alpha(b) = \alpha(a)$, then any value of c in $[a, b]$ gives the desired conclusion. Suppose $\alpha(b) > \alpha(a)$. By Theorem 34.4, f attains its maximum M and minimum m on $[a, b]$. For the partition $P = \{a, b\}$ we have

$$m[\alpha(b) - \alpha(a)] = L(f, P) \leq \int_a^b f \, d\alpha \leq U(f, P) = M[\alpha(b) - \alpha(a)]$$

Therefore,
$$m \leq \frac{\displaystyle\int_a^b f \, d\alpha}{\alpha(b) - \alpha(a)} \leq M$$

By the intermediate-value theorem (Theorem 45.6), there exists c in $[a, b]$ such that

$$f(c) = \frac{\displaystyle\int_a^b f \, d\alpha}{\alpha(b) - \alpha(a)} \qquad \blacksquare$$

Exercises

In Exercises 51.1 to 51.5 f is a bounded function on $[a, b]$ and α is an increasing function on $[a, b]$.

51.1 Prove that if P is a partition of $[a, b]$, then $L(f, P) \leq U(f, P)$.

51.2 Prove that if P and S are partitions of $[a, b]$ and S is a refinement of P, then $L(f, S) \geq L(f, P)$, thus completing the proof of Lemma 51.5.

51.3 Prove that $f \in \mathcal{R}_\alpha[a, b]$ if and only if for every $\varepsilon > 0$, there exist partitions P and S of $[a, b]$ such that
$$U(f, P) - L(f, S) < \varepsilon$$

51.4 Let $c \in (a, b)$. Prove that
$$\overline{\int_a^b} f \, d\alpha = \overline{\int_a^c} f \, d\alpha + \overline{\int_c^b} f \, d\alpha \qquad \text{and} \qquad \underline{\int_a^b} f \, d\alpha = \underline{\int_a^c} f \, d\alpha + \underline{\int_c^b} f \, d\alpha$$

51.5 Prove that
$$\overline{\int_a^b} f \, d\alpha = -\underline{\int_a^b} (-f) \, d\alpha$$

51.6 Use the results of this section to prove that
$$\int_0^1 x \, dx = \frac{1}{2} \qquad \text{and} \qquad \int_0^1 x^2 \, dx = \frac{1}{3}$$

51.7 Let
$$\alpha(x) = \begin{cases} x & 0 \le x \le 1 \\ 2+x & 1 < x \le 2 \end{cases}$$
Compute $\int_0^2 x \, d\alpha(x)$.

51.8 Let α be an increasing function on $[a, b]$ and let $c < 0$. Prove that if $f \in \mathcal{R}_\alpha[a, b]$, then $cf \in \mathcal{R}_\alpha[a, b]$ and
$$\int_a^b cf \, d\alpha = c \int_a^b f \, d\alpha$$
thus completing the proof of Theorem 51.13(i).

51.9 Let α be an increasing function on $[a, b]$. Prove that if $f, g \in \mathcal{R}_\alpha[a, b]$, then
$$\int_a^b (f + g) \, d\alpha \ge \int_a^b f \, d\alpha + \int_a^b g \, d\alpha$$
thus completing the proof of Theorem 51.13(ii).

51.10 (a) Prove that if f is a continuous function on $[a, b]$, there exists a point c in (a, b) such that
$$\int_a^b f(x) \, dx = f(c)(b - a)$$

(b) Give an example to show that for an arbitrary increasing function α on $[a, b]$, part (a) may fail.

51.11 Prove that if f is a continuous nonnegative function on $[a, b]$ and
$$\int_a^b f(x) \, dx = 0, \text{ then } f(x) = 0 \text{ for all } x \text{ in } [a, b].$$

51.12 Let
$$\alpha(x) = f(x) = \begin{cases} 0 & \text{if } 0 \le x < 1 \\ 1 & \text{if } 1 \le x \le 2 \end{cases}$$
$$g(x) = \begin{cases} 0 & \text{if } 0 \le x \le 1 \\ 1 & \text{if } 1 < x \le 2 \end{cases}$$

(a) Is $f \in \mathcal{R}_\alpha[0, 2]$? If so, compute $\int_0^2 f \, d\alpha$.

(b) Is $g \in \mathcal{R}_\alpha[0, 2]$? If so, compute $\int_0^2 g \, d\alpha$.

(c) Do (a) and (b) above with $\alpha(x) = x$.

51.13 Let α_1 and α_2 be increasing functions on $[a, b]$.
(a) Prove that $\alpha_1 + \alpha_2$ is increasing on $[a, b]$.
(b) Let $f \in \mathcal{R}_{\alpha_1}[a, b] \cap \mathcal{R}_{\alpha_2}[a, b]$. Prove that $f \in \mathcal{R}_{\alpha_1 + \alpha_2}[a, b]$ and
$$\int_a^b f \, d(\alpha_1 + \alpha_2) = \int_a^b f \, d\alpha_1 + \int_a^b f \, d\alpha_2$$

51.14 Let α be an increasing function on $[a, b]$ and let f and g be bounded functions on $[a, b]$. Prove that
$$\underline{\int_a^b} f \, d\alpha + \underline{\int_a^b} g \, d\alpha \le \underline{\int_a^b} (f + g) \, d\alpha \le \overline{\int_a^b} (f + g) \, d\alpha \le \overline{\int_a^b} f \, d\alpha + \overline{\int_a^b} g \, d\alpha$$

51.15   Let α be an increasing function on $[a, b]$. Prove that if $f, g \in \mathcal{R}_\alpha[a, b]$, then max $\{f, g\}$ and min $\{f, g\}$ are in $\mathcal{R}_\alpha[a, b]$.

51.16   Let α be an increasing function on $[a, b]$. Let $f \in \mathcal{R}_\alpha[a, b]$ and suppose that for some positive number M, $|f(x)| \geq M$ for all x in $[a, b]$. Prove that $1/f \in \mathcal{R}_\alpha[a, b]$.

51.17   Let f be a positive continuous function on $[a, b]$, and let M be the maximum of f on $[a, b]$. Prove that

$$\lim_{n \to \infty} \left[\int_a^b [f(x)]^n dx \right]^{1/n} = M$$

51.18   Give an example of an increasing function α on $[a, b]$ and a bounded function f on $[a, b]$ such that $|f| \in \mathcal{R}_\alpha[a, b]$, but $f \notin \mathcal{R}_\alpha[a, b]$.

51.19   Prove that if f is an increasing function on $[a, b]$, then $f \in \mathcal{R}[a, b]$.

51.20   Let f be a bounded function on $[a, b]$. Prove that any of the following conditions implies that $f \in \mathcal{R}[a, b]$.
   (a) f has a limit at each point of $[a, b]$.
   (b) f has a left-hand and a right-hand limit at each point of (a, b).
   (c) f has a left-hand limit at each point of (a, b).

52.   Riemann-Stieltjes Sums

We begin with a definition.

Definition 52.1   Let f and α be bounded functions on $[a, b]$. Let $P = \{x_0, x_1, \ldots, x_n\}$ be a partition of $[a, b]$. Let $T = \{t_1, t_2, \ldots, t_n\}$ where $t_i \in [x_{i-1}, x_i]$ for $i = 1, \ldots, n$. The sum

$$\sum_{i=1}^n f(t_i) \Delta\alpha_i$$

is called the *Riemann-Stieltjes sum for f with respect to α for the partition P and points T.* We denote this sum as

$$S(f, P, T)$$

Let f, α, and P be as in Definition 52.1, and assume that α is increasing. Clearly,

$$L(f, P) \leq S(f, P, T) \leq U(f, P)$$

for any points T. Thus if the upper and lower sums approximate the integral, we would expect that the Riemann-Stieltjes sums would also approximate the integral. The next theorem shows that this is the case.

Theorem 52.2   Let f be a bounded function on $[a, b]$ and let α be an increasing function on $[a, b]$. Then $f \in \mathcal{R}_\alpha[a, b]$ if and only if there exists a number I

having the property that for every $\varepsilon > 0$, there exists a partition P such that if P^* is a refinement of $P(P^* \supset P)$, then

$$|S(f, P^*, T) - I| < \varepsilon$$

for any points T.

If $f \in \mathscr{R}_\alpha[a, b]$, the number I is equal to $\int_a^b f \, d\alpha$.

Proof. Suppose $f \in \mathscr{R}_\alpha[a, b]$. Let $\varepsilon > 0$. By Theorem 51.8, there exists a partition P such that

$$U(f, P) - L(f, P) < \varepsilon$$

Let P^* be a refinement of P. Then

$$U(f, P^*) - L(f, P^*) \leq U(f, P) - L(f, P) < \varepsilon$$

Now $$L(f, P^*) \leq S(f, P^*, T) \leq U(f, P^*)$$

and $$L(f, P^*) \leq \int_a^b f \, d\alpha \leq U(f, P^*)$$

and hence $$\left| S(f, P^*, T) - \int_a^b f \, d\alpha \right| < \varepsilon \qquad \text{for any points } T$$

It follows that $I = \int_a^b f \, d\alpha$.

Now suppose the condition holds. We may assume $\alpha(b) > \alpha(a)$. Let $\varepsilon > 0$. There exists a partition $P = \{x_0, x_1, \ldots, x_n\}$ of $[a, b]$ such that

$$|S(f, P, T) - I| < \frac{\varepsilon}{4}$$

for all points T. For each i, $1 \leq i \leq n$, there exists $t_i^* \in [x_{i-1}, x_i]$ such that

$$M_i - \frac{\varepsilon}{4[\alpha(b) - \alpha(a)]} < f(t_i^*)$$

Thus $$U(f, P) - \frac{\varepsilon}{4} = \sum_{i=1}^n \left(M_i - \frac{\varepsilon}{4[\alpha(b) - \alpha(a)]} \right) \Delta\alpha_i \leq \sum_{i=1}^n f(t_i^*) \Delta\alpha_i$$

Let $T^* = \{t_1^*, \ldots, t_n^*\}$. Then

$$U(f, P) - \frac{\varepsilon}{4} \leq S(f, P, T^*)$$

Similarly, there exist points T^{**} such that

$$S(f, P, T^{**}) \leq L(f, P) + \frac{\varepsilon}{4}$$

Therefore,

$$U(f, P) - L(f, P) \leq \left[S(f, P, T^*) + \frac{\varepsilon}{4} \right] + \left[-S(f, P, T^{**}) + \frac{\varepsilon}{4} \right]$$

$$\leq |S(f, P, T^*) - S(f, P, T^{**})| + \frac{\varepsilon}{2}$$

$$\leq |S(f, P, T^*) - I| + |S(f, P, T^{**}) - I| + \frac{\varepsilon}{2}$$

$$< \frac{\varepsilon}{4} + \frac{\varepsilon}{4} + \frac{\varepsilon}{2} = \varepsilon$$

By Theorem 51.8, $f \in \mathscr{R}_\alpha[a, b]$. ∎

The condition "$P^* \supset P$" in Theorem 52.2 is analogous to the condition "$n \geq N$" in the definition of the limit of a sequence. These concepts can be treated under a unified theory, the theory of directed sets or filters. (See Kelley, 1955, Chapter 2.)

Theorem 52.2 shows that refining partitions is a way of forcing convergence of Riemann-Stieltjes sums. Another way to describe convergence of Riemann-Stieltjes sums is given in the next definition.

Definition 52.3   Let $P = \{x_0, x_1, \ldots, x_n\}$ be a partition of $[a, b]$. We define

$$\text{norm } P = \max \{|x_i - x_{i-1}| : 1 \leq i \leq n\}$$

Let f and α be bounded functions on $[a, b]$. We say that

$$\lim_{\text{norm } P \to 0} S(f, P, T) = I$$

if for every $\varepsilon > 0$, there exists $\delta > 0$ such that for any partition P of $[a, b]$ with norm $P < \delta$ and for any points T, we have

$$|S(f, P, T) - I| < \varepsilon$$

We cannot prove a theorem analogous to Theorem 52.2 for the kind of convergence described in Definition 52.3. That is, convergence of Riemann-Stieltjes sums as described in Definition 52.3 is not equivalent to the existence of the Riemann-Stieltjes integral. For example, let

$$f(x) = \begin{cases} 0 & \text{if } -1 \leq x < 0 \\ 1 & \text{if } 0 \leq x \leq 1 \end{cases}$$

and

$$\alpha(x) = \begin{cases} 0 & \text{if } -1 \leq x \leq 0 \\ 1 & \text{if } 0 < x \leq 1 \end{cases}$$

Let $P = \{-1, 0, 1\}$. Then $U(f, P) = L(f, P) = 1$, and thus $f \in \mathscr{R}_\alpha[-1, 1]$. However, $\lim_{\text{norm } P \to 0} S(f, P, T)$ does not exist.

If $P^* \supset P = \{-1, 0, 1\}$, then

$$1 = L(f, P) \le L(f, P^*) \le S(f, P^*, T) \le U(f, P^*) \le U(f, P) = 1$$

and therefore $S(f, P^*, T) = 1$. Thus there are partitions P^* of $[a, b]$ with arbitrarily small norm such that $S(f, P^*, T) = 1$. On the other hand, if

$$P^* = \left\{ -\frac{n}{n}, -\frac{n-1}{n}, \ldots, \frac{-2}{n}, -\frac{1}{n}, \frac{1}{n}, \frac{2}{n}, \ldots, \frac{n-1}{n}, \frac{n}{n} \right\}$$

and we choose $t = -(1/n)$ in the interval $[-(1/n), 1/n]$, we have

$$S(f, P^*, T) = 0$$

and norm $P^* = 2/n$. Thus there are partitions P^* of $[a, b]$ with arbitrarily small norm such that $S(f, P^*, T) = 0$. It follows that $\lim_{\text{norm } P \to 0} S(f, P, T)$ does not exist.

In spite of the previous example, we do have the following two results.

Theorem 52.4 Let f be a bounded function on $[a, b]$ and let α be an increasing function on $[a, b]$. If $\lim_{\text{norm } P \to 0} S(f, P, T)$ exists, then $f \in \mathscr{R}_\alpha[a, b]$ and

$$\lim_{\text{norm } P \to 0} S(f, P, T) = \int_a^b f \, d\alpha$$

Proof. Let $\varepsilon > 0$. There exists $\delta > 0$ such that if norm $P < \delta$, then

$$|S(f, P, T) - I| < \varepsilon$$

Let P be a partition of $[a, b]$ such that norm $P < \delta$. If $P^* \supset P$, then norm $P^* < \delta$ and thus

$$|S(f, P^*, T) - I| < \varepsilon$$

By Theorem 52.2, $f \in \mathscr{R}_\alpha[a, b]$ and $I = \int_a^b f \, d\alpha$. ∎

Theorem 52.5 Let α be an increasing function on $[a, b]$ and suppose that $f \in \mathscr{R}_\alpha[a, b]$. If either f or α is continuous on $[a, b]$, then

$$\int_a^b f \, d\alpha = \lim_{\text{norm } P \to 0} S(f, P, T)$$

Proof. First suppose that f is continuous on $[a, b]$. We may assume that $\alpha(b) > \alpha(a)$. Let $\varepsilon > 0$. Since f is continuous on $[a, b]$, f is uniformly continuous on $[a, b]$. Thus, there exists $\delta > 0$ such that if $|s - t| \le \delta$, then

$$|f(s) - f(t)| < \frac{\varepsilon}{2[\alpha(b) - \alpha(a)]}$$

Let $P = \{x_0, x_1, \ldots, x_n\}$ be a partition with norm $P < \delta$. Let

$$S(f, P, T) = \sum_{i=1}^{n} f(t_i)\Delta\alpha_i$$

be a Riemann-Stieltjes sum. By Theorems 51.11 and 51.14,

$$\int_a^b f\, d\alpha = \sum_{i=1}^{n} \int_{x_{i-1}}^{x_i} f\, d\alpha = \sum_{i=1}^{n} f(t_i^*)\Delta\alpha_i$$

where $t_i^* \in [x_{i-1}, x_i]$ for $i = 1, \ldots, n$. Therefore,

$$\left| S(f, P, T) - \int_a^b f\, d\alpha \right| = \left| \sum_{i=1}^{n} [f(t_i) - f(t_i^*)]\Delta\alpha_i \right|$$

$$\leq \sum_{i=1}^{n} |f(t_i) - f(t_i^*)|\, \Delta\alpha_i$$

$$\leq \frac{\varepsilon}{2[\alpha(b) - \alpha(a)]} \sum_{i=1}^{n} \Delta\alpha_i = \frac{\varepsilon}{2} < \varepsilon$$

Hence, $$\lim_{\text{norm } P \to 0} S(f, P, T) = \int_a^b f\, d\alpha$$

Now suppose that α is continuous on $[a, b]$. We first assume that $f(x) > 0$ for x in $[a, b]$. Let $\varepsilon > 0$. There exists a partition $Q = \{y_0, y_1, \ldots, y_n\}$ of $[a, b]$ such that

$$U(f, Q) < \int_a^b f\, d\alpha + \frac{\varepsilon}{4}$$

Let $M = \text{lub}\{f(x)\,|\,x \in [a, b]\}$. Now α is uniformly continuous on $[a, b]$ and thus there exists $\delta' > 0$ such that if $P = \{x_0, x_1, \ldots, x_j\}$ is a partition of $[a, b]$ with norm $P < \delta'$, we have

$$\Delta\alpha_i < \frac{\varepsilon}{4Mn}$$

for $i = 1, \ldots, j$. Let A denote the set of integers i for which $1 \leq i \leq j$ and (x_{i-1}, x_i) contains a point of Q. Let

$$B = \{1, \ldots, j\}\backslash A$$

Now $$U(f, P) = \sum_{i\in A} M_i\Delta\alpha_i + \sum_{i\in B} M_i\Delta\alpha_i$$

$$\leq (n - 1)M\frac{\varepsilon}{4Mn} + U(f, Q)$$

$$< \int_a^b f\, d\alpha + \frac{\varepsilon}{2}$$

Now let f be any function in $\mathscr{R}_\alpha[a, b]$ and let $\varepsilon > 0$. Choose a number k

such that $f(x) + k > 0$ on $[a, b]$. By the above argument, there exists $\delta_1 > 0$ such that if norm $P < \delta_1$, then

$$U(f + k, P) < \int_a^b (f + k)\, d\alpha + \frac{\varepsilon}{2}$$

Since $U(f + k, P) = U(f, P) + k(b - a)$ and

$$\int_a^b (f + k)\, d\alpha = \int_a^b f\, d\alpha + k(b - a)$$

we find that

$$U(f, P) < \int_a^b f\, d\alpha + \frac{\varepsilon}{2}$$

if norm $P < \delta_1$.

Applying the above argument to $-f$, we find that there exists $\delta_2 > 0$ such that if norm $P < \delta_2$, then

$$U(-f, P) < \int_a^b (-f)\, d\alpha + \frac{\varepsilon}{2}$$

Since $U(-f, P) = -L(f, P)$, we conclude that

$$L(f, P) > \int_a^b f\, d\alpha - \frac{\varepsilon}{2}$$

if norm $P < \delta_2$. Let $\delta = \min\{\delta_1, \delta_2\}$. If norm $P < \delta$, then

$$\int_a^b f\, d\alpha - \frac{\varepsilon}{2} < L(f, P) \le S(f, P, T) \le U(f, P) < \int_a^b f\, d\alpha + \frac{\varepsilon}{2}$$

and thus

$$\left| S(f, P, T) - \int_a^b f\, d\alpha \right| < \varepsilon$$

Therefore,
$$\lim_{\text{norm } P \to 0} S(f, P, T) = \int_a^b f\, d\alpha \quad \blacksquare$$

Corollary 52.6 If $f \in \mathscr{R}\,[a, b]$, then

$$\lim_{\text{norm } P \to 0} S(f, P, T) = \int_a^b f$$

Proof. Since $\alpha(x) = x$ is continuous, the conclusion follows immediately from Theorem 52.5. $\blacksquare$

Exercises

52.1 Let $f \in \mathscr{R}[0, 1]$. Prove that

$$\lim_{n \to \infty} \sum_{k=1}^{n} f\left(\frac{k}{n}\right) \frac{1}{n} = \int_0^1 f$$

52.2 Using Definition 52.3 to define what is meant by f being Riemann-Stieltjes integrable with respect to α on $[a, b]$, which we shall write $f \in RS[a, b]$, prove

(a) If $f, g \in RS[a, b]$, then $f + g \in RS[a, b]$ and

$$\int_a^b (f + g)\, d\alpha = \int_a^b f\, d\alpha + \int_a^b g\, d\alpha$$

(b) If $f \in RS[a, b]$, then $cf \in RS[a, b]$ and

$$\int_a^b cf = c \int_a^b f$$

(c) If $f, g \in RS[a, b]$ and $f(x) \le g(x)$ for all $x \in [a, b]$, then

$$\int_a^b f\, d\alpha \le \int_a^b g\, d\alpha$$

(d) Prove or disprove: if $f \in RS[a, b]$ and α is increasing on $[a, b]$, then $f \in RS[c, d]$ for every subinterval $[c, d]$ of $[a, b]$ and

$$\int_a^b f\, d\alpha = \int_a^c f\, d\alpha + \int_c^b f\, d\alpha$$

53. Riemann-Stieltjes Integration with Respect to an Arbitrary Integrator

In this section we extend the definition of the Riemann-Stieltjes integral $\int_a^b f\, d\alpha$ to arbitrary (not necessarily increasing) functions α. Motivated by Theorem 52.2, we make the following definition.

Definition 53.1 Let f and α be bounded functions on $[a, b]$. We say that f *is Riemann-Stieltjes integrable with respect to α on* $[a, b]$, and write $f \in \mathscr{R}_\alpha[a, b]$, if there exists a number I having the property that for every $\varepsilon > 0$, there exists a partition P of $[a, b]$ such that if P^* is a refinement of P ($P^* \supset P$), then

$$|S(f, P^*, T) - I| < \varepsilon$$

for any points T.

It can be shown (Exercise 53.4) that the number I of Definition 53.1 is unique. We denote this number I as $\int_a^b f\, d\alpha$. By Theorem 52.2, Definition 53.1 is consistent with Definition 51.3.

We now derive a result analogous to Theorem 51.13(i), and (ii).

Theorem 53.2 Let f, g, and α be bounded functions on $[a, b]$ and let $c \in \mathbf{R}$.

If $f, g \in \mathscr{R}_\alpha[a, b]$, then $cf, f + g \in \mathscr{R}_\alpha[a, b]$, and

$$\int_a^b (f + g)\, d\alpha = \int_a^b f\, d\alpha + \int_a^b g\, d\alpha, \qquad \int_a^b cf\, d\alpha = c \int_a^b f\, d\alpha$$

Proof. Let $\varepsilon > 0$. There exists a partition P_1 of $[a, b]$ such that if $P^* \supset P_1$, then

$$\left| S(f, P^*, T) - \int_a^b f\, d\alpha \right| < \frac{\varepsilon}{2}$$

There exists a partition P_2 of $[a, b]$ such that if $P^* \supset P_2$, then

$$\left| S(g, P^*, T) - \int_a^b g\, d\alpha \right| < \frac{\varepsilon}{2}$$

If we let $P = P_1 \cup P_2$ and $P^* \supset P$, then

$$\left| S(f + g, P^*, T) - \left(\int_a^b f\, d\alpha + \int_a^b g\, d\alpha \right) \right| < \varepsilon$$

Therefore $f + g \in \mathscr{R}_\alpha[a, b]$ and

$$\int_a^b (f + g)\, d\alpha = \int_a^b f\, d\alpha + \int_a^b g\, d\alpha$$

The second equation is proved in a similar way. ∎

Our next theorem demonstrates that the roles of f and α may be interchanged.

Theorem 53.3 (Integration by Parts) Let f and α be bounded functions on $[a, b]$. If $f \in \mathscr{R}_\alpha[a, b]$, then $\alpha \in \mathscr{R}_f[a, b]$ and

$$\int_a^b f\, d\alpha = f(b)\alpha(b) - f(a)\alpha(a) - \int_a^b \alpha\, df$$

Proof. Let $\varepsilon > 0$. There exists a partition P of $[a, b]$ such that if $P^* \supset P$, then

$$\left| S(f, P^*, T) - \int_a^b f\, d\alpha \right| < \varepsilon$$

Let

$$S(\alpha, P^*, T^{**}) = \sum_{i=1}^n \alpha(t_i)[f(x_i) - f(x_{i-1})]$$

be a Riemann-Stieltjes sum for α, where $P^* \supset P$. Now

$$f(b)\alpha(b) - f(a)\alpha(a) = \sum_{i=1}^n f(x_i)\alpha(x_i) - \sum_{i=1}^n f(x_{i-1})\alpha(x_{i-1})$$

so $f(b)\alpha(b) - f(a)\alpha(a) - S(\alpha, P^*, T^{**})$

$$= \sum_{i=1}^{n} f(x_i)[\alpha(x_i) - \alpha(t_i)] + \sum_{i=1}^{n} f(x_{i-1})[\alpha(t_i) - \alpha(x_{i-1})]$$

which is a Riemann-Stieltjes sum $S(f, P^{**}, T^{***})$, where $P^{**} = P^* \cup \{t_1, t_2, \ldots, t_n\}$. Now $P^{**} \supset P$ and we have

$$\left| S(\alpha, P^*, T^{**}) - \left[f(b)\alpha(b) - f(a)\alpha(a) - \int_a^b f\, d\alpha \right] \right|$$

$$= \left| \int_a^b f\, d\alpha - S(f, P^{**}, T^{***}) \right| < \varepsilon \quad\blacksquare$$

Corollary 53.4   Let f, α_1, and α_2 be bounded functions on $[a, b]$ and let $c \in \mathbf{R}$. If $f \in \mathcal{R}_{\alpha_1}[a, b] \cap \mathcal{R}_{\alpha_2}[a, b]$, then $f \in \mathcal{R}_{\alpha_1 + \alpha_2}[a, b]$ and $f \in \mathcal{R}_{c\alpha_1}[a, b]$ and

$$\int_a^b f\, d(\alpha_1 + \alpha_2) = \int_a^b f\, d\alpha_1 + \int_a^b f\, d\alpha_2, \qquad \int_a^b f\, d(c\alpha_1) = \int_a^b cf\, d\alpha_1$$

Proof.   By Theorem 53.3, α_1, $\alpha_2 \in \mathcal{R}_f[a, b]$. By Theorem 53.2, $\alpha_1 + \alpha_2 \in \mathcal{R}_f[a, b]$ and

$$\int_a^b (\alpha_1 + \alpha_2)\, df = \int_a^b \alpha_1\, df + \int_a^b \alpha_2\, df$$

Again by Theorem 53.3, $f \in \mathcal{R}_{\alpha_1 + \alpha_2}[a, b]$ and

$$\int_a^b f\, d(\alpha_1 + \alpha_2) = f(b)[\alpha_1(b) + \alpha_2(b)]$$

$$- f(a)[\alpha_1(a) + \alpha_2(a)] - \int_a^b (\alpha_1 + \alpha_2)\, df$$

$$= \left[f(b)\alpha_1(b) - f(a)\alpha_1(a) - \int_a^b \alpha_1\, df \right]$$

$$+ \left[f(b)\alpha_2(b) - f(a)\alpha_2(a) - \int_a^b \alpha_2\, df \right]$$

$$= \int_a^b f\, d\alpha_1 + \int_a^b f\, d\alpha_2$$

The second equation is proved in a similar way. $\blacksquare$

Let f and α be bounded functions on $[a, b]$. We consider the problem of imposing conditions on f and α so that $f \in \mathcal{R}_\alpha[a, b]$. Let $\varepsilon > 0$. We must be able to find a partition $P = \{x_0, x_1, \ldots, x_n\}$ such that if $P^* \supset P$, we have

$$|S(f, P^*, T^*) - I| < \varepsilon$$

In particular, we must have

$$|S(f, P, T) - S(f, P^*, T^*)| < 2\varepsilon$$

if $P^* \supset P$. We may write

$$S(f, P, T) - S(f, P^*, T^*) = \sum_{i=1}^{m} f(t_i)[\alpha(y_i) - \alpha(y_{i-1})]$$

$$- \sum_{i=1}^{m} f(t_i^*)[\alpha(y_i) - \alpha(y_{i-1})]$$

where $P^* = \{y_0, y_1, \ldots, y_m\}$ and t_i and t_i^* belong to the same interval $[x_{j-1}, x_j]$. Thus we must make the sum

$$\sum_{i=1}^{m} [f(t_i) - f(t_i^*)][\alpha(y_i) - \alpha(y_{i-1})] \qquad (53.1)$$

small. If f is continuous on $[a, b]$, we may use the uniform continuity of f to make $|f(t_i) - f(t_i^*)|$ small. If $\sum_{i=1}^{m} |\alpha(y_i) - \alpha(y_{i-1})|$ is bounded for every partition P^* of $[a, b]$, the sum (53.1) will be small. The next section is devoted to the study of functions α for which $\sum_{i=1}^{m} |\Delta\alpha_i|$ is bounded for every partition of $[a, b]$.

Exercises

53.1 Prove the second equation of Theorem 53.2.

53.2 Prove the second equation of Corollary 53.4.

53.3* Let f and α be bounded functions on $[a, b]$. Prove that the following are equivalent:
(a) $f \in \mathcal{R}_\alpha[a, b]$
(b) For every $\varepsilon > 0$, there exists a partition P of $[a, b]$ such that if $P^* \supset P$, then

$$|S(f, P^*, T^*) - S(f, P, T)| < \varepsilon$$

(c) For every $\varepsilon > 0$, there exists a partition P of $[a, b]$ such that if P^{**}, $P^* \supset P$, then

$$|S(f, P^{**}, T^{**}) - S(f, P^*, T^*)| < \varepsilon$$

53.4 Prove that the number I of Definition 53.1 is unique.

54. Functions of Bounded Variation

In this section we define *bounded variation*. Functions of bounded variation are close relatives of increasing functions (Theorem 54.9) and the results of this section will be used to extend the results of Section 51.

Definition 54.1 Let α be a function on $[a, b]$. If the set of all sums

$$\sum_{i=1}^{n} |\alpha(x_i) - \alpha(x_{i-1})|$$

where $\{x_0, x_1, \ldots, x_n\}$ is a partition of $[a, b]$, is bounded, we say that α is of *bounded variation* on $[a, b]$ and write $\alpha \in BV[a, b]$.

If $\alpha \in BV[a, b]$, we let

$$V_a^b\alpha = \text{lub} \left\{ \sum_{i=1}^{n} |\alpha(x_i) - \alpha(x_{i-1})| \right\}$$

where the least upper bound is taken over all partitions $\{x_0, x_1, \ldots, x_n\}$ of $[a, b]$. The number $V_a^b\alpha$ is called the *total variation* of α on $[a, b]$.

The next two theorems give examples of functions of bounded variation.

Theorem 54.2 If α is increasing (decreasing) on $[a, b]$, then $\alpha \in BV[a, b]$ and $V_a^b\alpha = \alpha(b) - \alpha(a)\ [= \alpha(a) - \alpha(b)]$.

Proof. Suppose α is increasing on $[a, b]$. (In case α is decreasing, the proof is similar.) If $\{x_0, x_1, \ldots, x_n\}$ is a partition of $[a, b]$, we have

$$\sum_{i=1}^{n} |\Delta\alpha_i| = \alpha(b) - \alpha(a)$$

and the conclusion follows immediately. ∎

Theorem 54.3 If α is continuous on $[a, b]$ and differentiable in (a, b) and α' is bounded on (a, b), then $\alpha \in BV[a, b]$.

Proof. Suppose $|\alpha'(x)| < M$ for x in (a, b). Let $\{x_0, x_1, \ldots, x_n\}$ be a partition of $[a, b]$. By the mean-value theorem (Theorem 49.6), we have

$$\Delta\alpha_i = \alpha'(t_i)(x_i - x_{i-1})$$

for some $t_i \in (x_{i-1}, x_i)$ for $i = 1, \ldots, n$. Thus

$$\sum_{i=1}^{n} |\Delta\alpha_i| = \sum_{i=1}^{n} |\alpha'(t_i)|(x_i - x_{i-1}) < M \sum_{i=1}^{n} (x_i - x_{i-1}) = M(b - a)$$

and therefore $\alpha \in BV[a, b]$. ∎

A continuous function on $[a, b]$ is not necessarily of bounded variation on $[a, b]$. Let

$$\alpha(x) = \begin{cases} \dfrac{1}{n} & \text{if } n \text{ is an even positive integer and } x = \dfrac{1}{n} \\ 0 & \text{if } n \text{ is an odd positive integer and } x = \dfrac{1}{n} \\ 0 & \text{if } x = 0 \text{ or } x = 1 \end{cases}$$

Let α be linear in between. (See Figure 54.1.) Then α is continuous on $[0, 1]$, but $\alpha \notin BV[0, 1]$. For the partition

$$\left\{0, \frac{1}{n}, \frac{1}{n-1}, \ldots, \frac{1}{2}, 1\right\}$$

where n is an odd positive integer, we have

$$\left|\alpha(0) - \alpha\left(\frac{1}{n}\right)\right| + \left|\alpha\left(\frac{1}{n}\right) - \alpha\left(\frac{1}{n-1}\right)\right| + \cdots$$

$$+ \left|\alpha\left(\frac{1}{3}\right) - \alpha\left(\frac{1}{2}\right)\right| + \left|\alpha\left(\frac{1}{2}\right) - \alpha(1)\right|$$

$$\geq \frac{1}{n-1} + \frac{1}{n-1} + \cdots + \frac{1}{4} + \frac{1}{4} + \frac{1}{2} + \frac{1}{2}$$

$$= 1 + \frac{1}{2} + \frac{1}{3} + \cdots + \frac{2}{n-1}$$

which is unbounded since the series $\sum_{n=1}^{\infty} 1/n$ diverges (Corollary 24.3). Therefore $\alpha \notin BV[0, 1]$.

The next theorems show that $BV[a, b]$ is closed under addition, multiplication by a constant, subtraction, and multiplication.

Theorem 54.4 If $\alpha, \beta \in BV[a, b]$ and $c \in \mathbf{R}$, then $\alpha + \beta, c\alpha \in BV[a, b]$ and

$$V_a^b(\alpha + \beta) \leq V_a^b\alpha + V_a^b\beta \qquad V_a^b(c\alpha) = |c|V_a^b\alpha$$

Proof. Let $\{x_0, x_1, \ldots, x_n\}$ be a partition of $[a, b]$. Then

$$\sum_{i=1}^{n} |[\alpha(x_i) + \beta(x_i)] - [\alpha(x_{i-1}) + \beta(x_{i-1})]|$$

$$\leq \sum_{i=1}^{n} |\Delta\alpha_i| + \sum_{i=1}^{n} |\Delta\beta_i| \leq V_a^b\alpha + V_a^b\beta$$

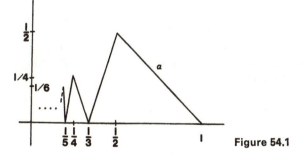

Figure 54.1

Therefore, $\alpha + \beta \in BV[a, b]$ and $V_a^b(\alpha + \beta) \le V_a^b\alpha + V_a^b\beta$.
The second equation is proved in a similar way. ■

Corollary 54.5   If α, $\beta \in BV[a, b]$, then $\alpha - \beta \in BV[a, b]$.

Proof.   By Theorem 54.4, we have $\alpha - \beta = \alpha + (-1)\beta \in BV[a, b]$. ■

Theorem 54.6   If α, $\beta \in BV[a, b]$, then $\alpha \cdot \beta \in BV[a, b]$.

Proof.   We first show that if $\alpha \in BV[a, b]$, then α is bounded on $[a, b]$. For, taking the partition $\{a, x, b\}$, where $a \le x \le b$, we have

$$|\alpha(x)| - |\alpha(a)| \le |\alpha(x) - \alpha(a)| + |\alpha(b) - \alpha(x)| \le V_a^b\alpha$$

and hence $$|\alpha(x)| \le V_a^b\alpha + |\alpha(a)|$$

for x in $[a, b]$ and thus α is bounded on $[a, b]$.

Let α, $\beta \in BV[a, b]$. There exist numbers M_1 and M_2 such that $|\alpha(x)| < M_1$ and $|\beta(x)| < M_2$ for x in $[a, b]$. Let $\{x_0, x_1, \ldots, x_n\}$ be a partition of $[a, b]$. Then

$$\sum_{i=1}^{n} |\alpha(x_i)\beta(x_i) - \alpha(x_{i-1})\beta(x_{i-1})|$$

$$= \sum_{i=1}^{n} |\alpha(x_i)\beta(x_i) - \alpha(x_i)\beta(x_{i-1}) + \alpha(x_i)\beta(x_{i-1}) - \alpha(x_{i-1})\beta(x_{i-1})|$$

$$\le \sum_{i=1}^{n} |\alpha(x_i)||\Delta\beta_i| + \sum_{i=1}^{n} |\beta(x_{i-1})||\Delta\alpha_i|$$

$$\le M_1 V_a^b\beta + M_2 V_a^b\alpha$$

Therefore, $\alpha \cdot \beta \in BV[a, b]$. ■

We next show that if $\alpha \in BV[a, b]$, the total variation is additive over the interval.

Theorem 54.7   If $\alpha \in BV[a, b]$, then α is of bounded variation on any closed subinterval of $[a, b]$ and if $c \in (a, b)$, we have

$$V_a^b\alpha = V_a^c\alpha + V_c^b\alpha$$

Proof.   Let $c \in (a, b)$. Let $P_1 = \{x_0, x_1, \ldots, x_n\}$ and $P_2 = \{y_0, y_1, \ldots, y_m\}$ be partitions of $[a, c]$ and $[c, b]$, respectively. Then $P_1 \cup P_2$ is a partition of $[a, b]$, and we have

$$\sum_{i=1}^{n} |\alpha(x_i) - \alpha(x_{i-1})| + \sum_{i=1}^{m} |\alpha(y_i) - \alpha(y_{i-1})| \le V_a^b\alpha$$

Thus for any partition $P_1 = \{x_0, x_1, \ldots, x_n\}$ of $[a, c]$, we have

$$\sum_{i=1}^{n} |\alpha(x_i) - \alpha(x_{i-1})| \leq V_a^b \alpha - \sum_{i=1}^{m} |\alpha(y_i) - \alpha(y_{i-1})|$$

Therefore, $\alpha \in BV[a, c]$ and

$$V_a^c \alpha \leq V_a^b \alpha - \sum_{i=1}^{m} |\alpha(y_i) - \alpha(y_{i-1})|$$

Thus for any partition $P_2 = \{y_0, y_1, \ldots, y_m\}$ of $[c, b]$, we have

$$\sum_{i=1}^{m} |\alpha(y_i) - \alpha(y_{i-1})| \leq V_a^b \alpha - V_a^c \alpha$$

It follows that $\alpha \in BV[c, b]$ and

$$V_a^c \alpha + V_c^b \alpha \leq V_a^b \alpha$$

Let $P = \{x_0, x_1, \ldots, x_n\}$ be a partition of $[a, b]$. Then $P \cup \{c\} = \{y_0, y_1, \ldots, y_m\}$ is a partition of $[a, b]$ and $y_j = c$ for some j, $1 \leq j \leq m - 1$. Now $\{y_0, y_1, \ldots, y_j\}$ and $\{y_j, \ldots, y_m\}$ are partitions of $[a, c]$ and $[c, b]$, respectively, and we have

$$\sum_{i=1}^{n} |\alpha(x_i) - \alpha(x_{i-1})| \leq \sum_{i=1}^{j} |\alpha(y_i) - \alpha(y_{i-1})| + \sum_{i=j+1}^{m} |\alpha(y_i) - \alpha(y_{i-1})|$$
$$\leq V_a^c \alpha + V_c^b \alpha$$

It follows that

$$V_a^b \alpha \leq V_a^c \alpha + V_c^b \alpha$$

and therefore,

$$V_a^b \alpha = V_a^c \alpha + V_c^b \alpha \qquad \blacksquare$$

As an application of Theorem 54.7, we compute the total variation of $\alpha(x) = |x|$ on $[-3, 1]$. Since α is increasing on $[0, 1]$, $V_0^1 \alpha = \alpha(1) - \alpha(0) = 1$. Similarly, $V_{-3}^0 \alpha = 3$. Thus

$$V_{-3}^1 \alpha = V_{-3}^0 \alpha + V_0^1 \alpha = 4$$

Theorem 54.8 Let $\alpha \in BV[a, b]$ and define $\beta(x) = V_a^x \alpha$ if $x \in (a, b]$ and let $\beta(a) = 0$. Then β and $\gamma = \beta - \alpha$ are increasing functions on $[a, b]$ and $\alpha = \beta - \gamma$.

Proof. Suppose $x, y \in [a, b]$ with $x < y$. By Theorem 54.7,

$$\beta(y) - \beta(x) = V_a^y \alpha - V_a^x \alpha = V_x^y \alpha \geq 0$$

and therefore β is increasing on $[a, b]$. For the partition $\{x, y\}$ of $[x, y]$ we have

$\alpha(y) - \alpha(x) \leq V_x^y \alpha$, and thus

$$\gamma(y) - \gamma(x) = V_x^y \alpha - [\alpha(y) - \alpha(x)] \geq 0$$

and therefore γ is increasing on $[a, b]$. ∎

Corollary 54.5 and Theorem 54.8 combine to give an elegant characterization of functions of bounded variation.

Theorem 54.9   Let α be a function on $[a, b]$. Then $\alpha \in BV[a, b]$ if and only if α is the difference of two increasing functions.

Proof.   The proof follows immediately from Corollary 54.5 and Theorem 54.8. ∎

Exercises

54.1   Show that the following functions are of bounded variation on the given interval and find the total variation.
(a) $\alpha(x) = x^2$ $[-2, 1]$
(b) $\alpha(x) = x^3 + x^2 - x + 1$ $[-2, 2]$

54.2   Find a formula for the total variation function $\beta(x) = V_a^x \alpha$ for the functions of Exercise 54.1.

54.3   Prove the second equation of Theorem 54.4.

54.4   Prove that if $\alpha \in BV[a, b]$, then $|\alpha| \in BV[a, b]$.

54.5   Prove that if $\alpha \in BV[a, b]$ and for some positive number M we have $|\alpha(x)| > M$ for x in $[a, b]$, then $1/\alpha \in BV[a, b]$.

54.6   Let $\alpha, \beta \in BV[a, b]$. Prove that max $\{\alpha, \beta\}$, min $\{\alpha, \beta\} \in BV[a, b]$.

54.7   Give an example of a function α which is continuous on $[a, b]$ and differentiable on (a, b) such that $\alpha \in BV[a, b]$, but α' is unbounded on (a, b).

54.8   (a)   Let $\alpha \in BV[a, b]$ and let $\beta(x) = V_a^x \alpha$. Prove that α is continuous at a point c in $[a, b]$ if and only if β is continuous at c.
(b)   Let α be a continuous function on $[a, b]$. Prove that $\alpha \in BV[a, b]$ if and only if α is the difference of two increasing continuous functions.

54.9   Let $\alpha \in BV[a, b]$ and let $\beta(x) = V_a^x \alpha$. Prove or disprove: If α is differentiable at a point c in (a, b), then β is differentiable at c.

54.10*   Let $\alpha \in BV[a, b]$. Prove that the set of discontinuities of α is countable.

54.11   If α is a function on $[a, b]$, α is said to satisfy a *uniform Lipschitz condition* of order $p > 0$ on $[a, b]$ if there exists a number $M > 0$ such that $|\alpha(x) - \alpha(y)| \leq M|x - y|^p$ for $x, y \in [a, b]$.
(a)   Let α be a function which satisfies a uniform Lipschitz condition of order p on $[a, b]$. Prove that if $p > 1$, α is constant. Prove that if $p = 1$, then $\alpha \in BV[a, b]$.

  (b) Give an example of a function α which satisfies a uniform Lipschitz condition of order $p < 1$ on $[a, b]$, but $\alpha \notin BV[a, b]$.

  (c) Give an example of a function $\alpha \in BV[a, b]$ which satisfies no uniform Lipschitz condition on $[a, b]$.

  (d) Prove that if $\alpha \in BV[a, b]$ and if g satisfies a uniform Lipschitz condition of order 1 on $[c, d] \supset \alpha([a, b])$, then $g \circ \alpha \in BV[a, b]$.

  (e) Use (d) above and an argument like that given in the proof of Theorem 51.13 to prove that if $\alpha, \beta \in BV[a, b]$, then $\alpha \cdot \beta$, $|\alpha|$, and $1/\beta$ (if β is bounded away from zero) are also in $BV[a, b]$.

55. Riemann-Stieltjes Integration with Respect to Functions of Bounded Variation

In this section we extend the results of Section 51 concerning integration with respect to increasing functions to integration with respect to functions of bounded variation. Our main tools are Theorem 54.8 which states that if $\alpha \in BV[a, b]$, then α is the difference $\beta - \gamma$ of the increasing functions $\beta(x) = V_a^x \alpha$ and $\gamma = \beta - \alpha$ and Theorem 55.2 which states that if $\alpha \in BV[a, b]$, and $f \in \mathcal{R}_\alpha[a, b]$, then $f \in \mathcal{R}_\beta[a, b] \cap \mathcal{R}_\gamma[a, b]$.

Theorem 55.1

(i) If f is continuous on $[a, b]$ and $\alpha \in BV[a, b]$, then $f \in \mathcal{R}_\alpha[a, b]$.

(ii) If $f \in BV[a, b]$ and α is continuous on $[a, b]$, then $f \in \mathcal{R}_\alpha[a, b]$.

Proof. Suppose that f is continuous on $[a, b]$ and $\alpha \in BV[a, b]$. By Theorem 54.8, $\alpha = \beta - \gamma$, where β and γ are increasing on $[a, b]$. By Theorem 51.10, $f \in \mathcal{R}_\beta[a, b] \cap \mathcal{R}_\gamma[a, b]$. By Corollary 53.4, $f \in \mathcal{R}_{\beta - \gamma}[a, b]$.

Part (ii) follows from (i) and Theorem 53.3 (integration by parts). ■

Let $\alpha \in BV[a, b]$ and suppose $f \in \mathcal{R}_\alpha[a, b]$. By Theorem 54.9, $\alpha = \beta^* - \gamma^*$, where β^* and γ^* are increasing on $[a, b]$. Now f may or may not be integrable with respect to β^* or γ^*. However, the next theorem shows that if we take $\beta^* = \beta$, where $\beta(x) = V_a^x \alpha$ and $\gamma^* = \gamma = \beta - \alpha$, then f is integrable with respect to β and γ.

Theorem 55.2   Let $\alpha \in BV[a, b]$ and suppose $f \in \mathcal{R}_\alpha[a, b]$. Let $\beta(x) = V_a^x \alpha$ and let $\gamma = \beta - \alpha$. Then $f \in \mathcal{R}_\beta[a, b] \cap \mathcal{R}_\gamma[a, b]$.

Proof. If $\beta(b) = 0$, then β is constant, and thus $f \in \mathcal{R}_\beta[a, b]$. Thus suppose $\beta(b) > 0$. Suppose $|f(x)| < M$ for all x in $[a, b]$.

Let $\varepsilon > 0$. There exists a partition P_1 of $[a, b]$ such that if $P^* \supset P_1$, we have

$$\left| S(f, P^*, T^*) - \int_a^b f \, d\alpha \right| < \frac{\varepsilon}{8}$$

There exists a partition $P_2 = \{y_0, y_1, \ldots, y_m\}$ of $[a, b]$ such that

$$\beta(b) - \frac{\varepsilon}{4M} = V_a^b\alpha - \frac{\varepsilon}{4M} < \sum_{i=1}^{m} |\alpha(y_i) - \alpha(y_{i-1})|$$

Let $P = P_1 \cup P_2 = \{x_0, x_1, \ldots, x_n\}$. Then

$$\left| S(f, P, T) - \int_a^b f\,d\alpha \right| < \frac{\varepsilon}{8}$$ (55.1)

$$\beta(b) - \sum_{i=1}^{n} |\alpha(x_i) - \alpha(x_{i-1})| < \frac{\varepsilon}{4M}$$

Now $\displaystyle\sum_{i=1}^{n} (M_i - m_i)\{[\beta(x_i) - \beta(x_{i-1})] - |\alpha(x_i) - \alpha(x_{i-1})|\}$

$$\leq 2M \sum_{i=1}^{n} \left\{ [\beta(x_i) - \beta(x_{i-1})] - \sum_{i=1}^{n} |\alpha(x_i) - \alpha(x_{i-1})| \right\}$$

$$= 2M\left(\beta(b) - \sum_{i=1}^{n} |\alpha(x_i) - \alpha(x_{i-1})| \right) < 2M\frac{\varepsilon}{4M} = \frac{\varepsilon}{2}$$ (55.2)

Let $\qquad\qquad A = \{i \mid 1 \leq i \leq n \text{ and } \alpha(x_i) - \alpha(x_{i-1}) \geq 0\}$

and $\qquad\qquad\qquad B = \{1, \ldots, n\}\backslash A$

Let $\delta = \varepsilon/(4\beta(b))$. If $i \in A$, choose $t_i, t_i^* \in [x_{i-1}, x_i]$ such that

$$f(t_i) - f(t_i^*) > M_i - m_i - \delta$$

If $i \in B$, choose $t_i, t_i^* \in [x_{i-1}, x_i]$ such that $f(t_i^*) - f(t_i) > M_i - m_i - \delta$. Then

$$\sum_{i=1}^{n} (M_i - m_i)|\alpha(x_i) - \alpha(x_{i-1})|$$

$$\leq \sum_{i \in A} [f(t_i) - f(t_i^*)]|\alpha(x_i) - \alpha(x_{i-1})|$$

$$+ \sum_{i \in B} [f(t_i^*) - f(t_i)]|\alpha(x_i) - \alpha(x_{i-1})| + \delta \sum_{i=1}^{n} |\alpha(x_i) - \alpha(x_{i-1})|$$

$$= \sum_{i=1}^{n} [f(t_i) - f(t_i^*)][\alpha(x_i) - \alpha(x_{i-1})] + \delta \sum_{i=1}^{n} |\alpha(x_i) - \alpha(x_{i-1})|$$ (55.3)

Let $T = \{t_1, \ldots, t_n\}$ and $T^* = \{t_1^*, \ldots, t_n^*\}$. By inequality (55.1), we have $S(f, P, T) - S(f, P, T^*) < \varepsilon/4$. Combining this inequality with (55.3), we have

$$\sum_{i=1}^{n} (M_i - m_i)|\alpha(x_i) - \alpha(x_{i-1})| < \frac{\varepsilon}{4} + \delta \sum_{i=1}^{n} |\alpha(x_i) - \alpha(x_{i-1})|$$

$$\leq \frac{\varepsilon}{4} + \frac{\varepsilon}{4\beta(b)}\beta(b) = \frac{\varepsilon}{2}$$ (55.4)

Inequalities (55.2) and (55.4) now give

$$U(f, P) - L(f, P) = \sum_{i=1}^{n} [M_i - m_i][\beta(x_i) - \beta(x_{i-1})] < \varepsilon$$

By Theorem 51.8, $f \in \mathscr{R}_{\beta}[a, b]$. By Corollary 53.4, $f \in \mathscr{R}_{\beta - \alpha}[a, b]$. ■

We are now ready to extend the results of Section 51.

Theorem 55.3 Let $\alpha \in BV[a, b]$ and suppose $f \in \mathscr{R}_{\alpha}[a, b]$. Then $f \in \mathscr{R}_{\alpha}[c, d]$ for every closed subinterval $[c, d]$ of $[a, b]$ and

$$\int_a^b f\, d\alpha = \int_a^c f\, d\alpha + \int_c^b f\, d\alpha$$

for all c in (a, b)

Proof. Let $\beta(x) = V_a^x \alpha$ and let $\gamma = \beta - \alpha$. By Theorem 55.2, $f \in \mathscr{R}_{\beta}[a, b] \cap \mathscr{R}_{\gamma}[a, b]$. By Theorem 51.11, $f \in \mathscr{R}_{\beta}[c, d] \cap \mathscr{R}_{\gamma}[c, d]$ for every closed subinterval $[c, d]$ of $[a, b]$ and

$$\int_a^b f\, d\beta = \int_a^c f\, d\beta + \int_c^b f\, d\beta$$

and

$$\int_a^b f\, d\gamma = \int_a^c f\, d\gamma + \int_c^b f\, d\gamma$$

for all c in (a, b). Now

$$
\begin{aligned}
\int_a^b f\, d\alpha &= \int_a^b f\, d(\beta - \gamma) = \int_a^b f\, d\beta - \int_a^b f\, d\gamma \\
&= \int_a^c f\, d\beta + \int_c^b f\, d\beta - \left(\int_a^c f\, d\gamma + \int_c^b f\, d\gamma \right) \\
&= \int_a^c f\, d(\beta - \gamma) + \int_c^b f\, d(\beta - \gamma) \\
&= \int_a^c f\, d\alpha + \int_c^b f\, d\alpha
\end{aligned}
$$

where we have used Corollary 53.4. ■

Theorem 55.4 Let $\alpha \in BV[a, b]$ and let $f \in \mathscr{R}_{\alpha}[a, b]$. Suppose that $m \leq f(x) \leq M$ for x in $[a, b]$. If g is continuous on $[m, M]$, then $g \circ f \in \mathscr{R}_{\alpha}[a, b]$.

Proof. The proof follows from Theorems 54.8, 55.2, and 51.12. ■

Theorem 55.5 Let $\alpha \in BV[a, b]$ and let $f, g \in \mathscr{R}_{\alpha}[a, b]$. Then $fg \in \mathscr{R}_{\alpha}[a, b]$.

Proof. The proof follows from Theorems 54.8, 55.2, and 51.13(iii). ■

Theorem 55.6   Let $\alpha \in BV[a, b]$ and let $\beta(x) = V_a^x \alpha$. If $f \in \mathscr{R}_\alpha[a, b]$, then $|f| \in \mathscr{R}_\alpha[a, b] \cap \mathscr{R}_\beta[a, b]$, and we have

$$\left| \int_a^b f \, d\alpha \right| \leq \int_a^b |f| \, d\beta$$

Proof.   Let $\gamma = \alpha - \beta$. By Theorem 55.2, $f \in \mathscr{R}_\beta[a, b] \cap \mathscr{R}_\gamma[a, b]$ and hence by Theorem 51.13(v), $|f| \in \mathscr{R}_\beta[a, b] \cap \mathscr{R}_\gamma[a, b]$. By Corollary 53.4, $|f| \in \mathscr{R}_\alpha[a, b]$. Let

$$p = \frac{\alpha + \beta}{2}, \qquad q = \frac{\beta - \alpha}{2}$$

By Corollary 53.4, $f, |f| \in \mathscr{R}_p[a, b] \cap \mathscr{R}_q[a, b]$. It is easy to verify that p and q are increasing on $[a, b]$. By Theorem 51.13(v) and Corollary 53.4, we have

$$
\begin{aligned}
\left| \int_a^b f \, d\alpha \right| &= \left| \int_a^b f \, d(p - q) \right| = \left| \int_a^b f \, dp - \int_a^b f \, dq \right| \\
&\leq \left| \int_a^b f \, dp \right| + \left| \int_a^b f \, dq \right| \leq \int_a^b |f| \, dp + \int_a^b |f| \, dq \\
&= \int_a^b |f| \, d(p + q) = \int_a^b |f| \, d\beta \qquad\blacksquare
\end{aligned}
$$

The last theorem of this section shows, that in certain cases, a Riemann-Stieltjes integral may be reduced to a Riemann integral.

Theorem 55.7   Let α be continuous and differentiable on $[a, b]$. If $f, \alpha' \in \mathscr{R}[a, b]$, then $f \in \mathscr{R}_\alpha[a, b]$ and

$$\int_a^b f \, d\alpha = \int_a^b f(x)\alpha'(x) \, dx$$

(Note since α' is bounded on $[a, b]$, then $\alpha \in BV[a, b]$.)

Proof.   Let $\beta(x) = x$. By Theorem 55.5, $f \cdot \alpha' \in \mathscr{R}[a, b]$. Suppose $|f(x)| < M$ for all x in $[a, b]$. In this proof we let S_γ denote the Riemann-Stieltjes sum for the integrator γ.

Let $\varepsilon > 0$. There exists a partition P of $[a, b]$ such that if $P^* \supset P$, then

$$\left| S_\beta(f\alpha', P^*, T^*) - \int_a^b f(x)\alpha'(x) \, dx \right| < \frac{\varepsilon}{2} \tag{55.5}$$

and

$$\left| S_\beta(\alpha', P^*, T^{**}) - \int_a^b \alpha'(x) \, dx \right| < \frac{\varepsilon}{4M} \tag{55.6}$$

Let $P^* = \{x_0, x_1, \ldots, x_n\} \supset P$ and let $t_i \in [x_{i-1}, x_i]$ for $i = 1, \ldots, n$.

Let $T = \{t_1, \ldots, t_n\}$. We show that

$$\left| S_\alpha(f, P^*, T) - \int_a^b f(x)\alpha'(x)\,dx \right| < \varepsilon$$

thus establishing the theorem.

By the mean-value theorem, there exist $s_i \in [x_{i-1}, x_i]$ such that $\Delta\alpha_i = \alpha(x_i) - \alpha(x_{i-1}) = \alpha'(s_i)(x_i - x_{i-1})$ for $i = 1, \ldots, n$. Let

$$v_i = \begin{cases} s_i & \text{if } \alpha'(s_i) - \alpha'(t_i) \geq 0 \\ t_i & \text{if } \alpha'(s_i) - \alpha'(t_i) < 0 \end{cases}$$

$$u_i = \begin{cases} t_i & \text{if } \alpha'(s_i) - \alpha'(t_i) \geq 0 \\ s_i & \text{if } \alpha'(s_i) - \alpha'(t_i) < 0 \end{cases}$$

for $i = 1, \ldots, n$. Let $T^* = \{v_1, \ldots, v_n\}$ and $T^{**} = \{u_1, \ldots, u_n\}$. Then

$$\begin{aligned}
\sum_{i=1}^n |\alpha'(s_i) - \alpha'(t_i)|(x_i - x_{i-1}) &= \sum_{i=1}^n [\alpha'(v_i) - \alpha'(u_i)](x_i - x_{i-1}) \\
&= S_\beta(\alpha', P^*, T^*) - S_\beta(\alpha', P^*, T^{**}) \\
&< \frac{\varepsilon}{2M}
\end{aligned} \tag{55.7}$$

by inequality (55.6). Now

$$\begin{aligned}
\left| S_\alpha(f, P^*, T) - \int_a^b f(x)\alpha'(x)\,dx \right| &= \left| \sum_{i=1}^n f(t_i)\,\Delta\alpha_i - \int_a^b f(x)\alpha'(x)\,dx \right| \\
&= \left| \sum_{i=1}^n f(t_i)\alpha'(t_i)(x_i - x_{i-1}) \right. \\
&\quad + \sum_{i=1}^n f(t_i)[\alpha'(s_i) - \alpha'(t_i)](x_i - x_{i-1}) - \left. \int_a^b f(x)\alpha'(x)\,dx \right| \\
&\leq \left| S_\beta(f\alpha', P^*, T) - \int_a^b f(x)\alpha'(x)\,dx \right| \\
&\quad + \sum_{i=1}^n |f(t_i)||\alpha'(s_i) - \alpha'(t_i)|(x_i - x_{i-1}) \\
&< \frac{\varepsilon}{2} + M \sum_{i=1}^n |\alpha'(s_i) - \alpha'(t_i)|(x_i - x_{i-1}) < \frac{\varepsilon}{2} + M\frac{\varepsilon}{2M} = \varepsilon
\end{aligned}$$

where we have used inequalities (55.5) and (55.7). ∎

We may deduce one part of the fundamental theorem of calculus which gives the standard method of evaluating Riemann integrals.

Corollary 55.8   Let f be continuous and differentiable on $[a, b]$. If $f' \in \mathscr{R}[a, b]$, then

$$\int_a^b f'(x)\, dx = f(b) - f(a)$$

Proof.   Take α of Theorem 55.7 to be f and take f of Theorem 55.7 to be the constant 1. Then

$$f(b) - f(a) = \int_a^b 1\, df = \int_a^b f'(x)\, dx \qquad\qquad \blacksquare$$

We adopt the standard notation

$$f(x)\Big|_a^b = f(b) - f(a)$$

Examples.

$$\int_0^2 x^2\, dx^2 = \int_0^2 x^2 2x\, dx = \int_0^2 2x^3\, dx = \frac{x^4}{2}\Big|_0^2 = 8$$

Let $[x]$ denote the greatest integer less than or equal to x.

$$\int_0^2 [x]\, dx^2 = \int_0^2 [x]\, 2x\, dx = \int_0^1 [x] 2x\, dx + \int_1^2 [x] 2x\, dx$$

$$= \int_0^1 0\, dx + \int_1^2 2x\, dx = x^2\Big|_1^2 = 3$$

The last integral may also be evaluated using integration by parts.

$$\int_0^2 [x]\, dx^2 = [x] x^2\Big|_0^2 - \int_0^2 x^2\, d[x] = 8 - [1 + 4] = 3$$

The integral $\int_0^2 x^2\, d[x]$ was evaluated using the method of Example 51.9.

Exercises

55.1   Let $\alpha \in BV[a, b]$. Give an example of increasing functions β^* and γ^* and a function $f \in \mathscr{R}_\alpha[a, b]$ such that $\alpha = \beta^* - \gamma^*$ and $f \notin \mathscr{R}_{\beta^*}[a, b]$.

55.2   Give the details of the proofs of Theorems 55.4 and 55.5.

55.3   Evaluate the following integrals:

(a) $\displaystyle\int_0^3 \sqrt{x}\, dx^3$

(b) $\displaystyle\int_1^4 \sqrt{x^2 + 1}\, d(x^2 + 3)$

(c) $\displaystyle\int_1^4 (x - [x])\, dx^2$

55.4 Let $\{a_n\}$ be a sequence. For $x \geq 1$, let

$$A(x) = \sum_{n=1}^{[x]} a_n$$

Suppose that f is continuous and differentiable in $[1, a]$. Prove that

$$\sum_{n \leq a} a_n f(n) = -\int_1^a A(x)f'(x)\, dx + A(a)f(a)$$

55.5 (Second mean-value theorem.) Suppose that α is continuous on $[a, b]$ and that f is increasing on $[a, b]$. Prove that there exists a point c in $[a, b]$ such that

$$\int_a^b f\, d\alpha = f(a)\int_a^c d\alpha + f(b)\int_c^b d\alpha$$

55.6 Let $\alpha \in BV[a, b]$ and let f be continuous on $[a, b]$. Prove that

$$\lim_{\text{norm } P \to 0} S(f, P, T) = \int_a^b f\, d\alpha$$

55.7 Let $\alpha \in BV[a, b]$ and let f be a bounded function on $[a, b]$. Prove that if $\lim_{\text{norm } P \to 0} S(f, P, T)$ exists, then $f \in \mathcal{R}_\alpha[a, b]$ and

$$\lim_{\text{norm } P \to 0} S(f, P, T) = \int_a^b f\, d\alpha$$

55.8 Let $\alpha \in BV[a, b]$ and suppose that α is continuous on $[a, b]$. Prove that if $f \in \mathcal{R}_\alpha[a, b]$, then

$$\lim_{\text{norm } P \to 0} S(f, P, T) = \int_a^b f\, d\alpha$$

55.9 Let $\alpha \in BV[a, b]$. Prove that if $f, g \in \mathcal{R}_\alpha[a, b]$, then max $\{f, g\}$ and min $\{f, g\}$ are in $\mathcal{R}_\alpha[a, b]$.

55.10 Let $\alpha \in BV[a, b]$. Suppose that $f \in \mathcal{R}_\alpha[a, b]$ and there exists a positive number M such that $|f(x)| \geq M$ for x in $[a, b]$. Prove that $1/f \in \mathcal{R}_\alpha[a, b]$.

56. The Riemann Integral

In this section we prove the fundamental theorem of integral calculus and a change of variable formula for the Riemann integral.

Theorem 56.1 (Fundamental Theorem of Integral Calculus)

(i) Let f be continuous and differentiable on $[a, b]$ and suppose $f' \in \mathcal{R}[a, b]$. Then

$$\int_a^b f'(x)\, dx = f(b) - f(a)$$

(ii) Let $f \in \mathcal{R}[a, b]$. Let

$$F(x) = \int_a^x f(t)\, dt$$

for x in $(a, b]$ and let $F(a) = 0$. Then F is continuous on $[a, b]$ and F is differentiable at every point at which f is continuous. If x is such a point, we have $F'(x) = f(x)$.

Proof.   Part (i) restates Corollary 55.8.

Suppose $|f(x)| < M$ for all x in $[a, b]$. Let $\varepsilon > 0$. Let $\delta = \varepsilon/M$. If $|x - y| < \delta$ with $x < y$, we have

$$|F(x) - F(y)| = \left| \int_x^y f(t)\, dt \right| \le \int_x^y |f(t)|\, dt \le \int_x^y M\, dt = M(y - x) < \varepsilon$$

and therefore F is continuous on $[a, b]$.

Suppose f is continuous at a point x in $[a, b]$. Let $\varepsilon > 0$. There exists $\delta > 0$ such that if $|y - x| < \delta$, then $|f(y) - f(x)| < \varepsilon$. Suppose $0 < |y - x| < \delta$. Without loss of generality, we assume $y > x$. (The case for $y < x$ is similar.) By Theorem 51.13(iv), since $f(x) - \varepsilon < f(z) < f(x) + \varepsilon$ for all z between x and y, we have

$$[f(x) - \varepsilon](y - x) \le F(y) - F(x) = \int_x^y f(t)\, dt \le [f(x) + \varepsilon](y - x)$$

Therefore, $$f(x) - \varepsilon \le \frac{F(y) - F(x)}{y - x} \le f(x) + \varepsilon$$

which implies that

$$\left| \frac{F(y) - F(x)}{y - x} - f(x) \right| \le \varepsilon$$

and thus $F'(x) = f(x)$. ∎

Corollary 56.2   Let f be a continuous function on $[a, b]$.

(i) There exists a function F such that F is continuous on $[a, b]$, differentiable on $[a, b]$, and $F'(x) = f(x)$ for all x in $[a, b]$.

(ii) If G is continuous on $[a, b]$, differentiable on $[a, b]$, and $G'(x) = f(x)$ for all x in $[a, b]$, then

$$\int_a^b f(x)\, dx = G(b) - G(a)$$

Proof.   Part (i) follows from Theorem 56.1(i).

Let G be as in part (ii). By Theorem 56.1(i),

$$\int_a^b f(x)\, dx = \int_a^b G'(x)\, dx = G(b) - G(a)$$ ∎

In our subsequent work we will need the following change of variable formula. Another change of variable formula where g is not assumed to be increasing is given in Exercise 56.3.

Theorem 56.3   Let $f \in \mathcal{R}[a, b]$ and let g be a strictly increasing function from $[c, d]$ onto $[a, b]$ such that g is differentiable on $[c, d]$ and $g' \in \mathcal{R}[c, d]$. Then $(f \circ g) \cdot g' \in \mathcal{R}[c, d]$ and

$$\int_a^b f(x)\, dx = \int_c^d f(g(t))g'(t)\, dt$$

Proof.   Suppose $|f(x)| < M$ for x in $[a, b]$. Let $\varepsilon > 0$. There exists a partition P_1 of $[a, b]$ such that if $P_1^* \supset P_1$, we have

$$\left| S(f, P_1^*, T_1) - \int_a^b f(x)\, dx \right| < \frac{\varepsilon}{3}$$

There exists a partition P_2 of $[c, d]$ such that if $P_2^* \supset P_2$, we have

$$\left| S(g', P_2^*, T_2) - \int_c^d g'(x)\, dx \right| < \frac{\varepsilon}{3M} \tag{56.1}$$

Since g is a strictly increasing function from $[c, d]$ onto $[a, b]$, it follows that $P_3 = g^{-1}(P_1)$ is a partition of $[c, d]$. Let $P = P_2 \cup P_3$. Let $P^* = \{y_0, y_1, \ldots, y_n\}$ be a partition of $[c, d]$ containing P. Let $s_i \in [y_{i-1}, y_i]$ for $i = 1, \ldots, n$ and let $T^* = \{s_1, s_2, \ldots, s_n\}$. We show that

$$\left| S((f \circ g) \cdot g', P^*, T^*) - \int_a^b f(x)\, dx \right| < \varepsilon$$

which will prove the theorem.

Let $x_i = g(y_i)$ for $i = 0, \ldots, n$ and $t_i = g(s_i)$ for $i = 1, \ldots, n$. Then

$$P_1^* = g(P^*) = \{x_0, x_1, \ldots, x_n\}$$

is a partition of $[a, b]$ which contains P_1 and $t_i \in [x_{i-1}, x_i]$ for $i = 1, \ldots, n$. Therefore,

$$\left| S(f, P_1^*, T_1) - \int_a^b f(x)\, dx \right| < \frac{\varepsilon}{3} \tag{56.2}$$

where $T_1 = \{t_1, \ldots, t_n\}$.

By the mean-value theorem $x_i - x_{i-1} = g(y_i) - g(y_{i-1}) = g'(u_i)(y_i - y_{i-1})$ where $u_i \in (y_{i-1}, y_i)$ for $i = 1, \ldots, n$. Now

$$S((f \circ g) \cdot g', P^*, T^*) - \int_a^b f(x)\, dx$$

$$= \sum_{i=1}^n f(g(s_i))g'(s_i)\, \Delta y_i - \int_a^b f(x)\, dx \tag{56.3}$$

$$= \sum_{i=1}^n f(g(s_i))g'(s_i)\, \Delta y_i - \sum_{i=1}^n f(t_i)\, \Delta x_i + \sum_{i=1}^n f(t_i)\, \Delta x_i - \int_a^b f(x)\, dx$$

$$= \sum_{i=1}^{n} f(g(s_i))g'(s_i)\,\Delta y_i - \sum_{i=1}^{n} f(g(s_i))g'(u_i)\,\Delta y_i + S(f, P_1^*, T_1)$$

$$- \int_a^b f(x)\,dx$$

$$= \sum_{i=1}^{n} f(g(s_i))[g'(s_i) - g'(u_i)]\,\Delta y_i + S(f, P_1^*, T_1) - \int_a^b f(x)\,dx$$

We have

$$\left| \sum_{i=1}^{n} f(g(s_i))[g'(s_i) - g'(u_i)]\,\Delta y_i \right| \le \sum_{i=1}^{n} |f(g(s_i))||g'(s_i) - g'(u_i)|\,\Delta y_i$$

$$\le M \sum_{i=1}^{n} |g'(s_i) - g'(u_i)|\,\Delta y_i$$

$$\le M\left(\frac{2\varepsilon}{3M}\right) = \frac{2\varepsilon}{3} \qquad (56.4)$$

where we have used inequality (56.1) and an argument like that given in the proof of Theorem 55.7. Inequalities (56.2), (56.3), and (56.4) combine to give

$$\left| S((f \circ g) \cdot g', P^*, T^*) - \int_a^b f(x)\,dx \right| < \varepsilon \qquad \blacksquare$$

For example, if $f \in \mathscr{R}[a, b]$ and we let $g(t) = t + k$, we have

$$\int_a^b f(x)\,dx = \int_{a-k}^{b-k} f(t + k)\,dt$$

Exercises

56.1 Let I be an interval of $\mathbf{R}$ and let f be a function on I. We say that f is *absolutely continuous* on I if for every $\varepsilon > 0$, there exists $\delta > 0$ such that if $\{(a_k, b_k)\}_{k=1}^{n}$ is a finite set of open subintervals of I with $(a_i, b_i) \cap (a_j, b_j) = \varnothing$ for $i \ne j$ and $\sum_{k=1}^{n} (b_k - a_k) < \delta$, then $\sum_{k=1}^{n} |f(b_k) - f(a_k)| < \varepsilon$.

    (a) Prove that if α is absolutely continuous on $[a, b]$, then α is continuous and of bounded variation on $[a, b]$.

    (b) Prove that if α is absolutely continuous on $[a, b]$, then $\alpha = \beta - \gamma$, where β and γ are increasing and absolutely continuous on $[a, b]$.

56.2 Let $\alpha \in BV[a, b]$ and suppose that $f \in \mathscr{R}_\alpha[a, b]$. Let

$$F(x) = \int_a^x f\,d\alpha$$

for x in $(a, b]$ and let $F(a) = 0$.

    (a) Prove that $F \in BV[a, b]$.

    (b) Prove that if α is continuous at x, then F is continuous at x.

    (c) Prove that if α is absolutely continuous on $[a, b]$, then F is absolutely continuous on $[a, b]$. (See Exercise 56.1 for the definition of *absolutely continuous*.)

(d) Prove that if α is increasing on $[a, b]$, then F is differentiable at each point at which α is differentiable and f is continuous; and that if x is such a point, we have $F'(x) = f(x)\alpha'(x)$.

56.3 Let g' be continuous on $[c, d]$ and let f be continuous on $g([c, d])$. Let $a = g(c)$, $b = g(d)$. Prove that

$$\int_a^b f(x)\, dx = \int_c^d f(g(t))g'(t)\, dt$$

56.4 Let g be continuous on $[a, b]$ and suppose that f is increasing on $[a, b]$.

(a) Prove that $\int_a^b f(x)g(x)\, dx = f(a)\cdot \int_a^c g(x)\, dx + f(b)\cdot \int_c^b g(x)\, dx$ for some c in $[a, b]$.

(b) (Bonnet's Theorem) Prove that if in addition to the hypotheses in part (a), f is nonnegative, then

$$\int_a^b f(x)g(x)\, dx = f(b)\cdot \int_c^b g(x)\, dx$$

for some c in $[a, b]$.

56.5 (a)* Prove the trapezoid rule: If f'' exists on $[c - h, c + h]$, then

$$\int_{c-h}^{c+h} f = h[f(c + h) + f(c - h)] - \frac{2f''(\xi)h^3}{3}$$

where $c - h < \xi < c + h$.

(b) Subdivide the interval $[0, 1]$ into 10 equal parts and use the trapezoid rule on each subdivision to estimate $\int_0^1 dx/(1 + x^2)$. Estimate the error. (We shall see later that $\int_0^1 dx/(1 + x^2) = \pi/4$.)

56.6 (a) Prove Simpson's rule: If $f^{(4)}$ exists on $[c - h, c + h]$, then

$$\int_{c-h}^{c+h} f = \frac{h}{3}[f(c + h) + 4f(c) + f(c - h)] - \frac{f^{(4)}(\xi)h^5}{90}$$

(b) Repeat Exercise 56.5(b) replacing the trapezoid rule with Simpson's rule.

56.7 (a) Derive the rectangle rule, which approximates $\int_{c-h}^{c+h} f(x)\, dx$ by $2hf(c)$, together with a remainder estimate similar to the trapezoid rule and Simpson's rule [see Exercises 56.5(a) and 56.6(a)].

(b) Repeat Exercise 56.5(b) replacing the trapezoid rule with the rectangle rule.

56.8 Prove that the following form of the integration by parts formula is valid for Riemann integrals: If f', $g' \in \mathscr{R}[a, b]$, then

$$\int_a^b f'g = f(b)g(b) - f(a)g(a) - \int_a^b fg'$$

56.9 Suppose that f is $(n + 1)$-times differentiable on $[c, d]$ and $f^{(n+1)} \in \mathscr{R}[c, d]$. Derive Taylor's theorem (Theorem 50.3) with integral remainder by beginning with the formula

$$f(d) - f(c) = \int_c^d f'$$

and proceeding by using integration by parts. The next step would be

$$\int_c^d f' = f'(t)(d-t)\Big|_c^d + \int_c^d f''(t)(d-t)\,dt$$

56.10* Suppose $f' \in \mathscr{R}[0, 1]$. Prove that

$$|f(\tfrac{1}{2})| \le \int_0^1 |f| + \tfrac{1}{2}\int_0^1 |f'|$$

57. Measure Zero

The next two sections of this chapter are devoted to deriving a necessary and sufficient condition for the existence of the Riemann integral. We will prove (Theorem 58.5) that $f \in \mathscr{R}[a, b]$ if and only if f is continuous "almost everywhere" in $[a, b]$. To define *almost everywhere*, we must begin with the concept of *measure zero*.

Definition 57.1 Let X be a subset of **R**. We say that X is of *measure zero* if for every $\varepsilon > 0$, there exists a countable family $\{I_n\}$ of open intervals such that $\bigcup_{n=1}^{\infty} I_n \supset X$ and $\sum_{n=1}^{\infty} |I_n| < \varepsilon$, where $|I|$ denotes the length of the interval I.

In other words, a subset X of **R** is of measure zero if X can be covered by a countable family of open intervals the sum of whose lengths can be made arbitrarily small.

Examples. A finite set $X = \{x_1, \ldots, x_n\}$ is of measure zero. For if $\varepsilon > 0$, the family

$$\left\{\left(x_i - \frac{\varepsilon}{3n}, x_i + \frac{\varepsilon}{3n}\right)\right\}_{i=1}^{n}$$

covers X and

$$\sum_{i=1}^{n}\left[\left(x_i + \frac{\varepsilon}{3n}\right) - \left(x_i - \frac{\varepsilon}{3n}\right)\right] = n\frac{2\varepsilon}{3n} = \frac{2\varepsilon}{3} < \varepsilon$$

and therefore X is of measure zero.

The set Q of rational numbers is of measure zero. Indeed any countable set is of measure zero. For suppose

$$X = \{x_1, x_2, \ldots\}$$

is a countable set. For each positive integer n, choose an open interval I_n containing x_n such that $|I_n| < \varepsilon/3^n$. Then

$$\bigcup_{n=1}^{\infty} I_n \supset X$$

and
$$\sum_{n=1}^{\infty} |I_n| \leq \varepsilon \sum_{n=1}^{\infty} \frac{1}{3^n} = \frac{\varepsilon}{2} < \varepsilon$$

Therefore, X is of measure zero.

The method of the last example may be used to prove that a countable union of sets of measure zero is of measure zero.

Theorem 57.2   If $\{X_n\}$ is a countable family of sets of measure zero, then $\bigcup_{n=1}^{\infty} X_n$ is of measure zero.

Proof.   Let $\varepsilon > 0$. For each positive integer n, there exists a countable family $\{I_i^{(n)}\}_{i=1}^{\infty}$ of open intervals such that $\bigcup_{i=1}^{\infty} I_i^{(n)} \supset X_n$ and $\sum_{i=1}^{\infty} |I_i^{(n)}| < \varepsilon/3^n$. The family

$$\{I_i^{(n)} \mid i,n \in \mathbf{P}\}$$

is countable (Theorem 9.5) and covers $\bigcup_{n=1}^{\infty} X_n$ and

$$\sum_{n=1}^{\infty} \sum_{i=1}^{\infty} |I_i^{(n)}| \leq \sum_{n=1}^{\infty} \frac{\varepsilon}{3^n} = \frac{\varepsilon}{2} < \varepsilon \qquad \blacksquare$$

Having given several examples of sets of measure zero, we now give an example of a set which is not of measure zero.

Theorem 57.3   The closed interval $[a, b]$ is not of measure zero.

Proof.   It is easy to show by induction that if $\{I_k\}_{k=1}^{n}$ is a *finite* set of open intervals which covers $[a, b]$, then

$$\sum_{k=1}^{n} |I_k| \geq b - a \qquad (57.1)$$

Suppose $[a, b]$ is of measure zero. Taking $\varepsilon = (b - a)/2$, there exists a countable family $\{I_k\}$ of open intervals which covers $[a, b]$ such that

$$\sum_{k=1}^{\infty} |I_k| < \frac{b - a}{2}$$

By the Heine-Borel theorem (Theorem 34.2), some finite subcollection $\{I_{k_j}\}_{j=1}^{n}$ covers $[a, b]$. Now

$$\sum_{j=1}^{n} |I_{k_j}| < \sum_{k=1}^{\infty} |I_k| < \frac{b - a}{2}$$

which contradicts (57.1). Therefore $[a, b]$ is not of measure zero. $\blacksquare$

Theorem 57.3 furnishes yet another proof that irrational numbers exist. For if $[0, 1]$ contains only rational numbers, by Theorem 57.2, $[0, 1]$ is countable, hence of measure zero. However, this contradicts Theorem 57.3.

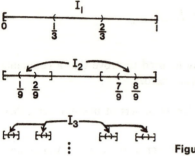

Figure 57.1

A remarkable fact is that uncountable sets of measure zero exist. The standard example is the Cantor set (see Figure 57.1).

Definition 57.4 From the closed interval $[0, 1]$ we remove the open middle third $I_1 = (\frac{1}{3}, \frac{2}{3})$ yielding the set

$$[0, \tfrac{1}{3}] \cup [\tfrac{2}{3}, 1]$$

From each of the closed intervals $[0, \frac{1}{3}]$ and $[\frac{2}{3}, 1]$ we remove the open middle thirds $(\frac{1}{9}, \frac{2}{9})$ and $(\frac{7}{9}, \frac{8}{9})$ yielding the set

$$[0, \tfrac{1}{9}] \cup [\tfrac{2}{9}, \tfrac{1}{3}] \cup [\tfrac{2}{3}, \tfrac{7}{9}] \cup [\tfrac{8}{9}, 1]$$

We let $I_2 = (\frac{1}{9}, \frac{2}{9}) \cup (\frac{7}{9}, \frac{8}{9})$. At the nth stage we have closed intervals

$$J_1, J_2, \ldots, J_{2^n}$$

From each of the closed intervals J_k we remove the open middle third $I^{(k)}$ for $k = 1, \ldots, 2^n$, yielding the set

$$(J_1 \backslash I^{(1)}) \cup (J_2 \backslash I^{(2)}) \cup \cdots \cup (J_{2^n} \backslash I^{(2^n)})$$

We let $I_n = \bigcup_{k=1}^{2^n} I^{(k)}$. The *Cantor set* C is the set

$$[0, 1] \Big\backslash \bigcup_{n=1}^{\infty} I_n$$

Some properties of the Cantor set are immediately obvious. Since C is the intersection

$$[0, 1] \cap \left(\bigcap_{n=1}^{\infty} I_n' \right)$$

of closed sets, C is closed (Theorem 38.8). Since $C \subset [0, 1]$, C is bounded. By Theorem 43.9, C is compact.

Theorem 57.5 The Cantor set is uncountable and of measure zero.

Proof. We first show that C is of measure zero. At the nth stage (see Definition 57.4), C is contained in the union of closed intervals

$$J_1 \cup J_2 \cup \cdots \cup J_{2^n}$$

where $|J_k| = 1/3^n$ for $k = 1, \ldots, 2^n$. Let $\varepsilon > 0$. Choose a positive integer n such that $(\frac{2}{3})^n < \varepsilon/2$. Cover each closed interval J_k by an open interval I_k such that $|I_k| < 1/3^n + \varepsilon/2^{n+1}$, for $k = 1, \ldots, 2^n$. Then $\{I_k\}_{k=1}^{2^n}$ is a finite set of open intervals which covers C and

$$\sum_{k=1}^{2^n} |I_k| < \sum_{k=1}^{2^n} \left(\frac{1}{3^n} + \frac{\varepsilon}{2^{n+1}} \right) = \left(\frac{2}{3} \right)^n + \frac{\varepsilon}{2} < \varepsilon$$

Therefore C is of measure zero.

Suppose that C is countable. Then

$$C = \{a_1, a_2, \ldots\}$$

Choose the interval $[0, \frac{1}{3}]$ or $[\frac{2}{3}, 1]$ to which a_1 does *not* belong. Call this interval $[c_1, d_1]$. Remove the open middle third from this interval $[c_1, d_1]$ yielding $[b_1, b_2] \cup [b_3, b_4]$. Choose the interval $[b_1, b_2]$ or $[b_3, b_4]$ to which a_2 does not belong. Call this interval $[c_2, d_2]$. Continue in this way. Then the sequence $\{c_k\}$ converges and if $c = \lim_{k \to \infty} c_k$, since C is closed, $c \in C$. But now $c \neq a_n$ for every positive integer n, and we have the desired contradiction. Therefore, C is uncountable. ∎

The argument given above to show that C is uncountable is similar to the proof of Theorem 9.8 which shows that **R** is uncountable. The fact that C is uncountable may also be deduced from Corollary 47.8

Definition 57.6 Let I be an interval in **R**. A property P is said to hold *almost everywhere in I* if the set of x in I for which P fails is of measure zero.

Examples. We will be concerned primarily with functions which are continuous almost everywhere. According to Definition 57.6, f is continuous almost everywhere on $[a, b]$ if the set

$$\{x \in [a, b] \mid f \text{ is not continuous at } x\}$$

is of measure zero.

Let $f(x) = 1$ for all x in $[a, b]$ and let

$$g(x) = \begin{cases} 1 & \text{if } x \text{ is irrational} \\ 0 & \text{if } x \text{ is rational} \end{cases}$$

Then $f(x) = g(x)$ almost everywhere in $[a, b]$ since

$$\{x \in [a, b] \mid f(x) \neq g(x)\}$$

is countable and hence of measure zero. Notice that g is continuous at no point of $[0, 1]$ and f is continuous at every point of $[0, 1]$. Therefore, $f = g$ almost everywhere in $[0, 1]$ implies nothing about continuity relationships.

Exercises

57.1 Let X be a subset of $\mathbf{R}$ which is of measure zero. Prove that every subset of X is of measure zero.

57.2 Give the inductive proof that if $\{I_k\}_{k=1}^n$ is a finite set of open intervals which covers $[a, b]$, then

$$\sum_{k=1}^{n} |I_k| \geq b - a$$

57.3 Prove that any interval of $\mathbf{R}$ is not of measure zero.

57.4 Prove that the Cantor set is the set of all numbers of the form $\sum_{n=1}^{\infty} a_n/3^n$ where a_n is either 0 or 2. Use this fact to deduce that C is uncountable.

57.5 Prove that the Cantor set is uncountable using Corollary 47.8.

57.6 For this exercise assume the following theorem.
Theorem If α is increasing on $[a, b]$, then α is differentiable almost everywhere in $[a, b]$.
(a) Rephrase this theorem using the term *measure zero*.
(b) Prove that if $\alpha \in BV[a, b]$, then α is differentiable almost everywhere in $[a, b]$.

57.7 Let f be continuous on $[a, b]$ and suppose that $f(x) = 0$ almost everywhere in $[a, b]$. Prove that $f(x) = 0$ for all x in $[a, b]$.

57.8 Prove that if $\alpha \in BV[a, b]$, then α is continuous almost everywhere in $[a, b]$.

57.9 Prove or disprove: If f and g are continuous almost everywhere in $[a, b]$, then $f + g$ is continuous almost everywhere in $[a, b]$.

58. A Necessary and Sufficient Condition for the Existence of the Riemann Integral

This section is devoted to proving that $f \in \mathcal{R}[a, b]$ if and only if f is continuous almost everywhere in $[a, b]$.

Definition 58.1 Let f be a bounded function on $[a, b]$. If I is an interval, we let

$$\omega_f(I) = \text{lub } \{f(x) \mid x \in I \cap [a, b]\} - \text{glb } \{f(x) \mid x \in I \cap [a, b]\}$$

We call $\omega_f(I)$ the *oscillation of f on I*. If $x \in [a, b]$, we let

$$\omega_f(x) = \text{glb } \{\omega_f(J)\}$$

where the greatest lower bound is taken over all open intervals J containing x. We call $\omega_f(x)$ the *oscillation of f at x.* (Since $\omega_f(J) \geq 0$ for every open interval J, glb $\{\omega_f(J)\}$ exists.)

The oscillation function enters into the theory of the integral in the following way. Let f be a bounded function on $[a, b]$ and let $P = \{x_0, x_1 \ldots, x_n\}$ be a partition of $[a, b]$. Let $I_k = [x_{k-1}, x_k]$ for $k = 1, \ldots, n$. Then we may write

$$U(f, P) - L(f, P) = \sum_{k=1}^{n} \omega_f(I_k)|I_k| \qquad (58.1)$$

If $f \in \mathscr{R}[a, b]$, by Theorem 51.8, the sum in (58.1) must be small. But for the sum in (58.1) to be small either $\omega_f(I_k)$ or $|I_k|$ must be small for $k = 1, \ldots, n$. The next lemma shows that the size of the oscillation is linked to continuity.

Lemma 58.2 Let f be a bounded function on $[a, b]$ and let $x \in [a, b]$. Then f is continuous at x if and only if $\omega_f(x) = 0$.

Proof. Suppose that f is continuous at x. Let $\varepsilon > 0$. There exists $\delta > 0$ such that if $|y - x| < \delta$, then $|f(y) - f(x)| < \varepsilon$. Thus if $y \in J = (x - \delta, x + \delta)$, then $f(x) - \varepsilon < f(y) < f(x) + \varepsilon$. Therefore, $\omega_f(J) \leq 2\varepsilon$ (verify). It follows that $0 \leq \omega_f(x) \leq 2\varepsilon$ for every $\varepsilon > 0$, and hence $\omega_f(x) = 0$.

Now suppose $\omega_f(x) = 0$. Let $\varepsilon > 0$. There exists an open interval J containing x such that $\omega_f(J) < \varepsilon$. There exists $\delta > 0$ such that $(x - \delta, x + \delta) \subset J$. Now if $|y - x| < \delta$, then $y \in J$, and hence

$$|f(y) - f(x)| \leq \omega_f(J) < \varepsilon \qquad \blacksquare$$

Lemma 58.3 Let f be a bounded function on $[a, b]$ and let $\varepsilon > 0$. Then $\{x \in [a, b] \mid \omega_f(x) < \varepsilon\}$ is open (in $[a, b]$).

Proof. Let

$$X = \{x \in [a, b] \mid \omega_f(x) < \varepsilon\}$$

Let $x \in X$. We must show that there exists $\delta > 0$ such that $(x - \delta, x + \delta) \cap [a, b] \subset X$. Since $\omega_f(x) < \varepsilon$, there exists an open interval J containing x such that $\omega_f(J) < \varepsilon$. Since $J \cap [a, b]$ is open, there exists $\delta > 0$ such that $(x - \delta, x + \delta) \cap [a, b] \subset J \cap [a, b]$. Let $y \in (x - \delta, x + \delta) \cap [a, b]$. Then $y \in J$; so $\omega_f(y) \leq \omega_f(J) < \varepsilon$. Therefore $y \in X$ and X is open. $\blacksquare$

Lemma 58.4 Let f be a bounded function on a closed interval J. If $\omega_f(x) < \varepsilon$ for every x in J, there exists a partition P of J such that

$$U(f, P) - L(f, P) < \varepsilon|J|$$

Proof. For each x in J, there exists an open interval I_x containing x such that $\omega_f(\bar{I}_x) < \varepsilon$. By the Heine-Borel theorem, the collection $\{I_x\}$ has a finite subcover $\{I_1, \ldots, I_n\}$. Let $P = \{x_0, x_1, \ldots, x_n\}$ be the set of end points of the intervals $\{I_1 \cap J, \ldots, I_n \cap J\}$. Then $\omega_f([x_{i-1}, x_i]) < \varepsilon$ for $i = 1, \ldots, n$, and we have

$$U(f, P) - L(f, P) = \sum_{i=1}^{n} \omega_f([x_{i-1}, x_i])(x_i - x_{i-1}) < \varepsilon |J| \qquad \blacksquare$$

We are now ready to prove the main theorem of this section.

Theorem 58.5 (Lebesgue) Let f be a bounded function on $[a, b]$. Then $f \in \mathcal{R}[a, b]$ if and only if f is continuous almost everywhere in $[a, b]$.

Proof. Suppose that $f \in \mathcal{R}[a, b]$. Let X be the set of points in $[a, b]$ at which f is not continuous and let

$$X_m = \left\{ x \in [a, b] \mid \omega_f(x) \geq \frac{1}{m} \right\}$$

for $m = 1, 2, \ldots$. By Lemma 58.2, $X = \bigcup_{m=1}^{\infty} X_m$. By Theorem 57.2 it suffices to prove that X_m is of measure zero for every positive integer m.

Let m be a positive integer. Let $\varepsilon > 0$. By Theorem 51.8, there exists a partition $P = \{x_0, x_1, \ldots, x_n\}$ of $[a, b]$ such that

$$\sum_{i=1}^{n} \omega_f(I_i)|I_i| < \frac{\varepsilon}{2m} \qquad (58.2)$$

where $I_i = [x_{i-1}, x_i]$ for $i = 1, \ldots, n$.

Let $X_m^* = X_m \cap P$ and let $X_m^{**} = X_m \backslash X_m^*$. If $x \in X_m^{**}$, then $x \in (x_{i-1}, x_i)$ for some positive integer i and $\omega_f([x_{i-1}, x_i]) \geq \omega_f(x) \geq 1/m$. Let $I_{i_1}, \ldots, I_{i_k}$ denote those intervals $[x_{i-1}, x_i]$ which contain a point of X_m^{**}. Then

$$\frac{1}{m} \sum_{j=1}^{k} |I_{i_j}| \leq \sum_{j=1}^{k} \omega_f(I_{i_j})|I_{i_j}| \leq \sum_{i=1}^{n} \omega_f(I_i)|I_i| < \frac{\varepsilon}{2m}$$

Therefore,
$$\sum_{j=1}^{k} |I_{i_j}| < \frac{\varepsilon}{2}$$

Now X_m^{**} is covered by the interiors of $I_{i_1}, \ldots, I_{i_k}$ the sum of whose lengths is less than $\varepsilon/2$. Since X_m^* is a finite set we may cover X_m^* by a finite number of open intervals the sum of whose lengths is less than $\varepsilon/2$. Therefore $X_m = X_m^* \cup X_m^{**}$ is covered by open intervals the sum of whose lengths is less than ε, and hence X_m is of measure zero.

Now suppose that f is continuous almost everywhere in $[a, b]$. If $\omega_f([a, b]) = 0$, then f is constant and hence $f \in \mathcal{R}[a, b]$; so we assume that $\omega_f([a, b]) > 0$. Let $\varepsilon > 0$. Choose a positive integer m such that $(b - a)/m < \varepsilon/2$.

If X_m is defined as above, then X_m is of measure zero. Thus there exists a countable set $\{I_n\}$ of open intervals such that $X_m \subset \bigcup_{n=1}^{\infty} I_n$ and

$$\sum_{n=1}^{\infty} |I_n| < \frac{\varepsilon}{2\omega_f([a, b])}$$

By Lemma 58.3 and Theorem 39.5, X_m is closed in $[a, b]$ and thus is compact. Hence there is a finite subcollection $\{I_{n_1}, \ldots, I_{n_k}\}$ of $\{I_n\}$ which covers X_m. Now

$$[a, b] \setminus \left(\bigcup_{j=1}^{k} I_{n_j} \right)$$

is a union of closed intervals $J_1, \ldots, J_p$. If $x \in J_i$, then $\omega_f(x) < 1/m$; so by Lemma 58.4, there exists a partition P_i of J_i such that

$$U(f, P_i) - L(f, P_i) < \frac{|J_i|}{m}$$

for $i = 1, \ldots, p$. Let $P = \bigcup_{j=1}^{p} P_j \cup \{a\} \cup \{b\}$. Now

$$U(f, P) - L(f, P) = \sum_{i=1}^{p} \{U(f, P_i) - L(f, P_i)\} + \sum_{i=1}^{k} \omega_f(\bar{I}_{n_i})|I_{n_i} \cap [a, b]|$$

$$< \frac{1}{m} \sum_{i=1}^{p} |J_i| + \omega_f([a, b]) \sum_{i=1}^{k} |I_{n_i}|$$

$$< \frac{b - a}{m} + \omega_f([a, b]) \frac{\varepsilon}{2\omega_f([a, b])} < \frac{\varepsilon}{2} + \frac{\varepsilon}{2} = \varepsilon$$

By Theorem 51.8, $f \in \mathscr{R}[a, b]$. ∎

Corollary 58.6 If $f \in \mathscr{R}[a, b]$ and $\int_a^b |f| = 0$, then $f(x) = 0$ almost everywhere on $[a, b]$.

Proof. We show that $f(x) = 0$ for each x in $[a, b]$ at which f is continuous. By Theorem 58.5, this implies that $f(x) = 0$ almost everywhere on $[a, b]$.

Suppose $f(x) \neq 0$ for some x in $[a, b]$ at which f is continuous. Then $|f(x)| > 0$ so there exists a closed interval $[c, d]$ containing x such that $|f(y)| > |f(x)|/2$ for all y in $[c, d]$. Now

$$0 = \int_a^b |f| = \int_a^c |f| + \int_c^d |f| + \int_d^b |f|$$

$$\geq \int_c^d |f| \geq \frac{|f(x)|}{2}(d - c) > 0$$

and we have a contradiction. Therefore, $f(x) = 0$ almost everywhere in $[a, b]$. ∎

Corollary 58.7   Let $f \in \mathscr{R}[a, b]$ and let

$$F(x) = \int_a^x f$$

for x in $(a, b]$ and let $F(a) = 0$. Then F is differentiable almost everywhere in $[a, b]$ and $F'(x) = f(x)$ almost everywhere in $[a, b]$.

Proof.   By Theorem 56.1(ii), $F'(x) = f(x)$ for each x in $[a, b]$ at which f is continuous. By Theorem 58.5, f is continuous almost everywhere in $[a, b]$ and the conclusion now follows.   ∎

Exercises

58.1   Let $f \in \mathscr{R}[a, b]$ and suppose that $\int_a^x f = 0$ for all x in $(a, b]$. Prove that $f(x) = 0$ almost everywhere in $[a, b]$.

58.2*   Let $f, g \in \mathscr{R}[a, b]$ and suppose that $f(x) = g(x)$ almost everywhere in $[a, b]$. Prove that $\int_a^b f = \int_a^b g$.

58.3   Using Theorem 58.5 prove that if $f \in BV[a, b]$, then $f \in \mathscr{R}[a, b]$.

58.4   State and prove an analogue of Theorem 58.5 valid for Riemann-Stieltjes integrals.

58.5   Using Theorem 58.5 prove that if $f, g \in \mathscr{R}[a, b]$, h is continuous on the range of f, and $c \in \mathbf{R}$, then $f + g$, cf, $f \cdot g$, and $h \circ f \in \mathscr{R}[a, b]$. Prove that if in addition $1/f$ is bounded on $[a, b]$, then $1/f \in \mathscr{R}[a, b]$.

59.   Improper Riemann-Stieltjes Integrals

So far in this chapter, all of our functions have been bounded and all of our integrals have been computed over closed, bounded intervals. In this section we relax these restrictions by defining improper Riemann-Stieltjes integrals.

Definition 59.1   Let f and α be functions defined on the interval $[a, \infty)$. Suppose that $f \in \mathscr{R}_\alpha[a, b]$ for every $b > a$. The *improper Riemann-Stieltjes integral (of the first kind)* $\int_a^\infty f \, d\alpha$ is the ordered pair (f, F), where

$$F(b) = \int_a^b f \, d\alpha$$

for $b > a$.

In a similar way, we define the improper integral $\int_{-\infty}^a f \, d\alpha$.

This definition is analogous to the definition (Definition 22.1) of an infinite series. The infinite series

$$\sum_{n=1}^{\infty} a_n$$

was defined to be the ordered pair $(\{a_n\}, \{s_n\})$, where $s_n = a_1 + \cdots + a_n$, $n = 1, 2, \ldots$. The function f of Definition 59.1 corresponds to the function a of Definition 22.1 and the function F of Definition 59.1 corresponds to the function s of Definition 22.1. The function F is sometimes called the *partial integral* in analogy with the partial sum of an infinite series. Convergence of an improper integral is defined in terms of convergence of the partial integral $F(b) = \int_a^b f \, d\alpha$.

Definition 59.2 Let $\int_a^\infty f \, d\alpha$ be an improper integral. If $\lim_{b \to \infty} \int_a^b f \, d\alpha$ exists we say that the improper integral $\int_a^\infty f \, d\alpha$ *converges*, and we write

$$\int_a^\infty f \, d\alpha = \lim_{b \to \infty} \int_a^b f \, d\alpha$$

If $\lim_{b \to \infty} \int_a^b f \, d\alpha$ does not exist, we say that the improper integral $\int_a^\infty f \, d\alpha$ *diverges*.

Convergence and divergence of the improper integral $\int_{-\infty}^a f \, d\alpha$ is defined in a similar way.

If the improper integrals $\int_{-\infty}^a f \, d\alpha$ and $\int_a^\infty f \, d\alpha$ are both convergent, we define

$$\int_{-\infty}^\infty f \, d\alpha = \int_{-\infty}^a f \, d\alpha + \int_a^\infty f \, d\alpha$$

Examples. The improper integral $\int_1^\infty (1/x^2) \, dx$ converges to 1 since

$$\lim_{a \to \infty} \int_1^a \frac{1}{x^2} \, dx = \lim_{a \to \infty} \left(1 - \frac{1}{a} \right) = 1$$

The improper integral $\int_1^\infty 1/\sqrt{x} \, dx$ diverges since

$$\lim_{a \to \infty} \int_1^a \frac{1}{\sqrt{x}} \, dx = \lim_{a \to \infty} 2(\sqrt{a} - 1) = \infty$$

Let $f(x) = x$. Then $\int_a^\infty f$ and $\int_{-\infty}^a f$ diverge (for every a), and hence the improper integral $\int_{-\infty}^\infty f$ diverges. However, $\lim_{a \to \infty} \int_{-a}^a f = 0$. We call $\lim_{a \to \infty} \int_{-a}^a f \, d\alpha$ (if this limit exists) the *Cauchy principal value* of the improper integral $\int_{-\infty}^\infty f \, d\alpha$. We have just seen that the Cauchy principal value may exist for a divergent integral. It is easy to show that if the improper integral $\int_{-\infty}^\infty f \, d\alpha$ converges, then the Cauchy principal value is equal to the number $\int_{-\infty}^\infty f \, d\alpha = \int_{-\infty}^a f \, d\alpha + \int_a^\infty f \, d\alpha$.

Many of the theorems for infinite series have analogues for improper integrals. The methods of proof are also often similar. For example, if the improper integrals $\int_a^\infty f \, d\alpha$ and $\int_a^\infty g \, d\alpha$ are convergent, then the improper integral $\int_a^\infty (f + g) \, d\alpha$ is convergent and

$$\int_a^\infty (f + g) \, d\alpha = \int_a^\infty f \, d\alpha + \int_a^\infty g \, d\alpha \qquad (59.1)$$

This follows from Theorem 53.2 (for ordinary Riemann-Stieltjes integrals) by taking limits.

Corresponding to an absolutely convergent series, we have the concept of an absolutely convergent improper integral.

Definition 59.3 We say that the improper integral $\int_a^\infty f \, d\alpha$ *converges absolutely* if $\int_a^\infty |f| \, d\alpha$ converges and $f \in \mathcal{R}_\alpha[a, b]$ for all $b > a$. We say that the improper integral $\int_a^\infty f \, d\alpha$ *converges conditionally* if $\int_a^\infty |f| \, d\alpha$ diverges, but $\int_a^\infty f \, d\alpha$ converges.

The next theorem parallels Theorem 24.1 for infinite series.

Theorem 59.4 Let α be an increasing function on $[a, \infty)$ and let f be a nonnegative function on $[a, \infty)$. Then the improper integral $\int_a^\infty f \, d\alpha$ converges if and only if the function $F(b) = \int_a^b f \, d\alpha$ is bounded on $[a, \infty)$.

Proof. Suppose F is bounded on $[a, \infty)$. Let

$$L = \text{lub} \, \{F(b) \mid b \in [a, \infty)\}.$$

Let $\varepsilon > 0$. There exists a number b such that $L - \varepsilon < F(b)$. If $x \geq b$, then

$$L - \varepsilon < F(b) \leq F(x) \leq L < L + \varepsilon$$

and thus $\int_a^\infty f \, d\alpha = L$.

Now suppose that $\int_a^\infty f \, d\alpha$ converges. If $b \in [a, \infty)$, we have

$$\int_a^b f \, d\alpha \leq \int_a^\infty f \, d\alpha$$

and therefore F is bounded on $[a, \infty)$. ∎

Theorem 59.5 (Comparison Test) Let α be an increasing function on $[a, \infty)$. Suppose that f and g are functions such that $0 \leq f(x) \leq g(x)$ for x in $[a, \infty)$. If $\int_a^\infty g \, d\alpha$ converges and $f \in \mathcal{R}_\alpha[a, b]$ for all $b > a$, then $\int_a^\infty f \, d\alpha$ converges and

$$\int_a^\infty f \, d\alpha \leq \int_a^\infty g \, d\alpha$$

Proof. The proof follows from Theorem 59.4 and the inequality

$$\int_a^b f \, d\alpha \le \int_a^b g \, d\alpha \le \int_a^\infty g \, d\alpha \qquad \blacksquare$$

We can now show that an absolutely convergent improper integral is convergent (for increasing integrators).

Theorem 59.6 Let α be an increasing function on $[a, \infty)$. If the improper integral $\int_a^\infty f \, d\alpha$ converges absolutely, then $\int_a^\infty f \, d\alpha$ converges.

Proof. If $x \ge a$, then $0 \le |f(x)| - f(x) \le 2|f(x)|$. By Theorem 59.5, $\int_a^\infty (|f| - f) \, d\alpha$ converges. By equation (59.1), $\int_a^\infty f \, d\alpha$ converges. $\blacksquare$

We now give an example of a conditionally convergent improper integral. Let

$$\alpha(n) = \begin{cases} 0 & \text{if } n \text{ is an odd positive integer} \\ 1 & \text{if } n = 2m \text{ is an even positive integer and } m \text{ is odd} \\ -1 & \text{if } n = 2m \text{ is an even positive integer and } m \text{ is even.} \end{cases}$$

Extend α linearly to $[1, \infty)$. (See Figure 59.1). Let $\beta(x) = \int_1^x \alpha(t) \, dt$ for $x \ge 1$. Then $0 \le \beta(x) \le 1$ for $x \ge 1$. Using integration by parts, we have

$$\int_1^b \frac{\alpha(x)}{x} \, dx = \frac{\beta(b)}{b} + \int_1^b \frac{\beta(x)}{x^2} \, dx$$

Since β is bounded on $[1, \infty)$, $\lim_{b \to \infty} (\beta(b)/b) = 0$. Since $0 \le (\beta(x)/x^2) \le 1/x^2$ for x in $[1, \infty)$ and $\int_1^\infty (1/x^2) \, dx$ is convergent, by Theorem 59.5, $\int_1^\infty (\beta(x)/x^2) \, dx$ is convergent. Therefore, $\int_1^\infty (\alpha(x)/x) \, dx$ is convergent.

Since $(1/x) \ge 1/(2k - 1)$ if $x \le 2k - 1$ and $\int_{2k-3}^{2k-1} |\alpha(x)| \, dx = 1$, we have

$$\int_1^{2n-1} \left| \frac{\alpha(x)}{x} \right| dx = \sum_{k=2}^n \int_{2k-3}^{2k-1} \left| \frac{\alpha(x)}{x} \right| dx \ge \sum_{k=2}^n \frac{1}{2k - 1}$$

for every positive integer n. The series $\sum_{k=2}^\infty 1/(2k - 1)$ diverges, and therefore the sequence of partial sums $\{\sum_{k=2}^n 1/(2k - 1)\}$ is unbounded (Theorem

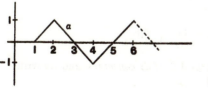

Figure 59.1

24.1). Therefore, the function $F(b) = \int_1^b |\alpha(x)/x| \, dx$ is unbounded. By Theorem 59.4, $\int_1^\infty |\alpha(x)/x| \, dx$ diverges. Thus the improper integral $\int_1^\infty (\alpha(x)/x) \, dx$ is conditionally convergent.

Our next theorem demonstrates a link between infinite series and improper integrals. See Exercise 59.10 for another result.

Theorem 59.7 (Integral Test) Let f be a nonnegative decreasing function on $[1, \infty)$. Then the infinite series $\sum_{n=1}^\infty f(n)$ converges if and only if the improper integral $\int_1^\infty f(x) \, dx$ converges.

Proof. Since $f(k + 1) \le f(x) \le f(k)$ for $k \le x \le k + 1$, we have

$$f(k + 1) \le \int_k^{k+1} f(x) \, dx \le f(k)$$

Therefore,
$$\sum_{k=1}^n f(k + 1) \le \int_1^{n+1} f \le \sum_{k=1}^n f(k)$$

The proof now follows from Theorems 24.1 and 59.4. ∎

We close this section by defining the improper integral of the second kind and giving two examples. Analogues of the theorems about improper integrals of the first kind are valid for improper integrals of the second kind (see Exercise 59.4).

Definition 59.8 Let f and α be functions defined on the half-open interval $[a, b)$. Suppose that $f \in \mathscr{R}_\alpha[a, c]$ for $a < c < b$. The *improper Riemann-Stieltjes integral (of the second kind)*

$$\int_a^{b^-} f \, d\alpha$$

is the ordered pair (f, F), where

$$F(c) = \int_a^c f \, d\alpha$$

for $a < c < b$. In a similar way we define the improper Riemann-Stieltjes integral

$$\int_{a^+}^b f \, d\alpha$$

If
$$\lim_{c \to b^-} F(c) = \lim_{c \to b^-} \int_a^c f \, d\alpha$$

exists, we say that the improper integral $\int_a^{b^-} f \, d\alpha$ *converges,* and we write

$$\int_a^{b^-} f\, d\alpha = \lim_{c \to b^-} F(c)$$

If $\lim_{c \to b^-} F(c)$ does not exist, we say that the improper integral $\int_a^{b^-} f\, d\alpha$ *diverges*. In a similar way, we define convergence and divergence of the improper integral $\int_{a^+}^b f\, d\alpha$.

Examples. The improper integral $\int_{0^+}^1 (1/\sqrt{x})\, dx$ converges to 2, since

$$\lim_{c \to 0^+} \int_c^1 \frac{1}{\sqrt{x}}\, dx = \lim_{c \to 0^+} 2 - 2\sqrt{c} = 2$$

The improper integral $\int_{0^+}^1 (1/x^2)\, dx$ diverges, since

$$\lim_{c \to 0^+} \int_c^1 \frac{1}{x^2}\, dx = \lim_{c \to 0^+} \frac{1}{c} - 1 = \infty$$

Exercises

59.1 Prove that the improper integral $\int_1^\infty (1/x^p)dx$ diverges if $p \le 1$ and converges if $p > 1$.

59.2 Determine the values of p and q for which the following improper integrals converge

$$\int_0^\infty \frac{x^{p-1}}{1+xq}\, dx \qquad \int_0^1 x^p(1-x^2)^q\, dx$$

59.3 Prove equation (59.1).

59.4 State and prove analogues of Theorems 59.4, 59.5, and 59.6 for improper integrals of the second kind.

59.5 Prove that if $f \in \mathscr{R}[a, b]$, we have

$$\int_{a^+}^b f = \int_a^b f = \int_a^{b^-} f$$

[For this reason, if $\alpha(x) = x$, we usually write

$$\int_{a^+}^b f = \int_a^b f = \int_a^{b^-} f$$

if either improper integral converges.]

59.6 Give an example of a function f such that the improper integral $\int_1^\infty f(x)\, dx$ converges, but $\lim_{x \to \infty} f(x)$ does not exist.

59.7 State and prove an integration-by-parts formula valid for improper integrals.

59.8 Prove an analogue of Theorem 26.4 valid for improper integrals.

59.9 State and prove analogues of Corollaries 28.4 and 28.7 valid for improper integrals.

59.10 Let f be a decreasing nonnegative function on $[1, \infty)$.

(a) Prove that if the series $\sum_{n=1}^{\infty} f(n)$ converges to L, then

$$\left| \sum_{n=1}^{m} f(n) - L \right| < \int_{m}^{\infty} f(x) \, dx$$

(b) Prove that if the improper integral $\int_{1}^{\infty} f(x) \, dx$ converges to L, then

$$\left| \int_{1}^{x} f(t) \, dt - L \right| < \sum_{n=[x]}^{\infty} f(n)$$

where $[x]$ denotes the greatest integer less than or equal to x.

59.11 Prove that if the improper integral $\int_{-\infty}^{\infty} f \, d\alpha$ converges, then the Cauchy principal value of $\int_{-\infty}^{\infty} f \, d\alpha$ exists and equals $\int_{-\infty}^{\infty} f \, d\alpha$.

59.12 (a) Given an example of functions f and α such that $\int_{1}^{\infty} f \, d\alpha$ converges absolutely, but $\int_{1}^{\infty} f \, d\alpha$ diverges.

(b) Suppose $\alpha \in BV[a, b]$ for every $b > a$. Let $\beta(x) = V_{a}^{x} \alpha$ for $x > a$. Prove that if the integrals $\int_{a}^{\infty} f \, d\alpha$ and $\int_{a}^{\infty} f \, d\beta$ converge absolutely, then $\int_{a}^{\infty} f \, d\alpha$ converges.

59.13 (a) Prove that the integral $\int_{0}^{\infty} t^{x-1} e^{-t} dt$ converges for positive x. The function $\Gamma(x) = \int_{0}^{\infty} t^{x-1} e^{-t} dt$, $0 < x < \infty$, is called the *gamma function*.

(b) Prove that $\Gamma(x + 1) = x\Gamma(x)$ for positive x. Deduce that $\Gamma(n + 1) = n!$ for $n = 1, 2, 3, \ldots$.

X

Sequences and Series of Functions

In this chapter we shall investigate convergence of sequences and series of functions. The results of this chapter will extend the results of Chapters IV and V. In particular, we shall be concerned with the following problem. If each element of a sequence $\{f_n\}$ of functions has a particular property, when does the limit function f have the same property? For example, we might ask whether the limit function f is continuous if each function f_n is continuous. We begin by defining convergence of a sequence of functions.

60. Pointwise Convergence and Uniform Convergence

We shall be concerned with sequences of real-valued functions defined on some set X. We shall denote a sequence as $\{f_n\}_{n=1}^{\infty}$ (or sometimes just $\{f_n\}$). Thus for each positive integer n we have a real-valued function f_n defined on a set X. The set X will not, in general, be a subset of the real numbers.

Definition 60.1 Let $\{f_n\}$ be a sequence of functions on a set X. Let f be a function on X. We say that $\{f_n\}$ *converges pointwise* to f on X if $\lim_{n \to \infty} f_n(x) = f(x)$ for every x in X.

Example 60.2 Let

$$f_n(x) = x^n \qquad 0 \le x \le 1$$

Then $\{f_n\}$ converges pointwise to f on $[0, 1]$, where

$$f(x) = \begin{cases} 0 & 0 \le x < 1 \\ 1 & x = 1 \end{cases}$$

Notice that in Example 60.2 each function f_n is continuous on $[0, 1]$ but that the pointwise limit function f is not continuous on $[0, 1]$. Thus continuity is *not* in general preserved under taking pointwise limits. Thus if we wish to preserve continuity we must seek a stronger definition of convergence of a sequence of functions. One of the most useful notions is that of uniform convergence. (See Theorems 60.4 and 61.1.)

Definition 60.3 Let $\{f_n\}$ be a sequence of functions on a set X. Let f be a function on X. We say that $\{f_n\}$ *converges uniformly* to f on X if for every $\varepsilon > 0$, there exists a positive integer N such that if $n \geq N$, then

$$|f_n(x) - f(x)| < \varepsilon$$

for every x in X.

The crucial part of Definition 60.3 is that $|f_n(x) - f(x)| < \varepsilon$ for *every* x in X (if $n \geq N$). That is, N depends on ε, but *not* on x.

It is clear that if a sequence of functions converges uniformly it also converges pointwise (verify).

We now prove that continuity is preserved under uniform convergence.

Theorem 60.4 Let $\{f_n\}$ be a sequence of functions which converges uniformly to a function f on a metric space M. If each f_n is continuous at a point a in M, then f is continuous at a.

Proof.   Let d be the metric for M. Let $\varepsilon > 0$. There exists a positive integer N such that

$$|f_N(x) - f(x)| < \frac{\varepsilon}{3}$$

for every $x \in M$. There exists $\delta > 0$ such that if $d(x, a) < \delta$, then

$$|f_N(x) - f_N(a)| < \frac{\varepsilon}{3}$$

Now if $d(x, a) < \delta$, then

$$|f(x) - f(a)| \leq |f(x) - f_N(x)| + |f_N(x) - f_N(a)| + |f_N(a) - f(a)|$$

$$< \frac{\varepsilon}{3} + \frac{\varepsilon}{3} + \frac{\varepsilon}{3} = \varepsilon \qquad\blacksquare$$

Corollary 60.5   If $\{f_n\}$ is a sequence of continuous functions which converges uniformly to a function f on a metric space M, then f is continuous.

It follows from Corollary 60.5 that the sequence $\{f_n\}$ of Example 60.2 does *not* converge uniformly to f. Thus a sequence may converge pointwise, but fail to converge uniformly.

The conclusion of Corollary 60.5 may also be written

$$\lim_{x \to a} \lim_{n \to \infty} f_n(x) = \lim_{n \to \infty} \lim_{x \to a} f_n(x)$$

For $\lim_{n \to \infty} f_n(x) = f(x)$ and since f is continuous, $\lim_{x \to a} f(x) = f(a)$. Also $\lim_{x \to a} f_n(x) = f_n(a)$ since each function f_n is continuous, and $\lim_{n \to \infty} f_n(a) = f(a)$. Therefore, Corollary 60.5 provides another example of a result which deals with interchanging limits. (See also Theorem 40.2 and the subsequent discussion.)

Example 60.6 We let $C[a, b]$ denote the set of continuous real-valued functions on $[a, b]$. We define a metric d on $C[a, b]$ by the formula

$$d(f, g) = \text{lub } \{|f(x) - g(x)| \mid x \in [a, b]\}$$

Since the function $f - g$ is continuous on $[a, b]$, it is bounded, and therefore the least upper bound of the above set exists. The rest of the proof that d is a metric on $C[a, b]$ is straightforward (Exercise 60.9). When we refer to $C[a, b]$, unless otherwise specified, we will assume that the metric on $C[a, b]$ is that defined above. This metric is often called the *metric of uniform convergence* because of the next theorem.

Theorem 60.7 Let $\{f_n\}$ be a sequence of functions in $C[a, b]$. Then $\lim_{n \to \infty} f_n = f$ in the metric space $C[a, b]$ if and only if f_n converges uniformly to f on $[a, b]$.

Proof. Suppose $\{f_n\}$ converges to f in the metric space $C[a, b]$. Let $\varepsilon > 0$. There exists a positive integer N such that if $n \geq N$, then $d(f_n, f) < \varepsilon$. This means $d(f_n, f) = \text{lub } \{|f_n(x) - f(x)| \mid x \in [a, b]\} < \varepsilon$, or equivalently, $|f_n(x) - f(x)| < \varepsilon$ for every x in $[a, b]$. By Definition 60.3, this says that f_n converges uniformly to f on $[a, b]$.

Conversely, suppose $\{f_n\}$ is a sequence in $C[a, b]$ which converges uniformly to a function f on $[a, b]$. By Corollary 60.5 we have that $f \in C[a, b]$. We must show that $\lim_{n \to \infty} f_n = f$ in the metric of $C[a, b]$.

Let $\varepsilon > 0$. Since f_n converges uniformly to f on $[a, b]$, there exists a positive integer N such that if $n \geq N$, then

$$|f_n(x) - f(x)| < \varepsilon/2 \text{ for every } x \in [a, b]$$

Thus if $n \geq N$, $\varepsilon/2$ is an upper bound for the set

$$\{|f_n(x) - f(x)| \mid x \in [a, b]\}$$

and so $d(f_n, f) \leq \varepsilon/2 < \varepsilon$. Therefore, $\lim_{n \to \infty} f_n = f$. ∎

Corollary 60.8 $C[a, b]$ is a complete metric space in the metric of Example 60.6.

Proof.   The proof is left as an exercise (Exercise 60.10).   ■

Exercises

60.1   Prove directly from Definition 60.3 that the sequence in Example 60.2 does not converge uniformly.

60.2   Let $f_n(x) = 1/(1 + n^2x^2)$ and $g_n(x) = nx(1 - x)^n$, $x \in [0, 1]$. Prove that $\{f_n\}$ and $\{g_n\}$ converge pointwise but not uniformly on $[0, 1]$.

60.3   Prove that if the sequence $\{f_n\}$ converges uniformly to the function f on a set X, then $\{f_n\}$ converges pointwise to f.

60.4   Give an example of a sequence of continuous functions (on a metric space) which converges pointwise, but not uniformly to a continuous function.

60.5   Let $\{f_n\}$ be a sequence of bounded functions on a set X. Prove that if $\{f_n\}$ converges uniformly to f on X, then f is bounded. Give an example to show that this statement is false if uniform convergence is replaced by pointwise convergence.

60.6   We say that a sequence $\{f_n\}$ is *uniformly bounded* on X if there exists M such that $|f_n(x)| < M$ for all $x \in X$ and for every positive integer n.

Suppose that $\{f_n\}$ and $\{g_n\}$ converge uniformly to f and g, respectively, on X.

  (a)   Prove that if each function f_n is bounded, then f is bounded and $\{f_n\}$ is uniformly bounded.

  (b)   Prove that $\{f_n + g_n\}$ converges uniformly to $f + g$ on X.

  (c)   Prove that if $c \in \mathbf{R}$, then $\{cf_n\}$ converges uniformly to cf on X.

  (d)   Prove that if $\{f_n\}$ and $\{g_n\}$ are uniformly bounded sequences on X, then $\{f_n g_n\}$ converges uniformly to $f \cdot g$.

  (e)   Give an example to show that statement (d) is false if the boundedness hypothesis is removed.

60.7   (*Cauchy condition for uniform convergence*) Let $\{f_n\}$ be a sequence of functions on a set X. Prove that there exists a function f such that $\{f_n\}$ converges uniformly to f on X if and only if for every $\varepsilon > 0$, there exists a positive integer N such that if $m, n \geq N$, then

$$|f_n(x) - f_m(x)| < \varepsilon \qquad \text{for every } x \text{ in } X$$

60.8*   Prove Dini's theorem. If a sequence $\{f_n\}$ of continuous functions converges pointwise to a continuous function f on a compact metric space M and if $f_n(x) \geq f_{n+1}(x)$ for every x in M and for every positive integer n, then $\{f_n\}$ converges uniformly to f on M.

60.9   Prove that $C[a, b]$ is a metric space with d defined in Example 60.6.

60.10*   Prove that $C[a, b]$ is a complete metric space with d defined in Example 60.6.

61. Integration and Differentiation of Uniformly Convergent Sequences

We first prove that the analogue of Corollary 60.5 holds for Riemann-Stieltjes integrable functions.

Theorem 61.1 Let α be a function of bounded variation on $[a, b]$ and let $\{f_n\}$ be a sequence of functions in $\mathscr{R}_\alpha[a, b]$ which converges uniformly to a function f. Then $f \in \mathscr{R}_\alpha[a, b]$ and

$$\lim_{n \to \infty} \int_a^b f_n \, d\alpha = \int_a^b f \, d\alpha$$

The conclusion of Theorem 61.1 may be written

$$\lim_{n \to \infty} \int_a^b f_n \, d\alpha = \int_a^b (\lim_{n \to \infty} f_n) \, d\alpha$$

where $\lim_{n \to \infty} f_n$ denotes the uniform limit of the sequence $\{f_n\}$.

Proof. First assume that α is increasing. If $\alpha(a) = \alpha(b)$, there is nothing to prove; so assume $\alpha(a) < \alpha(b)$. Let $\varepsilon > 0$. There exists a positive integer N such that

$$|f(x) - f_N(x)| < \frac{\varepsilon}{3[\alpha(b) - \alpha(a)]}$$

for all x in $[a, b]$. It follows that for any partition P of $[a, b]$

$$|U(f - f_N, P)| \le \frac{\varepsilon}{3} \qquad |L(f - f_N, P)| \le \frac{\varepsilon}{3} \tag{61.1}$$

Since $f_N \in \mathscr{R}_\alpha[a, b]$, by Theorem 51.8 there exists a partition P of $[a, b]$ such that

$$U(f_N, P) - L(f_N, P) < \frac{\varepsilon}{3} \tag{61.2}$$

Let

$$M_i = \text{lub} f \qquad \text{on} \quad [x_{i-1}, x_i]$$
$$M_i^* = \text{lub} f_N \qquad \text{on} \quad [x_{i-1}, x_i]$$
$$M_i^{**} = \text{lub} (f - f_N) \qquad \text{on} \quad [x_{i-1}, x_i]$$

If $x \in [x_{i-1}, x_i]$, then

$$f(x) = [f(x) - f_N(x)] + f_N(x) \le M_i^{**} + M_i^*$$

and hence

$$M_i \le M_i^{**} + M_i^*$$

It follows that

$$U(f, P) \leq U(f - f_N, P) + U(f_N, P) \qquad (61.3)$$

Similarly $\qquad L(f, P) \geq L(f - f_N, P) + L(f_N, P) \qquad (61.4)$

Using inequalities (61.1) through (61.4), we have

$$U(f, P) - L(f, P) \leq U(f - f_N, P) - L(f - f_N, P) + U(f_N, P) - L(f_N, P)$$

$$< |U(f - f_N, P)| + |L(f - f_N, P)| + \frac{\varepsilon}{3}$$

$$\leq \frac{\varepsilon}{3} + \frac{\varepsilon}{3} + \frac{\varepsilon}{3} = \varepsilon$$

By Theorem 51.8, $f \in \mathscr{R}_\alpha[a, b]$.
We next show that

$$\lim_{n \to \infty} \int_a^b f_n \, d\alpha = \int_a^b f \, d\alpha$$

(where α is increasing).

Let $\varepsilon > 0$. There exists a positive integer N such that if $n \geq N$, then

$$|f_n(x) - f(x)| < \frac{\varepsilon}{2[\alpha(b) - \alpha(a)]}$$

By Theorem 51.13(iv), if $n \geq N$, then

$$\int_a^b |f_n - f| \, d\alpha \leq \int_a^b \frac{\varepsilon}{2[\alpha(b) - \alpha(a)]} \, d\alpha = \frac{\varepsilon}{2}$$

Thus if $n \geq N$, then

$$\left| \int_a^b f \, d\alpha - \int_a^b f_n \, d\alpha \right| = \left| \int_a^b (f - f_n) \, d\alpha \right| \leq \int_a^b |f - f_n| \, d\alpha \leq \frac{\varepsilon}{2} < \varepsilon$$

If α is of bounded variation on $[a, b]$, we let $\alpha = \alpha_1 - \alpha_2$ be the decomposition of α as the difference of two increasing functions as in Theorem 54.8. By Theorem 55.2, $f_n \in \mathscr{R}_{\alpha_1}[a, b] \cap \mathscr{R}_{\alpha_2}[a, b]$ for every positive integer n. By our preceding result, $f \in \mathscr{R}_{\alpha_1}[a, b] \cap \mathscr{R}_{\alpha_2}[a, b]$, and hence $f \in \mathscr{R}_\alpha[a, b]$ by Corollary 53.4. Furthermore,

$$\lim_{n \to \infty} \int_a^b f_n \, d\alpha = \lim_{n \to \infty} \left(\int_a^b f_n \, d\alpha_1 - \int_a^b f_n \, d\alpha_2 \right)$$

$$= \lim_{n \to \infty} \int_a^b f_n \, d\alpha_1 - \lim_{n \to \infty} \int_a^b f_n \, d\alpha_2$$

$$= \int_a^b f \, d\alpha_1 - \int_a^b f \, d\alpha_2 = \int_a^b f \, d\alpha \qquad \blacksquare$$

If $\{f_n\}$ is a sequence of functions which converges pointwise to a function f on $[a, b]$ and each f_n is in $\mathscr{R}[a, b]$, it may happen that f fails to be in $\mathscr{R}[a, b]$. For example, let $\{r_1, r_2, \ldots\}$ be an enumeration of the rational numbers in $[a, b]$. Let

$$f_n(x) = \begin{cases} 0 & \text{if } x \in \{r_1, \ldots, r_n\} \\ 1 & \text{if } x \in \{r_1, \ldots, r_n\}' \cap [a, b] \end{cases}$$

Then $f_n \in \mathscr{R}[a, b]$ for every positive integer n, but the pointwise limit function

$$f(x) = \begin{cases} 0 & \text{if } x \text{ is rational} \\ 1 & \text{if } x \text{ is irrational} \end{cases}$$

is not in $\mathscr{R}[a, b]$. Some difficulties of this sort can be eliminated by considering a more general integral such as the Lebesgue integral (see Chapter XIV). See Exercises 61.1 to 61.4 for further examples.

No analogue of Corollary 60.5 and Theorem 61.1 holds for differentiation. If we let

$$f_n(x) = \frac{x^n}{n} \qquad 0 \le x \le 1$$

then $\{f_n\}$ converges uniformly to 0 on $[0, 1]$ (verify). However, $f_n'(1) = 1 \nrightarrow 0$. Other problems are also possible. We will show later (see Section 77) that there exists a sequence $\{f_n\}$ of infinitely differentiable functions on $[0, 1]$ which converges uniformly to a function f that is nowhere differentiable in $[0, 1]$.

The following theorem on differentiable functions is often quite useful.

Theorem 61.2 Let $\{f_n\}$ be a sequence of differentiable functions on (a, b). Suppose that

(i) f_n' is continuous on (a, b).
(ii) $\{f_n\}$ converges pointwise to f on (a, b).
(iii) $\{f_n'\}$ converges uniformly on (a, b).

Then f is differentiable on (a, b) and f_n' converges uniformly to f' on (a, b).

Proof. Let $c \in (a, b)$ and let $x \in (a, b)$. Let g denote the limit of $\{f_n'\}$. Then $f_n' \in \mathscr{R}[c, x]$ (or $\mathscr{R}[x, c]$) since f_n' is continuous. By Theorem 61.1,

$$\lim_{n \to \infty} \int_c^x f_n' = \int_c^x g \qquad (61.5)$$

By the fundamental theorem of calculus

$$\int_c^x f_n' = f_n(x) - f_n(c)$$

Using hypothesis (ii), we have

$$\lim_{n \to \infty} \int_c^x f_n' = \lim_{n \to \infty} [f_n(x) - f_n(c)] = f(x) - f(c) \qquad (61.6)$$

Combining equations (61.5) and (61.6), we have

$$\int_c^x g = f(x) - f(c)$$

Differentiating each side of this equation, we have

$$g(x) = f'(x) \qquad\qquad ■$$

Exercises 61.5 and 61.6 give other examples of phenomena that can occur with sequences of differentiable functions.

Exercise 61.7 gives a less restrictive theorem on limits of differentiable functions.

Exercises

61.1 Give an example of a sequence $\{f_n\}$ of functions in $\mathscr{R}[a, b]$ which converges pointwise but not uniformly to a function f on $[a, b]$ in $\mathscr{R}[a, b]$ such that
$$\lim_{n \to \infty} \int_a^b f_n = \int_a^b f$$

61.2 Give an example of a sequence $\{f_n\}$ of functions in $\mathscr{R}[a, b]$ which converges pointwise to a function f in $\mathscr{R}[a, b]$ but
$$\lim_{n \to \infty} \int_a^b f_n \neq \int_a^b f$$

61.3 Give an example of a sequence $\{f_n\}$ of functions and a function f such that
(a) $f_n \in \mathscr{R}[a, b]$ for every positive integer n
(b) $f \in \mathscr{R}[a, b]$
(c) $\lim_{n \to \infty} \int_a^b f_n = \int_a^b f$
(d) $\lim_{n \to \infty} f_n(x)$ does not exist for any $x \in [a, b]$.

61.4* Let $\{f_n\}$ be a uniformly bounded sequence of functions in $\mathscr{R}[a, b]$ which converges pointwise to a function f in $\mathscr{R}[a, b]$. Prove that
$$\lim_{n \to \infty} \int_a^b f_n = \int_a^b f$$

61.5 Give an example of a sequence $\{f_n\}$ of differentiable functions which converges uniformly on $\mathbf{R}$ to a differentiable function f such that $\{f_n'\}$ converges pointwise but not uniformly to f' on $\mathbf{R}$.

61.6 Let
$$f_n(x) = x^{1 + 1/(2n-1)} \qquad -1 < x < 1, n = 1, 2, \ldots$$
Prove that $\{f_n\}$ is a sequence of differentiable functions on $(-1, 1)$ which converges uniformly to $f(x) = |x|$ on $(-1, 1)$, but $f'(0)$ does not exist.

61.7 Let $\{f_n\}$ be a sequence of differentiable functions on (a, b). Suppose that the sequence $\{f_n(x)\}$ converges for some $x \in (a, b)$ and that there exists a function g such that $\{f_n'\}$ converges uniformly to g on (a, b). Prove that there exists a function f such that $\{f_n\}$ converges to f on (a, b) and that $f'(x)$ exists and equals $g(x)$ for every x in (a, b).

61.8 Use Theorems 58.5 and 60.4 to prove that if $\{f_n\}$ is a sequence of functions in $\mathscr{R}[a, b]$ which converges uniformly to f on $[a, b]$, then $f \in \mathscr{R}[a, b]$.

62. Series of Functions

If $\{u_n\}_{n=1}^{\infty}$ is a sequence of real-valued functions on a set X, we define the infinite series $\sum_{n=1}^{\infty} u_n$ to be the ordered pair $(\{u_n\}, \{s_n\})$, where

$$s_n = u_1 + \cdots + u_n$$

This definition is identical to the definition (Definition 22.1) of an infinite series of real numbers with "sequence of real numbers" replaced by "sequence of real-valued functions." As in Definition 22.1, we call $\{s_n\}$ the sequence of partial sums. Convergence of the series $\sum_{n=1}^{\infty} u_n$ is defined in terms of convergence of the sequence of partial sums.

Definition 62.1 Let $\{u_n\}$ be a sequence of real-valued functions on a set X and let f be a function on X. Let

$$s_n = u_1 + \cdots + u_n$$

We say that $\sum_{n=1}^{\infty} u_n$ *converges pointwise* to f on X if the sequence of partial sums $\{s_n\}$ converges pointwise to f on X.

We say that $\sum_{n=1}^{\infty} u_n$ *converges uniformly* to f on X if the sequence of partial sums $\{s_n\}$ converges uniformly to f on X.

Theorems 60.5, 61.1, and 61.2 may be used to give corresponding results about series.

Theorem 62.2 If $\{u_n\}$ is a sequence of continuous functions on a metric space M and if $\sum_{n=1}^{\infty} u_n$ converges uniformly to f on M, then f is continuous on M.

Proof. By Theorem 40.4, the sequence $\{s_n\}$ of partial sums of the series $\sum_{n=1}^{\infty} u_n$ is a sequence of continuous functions on M. By Corollary 60.5, f is continuous on M. ∎

Theorem 62.3 Let α be a function of bounded variation on $[a, b]$ and let $\{u_n\}$ be a sequence of functions in $\mathscr{R}_\alpha[a, b]$. If $\sum_{n=1}^{\infty} u_n$ converges uniformly to a function f on $[a, b]$, then $f \in \mathcal{R}_\alpha[a, b]$ and

$$\int_a^b f \, d\alpha = \sum_{n=1}^\infty \int_a^b u_n \, d\alpha$$

Proof.   Use Theorem 61.1.   ∎

Theorem 62.3 is often summarized by saying that a uniformly convergent series may be integrated term by term.

Theorem 62.4   Let $\{u_n\}$ be a sequence of differentiable functions on (a, b). Suppose that

(i) u_n' is continuous on (a, b).
(ii) $\sum_{n=1}^\infty u_n$ converges pointwise to a function f on (a, b).
(iii) $\sum_{n=1}^\infty u_n'$ converges uniformly on (a, b).

Then f is differentiable on (a, b) and

$$f'(x) = \sum_{n=1}^\infty u_n'(x)$$

for every x in (a, b).

Proof.   Use Theorem 61.2.   ∎

We may deduce a necessary and sufficient condition for uniform convergence of an infinite series from the Cauchy condition (Theorem 19.3).

Theorem 62.5   Let $\{u_n\}$ be a sequence of functions on a set X. Then $\sum_{n=1}^\infty u_n$ converges uniformly on X if and only if for every $\varepsilon > 0$, there exists a positive integer N such that if $n > m \geq N$, then

$$\left| \sum_{k=m+1}^n u_k(x) \right| < \varepsilon \tag{62.1}$$

for every x in X.

Proof.   Let $s_n = u_1 + \cdots + u_n$ denote the nth partial sum of the series $\sum_{n=1}^\infty u_n$.

First suppose that $\sum_{n=1}^\infty u_n$ converges uniformly to f on X. Let $\varepsilon > 0$. There exists a positive integer N such that if $n \geq N$, then

$$|s_n(x) - f(x)| < \frac{\varepsilon}{2}$$

for every $x \in X$. Thus if $n > m \geq N$, then

$$\left| \sum_{k=m+1}^{n} u_k(x) \right| = |s_n(x) - s_m(x)|$$

$$\leq |s_n(x) - f(x)| + |f(x) - s_m(x)|$$

$$< \frac{\varepsilon}{2} + \frac{\varepsilon}{2} = \varepsilon$$

for every $x \in X$.

Now suppose condition (62.1) holds. Let $\varepsilon > 0$. There exists a positive integer N such that if $n > m \geq N$, then

$$\left| \sum_{k=m+1}^{n} u_k(x) \right| < \frac{\varepsilon}{2}$$

for every $x \in X$. Since $\sum_{k=m+1}^{n} u_k(x) = s_n(x) - s_m(x)$, it follows that if $x \in X$, then $\{s_n(x)\}$ is a Cauchy sequence of real numbers. By Theorem 19.3, the sequence $\{s_n(x)\}$ converges for every $x \in X$. Let $f(x) = \lim_{n \to \infty} s_n(x)$, $x \in X$. We will show that $\{s_n\}$ converges uniformly to f on X. (So far we only know that $\{s_n\}$ converges pointwise to f on X.)

Let $x \in X$ and let $m \geq N$. Then

$$|s_m(x) - f(x)| = \left| \sum_{k=1}^{m} u_k(x) - \sum_{k=1}^{\infty} u_k(x) \right|$$

$$= \left| \sum_{k=m+1}^{\infty} u_k(x) \right| = \lim_{n \to \infty} \left| \sum_{k=m+1}^{n} u_k(x) \right|$$

$$\leq \frac{\varepsilon}{2} < \varepsilon$$

Thus if $m \geq N$, then

$$|s_m(x) - f(x)| < \varepsilon$$

for every $x \in X$. Therefore, $\{s_n\}$ converges uniformly to f on X. ∎

An important sufficient condition for uniform convergence of a series of functions is given in the next theorem. (See Exercises 62.3 and 62.4 for other conditions which yield uniform convergence.)

Theorem 62.6 (Weierstrass M-Test) Let $\{u_n\}$ be a sequence of functions on a set X. If there exists a sequence $\{M_n\}$ of positive numbers such that $\sum_{n=1}^{\infty} M_n$ converges and $|u_n(x)| \leq M_n$ for every $x \in X$ and for each positive integer n, then $\sum_{n=1}^{\infty} u_n$ converges uniformly on X.

Proof. We show that the series $\sum_{n=1}^{\infty} u_n$ satisfies the hypotheses of Theorem 62.5. Let $\varepsilon > 0$. Since $\sum_{n=1}^{\infty} M_n$ converges, there exists a positive integer N such that if $n > m \geq N$, then

$$M_{m+1} + \cdots + M_n < \varepsilon$$

If $n > m \geq N$ and $x \in X$, then

$$|u_{m+1}(x) + \cdots + u_n(x)| \leq |u_{m+1}(x)| + \cdots + |u_n(x)|$$
$$< M_{m+1} + \cdots + M_n < \varepsilon$$

and so by Theorem 62.5, $\sum_{n=1}^{\infty} u_n$ converges uniformly on X. ∎

As an example of the use of the results of this section, we will give an example of a function which is continuous on **R** but differentiable at no point of **R**.

Let $f_0(x)$ be the distance from x to the integer nearest x. (See Figure 62.1.)

We observe that f_0 is continuous on **R**, $f_0(x + N) = f_0(x)$ for every integer N, and $|f_0(x)| \leq \frac{1}{2}$ for every $x \in \mathbf{R}$. Let $f_k(x) = f_0(4^k x)/4^k$ for $x \in \mathbf{R}$ and $k = 1, 2, \ldots$ (See Figure 62.2.)

Then f_k is continuous on **R** and $|f_k(x)| \leq 1/(2 \cdot 4^k)$. By the Weierstrass M-test (Theorem 62.6), $\sum_{k=0}^{\infty} f_k$ converges uniformly on **R**. Since each f_k is continuous on **R**, the limit function, which we denote by F, is continuous on **R** (Theorem 62.2). We will show that F is nowhere differentiable.

Let us define a "roof" of f_m to be the graph of f_m over $[n/(2 \cdot 4^m), (n + 1)/(2 \cdot 4^m)]$ for some integer n. (See Figure 62.3.)

We note that if x and y are under a roof of f_m, then $(x, f_m(x))$ and $(y, f_m(y))$

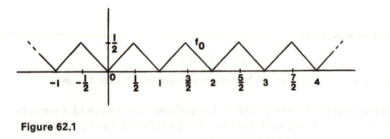

Figure 62.1

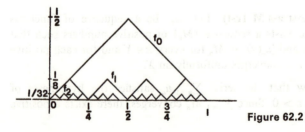

Figure 62.2

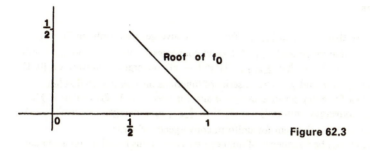

Figure 62.3

lie on a line segment which has slope either $+1$ or -1 and hence $[f_m(x) - f_m(y)]/(x - y) = \pm 1$. (See Figure 62.2.) Furthermore, in this case, x and y are under a roof of f_k for $k < m$, so that

$$\frac{f_k(x) - f_k(y)}{x - y} = \pm 1 \qquad \text{for} \quad k \le m$$

Let $a \in \mathbf{R}$. Let m be any positive integer. Then $a \in [n/(2 \cdot 4^{m-1}), (n + 1)/(2 \cdot 4^{m-1})]$ for some integer n. Since the length of this interval is $1/(2 \cdot 4^{m-1})$, by taking $h_m = +1/4^m$ or $-1/4^m$, we can assure that a and $h_m \in [n/(2 \cdot 4^{m-1}), (n + 1)/(2 \cdot 4^{m-1})]$. We showed above that, in this case,

$$\frac{f_k(a + h_m) - f_k(a)}{h_m} = \pm 1 \qquad \text{for} \quad k < m$$

If $k \ge m$,

$$f_k(a + h_m) - f_k(a) = \frac{f_0(4^k a + 4^k h_m) - f_0(4^k a)}{4^k}$$

$$= \frac{f_0(4^k a + N) - f_0(4^k a)}{4^k}$$

for some integer N. It follows that $f_k(a + h_m) - f_k(a) = 0$ for $k \ge m$. Therefore,

$$\frac{F(a + h_m) - F(a)}{h_m} = \sum_{k=0}^{\infty} \frac{f_k(a + h_m) - f_k(a)}{h_m}$$

$$= \sum_{k=0}^{m-1} \frac{f_k(a + h_m) - f_k(a)}{h_m} = \sum_{k=0}^{m-1} \pm 1$$

Thus $[F(a + h_m) - F(a)]/h_m$ is an odd integer if m is odd and $[F(a + h_m) - F(a)]/h_m$ is an even integer if m is even. It follows that $\lim_{m \to \infty} [F(a + h_m) - F(a)]/h_m$ does not exist. If F were differentiable at a, then this limit would exist. Consequently, F is differentiable at no point of the real line.

Exercises

62.1 Prove that the series $\sum_{n=1}^{\infty} 1/(n^2 + x^2)$ converges uniformly on $[0, \infty)$.

62.2 Prove that the series $\sum_{n=1}^{\infty} x^n (1 - x)$ converges pointwise but not uniformly on $[0, 1]$. Prove that $\sum_{n=1}^{\infty} (-1)^n x^n (1 - x)$ converges uniformly on $[0, 1]$.

62.3 Let $\sum_{n=1}^{\infty} u_n$ and $\sum_{n=1}^{\infty} v_n$ be series of functions on a set X such that $|u_n(x)| \leq |v_n(x)|$ for every positive integer n and for every $x \in X$. Prove that if $\sum_{n=1}^{\infty} |v_n|$ converges uniformly on X, then $\sum_{n=1}^{\infty} u_n$ converges uniformly on X.

62.4* Prove Dini's theorem for uniform convergence of series:

Let $\{u_n\}$ be a sequence of nonnegative continuous functions on a compact metric space M. If $\sum_{n=1}^{\infty} u_n$ converges pointwise to a continuous function f on M, then $\sum_{n=1}^{\infty} u_n$ converges uniformly to f on M.

62.5 Fill in the details of the following proof of the Tietze extension theorem for metric spaces.

Theorem Let f be a continuous function from a closed subset C of a metric space M into $[-1, 1]$. Then there exists a continuous extension of f from M into $[-1, 1]$, (that is, there exists a continuous function g from M into $[-1, 1]$ such that $g \mid C = f$).

Proof

(a) Let $A = \{x \in C \mid f(x) \leq -\frac{1}{3}\}$ and $B = \{x \in C \mid f(x) \geq \frac{1}{3}\}$. Modify Exercise 40.14(c) to conclude that there exists a continuous function f_1 from M into $[-\frac{1}{3}, \frac{1}{3}]$ such that $f_1 = -\frac{1}{3}$ on A and $\frac{1}{3}$ on B.

(b) Show that $|f(x) - f_1(x)| \leq \frac{2}{3}$ for all $x \in C$.

(c) Repeat (a) and (b) with f replaced by $f - f_1$ to find a continuous function f_2 from M into $[-\frac{2}{9}, \frac{2}{9}]$ such that f_2 is $-\frac{2}{9}$ on $\{x \in C \mid f(x) - f_1(x) \leq -\frac{2}{3}\}$ and f_2 is $\frac{2}{9}$ on $\{x \in C \mid f(x) - f_1(x) \geq \frac{2}{3}\}$ and
$$|f(x) - [f_1(x) + f_2(x)]| \leq \frac{4}{9} \qquad \text{for all } x \in C$$

(d) Show that there exists a sequence $\{f_n\}$ of continuous functions on M such that (i) the range of f_n is contained in $[-2^{n-1}/3^n, 2^{n-1}/3^n]$; (ii) f_{n+1} is $-2^n/3^{n+1}$ on $\{x \in C \mid f(x) - [f_1(x) + \cdots + f_n(x)] \leq -2^n/3^{n+1}\}$ and f_{n+1} is $2^n/3^{n+1}$ on $\{x \in C \mid f(x) - [f_1(x) + \cdots + f_n(x)] \geq 2^n/3^{n+1}\}$; and (iii) $|f(x) - [f_1(x) + \cdots + f_n(x)]| \leq (\frac{2}{3})^n$ for all $x \in C$.

(e) Prove that $\sum_{n=1}^{\infty} f_n$ converges uniformly to a continuous function on M.

(f) Prove that $g = \sum_{n=1}^{\infty} f_n$ is a continuous extension of f to M into $[-1, 1]$.

62.6 Let M be a metric space. Prove that M is compact if and only if every continuous real-valued function on M is bounded.

62.7 Suppose that $\sum_{n=1}^{\infty} u_n$ converges pointwise on a set X. Suppose also that $\{u_n\}$ converges uniformly to zero on X. Must $\sum_{n=1}^{\infty} u_n$ converge uniformly?

63. Applications to Power Series

We recall that a power series is a series of the form $\sum_{n=0}^{\infty} a_n(x - t)^n$ (see Sections 27 and 29). If R is the radius of convergence of a power series $\sum_{n=0}^{\infty} a_n(x - t)^n$, then $\sum_{n=0}^{\infty} a_n(x - t)^n$ converges absolutely if $|x - t| < R$ and diverges if $|x - t| > R$ (see Section 27). We will use the results of Section 62 to prove that a power series may be integrated or differentiated term by term for $|x - t| < R$.

Theorem 63.1 Let $\sum_{n=0}^{\infty} a_n(x - t)^n$ be a power series and let R be its radius of convergence. Let $f(x) = \sum_{n=0}^{\infty} a_n(x - t)^n$, $|x - t| < R$. Then

(i) If $0 < S < R$, then $\sum_{n=0}^{\infty} a_n(x - t)^n$ converges uniformly to f on $[t - S, t + S]$.

(ii) f is continuous on $(t - R, t + R)$.

(iii) If $a, b \in (t - R, t + R)$, $a < b$, and α is of bounded variation on $[a, b]$, then $f \in \mathscr{R}_\alpha[a, b]$, $\sum_{n=0}^{\infty} a_n \int_a^b (x - t)^n \, d\alpha(x)$ converges and

$$\int_a^b f \, d\alpha = \sum_{n=0}^{\infty} a_n \int_a^b (x - t)^n \, d\alpha(x)$$

(iv) f is differentiable on $(t - R, t + R)$, $\sum_{n=1}^{\infty} na_n(x - t)^{n-1}$ converges absolutely if $|x - t| < R$, and

$$f'(x) = \sum_{n=1}^{\infty} na_n(x - t)^{n-1}, \qquad |x - t| < R$$

Proof. (i) Suppose $0 < S < R$. If $|x - t| \leq S$, then $|a_n(x - t)^n| \leq |a_n|S^n$. Since $0 < S < R$, $\sum_{n=0}^{\infty} |a_n|S^n$ converges. By Theorem 62.6, $\sum_{n=0}^{\infty} a_n(x - t)^n$ converges uniformly on $[t - S, t + S]$.

(ii) Suppose $0 < S < R$. Each function $u_n(x) = a_n(x - t)^n$ is continuous on $[t - S, t + S]$. By Theorem 62.2, the uniform limit f is continuous on $[t - S, t + S]$. Since f is continuous on $[t - S, t + S]$ for every S, $0 < S < R$, it follows that f is continuous on $(t - R, t + R)$.

(iii) The proof follows immediately from Theorem 62.3.

(iv) By Theorem 27.2, the radius of convergence of the power series $\sum_{n=1}^{\infty} na_n(x - t)^{n-1}$ is

$$\frac{1}{\limsup_{n \to \infty} |na_n|^{1/n}}$$

Since $\lim_{n \to \infty} n^{1/n} = 1$, by Theorem 20.8, $\limsup_{n \to \infty} |na_n|^{1/n} = \limsup_{n \to \infty} |a_n|^{1/n}$. Thus

$$R = \frac{1}{\limsup_{n \to \infty} |a_n|^{1/n}} = \frac{1}{\limsup_{n \to \infty} |na_n|^{1/n}}$$

and hence $\sum_{n=1}^{\infty} na_n(x - t)^{n-1}$ converges absolutely if $|x - t| < R$.

Let $u_n(x) = a_n(x - t)^n$ and let $0 < S < R$. Then

(a) $u_n'(x) = na_n(x - t)^{n-1}$ is continuous on $(t - S, t + S)$.

(b) $\sum_{n=0}^{\infty} u_n$ converges pointwise (in fact uniformly) to f on $(t - S, t + S)$.

(c) $\sum_{n=1}^{\infty} u_n'$ converges uniformly on $(t - S, t + S)$.

In (b) and (c) we have used part (i) of this theorem. By Theorem 62.4, f is differentiable on $(t - S, t + S)$ and

$$f'(x) = \sum_{n=1}^{\infty} na_n(x - t)^{n-1}, \qquad |x - t| < S$$

Part (iv) now follows immediately. ■

Examples. Consider the geometric series

$$1 - x + x^2 - \cdots = \frac{1}{1 + x}, \qquad |x| < 1$$

By Theorem 63.1(iv), we have

$$-1 + 2x - 3x^2 + \cdots = D\left(\frac{1}{1 + x}\right) = \frac{-1}{(1 + x)^2}, \qquad |x| < 1$$

We will show in the next chapter that

$$\int_0^x \frac{1}{1 + t}\, dt = \log (1 + x), \qquad x > -1$$

By Theorem 63.1(iii),

$$\log (1 + x) = x - \frac{x^2}{2} + \frac{x^3}{3} - \cdots = \sum_{n=1}^{\infty} \frac{(-1)^{n+1} x^n}{n}, \qquad |x| < 1$$

Corollary 63.2 (Uniqueness of Power Series) Suppose the power series $\sum_{n=0}^{\infty} a_n(x - t)^n$ and $\sum_{n=0}^{\infty} b_n(x - t)^n$ converge for $|x - t| < R$. If $\sum_{n=0}^{\infty} a_n(x - t)^n = \sum_{n=0}^{\infty} b_n(x - t)^n$ for all x, $|x - t| < R$, then $a_n = b_n$ for every nonnegative integer n.

Proof. Taking $x = t$ in the equation $\sum_{n=0}^{\infty} a_n(x - t)^n = \sum_{n=0}^{\infty} b_n(x - t)^n$, we find that $a_0 = b_0$. By Theorem 63.1(iv), $\sum_{n=1}^{\infty} na_n(x - t)^{n-1} = \sum_{n=1}^{\infty} nb_n(x - t)^{n-1}$, $|x - t| < R$. Taking $x = t$ in this equation, we find that $a_1 = b_1$. Similarly, $a_2 = b_2$, $a_3 = b_3$, ... ■

Corollary 63.3 Suppose the power series $\sum_{n=0}^{\infty} a_n(x - t)^n$ has radius of convergence R. Let $f(x) = \sum_{n=0}^{\infty} a_n(x - t)^n$, $|x - t| < R$. Then

$$f^{(m)}(x) = \sum_{n=m}^{\infty} n(n-1)(n-2) \cdots (n-m+1)a_n(x-t)^{n-m}, \qquad |x-t| < R$$

for $m = 1, 2, \ldots$.

Proof. The proof follows from m applications of Theorem 63.1(iv). ∎

Corollary 63.4 (Taylor Series) Suppose the power series $\sum_{n=0}^{\infty} a_n(x-t)^n$ has radius of convergence R. Let $f(x) = \sum_{n=0}^{\infty} a_n(x-t)^n$, $|x-t| < R$. Then $a_n = f^{(n)}(t)/n!$, and hence

$$f(x) = \sum_{n=0}^{\infty} \frac{f^{(n)}(t)}{n!}(x-t)^n \qquad \text{for} \quad |x-t| < R$$

Proof. By Corollary 63.3,

$$f^{(m)}(t) = \sum_{n=m}^{\infty} n(n-1) \cdots (n-m+1)a_n 0^{n-m} = a_m m! \qquad ∎$$

Exercises

63.1 (a) Prove that the series $\sum_{n=0}^{\infty} (2^n/n!)x^n$ has radius of convergence ∞.

(b) Let $f(x) = \sum_{n=0}^{\infty} (2^n/n!)x^n$. Prove that $f'(x) = 2f(x)$.

63.2 Prove that

$$\int_0^x \frac{1}{1+t^2}\, dt = \sum_{n=0}^{\infty} \frac{(-1)^n x^{2n+1}}{2n+1} \qquad \text{for } |x| < 1$$

63.3 Suppose that the series $\sum_{n=0}^{\infty} a_n$ converges. Prove that the equation $f(x) = \sum_{n=0}^{\infty} a_n x^n$ defines a continuous function on $(-1, 1)$.

63.4 Prove that if the series $\sum_{n=0}^{\infty} a_n$ converges absolutely, then $\sum_{n=0}^{\infty} a_n x^n$ converges uniformly for $|x| < 1$.

63.5* Let $\sum_{n=0}^{\infty} a_n x^n$ be a power series with radius of convergence R. Let $f(x) = \sum_{n=0}^{\infty} a_n x^n$, $|x| < R$. Prove that if f is not constant the zeros of f are isolated [that is, prove that if $f(a) = 0$, $|a| < R$, there exists an open interval I such that $a \in I \subset (-R, R)$ and $f(x) \neq 0$ if $x \in I - \{a\}$].

63.6* Suppose the power series $\sum_{n=0}^{\infty} a_n x^n$ and $\sum_{n=0}^{\infty} b_n x^n$ converge for $|x| < R$ and that $\sum_{n=0}^{\infty} a_n x^n = \sum_{n=0}^{\infty} b_n x^n$ for all x belonging to an infinite subset of $[-R + \varepsilon, R - \varepsilon]$ for some $\varepsilon > 0$. Prove that $\sum_{n=0}^{\infty} a_n x^n = \sum_{n=0}^{\infty} b_n x^n$ for all x, $|x| < R$.

63.7 (Binomial Series) Let a be a real number. Let $\binom{a}{0} = 1$ and

$$\binom{a}{n} = \frac{a(a-1)\cdots(a-n+1)}{n!}$$

if $n \geq 1$ and n is a positive integer.

Notice that if $a \geq n$ and a is an integer, $\binom{a}{n}$ is the ordinary binomial coefficient.

(a)   Prove that the radius of convergence of the binomial series $\sum_{n=0}^{\infty} \binom{a}{n} x^n$ is 1 (assuming a is not a nonnegative integer).

(b)   Let $f(x) = \sum_{n=0}^{\infty} \binom{a}{n} x^n$, $|x| < 1$. Prove that $f(0) = 1$ and
$$af(x) = (1 + x)f'(x) \qquad |x| < 1$$

(c)* Let f be any function satisfying
$$f(0) = 1$$
$$af(x) = (1 + x)f'(x) \qquad |x| < 1$$
Prove that $f(x) = (1 + x)^a$, $|x| < 1$.

(d)   Deduce that $\sum_{n=0}^{\infty} \binom{a}{n} x^n = (1 + x)^a$ for any real number a and any x, $|x| < 1$.

(e)   Investigate the convergence of the binomial series for $x = \pm 1$.

64.   Abel's Limit Theorems

In Section 63, we showed that

$$\log (1 + x) = \sum_{n=1}^{\infty} \frac{(-1)^{n+1} x^n}{n}, \qquad |x| < 1$$

We will show in the next chapter that log is continuous in $(0, \infty)$ and thus

$$\lim_{x \to 1} \log (1 + x) = \log 2$$

If we substitute 1 for x in the power series $\sum_{n=1}^{\infty} (-1)^{n+1} x^n/n$, we get the convergent series $\sum_{n=1}^{\infty} (-1)^{n+1}/n$. We might ask whether

$$\log 2 = \sum_{n=1}^{\infty} \frac{(-1)^{n+1}}{n}$$

This is equivalent to asking whether

$$\lim_{x \to 1} \sum_{n=1}^{\infty} \frac{(-1)^{n+1} x^n}{n} = \sum_{n=1}^{\infty} \frac{(-1)^{n+1}}{n}$$

Corollary 64.3 states that this substitution is valid.

Theorem 64.1 (Abel's Test for Uniform Convergence)   Let $\{u_n\}$ be a sequence of functions on a set X such that $\sum_{n=1}^{\infty} u_n$ converges uniformly on X. Let $\{v_n\}$ be a uniformly bounded sequence on X such that $v_{n+1}(x) \leq v_n(x)$ for every x in X and for every positive integer n. Then $\sum_{n=1}^{\infty} u_n v_n$ converges uniformly on X.

Proof. We will use an analogue of the summation by parts formula

(Theorem 28.1), replacing $\sum_{k=1}^{n}$ by $\sum_{k=m+1}^{n}$ so that we can apply the Cauchy condition for uniform convergence (Theorem 62.5).

Let $\{a_n\}$ and $\{b_n\}$ be real sequences. Let

$$S_{j,k} = \sum_{n=j}^{k} a_n, \qquad j \le k$$

As in Theorem 28.1 we can prove the formula

$$\sum_{k=m+1}^{n} a_k b_k = \sum_{k=m+1}^{n} S_{m+1,k}(b_k - b_{k+1}) + S_{m+1,n}b_{n+1}, \qquad m < n \quad (64.1)$$

Let $x \in X$, let $a_n = u_n(x)$, and let $b_n = v_n(x)$. Let M be a positive number such that $|v_n(x)| < M$ for every $x \in X$ and for every positive integer n. Suppose $\varepsilon > 0$. There exists a positive integer N such that if $n > m \ge N$,

$$\left| \sum_{k=m+1}^{n} u_k(x) \right| < \frac{\varepsilon}{3M}$$

Now if $n > m \ge N$, then

$$\left| \sum_{k=m+1}^{n} u_k(x)v_k(x) \right| = \left| \sum_{k=m+1}^{n} S_{m+1,k}(x)[v_k(x) - v_{k+1}(x)] + S_{m+1,n}(x)v_{n+1}(x) \right|$$

$$\le \sum_{k=m+1}^{n} |S_{m+1,k}(x)||v_k(x) - v_{k+1}(x)| + |S_{m+1,n}(x)||v_{n+1}(x)|$$

$$\le \frac{\varepsilon}{3M} \sum_{k=m+1}^{n} [v_k(x) - v_{k+1}(x)] + \frac{\varepsilon}{3M} M$$

$$= \frac{\varepsilon}{3M}[v_{m+1}(x) - v_{n+1}(x)] + \frac{\varepsilon}{3}$$

$$< \frac{\varepsilon}{3M}(M + M) + \frac{\varepsilon}{3} = \varepsilon$$

By Theorem 62.5, $\sum_{n=1}^{\infty} u_n v_n$ converges uniformly on X. ∎

Corollary 64.2 (Abel's Theorem) If $\sum_{n=0}^{\infty} a_n$ is a convergent series, then the power series $\sum_{n=0}^{\infty} a_n x^n$ converges uniformly on $[0, 1]$.

Proof. Let $u_n(x) = a_n$ and $v_n(x) = x^n$, $x \in [0, 1]$ and then apply Theorem 64.1. ∎

Corollary 64.3 (Abel's Limit Theorem) Let $\sum_{n=0}^{\infty} a_n$ be a convergent series. Let

$$f(x) = \sum_{n=0}^{\infty} a_n x^n, \qquad 0 \le x \le 1$$

Then
$$\lim_{x \to 1} f(x) = f(1) = \sum_{n=0}^{\infty} a_n$$

Proof. By Theorems 64.2 and 61.5, f is continuous on $[0, 1]$. ∎

Corollary 64.4 Suppose that the power series $\sum_{n=0}^{\infty} a_n x^n$ converges for some $T \neq 0$. Let

$$f(x) = \sum_{n=0}^{\infty} a_n x^n, \qquad |x| < |T| \quad \text{or} \quad x = T$$

Then
$$\lim_{x \to T} f(x) = f(T) = \sum_{n=0}^{\infty} a_n T^n$$

Proof. Let

$$g(x) = \sum_{n=0}^{\infty} a_n T^n x^n, \qquad 0 \le x \le 1$$

By Corollary 64.2,

$$\lim_{x \to 1} g(x) = g(1) = \sum_{n=0}^{\infty} a_n T^n$$

Now $h(x) = x/T$ is continuous at $x = T$ and g is continuous at $x = 1$; so $f = g \circ h$ is continuous at $x = T$ and the conclusion now follows. ∎

Exercises

64.1 Prove equation (64.1).

64.2 Give an example of a divergent series $\sum_{n=0}^{\infty} a_n$ such that $\sum_{n=0}^{\infty} a_n x^n$ converges if $|x| < 1$ and $\lim_{x \to 1} \left(\sum_{n=0}^{\infty} a_n x^n \right)$ exists.

64.3* Let $\sum_{n=0}^{\infty} a_n$ and $\sum_{n=0}^{\infty} b_n$ be convergent series and let $\sum_{n=0}^{\infty} c_n$ denote the Cauchy product of $\sum_{n=0}^{\infty} a_n$ and $\sum_{n=0}^{\infty} b_n$. Prove that if $\sum_{n=0}^{\infty} c_n$ converges, then

$$\sum_{n=0}^{\infty} c_n = \left(\sum_{n=0}^{\infty} a_n \right) \left(\sum_{n=0}^{\infty} b_n \right)$$

Compare this result with Theorem 29.9.

64.4* Prove Dirichlet's test for uniform convergence:

Let $\{u_n\}$ be a sequence of functions on X. Suppose that the sequence $\{s_n\}$ of partial sums of the series $\sum_{n=1}^{\infty} u_n$ is uniformly bounded on X. Let $\{v_n\}$ be a sequence of functions on X such that $v_{n+1}(x) \le v_n(x)$ for every x in X and for every positive integer n. Suppose also that $\{v_n\}$ converges uniformly to 0. Then the series $\sum_{n=1}^{\infty} u_n v_n$ converges uniformly on X.

65. Summability Methods and Tauberian Theorems

Consider the divergent series $\sum_{n=0}^{\infty} (-1)^n$. If $|x| < 1$, then $\sum_{n=0}^{\infty} (-1)^n x^n$ converges and

$$\sum_{n=0}^{\infty} (-1)^n x^n = \frac{1}{1 + x}, \qquad |x| < 1$$

Since $\lim_{x \to 1} 1/(1 + x) = \frac{1}{2}$, we can assign $\sum_{n=0}^{\infty} (-1)^n$ the generalized sum of $\frac{1}{2}$. This summability method is known as Abel summability.

Definition 65.1 If $\sum_{n=0}^{\infty} a_n$ is an infinite series such that $\sum_{n=0}^{\infty} a_n x^n$ converges if $|x| < 1$ and $\lim_{x \to 1} \sum_{n=0}^{\infty} a_n x^n = T$ exists, we say that $\sum_{n=0}^{\infty} a_n$ is *Abel summable* (to T).

By Corollary 64.3, if $\sum_{n=0}^{\infty} a_n$ converges to L, then the Abel sum and the ordinary sum ($L = \sum_{n=0}^{\infty} a_n$) are the same. Thus Abel summability generalizes the ordinary method of summing an infinite series (Definition 22.2).

Another generalized sum is the Cesàro sum.

Definition 65.2 Let $\sum_{n=1}^{\infty} a_n$ be an infinite series. Let $\{s_n\}$ be the sequence of partial sums of the series $\sum_{n=1}^{\infty} a_n$. If $\lim_{n \to \infty} (s_1 + s_2 + \cdots + s_n)/n = T$ exists, we say that $\sum_{n=1}^{\infty} a_n$ is *Cesàro summable* (to T).

By Theorem 20.7, if $\sum_{n=1}^{\infty} a_n$ converges to L, then the Cesàro sum and the ordinary sum ($L = \sum_{n=1}^{\infty} a_n$) are the same. The divergent series $\sum_{n=0}^{\infty} (-1)^n$ has Cesàro sum $\frac{1}{2}$ (verify), and thus Cesàro summability also generalizes the ordinary method of summing an infinite series. Cesàro summability plays an important role in the theory of Fourier series (see Chapter XII).

It can be shown that if $\sum_{n=0}^{\infty} a_n$ is Cesàro summable to T, then $\sum_{n=0}^{\infty} a_n$ is Abel summable to T, but that there are Abel summable series which are not Cesàro summable (see Exercises 65.1 and 65.2).

Theorem 64.3 can be rephrased as follows: If $\sum_{n=0}^{\infty} a_n$ converges to L, then $\sum_{n=0}^{\infty} a_n$ is Abel summable to L. The series $\sum_{n=0}^{\infty} (-1)^n$ shows that the converse of this theorem is false. If certain additional hypotheses are imposed, it is often possible to deduce that $\sum_{n=0}^{\infty} a_n$ converges. Such a theorem is called a *Tauberian theorem* and the additional hypotheses are called *Tauberian conditions*. There are very complicated Tauberian theorems which play an important role in higher analysis. We shall give one example now of a Tauberian theorem and another later (Theorem 79.1). Other Tauberian theorems are given as exercises (Exercises 65.3 and 65.4).

Theorem 65.3 (Tauber's First Theorem) If $\sum_{n=0}^{\infty} a_n$ is Abel summable to T and $\lim_{n \to \infty} n a_n = 0$, then $\sum_{n=0}^{\infty} a_n$ converges to T.

Proof. Let

$$\sigma_n = \frac{|a_1| + 2|a_2| + \cdots + n|a_n|}{n}$$

$$f(x) = \sum_{n=0}^{\infty} a_n x^n \quad \text{for} \quad |x| < 1$$

Since $\lim_{n \to \infty} n|a_n| = 0$, by Theorem 20.7, $\lim_{n \to \infty} \sigma_n = 0$. By hypothesis $\lim_{n \to \infty} f(1 - (1/n)) = T$.

Let $\varepsilon > 0$. There exists a positive integer N such that if $n \geq N$, then

$$\left| f\left(1 - \frac{1}{n}\right) - T \right| < \frac{\varepsilon}{3}, \qquad \sigma_n < \frac{\varepsilon}{3}, \qquad n|a_n| < \frac{\varepsilon}{3}$$

Let $s_n = a_0 + a_1 + \cdots + a_n$. Then if $|x| < 1$,

$$s_n - T = f(x) - T + \sum_{k=0}^{n} a_k(1 - x^k) - \sum_{k=n+1}^{\infty} a_k x^k$$

Also, if $0 < x < 1$, then

$$(1 - x^k) = (1 - x)(1 + x + \cdots + x^{k-1}) \leq k(1 - x)$$

Therefore, if $n \geq N$ and $0 < x < 1$, then

$$|s_n - T| \leq |f(x) - T| + (1 - x) \sum_{k=0}^{n} k|a_k| + \frac{\varepsilon}{3n(1 - x)}$$

Now if $x = 1 - 1/n$ and $n \geq N$, then

$$|s_n - T| \leq \left| f\left(1 - \frac{1}{n}\right) - T \right| + \sigma_n + \frac{\varepsilon}{3}$$

$$< \frac{\varepsilon}{3} + \frac{\varepsilon}{3} + \frac{\varepsilon}{3} = \varepsilon \qquad \blacksquare$$

Exercises

65.1 Prove that the series $\sum_{n=1}^{\infty} (-1)^{n+1} n$ is Abel summable to $\frac{1}{4}$, but is not Cesàro summable.

65.2 Prove that if the series $\sum_{n=0}^{\infty} a_n$ is Cesàro summable to T, then $\sum_{n=0}^{\infty} a_n$ is Abel summable to T.

65.3 If $\sum_{n=0}^{\infty} a_n$ is Abel summable to T and $a_n \geq 0$ for every nonnegative integer n, prove that $\sum_{n=0}^{\infty} a_n$ converges to T.

65.4 Prove that an infinite series $\sum_{n=0}^{\infty} a_n$ converges to T if and only if $\sum_{n=0}^{\infty} a_n$ is Abel summable to T and

$$\lim_{N \to \infty} \frac{a_1 + 2a_2 + \cdots + Na_N}{N} = 0$$

65.5 Investigate the theorems of Chapter V where ordinary summability (or convergence) is replaced by Abel summability or Cesàro summability. (For example, if $\sum_{n=0}^{\infty} a_n$ and $\sum_{n=0}^{\infty} b_n$ are Abel summable to A and B, respectively, is $\sum_{n=0}^{\infty} (a_n + b_n)$ Abel summable to $A + B$?)

XI

Transcendental Functions

In this chapter we introduce some special real-valued functions on **R**.

66. The Exponential Function

Definition 66.1   For each $x \in \mathbf{R}$ we define

$$\exp x = \sum_{n=0}^{\infty} \frac{x^n}{n!}$$

By Theorem 26.6 we know that exp (x) converges absolutely for all x in **R**. We also know, from Corollary 29.10 on the Cauchy product, that for each $x, y \in \mathbf{R}$, we have

$$\exp x \exp y = \sum_{n=0}^{\infty} \frac{x^n}{n!} \sum_{n=0}^{\infty} \frac{y^n}{n!} = \sum_{n=0}^{\infty} \sum_{k=0}^{n} \frac{x^k}{k!} \frac{y^{n-k}}{(n-k)!}$$

$$= \sum_{n=0}^{\infty} \frac{(x+y)^n}{n!} = \exp (x+y) \qquad (66.1)$$

Clearly exp $0 = 1$ and since for each x in **R** we have [by equation (66.1)], $\exp x \cdot \exp (-x) = \exp (x - x) = \exp 0 = 1$, we conclude that

$$\exp x \neq 0 \qquad \text{for all } x \text{ in } \mathbf{R} \qquad (66.2)$$

and

$$\exp (-x) = \frac{1}{\exp x} \qquad \text{for all } x \text{ in } \mathbf{R} \qquad (66.3)$$

Let us investigate the differentiability of exp. We have by Theorem 63.1(iv) that $\exp' x$ exists for all $x \in \mathbf{R}$ and

$$\exp' x = \sum_{n=1}^{\infty} \frac{nx^{n-1}}{n!} = \sum_{n=0}^{\infty} \frac{x^n}{n!} = \exp x$$

Therefore, exp is differentiable for all x in **R** and

$$\exp' x = \exp x \qquad (66.4)$$

Therefore, $\exp x$ is continuous on **R**. This along with $\exp 0 = 1$ and $\exp x \neq 0$ for all $x \in \mathbf{R}$ and Corollary 45.6 implies that

$$\exp x > 0 \qquad \text{for all } x \in \mathbf{R} \qquad (66.5)$$

Theorem 66.2 $\exp 1 = e$.

Proof. From Theorem 16.6 we have $e = \lim_{n \to \infty} (1 + 1/n)^n$. Let $a_n = \sum_{k=0}^{n} 1/k!$ and $b_n = (1 + 1/n)^n$ for $n = 1, 2, \ldots$. We must show that $\lim_{n \to \infty} a_n = \lim_{n \to \infty} b_n$. By the binomial theorem we have

$$b_n = 1 + 1 + \frac{1}{2!}\left(1 - \frac{1}{n}\right)$$

$$+ \frac{1}{3!}\left(1 - \frac{1}{n}\right)\left(1 - \frac{2}{n}\right) + \cdots + \frac{1}{n!}\left(1 - \frac{1}{n}\right) \cdots \left(1 - \frac{n-1}{n}\right)$$

$$\leq 1 + 1 + \frac{1}{2!} + \frac{1}{3!} + \cdots + \frac{1}{n!} = a_n \qquad \text{for } n = 1, 2, \ldots$$

So

$$e = \lim_{n \to \infty} b_n \leq \lim_{n \to \infty} a_n = \sum_{n=0}^{\infty} \frac{(1)^n}{n!} = \exp 1.$$

We now prove the reverse inequality. For each $m \leq n$, positive integers, we have

$$b_n = 1 + 1 + \frac{1}{2!}\left(1 - \frac{1}{n}\right) + \cdots + \frac{1}{n!}\left(1 - \frac{1}{n}\right) \cdots \left(1 - \frac{n-1}{n}\right)$$

$$\geq 1 + 1 + \frac{1}{2!}\left(1 - \frac{1}{n}\right) + \cdots + \frac{1}{m!}\left(1 - \frac{1}{n}\right) \cdots \left(1 - \frac{m-1}{n}\right) = c_{m,n}$$

For fixed m we have

$$e = \lim_{n \to \infty} b_n \geq \lim_{n \to \infty} c_{m,n} = 1 + 1 + \frac{1}{2!} + \cdots + \frac{1}{m!} = a_m$$

Therefore $e \geq \lim_{m \to \infty} a_m = \exp 1$. ∎

For each positive integer n we have by equation (66.1),

$$e^n = e \cdot e \cdots e = \exp 1 \cdot \exp 1 \cdots \exp 1 = \exp(1 + 1 + \cdots + 1) = \exp n$$

$$(66.6)$$

and by equation (66.3),

$$e^{-n} = \frac{1}{e^n} = \frac{1}{\exp n} = \exp(-n) \qquad (66.7)$$

Therefore for any rational number $r = m/n$ (where $n > 0$), we have by equations (66.1) and (66.6) or (66.7)

$$(\exp r)^n = \exp r \cdots \exp r = \exp nr = e^{nr}$$

therefore, $\exp r = (e^{nr})^{1/n} = e^r$. We conclude that

$$\exp r = e^r \qquad \text{for any rational number } r \qquad (66.8)$$

Let α be any real number. Choose $\{r_n\}_{n=1}^{\infty}$ any increasing sequence of rational numbers such that

$$\lim_{n \to \infty} r_n = \alpha$$

By Section 17 and equation (66.8), we have

$$e\alpha = \lim_{n \to \infty} e^{r_n} = \lim_{n \to \infty} \exp r_n = \exp(\lim_{n \to \infty} r_n) = \exp \alpha$$

Therefore

$$\exp \alpha = e\alpha \qquad \text{for any real number } \alpha \qquad (66.9)$$

By equations (66.8) and (66.9) we have the following theorem.

Theorem 66.3 $\exp x = e^x$ for every $x \in \mathbf{R}$.

Since $e > 2$ and e^x is strictly increasing, we have $\lim_{x \to \infty} e^x = \infty$. By equation (66.3) $\lim_{x \to -\infty} e^x = 0$. By the intermediate value theorem (Corollary 45.6) it follows that the range of e^x is $(0, \infty)$.

In Figure 66.1 we have sketched the graph of e^x.

We now summarize the above results.

Theorem 66.4

(i) $e^x = \sum_{n=0}^{\infty} x^n/n!$ for all $x \in \mathbf{R}$.

(ii) e^x is continuous and differentiable for all x in $\mathbf{R}$.

(iii) $(e^x)' = e^x$ for all $x \in \mathbf{R}$.

(iv) e^x is strictly increasing on $\mathbf{R}$ and $e^x > 0$ for all $x \in \mathbf{R}$.

(v) $e^{x+y} = e^x \cdot e^y$ for all x and y in $\mathbf{R}$.

(vi) $\lim_{x \to \infty} e^x = \infty$ and $\lim_{x \to -\infty} e^x = 0$.

Exercises

66.1 Suppose $f: \mathbf{R} \to \mathbf{R}$ and $f'(x) = f(x)$ for all $x \in \mathbf{R}$ and $f(0) = 1$. Prove that $f(x) = e^x$ for all $x \in \mathbf{R}$.

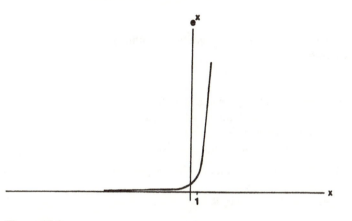

Figure 66.1

66.2 Prove $\lim_{n \to \infty} (1 + x/n)^n = e^x$ for all $x \in \mathbf{R}$.

66.3 Suppose $f(x) \cdot f(y) = f(x + y)$ for all x and y in $\mathbf{R}$. Assuming f is differentiable on $\mathbf{R}$ and not zero, prove that $f(x) = e^{cx}$, where c is some constant.

66.4 Do Exercise 66.3 with *differentiable* replaced by *continuous*.

66.5 Evaluate $\int_0^\infty e^{-(x-1/x)^2} dx$.

66.6 Prove that $\lim_{x \to \infty} x^n e^{-x} = 0$ for every positive integer n.

66.7* Suppose that $T'(t) = cT(t)$ for t in an interval containing zero. Show that $T(t) = T(0)e^{ct}$.

66.8* Suppose that $f'(x) \geq xf(x)$ for all real x. Prove that there exists a constant k such that $ke^x \leq f(x)$ for every real x.

66.9 Let
$$f(x) = \begin{cases} e^{-1/x^2} & x \neq 0 \\ 0 & x = 0 \end{cases}$$
Prove that $f^{(n)}(0) = 0$ for $n = 1, 2, \ldots$.

67. The Natural Logarithm Function

In Section 66 we showed that $e^x = \exp x$ is a strictly increasing function from $\mathbf{R}$ onto $(0, \infty)$. Therefore, exp has an inverse function from $(0, \infty)$ onto $\mathbf{R}$. This section is devoted to the study of this inverse function.

Definition 67.1 We define
$$\log x = \exp^{-1} x \qquad \text{for } x \text{ in } (0, \infty)$$
We call log the *natural logarithm function*.

Theorem 67.2

(i) log is a strictly increasing, continuous, differentiable function from $(0, \infty)$ onto $\mathbf{R}$.

(ii) $\log' x = 1/x$ for x in $(0, \infty)$.

(iii) $\log 1 = 0$ and $\log e = 1$.

(iv) $\log xy = \log x + \log y$ for x and y in $(0, \infty)$.

(v) $\log (x/y) = \log x - \log y$ for x and y in $(0, \infty)$.

(vi) $\log x^r = r \log x$ for x in $(0, \infty)$ and r in $\mathbf{R}$.

(vii) $\lim_{x \to 0+} \log x = -\infty$ and $\lim_{x \to \infty} \log x = \infty$.

Proof. For convenience of notation, in this proof we will denote $\log x$ by $L(x)$.

By Theorem 66.4 for each y in $\mathbf{R}$

$$\exp' y = \exp y > 0$$

By Theorem 49.10,

$$L'(\exp y) = \frac{1}{\exp' y} = \frac{1}{\exp y}$$

Let $x \in (0, \infty)$. There exists a unique y in $\mathbf{R}$ such that $\exp y = x$. Therefore, $L'(x) = 1/x$. Parts (i) and (ii) now follow.

Since $\exp 0 = 1$ and $\exp 1 = e$, part (iii) follows.

If $x, y \in (0, \infty)$, we have

$$\exp [L(x) + L(y)] = \exp [L(x)] \exp [L(y)] = xy = \exp [L(xy)]$$

and since exp is one to one, it follows that

$$L(x) + L(y) = L(xy)$$

If $x \in (0, \infty)$ and $r \in \mathbf{R}$, we have

$$\exp [L(x^r)] = x^r = (\exp [L(x)])^r = \exp [rL(x)]$$

and thus $L(x^r) = rL(x)$.

Part (v) follows from parts (iv) and (vi).

Part (vii) follows from Theorem 66.4(vi). ∎

The functions exp and log are examples from the classes of exponential and logarithm functions. (See Exercises 67.4 and 67.5).

Figure 67.1 is a sketch of the graph of $\log x$.

Exercises

67.1 Suppose $g(xy) = g(x) + g(y)$ for all x and y in $(0, \infty)$. Assuming g is continuous on $(0, \infty)$ and not constant prove that $g(x) = c \log x$ for some c in $\mathbf{R}$.

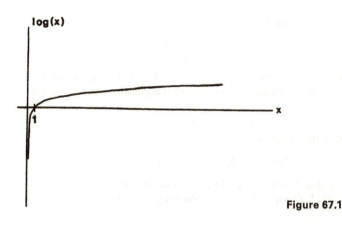

Figure 67.1

67.2 Prove that $\lim_{x \to \infty} x^{-n} \log x = 0$ for every positive integer n.

67.3 Prove that $\int_1^x (1/t)dt = \log x$ for x in $(0, \infty)$.

67.4 Let $a > 0$.
(a) Show that $a^x = e^{x \log a}$ for x in **R**.
(b) Prove that $(a^x)' = (\log a)a^x$.

67.5 Let $a > 0$ and $a \neq 1$. Define $\log_a$ to be the inverse function to a^x. We call $\log_a x$ the logarithm of x to the base a. Compute the derivative of $\log_a x$.

67.6 Let $a > 0$ and let $f(x) = x^a$ for x in $(0, \infty)$. Prove that $f'(x) = ax^{a-1}$.

67.7 (a) Let $f(x) = \log(1 + x)$. Apply the mean-value theorem to f on $[0, 1/n]$ to conclude that

$$n \log \left(1 + \frac{1}{n}\right) = \frac{1}{1 + c}$$

where $0 < c < 1/n$.

(b) Use (a) to prove that

$$\frac{1}{n+1} < \log \left(1 + \frac{1}{n}\right) < \frac{1}{n}$$

(c) Let $C_n = 1 + \frac{1}{2} + \cdots + \frac{1}{n} - \log n$

$$D_n = C_n - \frac{1}{n}$$

By considering $C_{n+1} - C_n$ and $D_{n+1} - D_n$ and (b), prove that $\{C_n\}$ is decreasing and $\{D_n\}$ is increasing.

(d) Prove that $\{C_n\}$ is bounded below and that $\{D_n\}$ is bounded above. Deduce that $\{C_n\}$ and $\{D_n\}$ converge to a common limit. This limit is known as *Euler's constant* and is denoted γ.

67.8 (a) Derive the following expansions valid for $|x| < 1$:

$$\log(1 + x) = \sum_{k=1}^{\infty} \frac{(-1)^{k+1}x^k}{k}$$

$$\log(1-x) = -\sum_{k=1}^{\infty} \frac{x^k}{k}$$

$$\log\left(\frac{1+x}{1-x}\right) = 2\sum_{k=1}^{\infty} \frac{x^{2k-1}}{2k-1}$$

(b) Show that the error $2\sum_{k=n+1}^{\infty} x^{2k-1}/(2k-1)$ in approximating log $((1+x)/(1-x))$ with n terms of the last expansion in part (a) is at most

$$\frac{2x^{2n-1}}{(2n+1)(1-x^2)}.$$

(c) Derive the expansion

$$\log 2 = 2\sum_{k=1}^{\infty} \frac{1}{3^{2k-1}(2k-1)}.$$

Show that if we take six terms of this expansion, the error is less than $\frac{1}{2} \times 10^{-6}$. Compare this result with Exercise 25.5(e).

68. The Trigonometric Functions

We begin by defining two power series functions, which turn out to be the functions sin x and cos x.

Definition 68.1   For each $x \in \mathbf{R}$ we define

$$S(x) = \sum_{n=0}^{\infty} (-1)^n \frac{x^{2n+1}}{(2n+1)!} = x - \frac{x^3}{3!} + \frac{x^5}{5!} - \frac{x^7}{7!} + \cdots \tag{68.1}$$

and

$$C(x) = \sum_{n=0}^{\infty} (-1)^n \frac{x^{2n}}{(2n)!} = 1 - \frac{x^2}{2!} + \frac{x^4}{4!} - \frac{x^6}{6!} + \cdots \tag{68.2}$$

These series converge absolutely for all values of x by the ratio test. Since the reader is probably familiar with the properties of the functions sin x and cos x, we will proceed in justifying Definition 68.5 by showing that $S(x)$ and $C(x)$ have all those familiar properties and that they are the unique functions with those properties.

From (68.1) and (68.2) we see that

$$C(0) = 1 \qquad S(0) = 0 \tag{68.3}$$

$$C(-x) = C(x) \qquad S(-x) = -S(x) \qquad \text{for all } x \in \mathbf{R} \tag{68.4}$$

and by Theorem 63.1(iv)

$$C'(x) = -S(x) \qquad S'(x) = C(x) \qquad \text{for all } x \in \mathbf{R} \tag{68.5}$$

Lemma 68.2 $[S(x)]^2 + [C(x)]^2 = 1$ for all $x \in \mathbf{R}$.

Proof.   Let $g(x) = [S(x)]^2 + [C(x)]^2$ for all $x \in \mathbf{R}$. By (68.5) and the chain rule

$$g'(x) = 2S(x)C(x) + 2C(x)(-S(x)) = 0 \qquad \text{for all } x \in \mathbf{R}$$

By Theorem 49.9, $g(x)$ is a constant function on $\mathbf{R}$; however,

$$g(0) = 1; \text{ so } g(x) = 1 \qquad \text{for all } x \in \mathbf{R} \qquad ■$$

Lemma 68.3 If f is a real-valued function on $\mathbf{R}$ satisfying $f''(x) = -f(x)$ for all $x \in \mathbf{R}$ and $f(0) = 0$ and $f'(0) = b$, then

$$f(x) = bS(x) \qquad \text{for all } x \in \mathbf{R}$$

Proof. If we let $U(x) = f(x)S(x) + f'(x)C(x)$

and $\qquad\qquad\qquad V(x) = f(x)C(x) - f'(x)S(x)$

it is easily verified that

$$U'(x) = 0 = V'(x) \qquad \text{for all } x \in \mathbf{R}$$

Therefore, we have

$$b = f(x)S(x) + f'(x)C(x)$$

and $\qquad\qquad f(x)C(x) = f'(x)S(x) \qquad \text{for all } x \in \mathbf{R}$

Using Lemma 68.2, we conclude that

$$\begin{aligned} bS(x) &= S(x)[f(x)S(x) + f'(x)C(x)] \\ &= f(x)\{1 - [C(x)]^2\} + f'(x)S(x)C(x) \\ &= f(x) - C(x)[f(x)C(x) - f'(x)S(x)] = f(x) \qquad \text{for all } x \in \mathbf{R} \qquad ■ \end{aligned}$$

Theorem 68.4 If f is a real-valued function on $\mathbf{R}$ satisfying $f''(x) = -f(x)$ for all $x \in \mathbf{R}$ and $f(0) = a$ and $f'(0) = b$, then

$$f(x) = aC(x) + bS(x) \qquad \text{for all } x \in \mathbf{R}$$

Proof. Let $g(x) = f(x) - aC(x)$. Then $g''(x) = -g(x)$ for all $x \in \mathbf{R}$ and $g(0) = 0$ and $g'(0) = b$; so by Lemma 68.3, we have

$$g(x) = bS(x) \qquad \text{for all } x \in \mathbf{R} \qquad ■$$

Theorem 68.4 justifies the following definition.

Definition 68.5 We define for all $x \in \mathbf{R}$ the functions

$$\sin x = S(x) \qquad \cos x = C(x)$$

and we call these functions the sine and cosine, respectively.

We proceed by proving many of the formulas involving $\sin x$ and $\cos x$.

Theorem 68.6 $\sin (x + y) = \sin x \cos y + \sin y \cos x$ for all x and y in **R**.

Proof. Fix y in **R** and let $f(x) = \sin (x + y)$ for all x in **R**. Then $f''(x) = -f(x)$ for all $x \in \mathbf{R}$ and $f(0) = \sin y$ and $f'(0) = \cos y$; so by Theorem 68.4 we have

$$f(x) = \sin y \cos x + \cos y \sin x \qquad \text{for all } x \text{ and } y \text{ in } \mathbf{R} \qquad \blacksquare$$

Theorem 68.7 $\cos (x + y) = \cos x \cos y - \sin x \sin y$ for all x and y in **R**.

Proof. Fix y in **R** and let $g(x) = \sin (x + y)$ for all $x \in \mathbf{R}$. By Theorem 68.6, $g(x) = \sin (x + y) = \sin x \cos y + \sin y \cos x$; so $g'(x) = \cos (x + y) = \cos x \cos y - \sin y \sin x$ for all x and y in **R**. $\blacksquare$

We now define the real number π.

It is easily verified (see Exercise 25.3) that since

$$\cos 2 = 1 - \frac{2^2}{2!} + \frac{2^4}{4!} - \cdots$$

we have

$$\left| \cos 2 - \left(1 - \frac{2^2}{2!} \right) \right| \leq \frac{2^4}{4!} = \frac{2}{3}$$

that is, $\cos 2$ is within $\frac{2}{3}$ of -1. This implies that $\cos 2 < 0$. Now since $\cos 0 = 1$ and $\cos x$ is continuous on **R**, we conclude by the intermediate-value theorem that $\cos x = 0$ for at least one x in $(0, 2)$. Let x_1 denote the smallest number in $(0, 2)$ such that $\cos x_1 = 0$. The existence of x_1 is guaranteed by the completeness of **R** and the continuity of $\cos x$. (Verify!)

Definition 68.8 We define

$$\pi = 2x_1$$

where x_1 is the smallest positive real number such that $\cos x_1 = 0$.

By this definition, $\cos (\pi/2) = 0$, and since

$[\cos (\pi/2)]^2 + [\sin (\pi/2)]^2 = 1$, we conclude that $\sin (\pi/2)$ is 1 or -1.

However, $\cos x > 0$ in $(0, \pi/2)$; so equation (68.5) implies that $\sin x$ is strictly increasing in $(0, \pi/2)$; therefore, $\sin (\pi/2) = 1$, since $\sin 0 = 0$. Using Theorems 68.6 and 68.7, we find that $\cos \pi = \cos (\pi/2 + \pi/2) = -1$, and similarly $\sin \pi = 0$, $\cos 2\pi = 1$ and $\sin 2\pi = 0$. Using these facts and Theorems 68.6 and 68.7, we have

$$\cos (x + 2\pi) = \cos x \qquad \sin (x + 2\pi) = \sin x \qquad \text{for all } x \in \mathbf{R} \qquad (68.6)$$

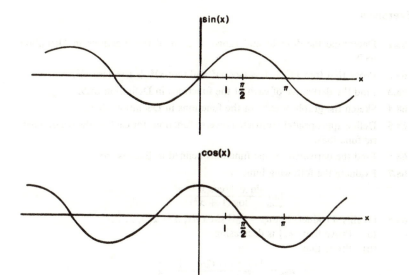

Figure 68.1

and

$$\cos\left(x + \frac{\pi}{2}\right) = -\sin x \qquad \sin\left(x + \frac{\pi}{2}\right) = \cos x \qquad \text{for all } x \in \mathbf{R} \quad (68.7)$$

The graphs of $\sin x$ and $\cos x$ are sketched in Figure 68.1.

We conclude this section with the definition of the other trigonometric functions. The formulas and interrelationships of these functions are left as exercises. (See Exercises 68.1 to 68.6.)

Definition 68.9 We define the following functions for each $x \in \mathbf{R}$, where the expression makes sense

$$\tan x = \frac{\sin x}{\cos x} \qquad \text{(tangent)}$$

$$\cot x = \frac{\cos x}{\sin x} \qquad \text{(cotangent)}$$

$$\sec x = \frac{1}{\cos x} \qquad \text{(secant)}$$

$$\csc x = \frac{1}{\sin x} \qquad \text{(cosecant)}$$

Exercises

68.1 Determine the domain and range of each of the functions in Definition 68.9.

68.2 Prove that $(\tan x)^2 + 1 = (\sec x)^2$ and $(\cot x)^2 + 1 = (\csc x)^2$.

68.3 Find the derivative of each of the functions in Definition 68.9.

68.4 Sketch the graph of each of the functions in Definition 68.9.

68.5 Define appropriately restricted inverse functions for each of the trigonometric functions.

68.6 Find the derivative of the functions defined in Exercise 68.5.

68.7 Evaluate the following limit:

$$\lim_{x \to 0} \frac{\sin x + \cos x - e^x}{\log (1 + x^2)}.$$

68.8 (Wallis' Product) Let $a_n = \int_0^{\pi/2} \sin^n x \, dx$, $n = 2, 3, \ldots$.

    (a) Prove that $\{a_n\}$ is decreasing.

    (b) Prove that

$$a_{2n} = \frac{1 \cdot 3 \cdot 5 \cdots (2n - 1)}{2 \cdot 4 \cdot 6 \cdots (2n)} \cdot \frac{\pi}{2}$$

and

$$a_{2n-1} = \frac{2 \cdot 4 \cdots (2n - 2)}{3 \cdot 5 \cdots (2n - 1)}$$

    (c) Prove that

$$\frac{2 \cdot 4 \cdots (2n)}{3 \cdot 5 \cdots (2n + 1)} < \frac{1 \cdot 3 \cdots (2n - 1)}{2 \cdot 4 \cdots (2n)} \cdot \frac{\pi}{2} < \frac{2 \cdot 4 \cdots (2n - 2)}{3 \cdot 5 \cdots (2n - 1)}$$

    and, hence, conclude that

$$\frac{2n}{2n + 1} \cdot \frac{\pi}{2} < \left(\frac{2 \cdot 4 \cdots (2n)}{1 \cdot 3 \cdots (2n - 1)} \right)^2 \frac{1}{2n + 1} < \frac{\pi}{2}$$

    (d) Derive Wallis' product:

$$\lim_{n \to \infty} \left(\frac{2 \cdot 4 \cdots (2n)}{1 \cdot 3 \cdots (2n - 1)} \right)^2 \frac{1}{2n + 1} = \frac{\pi}{2}$$

68.9 (Stirling's Formula)

    (a) Let $a_k = 1$ and $b_k = \log k$ in the summation by parts formula (Theorem 28.1) to derive

$$\sum_{k=1}^{n} \log k = - \sum_{k=1}^{n-1} k \log \left(1 + \frac{1}{k} \right) + n \log n$$

    (b) Take three terms in the expansion for $\log (1 + x)$ (see Exercise 67.8) to derive

$$\log \left(1 + \frac{1}{k} \right) = \frac{1}{k} - \frac{1}{2k^2} + \frac{c_k}{3k^3}$$

    where $0 < c_k < 1$. Combine the equations above to obtain

$$\sum_{k=1}^{n} \log k = n \log n - (n - 1) + \frac{1}{2} \sum_{k=1}^{n-1} \frac{1}{k} - \frac{1}{3} \sum_{k=1}^{n-1} \frac{c_k}{k^2}$$

(c) Using the result of Exercise 67.7, which established the existence of the limit $\lim_{k \to \infty} (\sum_{k=1}^{n} 1/k - \log n)$, prove that

$$\lim_{n \to \infty} \left(\sum_{k=1}^{n} \log k - n \log n - \frac{1}{2} \log n + n \right)$$

exists.

(d) Prove that

$$\lim_{n \to \infty} \frac{n! \, e^n}{n^n \sqrt{n}}$$

exists.

(e)* Let $a_n = n! e^n / (n^n \sqrt{n})$. Prove that

$$\frac{a_n^2}{a_{2n}} = \frac{2 \cdot 4 \cdots (2n)}{1 \cdot 3 \cdots (2n-1)} \frac{1}{\sqrt{2n+1}} \left[\frac{2n(2n+1)}{n^2} \right]^{1/2}$$

and deduce that

$$\lim_{n \to \infty} \frac{a_n^2}{a_{2n}} = \sqrt{2\pi}$$

(f) Prove Stirling's formula:

$$\lim_{n \to \infty} \frac{n! \, e^n}{n^n \sqrt{n}} = \sqrt{2\pi}$$

Stirling's formula states that $n!$ is approximately $n^n \sqrt{2n\pi}/e^n$ in the sense that the ratio of these two terms approaches 1. [The argument of (a) through (d) is due to Shohat (1933).]

68.10 (a) Derive the expansion

$$\arctan x = \sum_{k=0}^{\infty} (-1)^k \frac{x^{2k+1}}{2k+1}, \ |x| < 1$$

(b) Deduce from (a) that

$$\frac{\pi}{4} = \sum_{k=0}^{\infty} \frac{(-1)^k}{2k+1}$$

68.11 Derive a power series expansion for arcsin x similar to that given in Exercise 68.10 for arctan x.

68.12 Adapt the proof of Lemma 68.3 to prove that if $f''(x) = -cf(x)$ for some $c > 0$ on $[0, \pi]$ and $f(0) = f(\pi) = 0$, then $c = 1/n^2$ and $f(x) = k \sin nx$ for some constant k and some nonnegative integer n.

XII

Inner Product Spaces and Fourier Series

In many of the metric spaces we have studied it is possible to define addition and scalar multiplication in such a way that the spaces become vector spaces over the field of real numbers. (*Appendix: Vector Spaces* summarizes the definition of a vector space and some elementary results we will be using subsequently). For example, if we define addition and scalar multiplication pointwise in $\mathbf{R}^n$, l^2, and $\mathscr{R}[a, b]$, each of these sets becomes a vector space over $\mathbf{R}$. Thus these spaces (and others) have a much richer structure than merely being equipped with a distance function. In $\mathbf{R}^n$ it is possible to define the length or norm of a vector which corresponds to the geometric interpretation of length. It is possible to generalize the concept of the length of a vector to l^2 and $\mathscr{R}[a, b]$ (and to other spaces as well). A vector space with a length function is called a *normed linear space*. The vector space $\mathbf{R}^n$ has another important concept associated with it and that is the idea of perpendicularity or orthogonality of pairs of vectors in $\mathbf{R}^n$. This concept also generalizes to l^2 and $\mathscr{R}[a, b]$. A vector space with an orthogonality-measuring function is called an *inner product space*. In this chapter we will study inner product spaces and orthogonal expansion of vectors in inner product spaces with particular attention to $\mathscr{R}[a, a + 2\pi]$, where $a \in \mathbf{R}$.

69. Normed Linear Spaces

If we define

$$(x_1, \ldots, x_n) + (y_1, \ldots, y_n) = (x_1 + y_1, \ldots, x_n + y_n)$$

$$c(x_1, \ldots, x_n) = (cx_1, \ldots, cx_n)$$

where $(x_1, \ldots, x_n)$, $(y_1, \ldots, y_n) \in \mathbf{R}^n$ and $c \in \mathbf{R}$, then $\mathbf{R}^n$ becomes a real vector space. The zero vector of $\mathbf{R}^n$ is the n-tuple $(0, 0, \ldots, 0)$ and the additive inverse of $(x_1, \ldots, x_n)$ is $(-x_1, \ldots, -x_n)$. It is easy to verify that the other vector space axioms are also satisfied. We shall always denote the zero vector of a vector space by θ; this is to avoid confusion with the real number 0.

In $\mathbf{R}^n$ it is possible to define an "absolute-value-type" function. We let

$$\|x\| = \sqrt{x_1^2 + \cdots + x_n^2} \tag{69.1}$$

where $x = (x_1, \ldots, x_n) \in \mathbf{R}^n$. If $n = 1$, $\|x\|$ is precisely $|x|$, the absolute value of x. We call $\|x\|$ the *norm* of x. Geometrically, the norm of x is the length of the vector x. This norm function satisfies the following properties:

(i) $\|x\| = 0$ if and only if $x = \theta$.
(ii) $\|cx\| = |c| \, \|x\|$ for $c \in \mathbf{R}$, $x \in \mathbf{R}^n$.
(iii) $\|x + y\| \leq \|x\| + \|y\|$ for $x, y \in \mathbf{R}^n$.

Properties (i) and (ii) are easily verified. Property (iii) follows from the triangle inequality for the metric space $\mathbf{R}^n$. Indeed, if d denotes the usual metric for $\mathbf{R}^n$ and $x = (x_1, \ldots, x_n)$, $y = (y_1, \ldots, y_n) \in \mathbf{R}^n$, we have

$$\begin{aligned}
\|x + y\| &= \sqrt{(x_1 + y_1)^2 + \cdots + (x_n + y_n)^2} \\
&= d(x, -y) \\
&\leq d(x, \theta) + d(\theta, -y) \\
&= \sqrt{x_1^2 + \cdots + x_n^2} + \sqrt{y_1^2 + \cdots + y_n^2} \\
&= \|x\| + \|y\|
\end{aligned}$$

The preceding discussion motivates the next definition. We remark at this point that all vector spaces in this text are over the field of real numbers.

Definition 69.1 Let V be a vector space. A *norm* for V is a function $\| \cdot \|$ from V into $[0, \infty)$ which satisfies the following:

(i) $\|x\| = 0$ if and only if $x = \theta$.
(ii) $\|cx\| = |c| \, \|x\|$ for $c \in \mathbf{R}$, $x \in V$.
(iii) $\|x + y\| \leq \|x\| + \|y\|$ for $x, y \in V$.

The ordered pair $(V, \| \cdot \|)$ (or more simply, V) is called a *normed linear space* (or *normed vector space*).

Property (iii) of Definition 69.1 is called the *triangle inequality*.

We have already observed that equation (69.1) defines a norm on $\mathbf{R}^n$. To give further examples of normed linear spaces, we must first turn some of our familiar examples into vector spaces.

If $\{a_n\}$ and $\{b_n\}$ are real sequences and $c \in \mathbf{R}$, we define

$$\{a_n\} + \{b_n\} = \{a_n + b_n\} \qquad c\{a_n\} = \{ca_n\} \tag{69.2}$$

With these definitions of addition and scalar multiplication, l^1 and l^2 become vector spaces (verify).

Let $[a, b]$ be a closed interval and let $\mathscr{B}[a, b]$ be the set of all bounded functions on $[a, b]$. Under pointwise addition and scalar multiplication $\mathscr{B}[a, b]$ is a vector space (verify). The spaces $C[a, b]$ and $\mathscr{R}_\alpha[a, b]$, where $\alpha \in \mathscr{B}[a, b]$, are subspaces of $\mathscr{B}[a, b]$ (verify).

Theorem 69.2

(i) The equation

$$\|(x_1, \ldots, x_n)\| = \sqrt{x_1^2 + \cdots + x_n^2}$$

defines a norm on $\mathbf{R}^n$.

(ii) The equation

$$\|\{a_n\}\| = \sum_{n=1}^{\infty} |a_n|$$

defines a norm on l^1.

(iii) The equation

$$\|\{a_n\}\| = \sqrt{\sum_{n=1}^{\infty} a_n^2}$$

defines a norm on l^2.

(iv) The equation

$$\|f\| = \text{lub} \,\{|f(x)| : x \in [a, b]\}$$

defines a norm on $\mathscr{B}[a, b]$ and hence also on the subspaces $C[a, b]$ and $\mathscr{R}_\alpha[a, b]$, where $\alpha \in \mathscr{B}[a, b]$.

Proof. The proof of part (i) precedes Definition 69.1. The proofs of parts (ii) and (iii) are similar and thus are omitted. The proof of part (iv) is left as an exercise. ∎

The next theorem shows how it is possible to define a metric in a normed linear space.

Theorem 69.3   If $(V, \|\cdot\|)$ is a normed linear space, then the equation

$$d(x, y) = \|x - y\|$$

defines a metric on V.

Proof.   Let $x, y \in V$. Then $d(x, y) = 0$ if and only if $\|x - y\| = 0$ which by Definition 69.1(i) holds if and only if $x - y = \theta$. Thus property (i) of Definition 35.1 holds for d.

Let $x, y \in V$. Then

$$d(x, y) = \|x - y\| = \|(-1)(y - x)\| = |-1|\|y - x\| = \|y - x\| = d(y, x)$$

and thus property (ii) of Definition 35.1 holds for d.

Finally, let $x, y, z \in V$. Then

$$
\begin{aligned}
d(x, z) &= \|x - z\| \\
&= \|(x - y) + (y - z)\| \\
&\leq \|x - y\| + \|y - z\| \\
&= d(x, y) + d(y, z)
\end{aligned}
$$

and thus property (iii) of Definition 35.1 holds for d. Therefore, d is a metric for V. ∎

Consider the normed linear space $\mathbf{R}^n$, where the norm for $\mathbf{R}^n$ is defined by equation (69.1). The metric of Theorem 69.3 derived from this norm is

$$
\begin{aligned}
d[(x_1, \ldots, x_n), (y_1, \ldots, y_n)] &= \|(x_1, \ldots, x_n) - (y_1, \ldots, y_n)\| \\
&= \|(x_1 - y_1, \ldots, x_n - y_n)\| \\
&= \sqrt{(x_1 - y_1)^2 + \cdots + (x_n - y_n)^2}
\end{aligned}
$$

Thus the metric of Theorem 69.3 for $\mathbf{R}^n$ is identical to the usual metric for $\mathbf{R}^n$ (see Section 36). Similarly, it is easy to verify that the metric defined in Theorem 69.3 for l^1, l^2 or $C[a, b]$ with the norm given in Theorem 69.2 is identical to the metric defined previously for each of these spaces.

Whenever we apply any metric space definitions (such as continuity, limit of a sequence, open set, compact, complete, etc.) to a normed linear space V, unless otherwise specified, it will be assumed that the metric for V is that given in Theorem 69.3. For example, a sequence $\{x_n\}$ in a normed linear space V converges to x in V if for every $\varepsilon > 0$, there exists a positive integer N such that if $n \geq N$, we have

$$\|x_n - x\| = d(x_n, x) < \varepsilon$$

A complete normed linear space is called a *Banach space*. Thus each of the spaces $\mathbf{R}^n$, l^1, l^2, and $C[a, b]$ are Banach spaces. (See Section 46 and Corollary 60.8.)

Normed linear spaces will be considered in some detail in Chapter XIII. In the remainder of this chapter we will be concerned with a special kind of normed linear space known as an inner product space.

Exercises

69.1   Let S denote the set of all real sequences with addition and scalar multiplication defined by equation (69.2).

(a)   Prove that S is a vector space.

(b)   Prove that each of the spaces l^1, l^2, c_0, and l^∞ is a subspace of S.

(c)   Is H^∞ a subspace of S? Justify your answer.

69.2   (a)   Prove that $\mathscr{B}[a, b]$ is a vector space.

(b)   Prove that each of the spaces $C[a, b]$, $BV[a, b]$, and $\mathscr{R}_\alpha[a, b]$, where $\alpha \in \mathscr{B}[a, b]$, are subspaces of $\mathscr{B}[a, b]$.

69.3   Give the details of the proof of Theorem 69.2 (ii) to (iv).

69.4   Let V be a vector space and let d be a metric for V which satisfies

(a) $d(cx, cy) = |c| d(x, y)$ for $c \in \mathbf{R}$ and $x, y \in V$.

(b) $d(x, y) = d(x + z, y + z)$ for $x, y, z \in V$.

For x in V, define $\|x\| = d(x, \theta)$. Prove that $(V, \|\cdot\|)$ is a normed linear space.

69.5   Let $(V_1, \|\cdot\|_1)$ and $(V_2, \|\cdot\|_2)$ be normed linear spaces.

(a)   Prove that $V_1 \times V_2$ is a normed linear space if we define addition and scalar multiplication by the equations

$$(x_1, y_1) + (x_2, y_2) = (x_1 + x_2, y_1 + y_2) \qquad c(x, y) = (cx, cy)$$

and the norm by the equation

$$\|(x, y)\| = \|x\|_1 + \|y\|_2$$

(b)   Prove that if V_1 and V_2 are Banach spaces, then $V_1 \times V_2$ is a Banach space.

69.6   Let V be a normed linear space and let $y \in V$ and $c \in \mathbf{R}$. Prove that each of the following functions is continuous:

$$f(x) = x + y \qquad g(x) = cx \qquad h(t) = ty \qquad n(x) = \|x\|$$

69.7   Let V be the set of all real sequences $\{a_n\}$ such that for some positive integer N, $a_n = 0$ for all $n \geq N$.

(a)   Prove that V is a vector space.

(b)   Prove that the equation $\|\{a_n\}\| = \sum_{n=1}^\infty |a_n|$ defines a norm on V.

(c)   Prove that V is not a Banach space.

69.8*   Prove that every normed linear space is connected.

69.9   Prove that a normed linear space V is compact if and only if $V = \{\theta\}$.

69.10   Prove that $\|\{x_n\}\| = \text{lub } \{|x_n| \,|\, n \in \mathbf{P}\}$ defines a norm on l^∞ and, therefore, also on c_0. The metric of Exercise 35.6 is derived from this norm.

69.11   (a)   In analogy with Definition 46.6, define the completion of a normed linear space.

(b)*   Using your definition above, prove that every normed linear space V has a completion W such that $\bar{V} = W$.

70. The Inner Product Space $\mathbf{R}^3$

In $\mathbf{R}^3$ one encounters the *inner product* (or *dot product*) defined by the equation

$$(x, y) = x_1 y_1 + x_2 y_2 + x_3 y_3$$

where $x = (x_1, x_2, x_3)$, $y = (y_1, y_2, y_3) \in \mathbf{R}^3$. The norm of a vector x in $\mathbf{R}^3$ is related to the inner product by the equation

$$\|x\| = \sqrt{(x, x)}$$

Among the properties of the inner product useful in computations (all of which are easily verified) are

(i) $(cx + y, z) = c(x, z) + (y, z)$ for all $c \in \mathbf{R}$ and $x, y, z \in \mathbf{R}^3$.
(ii) $(x, cy + z) = c(x, y) + (x, z)$ for all $c \in \mathbf{R}$ and $x, y, z \in \mathbf{R}^3$.
(iii) $(x, y) = (y, x)$ for all $x, y \in \mathbf{R}^3$.
(iv) $(x, x) = 0$ if and only if $x = \theta$.

Properties (i) and (ii) state that the inner product is linear in each coordinate.

If x and y are unit vectors in $\mathbf{R}^3$, that is, if $\|x\| = 1 = \|y\|$, the inner product (x, y) has the geometric interpretation as $\cos \phi$, where ϕ is the angle between x and y (see Figure 70.1). The inner product gives a useful test for perpendicularity. The nonzero vectors x and y in $\mathbf{R}^3$ are perpendicular if and only if $\cos \phi = 0$, where ϕ is the angle between x and y. Thus unit vectors x and y are perpendicular if and only if $(x, y) = 0$. Arbitrary nonzero vectors x and y are perpendicular if and only if the associated unit vectors $x/\|x\|$ and $y/\|y\|$ are perpendicular. Thus nonzero vectors x and y are perpendicular if and only if $(x/\|x\|, y/\|y\|) = 0$. This equation holds (using the linearity of the inner product) if and only if $(x, y) = 0$. Thus nonzero vectors in $\mathbf{R}^3$ are perpendicular if and only if $(x, y) = 0$. If we adopt the standard assumption that any vector is perpendicular to the zero vector, we have the following theorem: vectors x and y are perpendicular in $\mathbf{R}^3$ if and only if $(x, y) = 0$. We will use the term *orthogonal* interchangably with *perpendicular*.

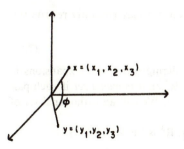

Figure 70.1

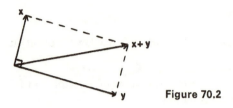

Figure 70.2

Many theorems from geometry may be proved using the inner product. We give a quick proof of the Pythagorean theorem. (See Figure 70.2.)

Theorem 70.1 If x and y are orthogonal vectors in $\mathbf{R}^3$, then $\|x + y\|^2 = \|x\|^2 + \|y\|^2$.

Proof. We compute using the linearity of the inner product and the fact that $(x, y) = (y, x) = 0$.

$$\begin{aligned}
\|x + y\|^2 &= (x + y, x + y) \\
&= (x, x + y) + (y, x + y) \\
&= (x, x) + (x, y) + (y, x) + (y, y) \\
&= (x, x) + (y, y) \\
&= \|x\|^2 + \|y\|^2 \qquad \blacksquare
\end{aligned}$$

Similar methods prove the generalized Pythagorean theorem.

Theorem 70.2 Let $\{x, y, z\}$ be a set of mutually orthogonal vectors in $\mathbf{R}^3$. Then

$$\|x + y + z\|^2 = \|x\|^2 + \|y\|^2 + \|z\|^2$$

Let $x = (x_1, x_2, x_3), y = (y_1, y_2, y_3) \in \mathbf{R}^3$. The Cauchy-Schwarz inequality (Theorem 36.1) states that

$$|x_1y_1 + x_2y_2 + x_3y_3| \le \sqrt{x_1^2 + x_2^2 + x_3^2}\sqrt{y_1^2 + y_2^2 + y_3^2}$$

Using the inner product and norm notation, we may elegantly rewrite this inequality as

$$|(x, y)| \le \|x\|\|y\| \tag{70.1}$$

We conclude this brief section by considering orthogonal expansions in $\mathbf{R}^3$. We call a set $\{x_1, x_2, x_3\}$ of vectors in $\mathbf{R}^3$ an *orthogonal set* if each pair $\{x_1, x_2\}$, $\{x_1, x_3\}$, and $\{x_2, x_3\}$ is orthogonal. We call an orthogonal set of unit vectors an *orthonormal set*.

One of the standard orthonormal sets in $\mathbf{R}^3$ is the set $\{\vec{\imath}, \vec{\jmath}, \vec{k}\}$,

where $\vec{\imath} = (1, 0, 0), \qquad \vec{\jmath} = (0, 1, 0), \qquad \vec{k} = (0, 0, 1)$

Any vector $x = (c_1, c_2, c_3) \in \mathbf{R}^3$ may be written as the unique linear combination

$$x = c_1 \vec{\imath} + c_2 \vec{\jmath} + c_3 \vec{k}$$

The coefficients are given by the equations

$$c_1 = (x, \vec{\imath}), \qquad c_2 = (x, \vec{\jmath}), \qquad c_3 = (x, \vec{k}). \qquad \text{(Verify)} \qquad (70.2)$$

We will generalize this situation to arbitrary orthonormal sets in $\mathbf{R}^3$. We begin by showing that if a vector in $\mathbf{R}^3$ has an orthonormal expansion, the coefficients are uniquely determined by equations like those in (70.2).

Theorem 70.3 Let $\{x_1, x_2, x_3\}$ be an orthonormal set in $\mathbf{R}^3$ and let $x \in \mathbf{R}^3$, where

$$x = c_1 x_1 + c_2 x_2 + c_3 x_3$$

Then $\qquad\qquad c_k = (x, x_k) \qquad$ for $k = 1, 2$ and 3

Proof. If $k = 1, 2$, or 3, then

$$(x, x_k) = (c_1 x_1 + c_2 x_2 + c_3 x_3, x_k)$$
$$= c_1(x_1, x_k) + c_2(x_2, x_k) + c_3(x_3, x_k) = c_k \qquad \blacksquare$$

We next show that any vector x in $\mathbf{R}^3$ has the orthonormal expansion

$$x = (x, x_1)x_1 + (x, x_2)x_2 + (x, x_3)x_3$$

where $\{x_1, x_2, x_3\}$ is an orthonormal set in $\mathbf{R}^3$.

Theorem 70.4 Let $\{x_1, x_2, x_3\}$ be an orthonormal set in $\mathbf{R}^3$. Then every vector x in $\mathbf{R}^3$ has the representation

$$x = (x, x_1)x_1 + (x, x_2)x_2 + (x, x_3)x_3$$

Proof. We first show that the set $\{x_1, x_2, x_3\}$ is linearly independent in $\mathbf{R}^3$ and hence a basis for $\mathbf{R}^3$.

Suppose we have scalars c_1, c_2, c_3 such that

$$c_1 x_1 + c_2 x_2 + c_3 x_3 = \theta$$

By Theorem 70.3, $c_k = (\theta, x_k) = 0$ for $k = 1, 2, 3$, and hence $\{x_1, x_2, x_3\}$ is linearly independent in $\mathbf{R}^3$.

Let $x \in \mathbf{R}^3$. Since $\{x_1, x_2, x_3\}$ is a basis for $\mathbf{R}^3$, there exist scalars c_1, c_2, c_3 such that

$$x = c_1 x_1 + c_2 x_2 + c_3 x_3$$

By Theorem 70.3, $c_k = (x, x_k)$ for $k = 1, 2, 3$. $\qquad \blacksquare$

The coefficients $c_k = (x, x_k)$, $k = 1, 2, 3$, of Theorem 70.4 are called the *Fourier coefficients* of x relative to the orthonormal set $\{x_1, x_2, x_3\}$. The sum

$$(x, x_1)x_1 + (x, x_2)x_2 + (x, x_3)x_3$$

is called the *Fourier series* of x relative to $\{x_1, x_2, x_3\}$. Theorem 70.4 states that the Fourier series of x is equal to x.

Parseval's theorem states that the sum of the squares of the Fourier coefficients of x is equal to $\|x\|^2$.

Theorem 70.5 (Parseval's Theorem for $\mathbf{R}^3$) Let $\{x_1, x_2, x_3\}$ be an orthonormal set in $\mathbf{R}^3$ and let $x \in \mathbf{R}^3$. Then

$$\|x\|^2 = \sum_{k=1}^{3} (x, x_k)^2$$

Proof. By Theorem 70.4,

$$x = \sum_{k=1}^{3} (x, x_k)x_k$$

Since $\{(x, x_1)x_1, (x, x_2)x_2, (x, x_3)x_3\}$ is a set of mutually orthogonal vectors, by Theorem 70.2

$$\left\| \sum_{k=1}^{3} (x, x_k)x_k \right\|^2 = \sum_{k=1}^{3} \|(x, x_k)x_k\|^2$$

and the theorem follows immediately. ∎

In the remainder of this chapter we will generalize the results of this section to other settings.

71. Inner Product Spaces

In this section we will define an abstract inner product space and give several important examples. Our definitions and results should be compared with those of Section 70 concerning $\mathbf{R}^3$.

Definition 71.1 Let V be a vector space. An *inner product* (sometimes called a *dot product* or *scalar product*) for V is a function (,) from $V \times V$ into $\mathbf{R}$ which satisfies the following:

(i) $(cx + y, z) = c(x, z) + (y, z)$ for all $c \in \mathbf{R}$ and all $x, y, z \in V$.
(ii) $(x, y) = (y, x)$ for all $x, y \in V$.
(iii) $(x, x) \geq 0$ for all $x \in V$.
(iv) If $(x, x) = 0$, then $x = \theta$.

The ordered pair $(V, (\ , \ ))$ (usually denoted by just V) is called an *inner product space*.

We now give three important examples of inner product spaces.

Theorem 71.2

(i) The equation

$$(x, y) = \sum_{k=1}^{n} x_k y_k$$

where $x = (x_1, \ldots, x_n)$, $y = (y_1, \ldots, y_n) \in \mathbf{R}^n$ defines an inner product on $\mathbf{R}^n$.

(ii) The equation

$$(x, y) = \sum_{n=1}^{\infty} x_n y_n$$

where $x = \{x_n\}$, $y = \{y_n\} \in l^2$ defines an inner product on l^2.

(iii) The equation

$$(f, g) = \int_a^b fg$$

where $f, g \in \mathcal{R}[a, b]$ defines an inner product on $\mathcal{R}[a, b]$ with the exception that property (iv) of Definition 71.1 is replaced by

(iv)$'$ If $(f, f) = 0$, then $f = \theta$ almost everywhere in $[a, b]$.

Proof. Part (i) easily verified.

The sum $\sum_{n=1}^{\infty} x_n y_n$ in part (ii) converges by Theorem 36.3. The fact that the equation in part (ii) defines an inner product on l^2 is easily established.

If $f, g \in \mathcal{R}[a, b]$, then $fg \in \mathcal{R}[a, b]$ by Theorem 51.13(iii) and thus $\int_a^b fg$ is defined. It is easy to verify that properties (i), (ii), and (iii) of Definition 71.1 hold for $\mathcal{R}[a, b]$. We verify (iv)$'$.

Let $f \in \mathcal{R}[a, b]$ and suppose $(f, f) = 0$. Then $\int_a^b f^2 = 0$; so by Corollary 58.6, $f^2 = 0$ almost everywhere in $[a, b]$. Therefore, $f = 0$ almost everywhere in $[a, b]$. (Note that θ in $\mathcal{R}[a, b]$ is $\theta(x) = 0$ for *all* $x \in [a, b]$.) ∎

If $n = 3$, the equation in Theorem 71.2(i) defines the usual inner product for $\mathbf{R}^3$ (see Section 70). The space $C[a, b]$ is a subspace of $\mathcal{R}[a, b]$ so the equation in Theorem 71.2(iii) also defines an inner product on $C[a, b]$. However, (iv)$'$ for $C[a, b]$ is equivalent to Definition 71.1(iv) for if a continuous function is zero almost everywhere in $[a, b]$, it is identically zero in $[a, b]$ (verify). The inner product space $\mathcal{R}[a, b]$ will be examined in some detail in Sections 72 to 76.

We next state an elementary theorem which will be useful in our subsequent computations.

Theorem 71.3 Let V be an inner product space. Then

(i) $(x, cy + z) = c(x, y) + (x, z)$ for $c \in \mathbf{R}$, and $x, y, z \in V$.

(ii) $(cx + dy, cx + dy) = c^2(x, x) + 2cd(x, y) + d^2(y, y)$ for $c, d \in \mathbf{R}$ and $x, y \in V$.

Let V be an inner product space. If we set

$$\|x\| = \sqrt{(x, x)} \qquad \text{for } x \in V$$

it is easy to verify that

(i) $\|x\| = 0$ if and only if $x = \theta$.

(ii) $\|cx\| = |c|\,\|x\|$ for all $c \in \mathbf{R}$ and $x \in V$.

In order to prove the triangle inequality, we need to prove the Cauchy-Schwarz inequality for arbitrary inner product spaces. The proof of Theorem 71.4 should be compared with the proof of Theorem 36.1 (which is the special case $V = \mathbf{R}^n$).

Theorem 71.4 (Cauchy-Schwarz Inequality for Inner Product Spaces) Let V be an inner product space and let

$$\|x\| = \sqrt{(x, x)} \qquad \text{for } x \in V$$

Then $|(x, y)| \le \|x\|\,\|y\|$ for all $x, y \in V$

Proof. If $y = \theta$, the conclusion is obvious; so we suppose $y \ne \theta$. Then $(y, y) > 0$. If c is any real number, then

$$0 \le (x - cy, x - cy) = (x, x) - 2c(x, y) + c^2(y, y)$$

Letting $c = (x, y)/(y, y)$, we have

$$0 \le (x, x) - 2\frac{(x, y)^2}{(y, y)} + \frac{(x, y)^2}{(y, y)}$$

which reduces to

$$0 \le (x, x) - \frac{(x, y)^2}{(y, y)}$$

and the inequality now follows. ∎

Corollary 71.5 Let V be an inner product space. The equation

$$\|x\| = \sqrt{(x, x)} \qquad x \in V$$

defines a norm on V.

Proof. Properties (i) and (ii) of Definition 69.1 are easily verified.

Let $x, y \in V$. Then

$$\|x + y\|^2 = (x + y, x + y) = (x, x) + 2(x, y) + (y, y)$$
$$\leq \|x\|^2 + 2\|x\|\|y\| + \|y\|^2$$
$$= (\|x\| + \|y\|)^2$$

where we have used the Cauchy-Schwarz inequality (Theorem 71.4). Therefore property (iii) of Definition 69.1 also holds. ∎

If V is either of $\mathbf{R}^n$ or l^2, Corollary 71.5 defines a norm identical to the norm on $\mathbf{R}^n$ or l^2 defined previously in Theorem 69.2. The Cauchy-Schwarz inequalities previously given for $\mathbf{R}^n$ and l^2 (Theorems 36.1 and 36.6) are identical to the Cauchy-Schwarz inequalities of Theorem 71.4 for $V = \mathbf{R}^n$ or $V = l^2$ (verify).

The norm defined by Corollary 71.5 for $\mathscr{R}[a, b]$ is given by the equation

$$\|f\| = \sqrt{\int_a^b f^2} \qquad \text{for } f \in \mathscr{R}[a, b]$$

This norm is called the L^2-*norm* for $\mathscr{R}[a, b]$. The distance between two functions f and g in $\mathscr{R}[a, b]$ is given by the equation

$$d(f, g) = \|f - g\| = \sqrt{\int_a^b (f - g)^2}$$

The Cauchy-Schwarz inequality for $\mathscr{R}[a, b]$ becomes

$$\left| \int_a^b fg \right| \leq \sqrt{\int_a^b f^2} \sqrt{\int_a^b g^2} \qquad \text{for all } f, g \in \mathscr{R}[a, b]$$

The subspace $C[a, b]$ of $\mathscr{R}[a, b]$ is now equipped with two norms, the L^2-norm and the uniform norm. The L^2-norm on $C[a, b]$ is given by the equation

$$\|f\|_2 = \sqrt{\int_a^b f^2}$$

and the uniform norm on $C[a, b]$ is given by

$$\|f\|_u = \text{lub } \{|f(x)| \mid x \in [a, b]\}$$

These norms are *not* the same (see Exercise 71.8).

A complete inner product space is called a *Hilbert space*, and thus $\mathbf{R}^n$ and l^2 furnish examples of Hilbert spaces. It can be shown that the space $\mathscr{R}[a, b]$ is not complete in the metric from the L^2-norm, and therefore $\mathscr{R}[a, b]$ is an inner product space which is not a Hilbert space. The accompanying chart shows the interrelationships of the various spaces we have defined.

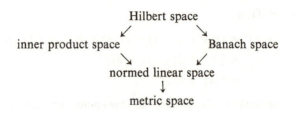

Exercises

71.1 Prove Theorem 71.2(i) and (ii).

71.2 Prove that properties (i), (ii), and (iii) of Definition 71.1 hold for the inner product of Theorem 71.2(iii).

71.3 Prove Theorem 71.3.

71.4 Prove that properties (i) and (ii) of Definition 69.1 hold for the norm of Corollary 71.5.

71.5 (a) Prove that equality holds in the Cauchy-Schwarz inequality (Theorem 71.4) if and only if there exist scalars c and d such that $cx = dy$.

(b) When does equality hold in Corollary 71.5?

71.6 Let V be the vector space of Exercise 69.7.

(a) Prove that the equation

$$(x, y) = \sum_{n=1}^{\infty} x_n y_n$$

defines an inner product on V.

(b) Prove that V is an inner product space which is not a Hilbert space.

71.7 (a) Let V be an inner product space. Prove the *polarization identity*
$$||x + y||^2 - ||x - y||^2 = 4(x, y) \qquad \text{for all } x, y \in V$$
and the *parallelogram law*
$$||x + y||^2 + ||x - y||^2 = 2(||x||^2 + ||y||^2) \qquad \text{for all } x, y \in V$$
Interpret the last equation geometrically in the plane.

(b) Show that the parallelogram law is not satisfied in any of the spaces l^1, l^∞, c_0, or $C[a, b]$ (where $C[a, b]$ is given the uniform norm). Deduce that there is no inner product which gives the norm for any of these spaces.

(c) Let V be a normed linear space in which the parallelogram law holds. Define (x, y) by the polarization identity. Prove that $(V, (,))$ is an inner product space and that $||x|| = \sqrt{(x, x)}$.

71.8 Define $f \in C[0, 1]$ by $f(x) = x$. Calculate $||f||_u$ and $||f||_2$.

71.9 Prove that $\mathscr{R}[a, b]$ is complete in the metric of the uniform norm.

71.10 (a) In analogy with Definition 46.6, define the completion of an inner product space.

(b) Using your definition above, prove that every inner product space V has a completion W such that $\bar{V} = W$.

72. Orthogonal Sets in Inner Product Spaces

We begin by defining orthogonality in abstract inner product spaces in terms of the inner product. (See Section 70 for motivation.)

Definition 72.1 Let V be an inner product space. We say that vectors x and y in V are *orthogonal* (or *perpendicular*) if $(x, y) = 0$. A set X of vectors in V is said to be an *orthogonal set* if any two distinct vectors in X are orthogonal. A set X of vectors in V is said to be an *orthonormal set* if X is an orthogonal set for which $\|x\| = 1$ for every $x \in X$.

Examples of orthonormal sets are provided by the next theorem.

Theorem 72.2

(i) The set

$$\{(1, 0, 0, \ldots, 0), (0, 1, 0, \ldots, 0), \ldots, (0, 0, \ldots, 0, 1)\}$$

is an orthonormal set in $\mathbf{R}^n$.

(ii) The set $\{\delta^{(n)}\}_{n=1}^{\infty}$ is an orthonormal set in l^2, where

$$\delta_m^{(n)} = \begin{cases} 1 & \text{if } n = m \\ 0 & \text{if } n \neq m \end{cases}$$

(iii) Let

$$\phi_0(x) = \frac{1}{\sqrt{2\pi}}, \qquad \phi_{2n-1}(x) = \frac{\cos nx}{\sqrt{\pi}}, \qquad \phi_{2n}(x) = \frac{\sin nx}{\sqrt{\pi}}$$

for $x \in \mathbf{R}$ and $n = 1, 2, \ldots$. Then $\{\phi_0, \phi_1, \ldots\}$ is an orthonormal set in $\mathcal{R}[a, a + 2\pi]$ for any real number a.

Proof. Parts (i) and (ii) are easily verified.

We first verify part (iii) for the special case $a = 0$. Now

$$\|\phi_0\|^2 = \int_0^{2\pi} \left(\frac{1}{\sqrt{2\pi}}\right)^2 = 1$$

$$\|\phi_{2n-1}\|^2 = \int_0^{2\pi} \frac{\cos^2 nx}{\pi}\, dx = \left. \frac{nx + \sin nx \cos nx}{2n\pi} \right|_0^{2\pi} = 1$$

Similarly, $$\|\phi_{2n}\|^2 = 1$$

Finally, we must show that any two distinct vectors in the set $\{\phi_0, \phi_1, \ldots\}$ are orthogonal. Now

$$(\phi_0, \phi_{2n-1}) = \int_0^{2\pi} \frac{\cos nx}{\sqrt{2\pi}}\, dx = \left. \frac{\sin nx}{n\sqrt{2\pi}} \right|_0^{2\pi} = 0$$

Similarly, $$(\phi_0, \phi_{2n}) = 0$$

If $n = m$,

$$(\phi_{2n-1}, \phi_{2m}) = \int_0^{2\pi} \frac{\cos nx \sin nx}{\pi} dx$$

$$= -\frac{1}{2\pi} \left\{ \frac{\cos 2nx}{2n} \right\}_0^{2\pi} = 0$$

If $n \neq m$,

$$(\phi_{2n-1}, \phi_{2m}) = \int_0^{2\pi} \frac{\cos nx \sin mx}{\pi} dx$$

$$= -\frac{1}{2\pi} \left\{ \frac{\cos [(m+n)x]}{m+n} - \frac{\cos [(m-n)x]}{m-n} \right\}_0^{2\pi} = 0$$

Similarly, one shows that if $n \neq m$,

$$(\phi_{2n-1}, \phi_{2m-1}) = 0 = (\phi_{2n}, \phi_{2m})$$

Therefore, the set $\{\phi_0, \phi_1, \ldots\}$ is an orthonormal set in $[0, 2\pi]$. Since all of the integrands appearing in our proof are periodic of period 2π, the values of the integrals are not altered by changing the limits of integration from $\int_0^{2\pi}$ to $\int_a^{a+2\pi}$. (We will give a formal proof of this fact in the next section.) Thus $\{\phi_0, \phi_1, \ldots\}$ is an orthonormal set in $\mathcal{R}[a, a + 2\pi]$. ∎

We will call the set $\{\phi_0, \phi_1, \ldots\}$ of Theorem 72.2(iii), the *trigonometric set*. We will study the inner product space $\mathcal{R}[a, a + 2\pi]$ and the trigonometric set in great detail in the remainder of this chapter.

Exercises

72.1   Prove Theorem 72.2(i) and (ii).

72.2   Let $\{\phi_0, \phi_1, \ldots\}$ be the trigonometric set.
    (a)   Prove that $\|\phi_{2n}\| = 1$.
    (b)   Prove that $(\phi_0, \phi_{2n}) = 0$.
    (c)   Prove the integration formula

$$\int \cos ax \sin bx \, dx = \frac{1}{2} \left\{ \frac{\cos [(a-b)x]}{a-b} - \frac{\cos [(a+b)x]}{a+b} \right\} + c$$

       for all $a \neq b$, $a \neq -b$
       used in the proof of Theorem 72.2.
    (d)   Complete the proof of Theorem 72.2 by showing that if $n \neq m$,

$$(\phi_{2n-1}, \phi_{2m-1}) = 0 = (\phi_{2n}, \phi_{2m})$$

72.3   Let X be an orthogonal set of nonzero vectors in an inner product space V. Prove that the set

$$\left\{ \frac{x}{\|x\|} \,\middle|\, x \in X \right\}$$

s an orthonormal set in V.

72.4 Let X be an orthogonal set of nonzero vectors in an inner product space V. Prove that X is a linearly independent set of vectors in V.

72.5 (Gram-Schmidt orthogonalization process) Let V be an inner product space. Let $\{x_1, x_2, \ldots\}$ be a linearly independent set of vectors in V. Let $y_1 = x_1$ and

$$y_m = x_m - \sum_{k=1}^{m-1} \frac{(x_m, y_k)}{\|y_k\|^2} y_k \qquad \text{for } m = 2, 3, \ldots$$

Prove that $\{y_1, y_2, \ldots\}$ is an orthogonal set in V. Interpret this process geometrically in $\mathbf{R}^3$.

72.6 Let $X = \{1, x, x^2, \ldots\}$.

   (a) Prove that X is a linearly independent set in the vector space $\mathscr{R}[-1, 1]$.

   (b) Let $\{\psi_0, \psi_1, \ldots\}$ be the orthogonal set obtained from X by the Gram-Schmidt orthogonalization process (see Exercise 72.5) for the inner product space $\mathscr{R}[-1, 1]$. Compute $\psi_0, \psi_1, \psi_2, \psi_3$.

Let

$$f_n(x) = (x^2 - 1)^n$$

for $x \in \mathbf{R}$ and $n = 0, 1, 2, \ldots$. Let

$$\theta_n(x) = \frac{1}{2^n n!} f_n^{(n)}(x)$$

for $x \in [-1, 1]$ and $n = 0, 1, 2, \ldots$. The polynomial θ_n is called the *Legendre polynomial* of order n.

   (c) Compute $\theta_0, \theta_1, \theta_2, \theta_3$ and compare your results with part (a).

   (d) Prove

$$\theta_{n+1}'(x) = x\theta_n'(x) + (n+1)\theta_n(x) \qquad \text{for } x \in \mathbf{R}$$

   (e) Prove

$$(n+1)\theta_{n+1}(x) = (n+1)x\theta_n(x) + (x^2 - 1)\theta_n'(x) \qquad \text{for } x \in \mathbf{R}$$

   (f) Prove

$$[(1 - x^2)(\theta_n(x)\theta_m'(x) - \theta_m(x)\theta_n'(x))]' = [n(n+1) - m(m+1)]\theta_m(x)\theta_n(x)$$

     for $x \in \mathbf{R}$.

   (g) Prove that the set $\{\theta_0, \theta_1, \ldots\}$ is orthogonal in $\mathscr{R}[-1, 1]$.

   (h) Let n be a nonnegative integer. Prove that there exist constants $c_0, c_1, \ldots, c_n$ such that

$$\theta_n = \sum_{k=0}^{n} c_k \psi_k$$

   (i)* Prove that there exist constants $k_0, k_1, \ldots$ such that $\theta_n = k_n \psi_n$, $n = 0, 1, \ldots$.

73. Periodic Functions

In this section we will first justify changing the limits of integration in the integral of a periodic function (Theorem 73.2). This fact was used in the proof of Theorem 72.2(iii). We will then briefly examine the problem of extending a given function to a periodic function.

Definition 73.1 Let f be a function on $\mathbf{R}$. We say that f is *periodic of period* $p > 0$ if

$$f(x + p) = f(x) \qquad \text{for all } x \text{ in } \mathbf{R}$$

Theorem 73.2 Let f be periodic of period p and suppose that $f \in \mathcal{R}[a, a + p]$. If $c \in \mathbf{R}$, then $f \in \mathcal{R}[c, c + p]$ and

$$\int_a^{a+p} f = \int_c^{c+p} f$$

Proof. Let n be an integer and let $g(t) = t - np$. Then g is a strictly increasing function from $[a + np, a + (n + 1)p]$ onto $[a, a + p]$; so by Theorem 56.3, $(f \circ g)g' \in \mathcal{R}[a + np, a + (n + 1)p]$. However,

$$[(f \circ g)g'](t) = f(t - np) = f(t)$$

and thus $f \in \mathcal{R}[a + np, a + (n + 1)p]$ for every integer n. It follows that f is Riemann integrable on every closed subinterval of $\mathbf{R}$ (why?).

We conclude the proof by showing that if $c \in \mathbf{R}$,

$$\int_c^{c+p} f = \int_0^p f$$

We first evaluate

$$\int_p^{c+p} f(x)\, dx$$

Letting $g(t) = t + p$ and using Theorem 56.3, we have

$$\int_p^{c+p} f(x)\, dx = \int_0^c f(t + p)\, dt = \int_0^c f(t)\, dt$$

Therefore $\displaystyle \int_c^{c+p} f = \int_c^p f + \int_p^{c+p} f = \int_c^p f + \int_0^c f = \int_0^p f$ ∎

We now consider the problem of extending a function f on $[a, a + p]$ to $\mathbf{R}$ so that f is periodic of period p. In case $f(a) = f(a + p)$, we may simply extend f by the rule

$$f(x) = f(x - np) \qquad a + np \le x < a + (n + 1)p$$

where n is an integer. (See Figure 73.1.)

Suppose $f \in \mathcal{R}[a, a + p]$. Define g by the rule

$$g(x) = \begin{cases} f(x) & a \le x < a + p \\ f(a) & x = a + p \end{cases} \tag{73.1}$$

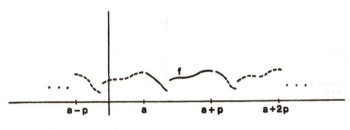

Figure 73.1

Then $g(x) = f(x)$ for $x \in [a, a + p)$; so

$$\int_a^{a+p} g = \int_a^{a+p} f$$

Since $g(a) = g(a + p)$, g may be extended to $\mathbf{R}$ so that g is periodic of period p. Therefore,

$$\int_c^{c+p} g = \int_a^{a+p} g = \int_a^{a+p} f \tag{73.2}$$

for any c in $\mathbf{R}$. Thus g serves as a kind of periodic extension of f, and so far as the integral is concerned [equation (73.2)], nothing is lost by redefining f at $x = a + p$.

Finally, let $f \in C[a, a + p]$. If $f(a) = f(a + p)$, we may extend f to a continuous function on $\mathbf{R}$ periodic of period p. If $f(a) \neq f(a + p)$, we may replace f by g [equation (73.1)] and extend g to a function on $\mathbf{R}$ periodic of period p. Equation (73.2) is still valid. However, we have introduced discontinuities. The extension g is discontinuous at $x = a + np$, $n = 0, \pm 1$, $\pm 2, \ldots$. (See Figure 73.2.)

We will let $CP[a, b]$ denote the set of functions continuous on $\mathbf{R}$ and periodic of period $b - a$. Our above discussion shows that $CP[a, b]$ can be identified with the set of continuous functions f on $[a, b]$ satisfying $f(a) = f(b)$. The set $CP[a, b]$ plays an important role in the theory of the inner product space $\mathscr{R}[a, b]$. (See Section 77.)

Exercises

73.1 Give an example of periodic functions f and g such that $f + g$ is not periodic.

73.2 Prove that $CP[a, b]$ is a vector space under pointwise addition and scalar multiplication.

73.3 A bounded, continuous, real-valued function on $\mathbf{R}$ is said to be *almost periodic* if for every $\varepsilon > 0$, there exists $L > 0$ such that in every interval of

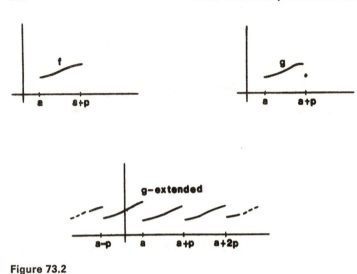

Figure 73.2

length L there exists a point a for which $|f(a + t) - f(t)| < \varepsilon$ for every $t \in \mathbf{R}$.
(a) Prove that any continuous periodic function is almost periodic.
(b) Prove that an almost periodic function is uniformly continuous on $\mathbf{R}$.
(c) Prove that if f and g are almost periodic functions, then $f + g$ is almost periodic. (Compare with Exercise 73.1).

74. Fourier Series: Definitions and Examples

We begin by extending Theorem 70.3 (which is concerned with $\mathbf{R}^3$) to an arbitrary inner product space. The proof of Theorem 74.1 is identical to the proof of Theorem 70.3.

Theorem 74.1 Let $\{x_1, \ldots, x_n\}$ be an orthonormal set in an inner product space V and let $x \in V$. Suppose there exist scalars $c_1, \ldots, c_n$ such that

$$x = \sum_{k=1}^{n} c_k x_k$$

Then $c_k = (x, x_k) \qquad \text{for } k = 1, \ldots, n$

   Proof.

$$(x, x_k) = \left(\sum_{j=1}^{n} c_j x_j, x_k \right) = \sum_{j=1}^{n} c_j (x_j, x_k) = c_k \qquad \blacksquare$$

Theorem 74.1 states that if a vector x in an inner product space V is a linear combination of the members of an orthonormal set $\{x_1, \ldots, x_n\}$, then x is the unique linear combination

$$x = \sum_{k=1}^{n} (x, x_k)x_k$$

Let $\{x_1, x_2, \ldots\}$ be a countable orthonormal set in an inner product space V and let $x \in V$. Motivated by Theorem 74.1, we can form the infinite series

$$\sum_{n=1}^{\infty} (x, x_n)x_n$$

and ask when

$$x = \sum_{n=1}^{\infty} (x, x_n)x_n$$

This is one of the central topics studied in the theory of Fourier series.

Definition 74.2   Let $X = \{x_1, x_2, \ldots\}$ be a countable orthonormal set in an inner product space V and let $x \in V$. The infinite series

$$\sum_{n=1}^{\infty} (x, x_n)x_n$$

is called the *Fourier series* of x (relative to X). The coefficient (x, x_n) is called the *nth Fourier coefficient* of x (relative to X).

The Fourier series of a function f in $\mathscr{R}[a, a + 2\pi]$ relative to the trigonometric set is

$$\sum_{n=0}^{\infty} (f, \phi_n)\phi_n(x) = \frac{1}{\sqrt{2\pi}} \int_a^{a+2\pi} \frac{f(t)}{\sqrt{2\pi}}\, dt + \sum_{n=1}^{\infty} \left(\frac{\cos nx}{\sqrt{\pi}} \int_a^{a+2\pi} f(t)\, \frac{\cos nt}{\sqrt{\pi}}\, dt \right.$$
$$\left. + \frac{\sin nx}{\sqrt{\pi}} \int_a^{a+2\pi} f(t)\, \frac{\sin nt}{\sqrt{\pi}}\, dt \right)$$

If we let

$$a_n = \frac{1}{\pi} \int_a^{a+2\pi} f(t) \cos nt\, dt, \qquad n = 0, 1, \ldots$$

$$b_n = \frac{1}{\pi} \int_a^{a+2\pi} f(t) \sin nt\, dt, \qquad n = 1, 2, \ldots$$

the Fourier series of f reduces to

$$\frac{a_0}{2} + \sum_{n=1}^{\infty} (a_n \cos nx + b_n \sin nx)$$

As an example, we calculate the Fourier series of $f(x) = x$, $-\pi \le x \le \pi$.

We have

$$a_n = \frac{1}{\pi} \int_{-\pi}^{\pi} t \cos nt \, dt = 0 \qquad n = 0, 1, \ldots$$

$$b_n = \frac{1}{\pi} \int_{-\pi}^{\pi} t \sin nt \, dt = \frac{2(-1)^{n+1}}{n}, \qquad n = 1, 2, \ldots$$

and thus the Fourier series of f is

$$\sum_{n=1}^{\infty} \frac{2(-1)^{n+1}}{n} \sin nx$$

Let $\{x_1, x_2, \ldots\}$ be an orthonormal set in an inner product space V and let $x \in V$. If we ask whether the Fourier series of x converges to x, we are asking whether the series

$$\sum_{n=1}^{\infty} (x, x_n)x_n$$

converges to x. In an arbitrary inner product space, the norm is given by $\|y\| = \sqrt{(y, y)}$, $y \in V$, and thus we are asking if the sequence of partial sums of the series $\sum_{n=1}^{\infty} (x, x_n)x_n$ converges to x in the inner product space norm. Therefore, the Fourier series of x converges to x in the inner product space norm if for every $\varepsilon > 0$, there exists a positive integer N such that if $k \geq N$, we have

$$\left\| x - \sum_{n=1}^{k} (x, x_n)x_n \right\| < \varepsilon \qquad (74.1)$$

In the inner product space $\mathscr{R}[a, a + 2\pi]$ convergence of the Fourier series of a function $f \in \mathscr{R}[a, a + 2\pi]$ to f in the inner product space norm means for every $\varepsilon > 0$, there exists a positive integer N such that if $k \geq N$,

$$\sqrt{\int_a^{a+2\pi} \left[f - \sum_{n=1}^{k} (f, \phi_n)\phi_n \right]^2} < \varepsilon$$

We have called such convergence L^2-convergence. We will show (Theorem 78.5) that the Fourier series of a function $f \in \mathscr{R}[a, a + 2\pi]$ converges to f in the L^2-sense.

The sequence of partial sums of the Fourier series of a function $f \in \mathscr{R}[a, a + 2\pi]$ is a sequence of real-valued functions. Thus we can also pose the questions: (1) Does the Fourier series of f converge pointwise? (2) Does the Fourier series of f converge uniformly?

Question (1) is very difficult. It has been known for some time that continuity of f at x does not imply convergence of the Fourier series of f at x. (See Hewitt and Stromberg, 1965.) It has recently been shown that if $f \in \mathscr{R}[a, a + 2\pi]$, then the Fourier series of f converges almost everywhere to f.

(See Carleson, 1966.) We will show (Corollary 76.8) that if f is differentiable at x, then the Fourier series of f converges to f at x.

Concerning question (2), we will prove (Theorem 77.5) that if $f \in CP[a, a + 2\pi]$, then the Fourier series of f is uniformly Cesàro summable to f and that if $f \in CP[a, a + 2\pi] \cap BV[a, a + 2\pi]$, then the Fourier series of f converges uniformly to f (Theorem 79.3).

Exercises

74.1 What is the Fourier series of an element of $\mathbf{R}^n$ or l^2 relative to the orthonormal sets defined in Theorem 72.2?

74.2 Let $\{x_1, x_2, \ldots\}$ be an orthonormal set in an inner product space V and let $x, y \in V$. Let $\sum_{n=1}^{\infty} c_n x_n$ and $\sum_{n=1}^{\infty} d_n x_n$ be the Fourier series of x and y, respectively. Prove that $\sum_{n=1}^{\infty} (c_n + d_n) x_n$ is the Fourier series of $x + y$.

74.3 Compute the Fourier series of the following functions:

(a) $f(x) = x \qquad 0 \le x \le 2\pi$

(b) $f(x) = x^2 \qquad -\pi \le x \le \pi$

(c) $f(x) = x^2 \qquad 0 \le x \le 2\pi$

(d) $f(x) = \begin{cases} 1 & 0 \le x \le \pi \\ 0 & -\pi \le x < 0 \end{cases}$

(e) $f(x) = \begin{cases} x & 0 \le x \le \pi \\ 0 & -\pi \le x < 0 \end{cases}$

(f) $f(x) = |x| \qquad -\pi \le x \le \pi$

Which of the above functions belong to $CP[0, 2\pi]$ or $CP[-\pi, \pi]$?

74.4 (a) Suppose that $f \in \mathscr{R}[-\pi, \pi]$ and $f(x) = f(-x)$ for $x \in [0, \pi]$. Prove that the Fourier series for f is $a_0/2 + \sum_{n=1}^{\infty} a_n \cos nx$, where $a_n = (2/\pi) \int_0^{\pi} f(t) \cos nt \, dt$.

(b) Suppose that $f \in \mathscr{R}[-\pi, \pi]$ and $f(x) = -f(-x)$ for $x \in [0, \pi]$. Prove that the Fourier series for f is $\sum_{n=1}^{\infty} b_n \sin nx$, where $b_n = (2/\pi) \int_0^{\pi} f(t) \sin nt \, dt$.

Let $f \in \mathscr{R}[0, \pi]$. Define $F(x) = f(-x)$ for $x \in [-\pi, 0)$ and $F(x) = f(x)$ for $x \in [0, \pi]$. Define $G(x) = -f(-x)$ for $x \in [-\pi, 0)$ and $G(x) = f(x)$ for $x \in [0, \pi]$. By (a) the Fourier series for F is $a_0/2 + \sum_{n=1}^{\infty} a_n \cos nx$, where $a_n = (2/\pi) \int_0^{\pi} f(t) \cos nt \, dt$, and by (b) the Fourier series for G is $\sum_{n=1}^{\infty} b_n \sin nx$, where $b_n = (2/\pi) \int_0^{\pi} f(t) \sin nt \, dt$. We call these series the *Fourier cosine* and *Fourier sine series* (respectively) for f on $[0, \pi]$.

74.5 The temperature $v(x, t)$, at location x and time t, of a rod of length π whose ends are kept at temperature zero is described by the boundary value problem:

(i) $\dfrac{\partial v(x, t)}{\partial t} = k \dfrac{\partial^2 v(x, t)}{\partial x^2} \qquad 0 < x < \pi, t > 0$

(ii) $v(0, t) = 0 = v(\pi, t) \qquad t \ge 0$

(iii) $v(x, 0) = f(x) \qquad 0 \le x \le \pi$

where $f(x)$ is the initial temperature distribution. (See Figure 74.1.)

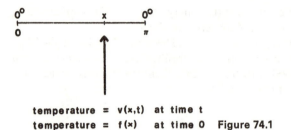

temperature = v(x,t) at time t
temperature = f(x)   at time 0   Figure 74.1

(a) Show that if $v_1(x, t)$ and $v_2(x, t)$ satisfy (i) and (ii) so does $c_1v_1 + c_2v_2$, where $c_1, c_2 \in \mathbf{R}$.

(b) Assuming that (i) has a solution of the form $v(x, t) = X(x)T(t)$, prove that

$$\frac{X''(x)}{X(x)} = \frac{T'(t)}{kT(t)}$$

and that, therefore, each side of this equation defines a constant function. Thus, we must have

(iv) $X''(x) = cX(x)$ and $T'(t) = ckT(t)$.

(c) Show that if X and T satisfy (iv), then $v(x, t) = X(x)T(t)$ satisfies (i) and that if further $X(0) = 0 = X(\pi)$, $v(x, t)$ satisfies (ii) also.

(d)* Show that any nontrivial solution of the system $X''(x) = cX(x)$, $T'(t) = ckT(t)$, $X(0) = 0 = X(\pi)$ is $X(x) = k_1 \sin nx$ and $T(t) = k_2 e^{-n^2kt}$ for some positive integer n and for constants k_1 and k_2.

By (a) the sum of solutions is a solution. Therefore, guided by (d), we try for a solution of the form

(v) $v(x, t) = \sum\limits_{n=1}^{\infty} C_n e^{-n^2kt} \sin nx$

In order to satisfy condition (iii), we must have

$$f(x) = v(x, 0) = \sum\limits_{n=1}^{\infty} C_n \sin nx$$

But this equation tells us that C_n should be the nth Fourier coefficient of the Fourier sine series of f. (See Exercise 74.4.) Moreover, if f is a reasonable function, the Fourier series of f will converge to f so that condition (iii) is satisfied. (Differentiability of f suffices by Corollary 76.8.) Under appropriate conditions series (v) converges and may be differentiated term by term so that conditions (i) and (ii) are also satisfied. (See Rabenstein, 1972.)

75. Orthonormal Expansions in Inner Product Spaces

In this section we will extend several of the theorems of Section 70 (concerning $\mathbf{R}^3$) to arbitrary inner product spaces. Our first goal is to extend Theorem 70.3 to the corresponding statement about a countable orthonormal set $\{x_1, x_2, \ldots\}$. If we could replace n by ∞ in the proof of Theorem 74.1, we would have a valid proof. Thus what we are demanding is that

$$\left(\sum_{k=1}^{\infty} y_k, z\right) = \sum_{k=1}^{\infty} (y_k, z)$$

or equivalently,

$$\left(\lim_{n \to \infty} \sum_{k=1}^{n} y_k, z\right) = \lim_{n \to \infty} \sum_{k=1}^{n} (y_k, z)$$

As usual, to justify the interchange of limit, we must establish continuity (in this case, of the inner product).

Theorem 75.1 Let V be an inner product space and let $z \in V$. The function f defined by the equation

$$f(x) = (x, z)$$

is continuous on V.

Proof. If $z = \theta, f$ is constant and hence continuous. Thus suppose $z \neq \theta$. Let $\varepsilon > 0$. Let $\delta = \varepsilon / \|z\|$. If $\|x - y\| < \delta$, then

$$|f(x) - f(y)| = |(x, z) - (y, z)|$$
$$= |(x - y, z)|$$
$$\leq \|x - y\| \|z\| < \delta \|z\| = \varepsilon$$

where we have used the Cauchy-Schwarz inequality. Therefore, f is continuous. ■

Corollary 75.2 Let V be an inner product space and let $z \in V$. Suppose that the series $\sum_{n=1}^{\infty} y_n$ converges to y in V. Then

$$(y, z) = \left(\sum_{n=1}^{\infty} y_n, z\right) = \sum_{n=1}^{\infty} (y_n, z)$$

Proof. We are given that the sequence $\{s_n\}, s_n = y_1 + \cdots + y_n$ converges to y in V. By Theorem 40.2, since $f(x) = (x, z)$ is continuous,

$$\lim_{n \to \infty} f(s_n) = f(y)$$

Now $$\lim_{n \to \infty} f(s_n) = \lim_{n \to \infty} \left(\sum_{k=1}^{n} y_k, z\right) = \lim_{n \to \infty} \sum_{k=1}^{n} (y_k, z)$$

$$= \sum_{k=1}^{\infty} (y_k, z)$$

On the other hand,

$$f(y) = (y, z) = \left(\sum_{k=1}^{\infty} y_k, z \right)$$

and the conclusion follows. ∎

We are now ready to prove the analogue of Theorems 70.3 and 74.1.

Theorem 75.3   Let $\{x_1, x_2, \ldots\}$ be an orthonormal set in an inner product space V and let $x \in V$. Suppose

$$x = \sum_{n=1}^{\infty} c_n x_n$$

for $c_n \in \mathbf{R}$ and $n = 1, 2, \ldots$. Then

$$c_n = (x, x_n) \qquad n = 1, 2, \ldots$$

Proof.   Using Corollary 75.2, we have

$$(x, x_n) = \left(\sum_{j=1}^{\infty} c_j x_j, x_n \right) = \sum_{j=1}^{\infty} c_j (x_j, x_n) = c_n$$

for $n = 1, 2, \ldots$. ∎

Theorem 75.3 states that if we wish to have $x = \sum_{n=1}^{\infty} c_n x_n$, we must take c_n to be the nth Fourier coefficient of x.

Let $\{x_1, x_2, \ldots\}$ be an orthonormal set in an inner product space V and let $x \in V$. We derive a lemma from which we prove that the best approximation to x using a linear combination of the form

$$c_1 x_1 + \cdots + c_n x_n$$

is achieved by taking c_k to be the kth Fourier coefficient of x. In view of Theorem 75.3 this is not too surprising.

Lemma 75.4   Let $\{x_1, \ldots, x_n\}$ be an orthonormal set in an inner product space V and let $x \in V$. Let $c_1, \ldots, c_n$ be n scalars and set

$$y = \sum_{k=1}^{n} c_k x_k$$

Then $\|x - y\|^2 = \|x\|^2 - \sum_{k=1}^{n} (x, x_k)^2 + \sum_{k=1}^{n} [(x, x_k) - c_k]^2$

Proof.   The proof is a straightforward computation.

$$\|x - y\|^2 = (x - y, x - y) = (x, x) - 2(x, y) + (y, y)$$

$$= \|x\|^2 - 2\left(x, \sum_{k=1}^{n} c_k x_k\right) + \left(\sum_{k=1}^{n} c_k x_k, \sum_{j=1}^{n} c_j x_j\right)$$

$$= \|x\|^2 - 2\sum_{k=1}^{n} c_k(x, x_k) + \sum_{k=1}^{n} c_k\left(x_k, \sum_{j=1}^{n} c_j x_j\right)$$

$$= \|x\|^2 - 2\sum_{k=1}^{n} c_k(x, x_k) + \sum_{k=1}^{n} c_k\left[\sum_{j=1}^{n} c_j(x_k, x_j)\right]$$

$$= \|x\|^2 - 2\sum_{k=1}^{n} c_k(x, x_k) + \sum_{k=1}^{n} c_k^2$$

$$= \|x\|^2 - \sum_{k=1}^{n} (x, x_k)^2 + \sum_{k=1}^{n} [(x, x_k) - c_k]^2 \qquad \blacksquare$$

Theorem 75.5 Let $\{x_1, x_2, \ldots\}$ be an orthonormal set in an inner product space V and let $x \in V$. Let $c_1, \ldots, c_n$ be n scalars and set

$$y_n = \sum_{k=1}^{n} c_k x_k \qquad z_n = \sum_{k=1}^{n} (x, x_k) x_k$$

Then
$$\|x - z_n\| \le \|x - y_n\|$$

Proof. Taking $c_k = (x, x_k)$, $k = 1, \ldots, n$, in Lemma 75.4, we have

$$\|x - z_n\|^2 = \|x\|^2 - \sum_{k=1}^{n} (x, x_k)^2$$

Also by Lemma 75.4, we have

$$\|x - y_n\|^2 = \|x\|^2 - \sum_{k=1}^{n} (x, x_k)^2 + \sum_{k=1}^{n} [(x, x_k) - c_k]^2$$

and since $\sum_{k=1}^{n} [(x, x_k) - c_k]^2 \ge 0$ the inequality follows. $\blacksquare$

We next prove a weakened generalization of Parseval's theorem (Theorem 70.5) known as Bessel's inequality. We will again use Lemma 75.4.

Theorem 75.6 (Bessel's Inequality) Let $\{x_1, x_2, \ldots\}$ be an orthonormal set in an inner product space V and let $x \in V$. Then the series $\sum_{n=1}^{\infty} (x, x_n)^2$ converges and

$$\sum_{n=1}^{\infty} (x, x_n)^2 \le \|x\|^2$$

Proof. Taking $c_k = (x, x_k)$ in Lemma 75.4, we have

$$\|x\|^2 - \sum_{k=1}^{n} (x, x_k)^2 = \|x - y\|^2 \ge 0$$

Therefore for every positive integer n,

$$\sum_{k=1}^{n} (x, x_k)^2 \leq \|x\|^2$$

and the conclusions now follow. ∎

Theorem 75.6 for the special case $V = \mathscr{R}[a, a + 2\pi]$ states that if $f \in \mathscr{R}[a, a + 2\pi]$ and

$$\frac{a_0}{2} + \sum_{n=1}^{\infty} (a_n \cos nx + b_n \sin nx)$$

is the Fourier series for f, then

$$\frac{a_0^2}{2} + \sum_{n=1}^{\infty} (a_n^2 + b_n^2) \leq \frac{1}{\pi} \int_a^{a+2\pi} f^2 \qquad \text{(verify)} \qquad (75.1)$$

If $f(x) = x$, $-\pi \leq x \leq \pi$, inequality (75.1) becomes

$$\sum_{n=1}^{\infty} \frac{4}{n^2} \leq \frac{2}{3}\pi^2 \qquad (75.2)$$

(see Section 72). We will later show [Corollary 78.6(ii)] that equality holds in equation (75.1), and hence (75.2) will become

$$\sum_{n=1}^{\infty} \frac{4}{n^2} = \frac{2}{3}\pi^2$$

which reduces to

$$\sum_{n=1}^{\infty} \frac{1}{n^2} = \frac{\pi^2}{6}$$

Corollary 75.7   Let $\{x_1, x_2, \ldots\}$ be an orthonormal set in an inner product space V and let $x \in V$. Then

$$\lim_{n \to \infty} (x, x_n) = 0$$

Proof.   The proof follows immediately from Theorems 75.6 and 22.3. ∎

The special case of Corollary 75.7 for $V = \mathscr{R}[a, a + 2\pi]$ is important enough to isolate as a theorem.

Theorem 75.8 (Riemann-Lebesgue Lemma)   Let $f \in \mathscr{R}[a, a + 2\pi]$. Then

$$\lim_{n \to \infty} \int_a^{a+2\pi} f(t) \sin nt \, dt = 0 = \lim_{n \to \infty} \int_a^{a+2\pi} f(t) \cos nt \, dt$$

We now give a condition which will allow us to deduce Parseval's theorem as well as several other interesting consequences.

Definition 75.9 Let $X = \{x_1, x_2, \ldots\}$ be an orthonormal set in an inner product space V. If

$$x = \sum_{n=1}^{\infty} (x, x_n)x_n$$

for every x in V (convergence is relative to the inner product space norm), the orthonormal set X is said to be *complete*.

A word of caution is in order at this point. *Complete orthonormal set* is a different concept than *complete metric space*.

It is easily verified that the orthonormal sets defined in Theorem 72.2 for $\mathbf{R}^n$ and l^2 are complete in those spaces. We will show later (Theorem 78.5) that the trigonometric set is complete in $\mathscr{R}[a, a + 2\pi]$.

Theorem 75.10 Let $X = \{x_1, x_2, \ldots\}$ be a complete orthonormal set in an inner product space V. Then

(i) $(x, y) = \sum_{n=1}^{\infty} (x, x_n)(y, x_n)$ for all $x, y \in V$.
(ii) (Parseval's theorem) $\|x\|^2 = \sum_{n=1}^{\infty} (x, x_n)^2$ for all $x \in V$.
(iii) If a vector x in V is orthogonal to x_n for every positive integer n, then $x = \theta$.
(iv) If $x \notin X$, the set $X \cup \{x\}$ is not orthonormal (that is, X is a *maximal orthonormal set*).

Proof. Let $x, y \in V$. Then

$$x = \sum_{n=1}^{\infty} (x, x_n)x_n, \qquad y = \sum_{m=1}^{\infty} (y, x_m)x_m$$

Using Theorem 75.2, we have

$$(x, y) = \left(\sum_{n=1}^{\infty} (x, x_n)x_n, \sum_{m=1}^{\infty} (y, x_m)x_m \right)$$

$$= \sum_{n=1}^{\infty} (x, x_n)\left(x_n, \sum_{m=1}^{\infty} (y, x_m)x_m \right)$$

$$= \sum_{n=1}^{\infty} (x, x_n)\left[\sum_{m=1}^{\infty} (y, x_m)(x_n, x_m) \right]$$

$$= \sum_{n=1}^{\infty} (x, x_n)(y, x_n)$$

Part (ii) follows from part (i) by taking $x = y$.

If $(x, x_n) = 0$, $n = 1, 2, \ldots$, by part (ii),

$$\|x\|^2 = \sum_{n=1}^{\infty} (x, x_n)^2 = 0$$

Finally, suppose $x \notin X$ and $X \cup \{x\}$ is an orthonormal set. Then $(x, x_n) = 0$, $n = 1, 2, \ldots$, and by part (iii), $x = \theta$, which is impossible. ∎

Exercises

75.1 Prove the generalized Pythagorean theorem. If $\{x_1, \ldots, x_n\}$ is a set of mutually orthogonal vectors in an inner product space V, then

$$\left\| \sum_{k=1}^{n} x_k \right\|^2 = \sum_{k=1}^{n} \|x_k\|^2$$

75.2 Verify inequality (75.1).

75.3 Find a necessary and sufficient condition for equality in Theorem 75.5.

75.4 Deduce the Cauchy-Schwarz inequality (Theorem 71.4) from Bessel's inequality (Theorem 75.6).

75.5 Prove that an orthonormal set containing n elements is complete in $\mathbf{R}^n$. Can you generalize to l^2?

75.6* Let $X = \{x_1, x_2, \ldots\}$ be an orthonormal set in a Hilbert space V. Suppose that X is a maximal orthonormal set (that is, if $x \notin X$, then $X \cup \{x\}$ is not orthonormal). Prove that X is a complete orthonormal set. This exercise together with the proof of Theorem 75.10 shows that in a Hilbert space, properties (i) to (iv) of Theorem 75.10 are equivalent to X being a complete orthonormal set.

75.7 Prove the following generalization of Bessel's inequality. Let $\{x_1, x_2, \ldots\}$ be a sequence in an inner product space V. (This sequence is not assumed to be orthonormal.) Let $x \in V$. Then

$$\sum_{n=1}^{N} |(x, x_n)|^2 \leq \|x\|^2 \max_{1 \leq n \leq N} \sum_{m=1}^{N} |(x_n, x_m)|$$

75.8* Let V be a Hilbert space and let W be a dense subspace. Prove that if $X = \{x_1, x_2, \ldots\}$ is a complete orthonormal set in W, then X is also a complete orthonormal set in V.

76. Pointwise Convergence of Fourier Series in $\mathscr{R}[a, a + 2\pi]$

In this section we begin specializing the general theory of Fourier series to the inner product space $\mathscr{R}[a, a + 2\pi]$ by considering the problem of when the Fourier series of a function f in $\mathscr{R}[a, a + 2\pi]$ converges pointwise to f. We will prove (Corollary 76.8) that a sufficient condition for convergence is that f be differentiable.

To show that the Fourier series of a function f in $\mathcal{R}[a, a + 2\pi]$ converges at x, we must show that the limit of the nth partial sum of the Fourier series of f at x, $s_n(x)$ exists. We begin by deriving an integral representation for $s_n(x)$. Now

$$s_n(x) = \frac{1}{2\pi} \int_a^{a+2\pi} f(t)\, dt + \frac{1}{\pi} \sum_{k=1}^n \left[\cos kx \int_a^{a+2\pi} f(t) \cos kt\, dt \right.$$

$$\left. + \sin kx \int_a^{a+2\pi} f(t) \sin nt\, dt \right]$$

$$= \frac{1}{2\pi} \int_a^{a+2\pi} f(t)\, dt + \frac{1}{\pi} \sum_{k=1}^n \int_a^{a+2\pi} f(t)[\cos kx \cos kt$$

$$+ \sin kx \sin kt]\, dt$$

$$= \frac{1}{2\pi} \int_a^{a+\pi 2} f(t)\, dt + \frac{1}{\pi} \sum_{k=1}^n \int_a^{a+2\pi} f(t) \cos [k(x - t)]\, dt$$

Suppose that $f(a) = f(a + 2\pi)$, so that we can extend f to a periodic function on $\mathbf{R}$ of period 2π. [If $f(a) \neq f(a + 2\pi)$, we may redefine f at $a + 2\pi$ so that $f(a) = f(a + 2\pi)$. This does not affect the Fourier series of f. See Section 73.] Making the change of variable $u = x - t$ (Theorem 56.3), our formula becomes

$$s_n(x) = \frac{1}{2\pi} \int_{x-(a+2\pi)}^{x-a} f(x - u)\, du + \frac{1}{\pi} \sum_{k=1}^n \int_{x-(a+2\pi)}^{x-a} f(x - u) \cos ku\, du$$

$$= \frac{1}{2\pi} \int_{-\pi}^{\pi} f(x - t)\, dt + \frac{1}{\pi} \sum_{k=1}^n \int_{-\pi}^{\pi} f(x - t) \cos kt\, dt$$

since $[x - a] - [x - (a + 2\pi)] = 2\pi$ (Theorem 73.2). Thus if we let

$$D_n(t) = \frac{1}{2} + \sum_{k=1}^n \cos kt$$

we have the elegant integral representation

$$s_n(x) = \frac{1}{\pi} \int_{-\pi}^{\pi} f(x - t) D_n(t)\, dt$$

for $s_n(x)$. Thus the problem of convergence of $\{s_n(x)\}$ involves the study of the functions D_n.

Definition 76.1 The *Dirichlet kernel* D_n is defined by the equations

$$D_0(t) = \frac{1}{2} \qquad D_n(t) = \frac{1}{2} + \sum_{k=1}^n \cos kt \qquad \text{for } n = 1, 2, \ldots$$

We summarize the discussion preceding Definition 76.1 in the following theorem.

Theorem 76.2   Let f be periodic of period 2π on $\mathbf{R}$ and suppose $f \in \mathcal{R}[a, a + 2\pi]$. Let s_n be the nth partial sum of the Fourier series of f. Then

$$s_n(x) = \frac{1}{\pi} \int_{-\pi}^{\pi} f(x - t) D_n(t)\, dt$$

Proof.   The proof precedes Definition 76.1.   ∎

In order to study the functions D_n, it is convenient to derive a more explicit formula for D_n. We do this in the next theorem.

Theorem 76.3

(i) $D_n(t) = \dfrac{\sin\left[(n + \frac{1}{2})t\right]}{2 \sin (t/2)}$

for all $t \in \mathbf{R}$ with $\sin (t/2) \neq 0$.

(ii) $\dfrac{1}{\pi} \displaystyle\int_{-\pi}^{\pi} D_n(t)\, dt = 1$

Proof.   We use the identity

$$\sin (u + v) - \sin (u - v) = 2 \sin v \cos u$$

Taking $u = kt$ and $v = t/2$, we have

$$\sin\left[\left(k + \frac{1}{2}\right)t\right] - \sin\left[\left(k - 1 + \frac{1}{2}\right)t\right] = 2 \sin\left(\frac{t}{2}\right) \cos kt$$

We compute the telescoping sum

$$\sin\left[\left(n + \frac{1}{2}\right)t\right] - \sin\left(\frac{t}{2}\right)$$

$$= \sum_{k=1}^{n} \left\{ \sin\left[\left(k + \frac{1}{2}\right)t\right] - \sin\left[\left(k - 1 + \frac{1}{2}\right)t\right] \right\}$$

$$= \sum_{k=1}^{n} 2 \sin\left(\frac{t}{2}\right) \cos kt$$

and equation (i) now follows.

Equation (ii) follows immediately from the definition of D_n.   ∎

Let f be periodic of period 2π and suppose $f \in \mathcal{R}[a, a + 2\pi]$. The Fourier series of f at x will converge if

$$\lim_{n \to \infty} s_n(x) = \lim_{n \to \infty} \frac{1}{\pi} \int_{-\pi}^{\pi} f(x - t) D_n(t)\, dt \qquad (76.1)$$

exists. The Riemann localization theorem (Theorem 76.4) shows that in any case if $0 < \delta < \pi$,

$$\lim_{n\to\infty} \frac{1}{\pi}\left[\int_{-\pi}^{-\delta} f(x-t)D_n(t)\,dt + \int_{\delta}^{\pi} f(x-t)D_n(t)\,dt\right] = 0$$

Thus the limit in equation (76.1) will exist if and only if

$$\lim_{n\to\infty} \frac{1}{\pi}\int_{-\delta}^{\delta} f(x-t)D_n(t)\,dt \qquad (76.2)$$

exists for some δ, $0 < \delta < \pi$. If either of the limits in (76.1) or (76.2) exists, then the other does also and they are equal.

In (76.2), the variable t satisfies $-\delta \le t \le \delta$, so that $u = x - t$ satisfies $x - \delta \le u \le x + \delta$. Thus the behavior of the integral in (76.2) and hence also in (76.1) depends only on values of f in a small open interval containing x. This is the reason for the name *localization theorem*. In summary, convergence of the Fourier series of f at x depends on the behavior of f in a neighborhood of x. This fact should be contrasted with the situation for power series (see Theorem 27.2).

Theorem 76.4 (Riemann Localization Theorem) Let $g \in \mathcal{R}[-\pi, \pi]$ and let δ satisfy $0 < \delta < \pi$. Then

$$\lim_{n\to\infty}\left[\int_{-\pi}^{-\delta} g(t)D_n(t)\,dt + \int_{\delta}^{\pi} g(t)D_n(t)\,dt\right] = 0$$

Proof. Let

$$g_1(t) = \begin{cases} 0 & -\delta < t < \delta \\ g(t) & \delta \le |t| \le \pi \end{cases}$$

$$g_2(t) = \begin{cases} 0 & -\delta < t < \delta \\ g(t)\cot(t/2) & \delta \le |t| \le \pi \end{cases}$$

Then $g_1, g_2 \in \mathcal{R}[-\pi, \pi]$ and by Theorem 75.8,

$$\lim_{n\to\infty}\int_{-\pi}^{\pi} g_1(t)\cos nt\,dt = 0 = \lim_{n\to\infty}\int_{-\pi}^{\pi} g_2(t)\sin nt\,dt \qquad (76.3)$$

By Theorem 76.3(i), if $\delta \le |t| \le \pi$,

$$D_n(t) = \frac{\sin[(n+\frac{1}{2})t]}{2\sin(t/2)}$$

Thus

$$\left(\int_{-\pi}^{-\delta} + \int_{\delta}^{\pi}\right) g(t) D_n(t)\, dt$$

$$= \left(\int_{-\pi}^{-\delta} + \int_{\delta}^{\pi}\right) g(t) \frac{\sin nt \cos (t/2) + \sin (t/2) \cos nt}{2 \sin (t/2)}\, dt$$

$$= \frac{1}{2}\left(\int_{-\pi}^{-\delta} + \int_{\delta}^{\pi}\right) g(t) \left(\cot \frac{t}{2} \sin nt + \cos nt\right) dt$$

$$= \frac{1}{2}\left[\int_{-\pi}^{\pi} g_2(t) \sin nt\, dt + \int_{-\pi}^{\pi} g_1(t) \cos nt\, dt\right]$$

and the theorem now follows from equation (76.3). ∎

We are now in a position to prove that differentiability of f at x implies that the Fourier series of f converges to f at x. Actually we will prove that a slightly weaker condition known as the *Lipschitz condition* suffices.

Definition 76.5 A function f is said to satisfy a *Lipschitz condition* at x if for some $M > 0$ and $\delta > 0$,

$$|f(x) - f(y)| \le M|x - y|$$

whenever $|x - y| \le \delta$.

Theorem 76.6 If f is differentiable at x, f satisfies a Lipschitz condition at x.

Proof. Suppose

$$\lim_{y \to x} \frac{f(y) - f(x)}{y - x} = f'(x)$$

Taking $\varepsilon = 1$, there exists $\delta > 0$ such that if $0 < |y - x| \le \delta$, then

$$\left|\frac{f(y) - f(x)}{y - x} - f'(x)\right| < 1$$

Thus if $|y - x| \le \delta$,

$$|f(y) - f(x)| \le (1 + |f'(x)|)|y - x| \qquad\qquad ∎$$

Theorem 76.7 Let f be periodic of period 2π and suppose $f \in \mathcal{R}[a, a + 2\pi]$. If f satisfies a Lipschitz condition at x, then the Fourier series of f converges to f at x.

Proof. By Theorem 76.2, it suffices to prove that

$$\lim_{n \to \infty} \frac{1}{\pi} \int_{-\pi}^{\pi} f(x - t) D_n(t)\, dt = f(x)$$

By Theorem 76.3(ii)

$$f(x) = \frac{1}{\pi} \int_{-\pi}^{\pi} f(x) D_n(t)\, dt$$

and hence we must show that

$$\lim_{n \to \infty} \frac{1}{\pi} \int_{-\pi}^{\pi} [f(x - y) - f(x)] D_n(t)\, dt = 0$$

Let $\varepsilon > 0$. Since

$$\lim_{t \to 0} \frac{2 \sin (t/2)}{t} = 1$$

there exist constants k and δ_1, $0 < \delta_1 < \pi$, such that

$$\left| \frac{t}{2 \sin (t/2)} \right| < k \qquad \text{for all } 0 < |t| \le \delta_1$$

Thus for any positive integer n, if $0 < |t| \le \delta_1$, we have

$$|t D_n(t)| = \left| \frac{t}{2 \sin (t/2)} \sin \left[\left(n + \frac{1}{2} \right) t \right] \right| < k$$

If $t = 0$, $t D_n(t) = 0$, so that

$$|t D_n(t)| < k \qquad \text{for all } |t| \le \delta_1 \text{ and } n \text{ a positive integer}$$

There exists $\delta_2 > 0$ and $M > 0$ such that

$$|f(x) - f(y)| \le M|x - y| \qquad \text{for } |x - y| \le \delta_2$$

Let $\delta = \min \{\delta_1, \delta_2, \varepsilon/(4Mk)\}$. Then for any positive integer n,

$$\left| \int_{-\delta}^{\delta} [f(x - t) - f(x)] D_n(t)\, dt \right| \le \int_{-\delta}^{\delta} |f(x - t) - f(x)| |D_n(t)|\, dt$$

$$\le \int_{-\delta}^{\delta} M|t| |D_n(t)|\, dt$$

$$\le \int_{-\delta}^{\delta} Mk\, dt = 2Mk\delta \le \frac{\varepsilon}{2}$$

Let $\quad I_n = \int_{-\pi}^{-\delta} [f(x - t) - f(x)] D_n(t)\, dt + \int_{\delta}^{\pi} [f(x - t) - f(x)] D_n(t)\, dt$

By Theorem 76.4, there exists a positive integer N such that

$$|I_n| < \frac{\varepsilon}{2} \qquad \text{for all } n \ge N$$

Thus, if $n \geq N$,

$$\left| \int_{-\pi}^{\pi} [f(x - t) - f(x)] D_n(t) \, dt \right|$$

$$\leq \left| \int_{-\delta}^{\delta} [f(x - t) - f(x)] D_n(t) \, dt \right| + |I_n| < \varepsilon \qquad \blacksquare$$

Corollary 76.8 Let f be periodic of period 2π and suppose $f \in \mathcal{R}[a, a + 2\pi]$. If f is differentiable at x, then the Fourier series of f converges to f at x.

Proof. The result follows from Theorems 76.6 and 76.7. $\blacksquare$

If $f \in \mathcal{R}[a, a + 2\pi]$, and f is differentiable at a point x in the open interval $(a, a + 2\pi)$, then the Fourier series of f converges to f at x. For we may alter f at $a + 2\pi$ so that f may be extended to a periodic function on **R** of period 2π without altering either the Fourier series of f or the value of f at x.

Example. Let

$$f(x) = \begin{cases} x & -\pi < x < \pi \\ 0 & x = -\pi \text{ or } x = \pi \end{cases}$$

Then f is differentiable in $(-\pi, \pi)$; so by Corollary 76.8,

$$x = \sum_{n=1}^{\infty} \frac{2}{n} (-1)^{n+1} \sin nx, \qquad -\pi < x < \pi$$

If we take $x = \pi/2$, this reduces to

$$\frac{\pi}{4} = \sum_{n=1}^{\infty} \frac{(-1)^{n+1}}{2n - 1}$$

We have previously derived this formula by another method. (See Exercise 68.10.)

Exercises

76.1 Investigate the pointwise convergence of the Fourier series of Exercise 74.3.

76.2 Prove that if f satisfies a Lipschitz condition at x, then f is continuous at x. Give an example to show that the converse of this statement is false.

76.3 Prove that if f is left and right differentiable at x, then f satisfies a Lipschitz condition at x.

76.4 Derive a formula for $D_n(t)$ in case $\sin(t/2) = 0$.

76.5 Let f be periodic of period 2π on **R** and suppose $f \in \mathcal{R}[a, a + 2\pi]$. Let s_n

be the nth partial sum of the Fourier series of f. Derive the integral representation

$$s_n(x) = \frac{4}{\pi} \int_0^{\pi/2} \frac{f(x + 2t) + f(x - 2t)}{2} D_n(2t) \, dt$$

76.6 Prove Theorem 76.3(i) by induction.

76.7* Prove that the series $\sum_{k=1}^{\infty} (\cos kt/k)$ converges pointwise except when t is a multiple of 2π. When is convergence uniform?

77. Cesàro Summability of Fourier Series

We remarked earlier, that it is known that the Fourier series of a continuous function may diverge at some points. Fejér showed how this difficulty could be surmounted if instead of considering the sequence $\{s_n\}$ of partial sums of the Fourier series one instead considers the sequence $\{\sigma_n\}$

$$\sigma_n = \frac{s_0 + \cdots + s_n}{n + 1}$$

of Cesàro sums. Fejér proved that if $f \in CP[a, a + 2\pi]$, then $\{\sigma_n\}$ converges uniformly to f. In this section we will prove Fejér's theorem (Theorem 77.5) and derive several consequences.

We begin by deriving an integral representation of σ_n. This development parallels that for s_n of Section 76.

Let f be periodic of period 2π and suppose $f \in \mathcal{R}[a, a + 2\pi]$. Then

$$\sigma_n(x) = \frac{s_0(x) + \cdots + s_n(x)}{n + 1}$$

$$= \frac{1}{n + 1} \sum_{k=0}^{n} \frac{1}{\pi} \int_{-\pi}^{\pi} f(x - t) D_k(t) \, dt$$

$$= \frac{1}{\pi} \int_{-\pi}^{\pi} f(x - t) \left[\frac{\sum_{k=0}^{n} D_k(t)}{n + 1} \right] dt$$

Thus if we set $K_n(t) = [1/(n + 1)] \sum_{k=0}^{n} D_k(t)$, we have the integral representation

$$\sigma_n(x) = \frac{1}{\pi} \int_{-\pi}^{\pi} f(x - t) K_n(t) \, dt$$

Definition 77.1 The *Fejér kernel* K_n is defined by the equation

$$K_n(t) = \frac{1}{n + 1} \sum_{k=0}^{n} D_k(t) \qquad \text{for all } t \in \mathbf{R} \text{ and } n = 0, 1, \ldots$$

Theorem 77.2   Let f be periodic of period 2π and suppose $f \in \mathcal{R}[a, a + 2\pi]$. Let $\{s_n\}$ be the sequence of partial sums of the Fourier series of f and set

$$\sigma_n = \frac{1}{n+1} \sum_{k=0}^{n} s_k$$

Then $$\sigma_n(x) = \frac{1}{\pi} \int_{-\pi}^{\pi} f(x - t) K_n(t)\, dt$$

We next find a convenient formula for the Fejér kernel.

Theorem 77.3

$$K_n(t) = \frac{1}{2(n+1)} \left[\frac{\sin\left[(n+1)t/2\right]}{\sin(t/2)} \right]^2$$

for all t with $\sin(t/2) \neq 0$.

Proof.   By Theorem 76.3(i),

$$K_n(t) = \frac{1}{n+1} \sum_{k=0}^{n} D_k(t)$$

$$= \frac{1}{n+1} \sum_{k=0}^{n} \frac{\sin\left[(k + \tfrac{1}{2})t\right]}{2\sin(t/2)}$$

$$= \frac{1}{2(n+1)\sin^2(t/2)} \sum_{k=0}^{n} \sin\left[\left(k + \frac{1}{2}\right)t\right] \sin\frac{t}{2}, \qquad \text{for all } t$$

with $\sin(t/2) \neq 0$.

We use the identity

$$\cos(u - v) - \cos(u + v) = 2\sin u \sin v$$

Taking $u = (k + \tfrac{1}{2})t$ and $v = t/2$, we have

$$\frac{\cos kt - \cos\left[(k + 1)t\right]}{2} = \sin\left[\left(k + \frac{1}{2}\right)t\right] \sin\frac{t}{2}$$

We compute the telescoping sum

$$\frac{1 - \cos\left[(n+1)t\right]}{2} = \sum_{k=0}^{n} \frac{\cos kt - \cos\left[(k+1)t\right]}{2}$$

$$= \sum_{k=0}^{n} \sin\left[\left(k + \frac{1}{2}\right)t\right] \sin\frac{t}{2}$$

By the half-angle formula

$$\frac{1 - \cos\left[(n+1)t\right]}{2} = \sin^2\left[(n+1)\frac{t}{2}\right]$$

and the formula for K_n follows. ∎

We will make use of the following facts about K_n in proving Fejér's theorem.

Theorem 77.4

(i) $K_n(t) \geq 0$ for all t

(ii) $\dfrac{1}{\pi} \displaystyle\int_{-\pi}^{\pi} K_n(t)\, dt = 1$

(iii) $K_n(t) \leq \dfrac{1}{2(n+1)\sin^2(\delta/2)}$ for $0 < \delta \leq |t| \leq \pi$

Proof. (i) follows from Theorem 77.3.

(ii) follows from Theorem 76.3(ii) and Definition 77.1.

If $0 < \delta \leq |t| \leq \pi$, then $\sin^2(\delta/2) \leq \sin^2(t/2)$, and (iii) now follows from Theorem 77.3. ∎

Theorem 77.5 (Fejér) Let $f \in CP[a, a + 2\pi]$. Let $\{s_n\}$ be the sequence of partial sums of the Fourier series of f. Let

$$\sigma_n = \frac{s_0 + \cdots + s_n}{n+1}$$

Then $\{\sigma_n\}$ converges uniformly to f on **R**.

Proof. Let $\varepsilon > 0$. Choose M such that $|f(x)| \leq M$ for all $x \in \mathbf{R}$. Since f is uniformly continuous, there exists δ with $\pi \geq \delta > 0$ such that

$$|f(x) - f(y)| < \frac{\varepsilon}{4} \quad \text{for} \quad |x - y| \leq \delta$$

By Theorem 77.4(iii), there exists a positive integer N such that if $n \geq N$ and $\delta \leq |t| \leq \pi$,

$$K_n(t) \leq \frac{\varepsilon}{16M}$$

By Theorems 77.2 and 77.4(ii),

$$\sigma_n(x) - f(x) = \frac{1}{\pi}\int_{-\pi}^{\pi} f(x-t)K_n(t)\, dt - \frac{1}{\pi}\int_{-\pi}^{\pi} f(x)K_n(t)\, dt$$

$$= \frac{1}{\pi}\int_{-\pi}^{\pi} [f(x-t) - f(x)]K_n(t)\, dt$$

Thus $|\sigma_n(x) - f(x)| \leq \dfrac{1}{\pi}\displaystyle\int_{-\pi}^{\pi} |f(x-t) - f(x)|K_n(t)\, dt$

where we have used Theorem 77.4(i).

If $|t| \leq \delta$, then $|(x - t) - x| \leq \delta$

so
$$\frac{1}{\pi} \int_{-\delta}^{\delta} |f(x - t) - f(x)| K_n(t) \, dt \leq \frac{\varepsilon}{4\pi} \int_{-\delta}^{\delta} K_n(t) \, dt$$

$$\leq \frac{\varepsilon}{4\pi} \int_{-\pi}^{\pi} K_n(t) \, dt$$

$$= \frac{\varepsilon}{4}, \qquad n = 1, 2, \ldots$$

where we have used Theorem 77.4(i) and (ii).

If $n \geq N$ and $\delta \leq |t| \leq \pi$, then

$$K_n(t) \leq \frac{\varepsilon}{16M}$$

so $\quad \dfrac{1}{\pi} \left(\int_{-\pi}^{-\delta} + \int_{\delta}^{\pi} \right) |f(x - t) - f(x)| K_n(t) \, dt \leq \dfrac{\varepsilon}{16M\pi} \int_{-\pi}^{\pi} 2M \, dt = \dfrac{\varepsilon}{4}$

Thus if $n \geq N$ and $x \in \mathbf{R}$,

$$|\sigma_n(x) - f(x)| \leq \frac{\varepsilon}{2} < \varepsilon \qquad\qquad \blacksquare$$

Corollary 77.6 (Uniqueness of Fourier Series in $CP[a, a + 2\pi]$) Let $f, g \in CP[a, a + 2\pi]$. If f and g have the same Fourier series, then $f(x) = g(x)$ for every x in $\mathbf{R}$.

Proof. Let $\{\sigma_n\}$ be the sequence of Cesàro sums of the Fourier series of f and g. By Theorem 77.5, $\{\sigma_n\}$ converges uniformly to f and g and thus $f(x) = g(x)$ for every x in $\mathbf{R}$. $\quad\blacksquare$

Corollary 77.7 The trigonometric set is complete in $CP[a, a + 2\pi]$.

Proof. Let $f \in CP[a, a + 2\pi]$. Let $\{s_n\}$ be the sequence of partial sums of the Fourier series of f and let $\{\sigma_n\}$ be the sequence of Cesàro sums.

Let $\varepsilon > 0$. By Theorem 77.5, there exists a positive integer N such that if $n \geq N$, then

$$|f(x) - \sigma_n(x)| < \frac{\sqrt{\varepsilon}}{2\sqrt{\pi}} \qquad \text{for all } x \in [a, a + 2\pi]$$

Thus if $n \geq N$,

$$\|f - \sigma_n\|^2 = \int_a^{a+2\pi} (f - \sigma_n)^2 \leq \int_a^{a+2\pi} \frac{\varepsilon}{4\pi} = \frac{\varepsilon}{2}$$

It follows from the definition of σ_n that σ_n is a linear combination of $\phi_0, \ldots,$

ϕ_{2n}. By Theorem 75.5, if $n \geq N$,

$$\|f - s_n\|^2 \leq \|f - \sigma_n\|^2 \leq \frac{\varepsilon}{2} < \varepsilon$$ ∎

Corollary 77.8 Let $f, g \in CP[a, a + 2\pi]$. Suppose

$$\frac{a_0}{2} + \sum_{n=1}^{\infty} (a_n \cos nx + b_n \sin nx)$$

is the Fourier series of f and

$$\frac{c_0}{2} + \sum_{n=1}^{\infty} (c_n \cos nx + d_n \sin nx)$$

is the Fourier series of g. Then

(i) $\dfrac{a_0 c_0}{2} + \sum\limits_{n=1}^{\infty} (a_n c_n + b_n d_n) = \dfrac{1}{\pi} \displaystyle\int_a^{a+2\pi} fg$

(ii) $\dfrac{a_0^2}{2} + \sum\limits_{n=1}^{\infty} (a_n^2 + b_n^2) = \dfrac{1}{\pi} \displaystyle\int_a^{a+2\pi} f^2$

(iii) If $\displaystyle\int_a^{a+2\pi} f(x) \cos nx \, dx = 0 = \int_a^{a+2\pi} f(x) \sin nx \, dx$

for $n = 0, 1, 2, \ldots$, then $f(x) = 0$ for every $x \in \mathbf{R}$.

(iv) If $h \in CP[a, a + 2\pi]$ and $h \notin \{\phi_0, \phi_1, \ldots\}$, then $\{h, \phi_0, \phi_1, \ldots\}$ is not orthonormal in $CP[a, a + 2\pi]$.

Proof. The proof follows immediately from Theorems 77.7 and 75.10. ∎

Example. Let $f(x) = |x|$ for $-\pi \leq x \leq \pi$. Then f is in $CP[-\pi, \pi]$. The Fourier series of f is easily shown to be

$$\frac{\pi}{2} + \sum_{n=1}^{\infty} \frac{-4}{\pi(2n - 1)^2} \cos [(2n - 1)x]$$

By Corollary 77.8(ii),

$$\frac{\pi^2}{2} + \sum_{n=1}^{\infty} \frac{16}{\pi^2(2n - 1)^4} = \frac{1}{\pi} \int_{-\pi}^{\pi} x^2 \, dx$$

This equation reduces to

$$\sum_{n=1}^{\infty} \frac{1}{(2n - 1)^4} = \frac{\pi^4}{96}$$

Corollary 77.9 (Weierstrass Approximation Theorem, Special Case) Let f be continuous on $[-\pi/2, \pi/2]$. Let $\varepsilon > 0$. There exists a polynomial p such that

$$|f(x) - p(x)| < \varepsilon \quad \text{for all } x \in \left[-\frac{\pi}{2}, \frac{\pi}{2}\right]$$

Proof. Extend f as a continuous function on $\mathbf{R}$ periodic of period 2π. One possible extension of f is sketched in Figure 77.1. Let $\varepsilon > 0$. Let $\{\sigma_n\}$ be the sequence of Cesàro sums of the Fourier series of f. By Theorem 77.5, $\{\sigma_n\}$ converges uniformly to f on $[-\pi/2, \pi/2]$. Thus there exists a positive integer N_1 such that

$$|\sigma_{N_1}(x) - f(x)| < \frac{\varepsilon}{2} \quad \text{for all } x \in \left[-\frac{\pi}{2}, \frac{\pi}{2}\right]$$

Since σ_{N_1} is a linear combination of

$$\{\sin nx, \cos nx \mid n = 0, 1, \ldots, M\}$$

where
$$M = \begin{cases} \dfrac{N_1}{2} & \text{for } N_1 \text{ even} \\[2mm] \dfrac{N_1 - 1}{2} & \text{for } N_1 \text{ odd} \end{cases}$$

it follows that σ_{N_1} has a power series representation

$$\sigma_{N_1}(x) = \sum_{n=0}^{\infty} a_n x^n$$

valid for all x in $\mathbf{R}$. By Theorem 63.1(i), $\sum_{n=0}^{\infty} a_n x^n$ converges uniformly to σ_{N_1} on $[-\pi/2, \pi/2]$. Thus there exists a positive integer N such that

$$\left| \sigma_{N_1}(x) - \sum_{n=0}^{N} a_n x^n \right| < \frac{\varepsilon}{2}, \quad \text{for all } x \in \left[-\frac{\pi}{2}, \frac{\pi}{2}\right]$$

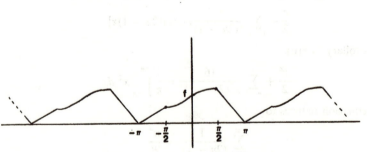

Figure 77.1

If we let $p(x) = \sum_{n=0}^{N} a_n x^n$, then

$$|p(x) - f(x)| \le |p(x) - \sigma_{N_1}(x)| + |\sigma_{N_1}(x) - f(x)| < \varepsilon$$

for all $x \in [-\pi/2, \pi/2]$. ∎

Corollary 77.10 (Weierstrass Approximation Theorem) Let f be continuous on $[a, b]$ and let $\varepsilon > 0$. Then there exists a polynomial p such that

$$|f(x) - p(x)| < \varepsilon$$

for all x in $[a, b]$.

Proof. Let g be the "change of scale" function defined by the equation

$$g(x) = b + \frac{b - a}{\pi}\left(x - \frac{\pi}{2}\right)$$

The graph of g is sketched in Figure 77.2. Then g is a one-to-one continuous function from $[-\pi/2, \pi/2]$ onto $[a, b]$ with a continuous inverse. Since $h = f \circ g$ is a continuous function on $[-\pi/2, \pi/2]$, by Corollary 77.9, if $\varepsilon > 0$ there exists a polynomial q such that

$$|h(x) - q(x)| < \varepsilon \qquad \text{for all } x \in \left[-\frac{\pi}{2}, \frac{\pi}{2}\right]$$

Let $x \in [a, b]$. Then $g^{-1}(x) \in [-\pi/2, \pi/2]$, and hence

$$|h(g^{-1}(x)) - q(g^{-1}(x))| < \varepsilon$$

Since $h \circ g^{-1} = (f \circ g) \circ g^{-1} = f$, we have

$$|f(x) - (q \circ g^{-1})(x)| < \varepsilon \qquad \text{for all } x \in [a, b]$$

Therefore, if we let $p = q \circ g^{-1}$, p satisfies the conclusion of the corollary. ∎

Example 77.11 We will describe a sequence of infinitely differentiable

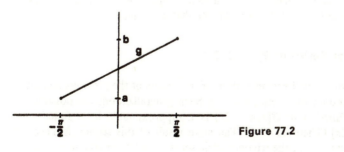

Figure 77.2

functions $\{p_n\}$ on $[a, b]$ which converges uniformly on $[a, b]$ to a continuous nowhere differentiable function.

Let f be a continuous nowhere differentiable function on $[a, b]$. By Corollary 77.10, for every positive integer n, there exists a polynomial p_n such that

$$|p_n(x) - f(x)| < \frac{1}{n} \quad \text{for all } x \in [a, b]$$

It follows that $\{p_n\}$ converges uniformly to f on $[a, b]$. Since a polynomial is infinitely differentiable, we have the desired example.

Exercises

77.1 Prove Theorem 77.3 by induction.

77.2 Apply Corollary 77.8(ii) to the functions $f(x) = x^2$ and $g(x) = |x|$ on $[-\pi, \pi]$.

77.3 Let f be continuous on $[a, b]$. Suppose that

$$\int_a^b f(x)x^n dx = 0$$

for $n = 0, 1, 2, \ldots$. Prove that $f(x) = 0$ for all $x \in [a, b]$.

77.4 Where does the proof of Theorem 77.5 break down if K_n is replaced by D_n?

77.5 Let f be periodic of period 2π and suppose $f \in \mathscr{R}[a, a + 2\pi]$. Let σ_n be defined as in Theorem 77.2. Derive the integral representation

$$\sigma_n(x) = \frac{4}{\pi} \int_0^{\pi/2} \frac{f(x + 2t) + f(x - 2t)}{2} K_n(2t)\, dt$$

77.6* Let f be periodic of period 2π and suppose $f \in \mathscr{R}[a, a + 2\pi]$. Let x be a point in $[a, a + 2\pi]$ for which

$$\lim_{t \to 0+} \frac{f(x + t) + f(x - t)}{2} = L$$

exists. Prove that the Fourier series of f at x is Cesàro summable to L. Deduce that if f is continuous at x, then the Fourier series of f at x is Cesàro summable to $f(x)$.

77.7* Prove that $C[a, b]$ in the uniform norm is separable, that is, that there exists a countable subset X of $C[a, b]$ such that $\bar{X} = C[a, b]$.

78. Fourier Series in $\mathscr{R}[a, a + 2\pi]$

In this section we will extend certain of the results of the previous section to the inner product space $\mathscr{R}[a, a + 2\pi]$. Such generalizations are possible because any function in $\mathscr{R}[a, a + 2\pi]$ is L^2-approximable by a function in $CP[a, a + 2\pi]$ (Theorem 78.4). The main result of this section is Theorem 78.5 which states that the trigonometric set is complete in $\mathscr{R}[a, a + 2\pi]$.

Definition 78.1 Let $[a, b]$ be a closed interval. A *step function f* on $[a, b]$ is a function on $[a, b]$ which is constant on each open subinterval (x_{i-1}, x_i), $i = 1, \ldots, n$ of some partition $\{x_0, x_1, \ldots, x_n\}$ of $[a, b]$.

A step function is pictured in Figure 78.1.

Throughout this section the norm on $\mathscr{R}[a, b]$ is the L^2-norm,

$$\|f\| = \sqrt{\int_a^b f^2}$$

We begin by showing that any function in $\mathscr{R}[a, b]$ is L^2 approximable by a step function.

Lemma 78.2 Let $f \in \mathscr{R}[a, b]$. If $\varepsilon > 0$, there exists a step function g on $[a, b]$ such that

$$\|f - g\| < \varepsilon$$

Proof. Choose M such that $|f(x)| \le M$ for all x in $[a, b]$. Let $\varepsilon > 0$. By definition, there exists a partition

$$P = \{x_0, x_1, \ldots, x_n\}$$

of $[a, b]$ such that

$$\int_a^b f - L(f, P) < \frac{\varepsilon^2}{2M}$$

Let $\quad g(x) = \begin{cases} \text{glb } \{f(t) \mid x_{i-1} \le t \le x_i\} & \text{if } x_{i-1} < x < x_i \\ f(x_i) & \text{if } x = x_i \end{cases}$

Then g is a step function on $[a, b]$; $|g(x)| \le M$ for all $x \in [a, b]$; and

$$L(f, P) = \int_a^b g$$

Now $\quad \int_a^b (f - g) = \int_a^b f - \int_a^b g = \int_a^b f - L(f, P) < \frac{\varepsilon^2}{2M}$

                a b **Figure 78.1**

so that

$$
\int_a^b (f-g)^2 = \int_a^b (f^2 - 2fg + g^2)
$$

$$
= \int_a^b f(f-g) + \int_a^b g(g-f)
$$

$$
\le \int_a^b |f||f-g| + \int_a^b |g||g-f|
$$

$$
= \int_a^b |f|(f-g) + \int_a^b |g|(f-g)
$$

$$
\le 2M \int_a^b (f-g) < 2M \frac{\varepsilon^2}{2M} = \varepsilon^2
$$

and hence $\|f - g\| < \varepsilon$. ∎

Lemma 78.3 Let g be a step function on $[a, b]$ and let $\varepsilon > 0$. Then there exists $h \in CP[a, b]$ such that

$$
\|g - h\| < \varepsilon
$$

Proof. Let $\varepsilon > 0$. First suppose that the partition associated with g is $\{a, b\}$. Then $g(x) = M, a < x < b$. Choose K such that $|g(x)| \le K, x \in [a, b]$. Let

$$
h(x) = \begin{cases} 0 & x = a \text{ or } x = b \\ M & a + \dfrac{\varepsilon^2}{32K^2} \le x \le b - \dfrac{\varepsilon^2}{32K^2} \end{cases}
$$

Extend h linearly to $[a, b]$ (see Figure 78.2).

Now $|g(x) - h(x)| \le |g(x)| + |h(x)| \le K + |M| \le 2K$, for all $x \in [a, b]$; so $[g(x) - h(x)]^2 \le 4K^2$, for all $x \in [a, b]$. Therefore,

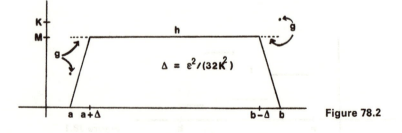

Figure 78.2

$$\int_a^b [g - h]^2 = \int_a^{a+\varepsilon^2/32K^2} (g - h)^2 + \int_{b-\varepsilon^2/32K^2}^b (g - h)^2$$

$$\leq 4K^2 \frac{\varepsilon^2}{32K^2} + 4K^2 \frac{\varepsilon^2}{32K^2} = \frac{\varepsilon^2}{4}$$

and hence $\|g - h\| \leq \varepsilon/2 < \varepsilon$.

Now let g be an arbitrary step function on $[a, b]$ and let $P = \{x_0, x_1, \ldots, x_n\}$ be the partition associated with g. By the first part of our proof, for each positive integer i, $1 \leq i \leq n$, there exists a continuous function h_i on $[x_{i-1}, x_i]$ such that $h_i(x_{i-1}) = 0 = h_i(x_i)$ and

$$\int_{x_{i-1}}^{x_i} (g - h_i)^2 < \frac{\varepsilon^2}{n}$$

Let $h(x) = h_i(x)$, for all $x_{i-1} \leq x \leq x_i, i = 1, \ldots, n$. Then h is continuous on $[a, b]$ and since $h(a) = 0 = h(b)$, $h \in CP[a, b]$. Now

$$\int_a^b (g - h)^2 = \sum_{i=1}^n \int_{x_{i-1}}^{x_i} (g - h_i)^2 < n\frac{\varepsilon^2}{n} = \varepsilon^2$$

and thus $\|g - h\| < \varepsilon$. ∎

Theorem 78.4 Let $f \in \mathscr{R}[a, b]$ and let $\varepsilon > 0$. Then there exists $h \in CP[a, b]$ such that

$$\|f - h\| < \varepsilon$$

Proof. By Lemma 78.2 there exists a step function g on $[a, b]$ such that

$$\|f - g\| < \frac{\varepsilon}{2}$$

By Lemma 78.3, there exists $h \in CP[a, b]$ such that

$$\|g - h\| < \frac{\varepsilon}{2}$$

Therefore $\|f - h\| \leq \|f - g\| + \|g - h\| < \frac{\varepsilon}{2} + \frac{\varepsilon}{2} = \varepsilon$ ∎

We can now generalize Corollary 77.7 to $\mathscr{R}[a, a + 2\pi]$.

Theorem 78.5 The trigonometric set is complete in $\mathscr{R}[a, a + 2\pi]$.

Proof. Let $f \in \mathscr{R}[a, a + 2\pi]$ and let $\{s_n\}$ be the sequence of partial sums of the Fourier series of f. Let $\varepsilon > 0$. By Theorem 78.4 there exists $h \in CP[a,$

$a + 2\pi]$ such that

$$\|f - h\| < \frac{\varepsilon}{2}$$

Let $\{t_n\}$ be the sequence of partial sums of the Fourier series of h. By Corollary 77.7, $\lim_{n\to\infty} \|h - t_n\| = 0$. Thus there exists a positive integer N such that if $n \geq N$,

$$\|h - t_n\| < \frac{\varepsilon}{2}$$

By Theorem 75.5,

$$\|f - s_n\| \leq \|f - t_n\|$$

Thus if $n \geq N$,

$$\|f - s_n\| \leq \|f - t_n\| \leq \|f - h\| + \|h - t_n\| < \frac{\varepsilon}{2} + \frac{\varepsilon}{2} = \varepsilon \qquad \blacksquare$$

Corollary 78.6   Let $f, g \in \mathcal{R}[a, a + 2\pi]$. Suppose

$$\frac{a_0}{2} + \sum_{n=1}^{\infty} (a_n \cos nx + b_n \sin nx)$$

is the Fourier series of f and

$$\frac{c_0}{2} + \sum_{n=1}^{\infty} (c_n \cos nx + d_n \sin nx)$$

is the Fourier series of g. Then

(i) $\dfrac{a_0 c_0}{2} + \displaystyle\sum_{n=1}^{\infty} (a_n c_n + b_n d_n) = \dfrac{1}{\pi} \int_a^{a+2\pi} fg$

(ii) (Parseval's theorem)

$$\frac{a_0^2}{2} + \sum_{n=1}^{\infty} (a_n^2 + b_n^2) = \frac{1}{\pi} \int_a^{a+2\pi} f^2$$

(iii) If

$$\int_a^{a+2\pi} f(x) \cos nx \, dx = 0 = \int_a^{a+2\pi} f(x) \sin nx \, dx \qquad \text{for } n = 0, 1, 2, \ldots$$

then $f = 0$ almost everywhere in $[a, a + 2\pi]$.

Proof.   The proof follows immediately from Theorems 78.5 and 75.10. $\blacksquare$

The uniqueness of Fourier series follows almost immediately from Corollary 78.6(iii).

Corollary 78.7   Let f, $g \in \mathscr{R}[a, a + 2\pi]$. If f and g have the same Fourier series, then $f = g$ almost everywhere in $[a, a + 2\pi]$.

Proof.   If f and g have the same Fourier series, then

$$\int_a^b f(x) \cos nx \, dx = \int_a^b g(x) \cos nx \, dx$$

and

$$\int_a^b f(x) \sin nx \, dx = \int_a^b g(x) \sin nx \, dx$$

for $n = 0, 1, 2, \ldots$.
Thus

$$\int_a^b [f(x) - g(x)] \cos nx \, dx = 0 = \int_a^b [f(x) - g(x)] \sin nx \, dx$$

for $n = 0, 1, \ldots$. By Corollary 77.8(iii), $f - g = 0$ almost everywhere in $[a, a + 2\pi]$. ∎

Since the values of the integrals

$$\frac{1}{\pi} \int_a^b f(x) \cos nx \, dx, \qquad \frac{1}{\pi} \int_a^b f(x) \sin nx \, dx$$

are not changed by altering f at a single point, it is clear that we cannot conclude $f = g$ *everywhere* in $[a, b]$.

Example.   Let $f(x) = x$, $-\pi \le x \le \pi$. The Fourier series of f is

$$\sum_{n=1}^{\infty} \frac{2(-1)^{n+1}}{n} \sin nx$$

By Corollary 78.6(ii),

$$\sum_{n=1}^{\infty} \frac{4}{n^2} = \frac{1}{\pi} \int_{-\pi}^{\pi} x^2 \, dx$$

This equation reduces to

$$\sum_{n=1}^{\infty} \frac{1}{n^2} = \frac{\pi^2}{6}$$

This example suggests the following method for summing an infinite series $\sum_{n=0}^{\infty} a_n$ whose terms are nonnegative. Find a function $f \in \mathscr{R}[a, a + 2\pi]$, such that $(f, \phi_{n_k}) = \sqrt{a_k}$ and $(f, \phi_n) = 0$ otherwise. Then by Corollary 78.6,

$$\int_a^{a+2\pi} f^2 = \sum_{n=0}^{\infty} (f, \phi_n)^2 = \sum_{n=0}^{\infty} a_n$$

Our example shows that it is possible to find such a function f for the series $\sum_{n=1}^{\infty} 1/n^2$.

Unfortunately, there are series for which no such function in $\mathcal{R}[a, a + 2\pi]$ exists which will sum the series. This difficulty can be overcome if we extend the class $\mathcal{R}[a, b]$ to the Lebesgue integrable functions on $[a, b]$ to the extent that if $\sum_{n=0}^{\infty} a_n^2$ converges, there exists a Lebesgue integrable function f such that $(f, \phi_n) = a_n$ (see Theorem 91.9).

We conclude this section by proving a theorem about Fourier series analogous to Theorem 63.1(iii) for power series. Our theorem states that whether or not the Fourier series of f converges, the Fourier series may be integrated term by term with the correct result.

Theorem 78.8   Let $f \in \mathcal{R}[a, a + 2\pi]$. Let

$$\frac{a_0}{2} + \sum_{n=1}^{\infty} (a_n \cos nx + b_n \sin nx)$$

be the Fourier series of f. If $a \le c \le a + 2\pi$, then the series

$$\int_c^x \frac{a_0}{2} dt + \sum_{n=1}^{\infty} \int_c^x (a_n \cos nt + b_n \sin nt) dt$$

converges uniformly to $\int_c^x f$ on $[a, a + 2\pi]$.

Proof.   Let $\{s_n\}$ be the sequence of partial sums of the Fourier series of f. We must show that

$$\left\{ \int_c^x s_n \right\}$$

converges uniformly to $\int_c^x f$ on $[a, a + 2\pi]$. This is equivalent to showing that $\{\int_c^x (s_n - f)\}$ converges uniformly to zero on $[a, a + 2\pi]$.

Let $\varepsilon > 0$. By Theorem 78.5, there exists a positive integer N such that if $n \ge N$,

$$\|s_n - f\| < \frac{\varepsilon}{\sqrt{2\pi}}$$

If $x \in [a, a + 2\pi]$ and $n \ge N$, then

$$\left| \int_c^x (s_n - f) \right| \le \sqrt{\left| \int_c^x (s_n - f)^2 \right|} \sqrt{\left| \int_c^x 1^2 \right|}$$

$$\le \sqrt{\int_a^{a+2\pi} (s_n - f)^2} \sqrt{\int_a^{a+2\pi} 1}$$

$$= \|s_n - f\| \sqrt{2\pi} < \varepsilon$$

where we have used the Cauchy-Schwarz inequality.   ∎

Example. The Fourier series for $f(x) = x$ for $-\pi \le x \le \pi$ is

$$\sum_{n=1}^{\infty} \frac{2(-1)^{n+1}}{n} \sin nx$$

By Theorem 78.8,

$$\int_{\pi}^{x} t \, dt = \sum_{n=1}^{\infty} \frac{2(-1)^{n+1}}{n} \int_{\pi}^{x} \sin nt \, dt \qquad \text{for } -\pi \le x \le \pi \qquad (78.1)$$

and convergence is uniform on $[-\pi, \pi]$.

It is easily verified that

$$\sum_{n=1}^{\infty} \frac{2(-1)^{n+1}}{n} \int_{\pi}^{x} \sin nt \, dt = \sum_{n=1}^{\infty} \frac{2(-1)^{n}}{n^2} \cos nx - 2 \sum_{n=1}^{\infty} \frac{1}{n^2}$$

Using the fact that $\sum_{n=1}^{\infty} 1/n^2 = \pi^2/6$, equation (78.1) becomes

$$x^2 = \frac{\pi^2}{3} + \sum_{n=1}^{\infty} \frac{4(-1)^{n}}{n^2} \cos nx \qquad \text{for } -\pi \le x \le \pi \qquad (78.2)$$

Taking $x = \pi$, we again derive $\sum_{n=1}^{\infty} 1/n^2 = \pi^2/6$. Taking $x = 0$, we have

$$\sum_{n=1}^{\infty} \frac{(-1)^{n+1}}{n^2} = \frac{\pi^2}{12}$$

By Theorem 75.3, equation (78.2) gives the Fourier series for $g(x) = x^2$ on $[-\pi, \pi]$. Thus by Corollary 78.6(ii)

$$\frac{1}{\pi} \int_{-\pi}^{\pi} x^4 \, dx = \frac{\frac{4}{9}\pi^4}{2} + \sum_{n=1}^{\infty} \frac{16}{n^4}$$

This equation reduces to

$$\sum_{n=1}^{\infty} \frac{1}{n^4} = \frac{\pi^4}{90}$$

Exercises

78.1   Let $f \in \mathscr{R}[a, b]$ and let $\varepsilon > 0$. Prove that there exists a polynomial p such that

$$\|f - p\| < \varepsilon$$

78.2   Let $f \in C[a, b]$ and let $\varepsilon > 0$. Prove that there exists a step function g on $[a, b]$ such that

$$|g(x) - f(x)| < \varepsilon \qquad \text{for all } x \in [a, b]$$

78.3 Let $f \in \mathcal{R}[a, b]$. Suppose that

$$\int_a^b f(x) x^n \, dx = 0$$

for $n = 0, 1, 2, \ldots$. Prove that $f = 0$ almost everywhere in $[a, b]$.

78.4 Apply Corollary 78.6(ii) to the functions of Exercise 74.3.

78.5 Prove the following expansions:

(a) $\dfrac{x^2}{2} = \pi x - \dfrac{\pi^2}{3} + 2 \displaystyle\sum_{n=1}^{\infty} \dfrac{\cos nx}{n^2}$ for all $0 \le x \le 2\pi$

(b) $x = \dfrac{\pi}{2} - \dfrac{4}{\pi} \displaystyle\sum_{n=1}^{\infty} \dfrac{\cos (2n - 1)x}{(2n - 1)^2}$ for all $0 \le x \le \pi$

(c) $x \sin x = 1 - \dfrac{1}{2} \cos x - 2 \displaystyle\sum_{n=2}^{\infty} \dfrac{(-1)^n \cos nx}{n^2 - 1}$ for all $-\pi \le x \le \pi$

When is convergence uniform?

78.6 Prove that $\displaystyle\sum_{n=1}^{\infty} \dfrac{1}{n^6} = \dfrac{\pi^6}{945}$

78.7 (a)* Prove that for no positive integer m is it true that

$$\sum_{n=1}^{\infty} \frac{1}{n^3} = \frac{\pi^3}{m}$$

(b) State and prove a generalization for the series $\sum_{n=1}^{\infty} (1/n^k)$, where k is an odd positive integer.

78.8 Let $f \in \mathcal{R}[a, a + 2\pi]$. Let

$$\frac{a_0}{2} + \sum_{n=1}^{\infty} (a_n \cos nx + b_n \sin nx)$$

be the Fourier series of f. Let $c \in [a, a + 2\pi]$.

(a)* Prove that the series $\sum_{n=1}^{\infty} [(b_n/n) \cos nc - (a_n/n) \sin nc]$ converges absolutely.

(b) Let

$$K = \sum_{n=1}^{\infty} \left(\frac{b_n}{n} \cos nc - \frac{a_n}{n} \sin nc \right)$$

and $g(x) = \displaystyle\int_c^x f - \dfrac{a_0 x}{2} - K$ for all $x \in [a, b]$

Prove that

$$\frac{-ca_0}{2} + \sum_{n=1}^{\infty} \left(\frac{a_n}{n} \sin nx - \frac{b_n}{n} \cos nx \right)$$

is the Fourier series of g and that the Fourier series of g converges uniformly to g on $[a, a + 2\pi]$.

(c) Prove that

$$\frac{1}{\pi} \int_a^{a+2\pi} g^2 = \frac{c^2 a_0^2}{2} + \sum_{n=1}^{\infty} \frac{a_n^2 + b_n^2}{n^2}$$

78.9 Compare the properties of Fourier series and power series.

78.10 Is it possible to sum the series $\sum_{n=1}^{\infty} (1/n^3)$ by the methods of this section?

79.   A Tauberian Theorem and an Application to Fourier Series

Fejér's theorem (Theorem 77.5) states that the Fourier series of a function f in $CP[a, a + 2\pi]$ is uniformly Cesàro summable to f. Thus a Tauberian theorem of the form, "If $\sum_{n=1}^{\infty} f_n$ is uniformly Cesàro summable to f and a Tauberian condition is satisfied, then $\sum_{n=1}^{\infty} f_n$ converges uniformly to f," will have an application to Fourier series. The Tauberian condition is that $\{nf_n\}$ is uniformly bounded. The theorem is due to G. H. Hardy.

Theorem 79.1 (Hardy)   If $\sum_{n=1}^{\infty} f_n$ is uniformly Cesàro summable to f on a set X and if the sequence $\{nf_n\}$ is uniformly bounded on X, then $\sum_{n=1}^{\infty} f_n$ converges uniformly to f on X.

Proof.   There exists a number M such that $|nf_n(x)| < M$ for every positive integer n and for every x in X.
Let

$$s_n(x) = \sum_{k=1}^{n} f_k(x)$$

and let

$$\sigma_n(x) = \frac{1}{n} \sum_{k=1}^{n} s_k(x)$$

for x in X. Then if $m > n$, we have

$$(m - n)s_m(x) - (m - n)f(x) = m[\sigma_m(x) - f(x)] - n[\sigma_n(x) - f(x)]$$
$$+ \sum_{k=n+2}^{m} (k - 1 - n)f_k(x) \quad \text{(verify)} \quad (79.1)$$

Let $\varepsilon > 0$. Choose $\varepsilon^* > 0$ such that

$$\varepsilon^* < 1 \quad \text{and} \quad \varepsilon^* + 2\sqrt{\varepsilon^*} + 2M\sqrt{\varepsilon^*} < \varepsilon$$

Since $\{\sigma_n\}$ converges uniformly to f on X, there exists a positive integer N^* such that if $n \geq N^*$, we have $|\sigma_n(x) - f(x)| < \varepsilon^*$ for every x in X. If $m > n \geq N^*$, we have, by (79.1),

$$|s_m(x) - f(x)| \leq \frac{m}{m - n}|\sigma_m(x) - f(x)| + \frac{n}{m - n}|\sigma_n(x) - f(x)|$$

$$+ \frac{1}{m - n} \sum_{k=n+2}^{m} (k - 1 - n)|f_k(x)|$$

$$< \frac{m}{m - n}\varepsilon^* + \frac{n}{m - n}\varepsilon^* + \frac{1}{m - n} \sum_{k=n+2}^{m} (k - 1 - n)\frac{M}{k}$$

$$\leq \frac{m + n}{m - n}\varepsilon^* + \frac{1}{m - n}\frac{M}{n} \sum_{k=1}^{m-n-1} k$$

$$= \frac{m+n}{m-n}\varepsilon^* + \frac{M}{n(m-n)} \frac{(m-n-1)(m-n)}{2}$$

$$< \frac{m+n}{m-n}\varepsilon^* + M\left(\frac{m-n}{n}\right)$$

$$= \varepsilon^* + \frac{2\varepsilon^*}{(m/n-1)} + M\left(\frac{m}{n}-1\right) \tag{79.2}$$

for every x in X.

Choose a positive integer $N > \max\{6/\sqrt{\varepsilon^*}, (1+2\sqrt{\varepsilon^*})N^*\}$ and suppose that $m \geq N$. Now

$$\frac{m}{1+2\sqrt{\varepsilon^*}} < \frac{m}{1+\sqrt{\varepsilon^*}} \tag{79.3}$$

Since $\varepsilon^* < 1$, it follows that $1 + 3\sqrt{\varepsilon^*} + 2\varepsilon^* < 6$, and therefore

$$\frac{m}{1+\sqrt{\varepsilon^*}} - \frac{m}{1+2\sqrt{\varepsilon^*}} = \frac{m\sqrt{\varepsilon^*}}{1+3\sqrt{\varepsilon^*}+2\varepsilon^*} > \frac{m\sqrt{\varepsilon^*}}{6} \geq \frac{N\sqrt{\varepsilon^*}}{6} > 1 \tag{79.4}$$

Inequalities (79.3) and (79.4) imply that there exists a positive integer n satisfying

$$\frac{m}{1+2\sqrt{\varepsilon^*}} < n < \frac{m}{1+\sqrt{\varepsilon^*}} \tag{79.5}$$

Now $N^* < \dfrac{N}{1+2\sqrt{\varepsilon^*}} \leq \dfrac{m}{1+2\sqrt{\varepsilon^*}} < n < \dfrac{m}{1+\sqrt{\varepsilon^*}} < m$

and thus (79.2) holds. Therefore

$$|s_m(x) - f(x)| < \varepsilon^* + \frac{2\varepsilon^*}{(m/n-1)} + M\left(\frac{m}{n}-1\right) \tag{79.6}$$

for every x in X. Inequality (79.5) may be rewritten

$$1 + \sqrt{\varepsilon^*} < \frac{m}{n} < 1 + 2\sqrt{\varepsilon^*}$$

and thus $\dfrac{m}{n} - 1 < 2\sqrt{\varepsilon^*}$ and $\dfrac{1}{(m/n-1)} < \dfrac{1}{\sqrt{\varepsilon^*}}$ \tag{79.7}

Therefore, if $m \geq N$ we have combining inequalities (79.6) and (79.7)

$$|s_m(x) - f(x)| < \varepsilon^* + 2\varepsilon^* \frac{1}{\sqrt{\varepsilon^*}} + M2\sqrt{\varepsilon^*}$$

$$= \varepsilon^* + 2\sqrt{\varepsilon^*} + 2M\sqrt{\varepsilon^*} < \varepsilon$$

for every x in X. ∎

For ordinary infinite series, Hardy's theorem takes the following form: If $\sum_{n=1}^{\infty} a_n$ is Cesàro summable to L and $\{na_n\}$ is bounded, then $\sum_{n=1}^{\infty} a_n$ converges to L. Hardy conjectured, but was unable to prove, that if $\sum_{n=1}^{\infty} a_n$ is Abel summable to L and $\{na_n\}$ is bounded, then $\sum_{n=1}^{\infty} a_n$ converges to L. This much deeper result was first proved by Hardy's colleague, Littlewood. Littlewood's theorem was a significant generalization of Tauber's original theorem (Theorem 65.3) for the Tauberian condition $\lim_{n \to \infty} na_n = 0$ of Theorem 65.3 implies that $\{na_n\}$ is bounded. Littlewood's theorem also generalizes Hardy's theorem since, as we have pointed out before, Cesàro summability implies Abel summability. (See Hardy, 1949.)

Hardy's theorem has a nice application to Fourier series since, as we will show, the terms of the Fourier series of a function of bounded variation satisfy Hardy's Tauberian condition.

Theorem 79.2   Let $f \in BV[a, a + 2\pi]$. Then the terms of the Fourier series of f are uniformly bounded on $\mathbf{R}$.

Proof.   First suppose that f is increasing on $[a, a + 2\pi]$. Let

$$\frac{a_0}{2} + \sum_{n=1}^{\infty} (a_n \sin nx + b_n \cos nx)$$

be the Fourier series of f. Then

$$na_n = \frac{n}{\pi} \int_a^{a+2\pi} f(x) \sin nx \, dx = -\frac{n}{\pi} \int_a^{a+2\pi} f(x) \, d\left(\frac{\cos nx}{n}\right)$$

$$= -\frac{1}{\pi}\left[f(x) \cos nx \Big|_a^{a+2\pi} - \int_a^{a+2\pi} \cos nx \, df(x) \right]$$

where we have used Theorems 55.7 and 53.3. Therefore,

$$|na_n| \leq \frac{1}{\pi}\left[|f(a + 2\pi)| + |f(a)| + \int_a^{a+2\pi} |\cos nx| \, df(x) \right]$$

$$\leq \frac{1}{\pi}\left[|f(a + 2\pi)| + |f(a)| + \int_a^{a+2\pi} df(x) \right]$$

$$= \frac{1}{\pi}[|f(a + 2\pi)| + |f(a)| + f(a + 2\pi) - f(a)]$$

The same inequality holds with a_n replaced by b_n, and thus the theorem is established in case f is increasing on $[a, a + 2\pi]$. The general case follows from Theorem 54.9. ∎

We may combine Fejér's theorem (Theorem 77.5) and the results of this section in the following theorem.

Theorem 79.3   If $f \in CP[a, a + 2\pi] \cap BV[a, a + 2\pi]$, then the Fourier series of f converges uniformly to f on $\mathbf{R}$.

Proof.   By Theorem 77.5, the Fourier series of f is uniformly Cesàro summable to f on $\mathbf{R}$. By Theorem 79.2, the terms of the Fourier series of f satisfy Hardy's Tauberian condition. By Theorem 79.1, the Fourier series of f converges uniformly to f on $\mathbf{R}$.   ■

Exercises

79.1   Complete the proof of Theorem 79.2.

79.2   Prove the following result. Let $f \in BV[a, a + 2\pi]$. Let x be a point in $(a, a + 2\pi)$ such that

$$\lim_{t \to 0^+} \frac{f(x + t) + f(x - t)}{2} = L$$

exists. Prove that the Fourier series of f at x converges to L. Deduce that the Fourier series of a function f of bounded variation on $[a, a + 2\pi]$ converges to f pointwise on $[a, a + 2\pi]$ except possibly on a countable set.

XIII

Normed Linear Spaces and the Riesz Representation Theorem

In this chapter we will present a brief introduction to functional analysis and then prove the Riesz representation theorem (Theorem 84.1), which states that if T is a continuous linear transformation (see Appendix: Vector Spaces) from $C[a, b]$ into $\mathbf{R}$, then there exists $\alpha \in BV[a, b]$ such that $T(f) = \int_a^b f \, d\alpha$ for all $f \in C[a, b]$.

80. Normed Linear Spaces and Continuous Linear Transformations

Functional analysis can be described as the study of normed linear spaces (and other more general vector spaces) in terms of the continuous linear transformations on these spaces.

We recall (see Section 69) that a normed linear space is a (real) vector space V together with a norm $\|\cdot\|$ on V. The metric on a normed linear space is given by

$$d(x, y) = \|x - y\| \qquad x, y \in V$$

and metric space concepts applied to V are relative to this metric.

When dealing with several normed linear spaces and when there is no possibility of confusion, we follow the usual practice of using the same notation $\|\cdot\|$ to denote any of the norms under consideration. For example, in Theorem 80.1, T is a linear transformation from V into W, where V and W are normed linear spaces. In part (iv) of this theorem we find the inequality

$$\|T(x)\| \le K\|x\| \qquad x \in V$$

Obviously, $\|T(x)\|$ is the norm in W and $\|x\|$ is the norm in V.

We begin by deriving several conditions equivalent to continuity for a linear transformation. Our theorem shows that linearity and continuity together form a rather strong condition.

Theorem 80.1   Let V and W be normed linear spaces and let T be a linear transformation from V into W. The following statements are equivalent:

(i) T is continuous on V.

(ii) T is continuous at some point a in V.

(iii) T is uniformly continuous on V.

(iv) There exists a constant K such that

$$\|T(x)\| \leq K\|x\| \qquad \text{for all } x \in V$$

(v) The set

$$A = \{\|T(x)\| \mid x \in V \text{ and } \|x\| \leq 1\}$$

is bounded.

(vi) The set

$$B = \{\|T(x)\| \mid x \in V \text{ and } \|x\| = 1\}$$

is bounded.

Proof.   Obviously (i) implies (ii).

Assume that T is continuous at a. Let $\varepsilon > 0$. Then there exists $\delta > 0$ such that if $\|x - a\| < \delta$, then $\|T(x) - T(a)\| < \varepsilon$. If $\|x - y\| < \delta$, then $\|(x + a - y) - a\| = \|x - y\| < \delta$; so $\|T(x + a - y) - T(a)\| < \varepsilon$. Since $T(x + a - y) - T(a) = T(x) - T(y)$, (ii) implies (iii).

If T is uniformly continuous on V, then T is continuous at θ, and thus taking $\varepsilon = 1$, there exists $\delta > 0$ such that if $\|x\| < \delta$, then $\|T(x)\| < 1$. Let $x \in V$, $x \neq \theta$. Then

$$\left\| \frac{\delta}{2\|x\|} x \right\| = \frac{\delta}{2} < \delta$$

so

$$\left\| T\left(\frac{\delta}{2\|x\|} x \right) \right\| < 1$$

This inequality reduces to

$$\|T(x)\| < \frac{2}{\delta} \|x\| \qquad \text{for } x \in V, x \neq \theta$$

Therefore, if we take $K = 2/\delta$, then $\|T(x)\| \leq K\|x\|$ for $x \in V$, $x \neq \theta$. However if $x = \theta$, then $\|T(x)\| = 0 \leq K\|x\|$ and thus (iii) implies (iv).

Assume (iv) holds. Let $x \in V$, $\|x\| \leq 1$. By (iv),

$$\|T(x)\| \leq K\|x\| \leq K$$

and thus A is bounded by K. Therefore (iv) implies (v).

Since $B \subset A$, if A is bounded, then B is bounded, and thus (v) implies (vi).

Assume B is bounded by K. Let $\varepsilon > 0$. Let $\delta = \varepsilon/(K + 1)$. Suppose $\|x - y\| < \delta$. If $x \neq y$, $\| (x - y)/\|x - y\| \| = 1$, and thus

$$\left\| T\left(\frac{x - y}{\|x - y\|}\right) \right\| \leq K$$

Thus $\qquad \|T(x) - T(y)\| = \|T(x - y)\| \leq K\|x - y\| \leq K\delta < \varepsilon$

On the other hand, if $x = y$, then $\|T(x) - T(y)\| = 0 < \varepsilon$, and therefore (vi) implies (i). ■

Theorem 80.1 states that the linear transformation T is continuous if and only if T is bounded on the unit ball

$$\{x \in V \mid \|x\| \leq 1\}$$

For this reason, continuous linear transformations (on normed linear spaces) are sometimes called *bounded linear transformations*.

The norm for **R** is the absolute value, and thus by Theorem 80.1 a linear transformation T from a normed linear space V into **R** is continuous if and only if there exists a constant K such that

$$|T(x)| \leq K\|x\| \qquad \text{for all } x \in V$$

We will use this condition to verify continuity in each of the following examples. We leave to the reader the (easy) task of verifying that each of our functions is linear.

Example 80.2 Define $T: \mathbf{R}^n \to \mathbf{R}$ by the equation

$$T(\mathbf{x}) = x_1$$

where $\mathbf{x} = (x_1, \ldots, x_n) \in \mathbf{R}^n$. Then

$$|T(\mathbf{x})| = |x_1| \leq \sqrt{x_1^2 + \cdots + x_n^2} = \|\mathbf{x}\|$$

and thus T is continuous on $\mathbf{R}^n$.

Example 80.3 Let $\{a_n\} \in l^2$. Define $T: l^2 \to \mathbf{R}$ by the equation

$$T(\mathbf{x}) = \sum_{n=1}^{\infty} a_n x_n$$

where $\mathbf{x} = \{x_n\} \in l^2$. By the Cauchy-Schwarz inequality (Theorem 36.6) we have

$$|T(\mathbf{x})| = \left| \sum_{n=1}^{\infty} a_n x_n \right| \leq \sqrt{\sum_{n=1}^{\infty} a_n^2} \sqrt{\sum_{n=1}^{\infty} x_n^2} = K\|\mathbf{x}\|$$

where $K = \sqrt{\sum_{n=1}^{\infty} a_n^2}$ and thus T is continuous on l^2.

Throughout this chapter, the norm on $C[a, b]$ is the sup or uniform norm (see Section 60 and Theorem 69.2)

$$\|f\| = \text{lub} \{|f(x)| \mid x \in [a, b]\}$$

for all $f \in C[a, b]$.

Example 80.4   We come now to the central example in this chapter. Let $\alpha \in BV[a, b]$ and let $\beta(x) = V_a^x \alpha$ be the total variation function (see Section 54). Define $T: C[a, b] \to \mathbf{R}$ by the equation

$$T(f) = \int_a^b f \, d\alpha \qquad \text{for all } f \text{ in } C[a, b]$$

Let $f \in C[a, b]$. The function $|f|$ is in $\mathbf{R}_\beta[a, b]$ (Theorem 55.6) and since $|f(x)| \leq \|f\|$ for x in $[a, b]$, we have [Theorem 51.13(iv)]

$$\int_a^b |f| \, d\beta \leq \int_a^b \|f\| \, d\beta = \|f\| \int_a^b d\beta = \|f\|[\beta(b) - \beta(a)] = (V_a^b \alpha)\|f\|$$

By Theorem 55.6, we have

$$|T(f)| = \left| \int_a^b f \, d\alpha \right| \leq \int_a^b |f| \, d\beta \leq (V_a^b \alpha)\|f\|$$

and thus T is continuous on $C[a, b]$.

Exercises

80.1   Let V be a normed linear space. Let $\{x_n\}$ and $\{y_n\}$ be sequences in V which converge to x and y, respectively. Let $\{c_n\}$ be a real sequence which converges to c. Prove that $\{c_n x_n + y_n\}$ converges to $cx + y$. Prove that $\{\|c_n x_n\|\}$ converges to $\|cx\|$.

80.2   Let $c \in [a, b]$. Prove that the equation

$$T(f) = f(c)$$

defines a continuous linear transformation on $C[a, b]$. Prove that there exists $\alpha \in BV[a, b]$ such that

$$T(f) = \int_a^b f \, d\alpha$$

80.3*   Prove that any linear transformation from $\mathbf{R}^n$ into $\mathbf{R}$ is continuous.

80.4   Let $\{a_n\} \in l^1$. Define $T: l^\infty \to \mathbf{R}$ by the equation

$$T(\mathbf{x}) = \sum_{n=1}^\infty a_n x_n$$

where $\mathbf{x} = \{x_n\} \in l^\infty$. Prove that T is continuous on l^∞.

80.5   Let c be the set of all convergent (real) sequences. If $\mathbf{x} = \{x_n\} \in c$, let $\|\mathbf{x}\| = \text{lub} \{|x_n| \mid n \in \mathbf{P}\}$.

(a) Prove that c is a normed linear space.

(b) Prove that the equation

$$T(\mathbf{x}) = \lim_{n \to \infty} x_n$$

$\mathbf{x} = \{x_n\} \in c$ defines a continuous linear transformation on c.

80.6* Let $\mathscr{P}[0, 1]$ be the subspace of $C[0, 1]$ consisting of all polynomials on $[0, 1]$. Prove that the linear transformation T from $\mathscr{P}[0, 1]$ into $\mathbf{R}$ defined by

$$T(p) = p(2)$$

is not continuous on $\mathscr{P}[0, 1]$.

80.7 Let V and W be normed linear spaces and let T be a linear transformation from V into W. Prove that T is continuous on V if and only if the set

$$\{\|T(x)\| \mid x \in V \text{ and } \|x\| < 1\}$$

is bounded.

80.8 Prove that the map $T: \mathbf{R}^3 \to C[0, 1]$ defined by $T(\alpha, \beta, \gamma) = \alpha x^2 + \beta x + \gamma$ is a one-to-one, continuous linear map. Is the mapping $T^{-1}: T(C[0, 1]) \to \mathbf{R}^3$ continuous?

81. The Normed Linear Space of Continuous Linear Transformations

Let V and W be vector spaces. Let $\mathscr{L}'(V, W)$ denote the set of all linear transformations from V into W. The set $\mathscr{L}'(V, W)$ is itself a vector space if we define operations pointwise

$$(T + U)(x) = T(x) + U(x) \qquad (cT)(x) = c(T(x))$$

$$\text{for all } T, U \in \mathscr{L}'(V, W), \, x \in V, \, c \in \mathbf{R}$$

If V and W are finite-dimensional vector spaces of dimensions n and m, respectively, $\mathscr{L}'(V, W)$ is identified with the space of all $m \times n$ matrices.

If V and W are normed linear spaces, we may consider the subset $\mathscr{L}(V, W)$ of $\mathscr{L}'(V, W)$ of all *continuous* linear transformations of V into W. The next theorem states that $\mathscr{L}(V, W)$ is a subspace of $\mathscr{L}'(V, W)$.

Theorem 81.1 Let V and W be normed linear spaces. Then $\mathscr{L}(V, W)$ is a vector space where the operations are defined pointwise.

Proof. We verify that $\mathscr{L}(V, W)$ is a subspace of $\mathscr{L}'(V, W)$. It suffices to show that if $T, U \in \mathscr{L}(V, W)$ and $c \in \mathbf{R}$, then $cT + U \in \mathscr{L}(V, W)$.

Let $T, U \in \mathscr{L}(V, W)$ and $c \in \mathbf{R}$. By Theorem 80.1, there exist constants K_1 and K_2 such that

$$\|T(x)\| \le K_1 \|x\| \qquad \|U(x)\| \le K_2 \|x\| \qquad \text{for all } x \in V$$

Thus if $x \in V$,

$$\|(cT + U)(x)\| = \|cT(x) + U(x)\| \leq |c|\|T(x)\| + \|U(x)\|$$
$$\leq |c|K_1\|x\| + K_2\|x\|$$
$$= (|c|K_1 + K_2)\|x\|$$

and by Theorem 80.1, $cT + U$ is continuous. ∎

We next show how to define a norm on $\mathscr{L}(V, W)$.

Theorem 81.2   Let V and W be normed linear spaces. Then the equation

$$\|T\| = \text{lub } \{\|T(x)\| \mid x \in V \text{ and } \|x\| = 1\}$$

defines a norm on $\mathscr{L}(V, W)$.

Proof.   If $T \in \mathscr{L}(V, W)$ let

$$A_T = \{\|T(x)\| \mid x \in V \text{ and } \|x\| = 1\}$$

By Theorem 80.1, A_T is bounded, and hence lub A_T exists.

If $T = 0$, then $A_T = \{0\}$, and hence $\|T\| = 0$. If $\|T\| = 0$, then $A_T = \{0\}$, and hence $T(x) = \theta$ if $x \in V$ and $\|x\| = 1$. If $x \in V$ and $x \neq \theta$, then $\|x/\|x\| \| = 1$, and hence $\|T(x/\|x\|)\| = 0$. Thus $T(x) = \theta$ if $x \in V$ and $x \neq \theta$, and it follows that $T = 0$. We have shown that $\|T\| = 0$ if and only if $T = 0$.

Let $T \in \mathscr{L}(V, W)$ and $c \in \mathbf{R}$. If $c = 0$, then

$$\|cT\| = 0 = |c|\|T\|$$

and thus we suppose $c \neq 0$. If $x \in V$ and $\|x\| = 1$, then

$$\|(cT)(x)\| = |c|\|T(x)\| \leq |c|\|T\|$$

and hence $$\|cT\| \leq |c|\|T\|$$

On the other hand, if $x \in V$ and $\|x\| = 1$, then

$$|c|\|T(x)\| = \|(cT)(x)\| \leq \|cT\|$$

Therefore, $$\|T(x)\| \leq \frac{1}{|c|}\|cT\| \qquad \text{for } x \in V \text{ with } \|x\| = 1$$

and hence $$\|T\| \leq \frac{1}{|c|}\|cT\|$$

It now follows that $\|cT\| = |c|\|T\|$.

Finally, let $T, U \in \mathscr{L}(V, W)$. If $x \in V$ and $\|x\| = 1$, then

$$\|T(x)\| \leq \|T\| \qquad \|U(x)\| \leq \|U\|$$

Therefore,

$$\|(T + U)(x)\| = \|T(x) + U(x)\| \leq \|T(x)\| + \|U(x)\| \leq \|T\| + \|U\|$$

and it follows that

$$\|T + U\| \leq \|T\| + \|U\| \qquad\blacksquare$$

We next give two other characterizations of the norm of a continuous linear transformation.

Theorem 81.3 Let V and W be normed linear spaces and let $T \in \mathscr{L}(V, W)$. Then

$$\|T\| = \begin{cases} \text{lub } \{\|T(x)\| \mid x \in V \text{ and } \|x\| \leq 1\} \\ \text{glb } \{K \mid \|T(x)\| \leq K\|x\|, x \in V\} \end{cases}$$

Proof. Let

$$A = \{\|T(x)\| \mid x \in V \text{ and } \|x\| \leq 1\}$$
$$B = \{\|T(x)\| \mid x \in V \text{ and } \|x\| = 1\}$$
$$C = \{K \mid \|T(x)\| \leq K\|x\|, x \in V\}$$

Since $B \subset A$,

$$\|T\| = \text{lub } B \leq \text{lub } A \qquad (81.1)$$

Let $K \in C$. If $x \in V$ and $\|x\| \leq 1$, then

$$\|T(x)\| \leq K\|x\| \leq K$$

and thus K is an upper bound for A. Therefore, lub $A \leq K$. This inequality states that lub A is a lower bound for C, and hence

$$\text{lub } A \leq \text{glb } C \qquad (81.2)$$

If $x \in V$ and $x \neq 0$, then $\|x/\|x\| \| = 1$, and hence $\|T(x/\|x\|)\| \leq \|T\|$. This reduces to $\|T(x)\| \leq \|T\| \|x\|$ for $x \neq \theta$. This inequality is also valid if $x = \theta$, and thus

$$\|T(x)\| \leq \|T\| \|x\| \qquad x \in V$$

Therefore $\|T\| \in C$, and hence

$$\text{glb } C \leq \|T\| \qquad (81.3)$$

Inequalities (81.1) to (81.3) now give the desired conclusion. $\blacksquare$

In the course of the proof of Theorem 81.3 we proved the following important fact.

Corollary 81.4 Let V and W be normed linear spaces and let $T \in \mathscr{L}(V, W)$. Then

$$\|T(x)\| \leq \|T\| \, \|x\| \qquad \text{for all } x \in V$$

Proof. See the proof of Theorem 81.3. ■

Examples. Let T be the linear transformation from $\mathbf{R}^n$ into $\mathbf{R}$ defined by

$$T(\mathbf{x}) = x_1$$

where $\mathbf{x} = (x_1, \ldots, x_n) \in \mathbf{R}^n$. We proved (see Example 80.2) that T is continuous and that

$$|T(\mathbf{x})| \leq 1 \cdot \|\mathbf{x}\| \qquad \text{for all } \mathbf{x} \in \mathbf{R}^n$$

By Theorem 81.3, $\|T\| \leq 1$.

Let $\delta^{(1)} = (1, 0, \ldots, 0)$. Then $\|\delta^{(1)}\| = 1$ and since $T(\delta^{(1)}) = 1$, by Theorem 81.2 $T(\delta^{(1)}) = 1 \leq \|T\|$. Therefore, $\|T\| = 1$.

Let $\{a_n\} \in l^2$. Let T be the linear transformation from l^2 into $\mathbf{R}$ defined by

$$T(\mathbf{x}) = \sum_{n=1}^{\infty} a_n x_n$$

where $\mathbf{x} = \{x_n\} \in l^2$. We proved (see Example 80.3) that T is continuous and that

$$|T(\mathbf{x})| \leq \|a\| \, \|\mathbf{x}\| \qquad \text{for all } \mathbf{x} \in l^2$$

By Theorem 81.3, $\|T\| \leq \|a\|$. If $a = \theta$, we have $\|T\| = 0 = \|a\|$. If $a \neq \theta$, then $T(a) = \|a\|^2$ and thus $T(a/\|a\|) = \|a\|$. Since $\|a/\|a\| \, \| = 1$, we have $\|a\| = T(a/\|a\|) \leq \|T\|$, and therefore $\|T\| = \|a\|$.

Exercises

81.1 Let V and W be normed linear spaces and let $T \in \mathscr{L}(V, W)$. Prove that
$$\|T\| = \text{lub } \{\|T(x)\| \mid x \in V \text{ and } \|x\| < 1\}$$

81.2 What is the norm of the linear transformation of Exercise 80.2? Compare this value with $V_a^b \alpha$.

81.3 What is the norm of the linear transformation of Exercise 80.4?

81.4 What is the norm of the linear transformation of Exercise 80.5?

81.5 Let V, W, and X be normed linear spaces and let $T \in \mathscr{L}(W, X)$ and $S \in \mathscr{L}(V, W)$. Prove that $T \circ S \in \mathscr{L}(V, X)$ and
$$\|T \circ S\| \leq \|T\| \, \|S\| \tag{81.4}$$

81.6 (a) Prove that any linear transformation from $\mathbf{R}^n$ into $\mathbf{R}^m$ is continuous.

   (b) Let T be an invertible linear transformation from $\mathbf{R}^n$ into $\mathbf{R}^n$. Prove that $1 \leq \|T\| \, \|T^{-1}\|$. Give examples to show that either $<$ or $=$ may occur.

(c) Let $T \in \mathscr{L}(\mathbf{R}^n, \mathbf{R}^n)$. Let $c_1, \ldots, c_n$ be the (possibly complex) eigenvalues of $T^t T$. (^{t} denotes transpose.)
Prove that

$$\|T\| = \max \{ \sqrt{|c_1|}, \sqrt{|c_2|}, \ldots, \sqrt{|c_n|} \}$$

81.7 A linear algebra which is a normed linear space whose norm satisfies equation (81.4) is called a *normed linear algebra*. Prove that $\mathbf{R}^n$, $C[a, b]$, l^2, c_0, and l^∞ are normed algebras (multiplication is defined pointwise). A complete normed algebra is called a *Banach algebra*, and thus each of the spaces above furnishes an example of a Banach algebra.

81.8* Prove that the set of power series $\sum_{n=0}^\infty a_n x^n$ is a vector subspace of $C[a, b]$. Is this subspace complete?

82. The Dual Space of a Normed Linear Space

The dual space of a normed linear space V is the set of all continuous linear transformations from V into $\mathbf{R}$ with the norm defined as in Theorem 81.2.

Definition 82.1 Let V be a normed linear space. The *dual space* V^* of V is the space $\mathscr{L}(V, \mathbf{R})$ of all continuous linear transformations from V into $\mathbf{R}$. The norm on V^* is given by

$$\|T\| = \text{lub} \{ |T(x)| \mid x \in V \text{ and } \|x\| = 1 \} \qquad \text{for each } T \in V^*$$

The elements of V^* are called *continuous linear functionals*.

An interesting problem in functional analysis is to identify the dual space of a given normed linear space. As examples, we show that $(\mathbf{R}^n)^*$ may be identified with $\mathbf{R}^n$ and c_0^* may be identified with l^1. The Riesz representation theorem states that $(C[a, b])^*$ may be identified with $BV[a, b]$.

Theorem 82.2 Let $T \in (\mathbf{R}^n)^*$. Then there exists $\mathbf{a} = (a_1, \ldots, a_n) \in \mathbf{R}^n$ such that

$$T(\mathbf{x}) = \sum_{k=1}^n x_k a_k \qquad \text{for all } \mathbf{x} = (x_1, \ldots, x_n) \in \mathbf{R}^n$$

and $\|T\| = \|\mathbf{a}\|$.

Proof. Let $T \in (\mathbf{R}^n)^*$. Let

$$\delta^{(k)} = (0, 0, \ldots, 0, 1, 0, \ldots, 0)$$

where the number 1 appears in the kth coordinate. Set

$$a_k = T(\delta^{(k)}) \qquad \text{for } k = 1, \ldots, n$$

and let $\mathbf{a} = (a_1, \ldots, a_n)$.

If $\mathbf{x} = (x_1, \ldots, x_n) \in \mathbf{R}^n$, then

$$T(\mathbf{x}) = T\left(\sum_{k=1}^{n} x_k \delta^{(k)}\right) = \sum_{k=1}^{n} x_k T(\delta^{(k)}) = \sum_{k=1}^{n} x_k a_k$$

Now
$$|T(\mathbf{x})| = \left|\sum_{k=1}^{n} x_k a_k\right| \leq \sqrt{\sum_{k=1}^{n} x_k^2} \sqrt{\sum_{k=1}^{n} a_k^2} = \|\mathbf{x}\| \|\mathbf{a}\|$$

so by Theorem 81.3, $\|T\| \leq \|\mathbf{a}\|$.

If $\mathbf{a} = \theta$, $\|T\| = 0 = \|\mathbf{a}\|$. If $\mathbf{a} \neq \theta$, we argue as in the last example in Section 81 to prove that $\|T\| \geq \|\mathbf{a}\|$. Therefore $\|T\| = \|\mathbf{a}\|$. ■

Similar methods identify the dual space of c_0.

Theorem 82.3　Let $T \in c_0^*$. Then there exists $\mathbf{a} = \{a_n\} \in l^1$ such that

$$T(\mathbf{x}) = \sum_{n=1}^{\infty} x_n a_n \qquad \text{for all } \mathbf{x} = \{x_n\} \in c_0$$

and $\|T\| = \|\mathbf{a}\|$.

Proof.　Let $T \in c_0^*$. Let

$$\delta_m^{(n)} = \begin{cases} 1 & n = m \\ 0 & n \neq m \end{cases}$$

and set
$$a_n = T(\delta^{(n)})$$

Let $\mathbf{x} = \{x_n\} \in c_0$ and let $y_n = \sum_{k=1}^{n} x_k \delta^{(k)}$. Then

$$T(y_n) = T\left(\sum_{k=1}^{n} x_k \delta^{(k)}\right) = \sum_{k=1}^{n} x_k T(\delta^{(k)}) = \sum_{k=1}^{n} x_k a_k$$

Now
$$y_n - \mathbf{x} = (0, 0, \ldots, 0, x_{n+1}, x_{n+2}, \ldots)$$

and thus
$$\|y_n - \mathbf{x}\| = \text{lub}\,\{|x_k| \mid k > n\}$$

Hence
$$\lim_{n \to \infty} \|y_n - \mathbf{x}\| = \lim_{n \to \infty} \text{lub}\,\{|x_k| \mid k > n\} = \lim_{n \to \infty} \sup |x_n| = 0$$

since $\{x_n\} \in c_0$. Thus $\lim_{n \to \infty} y_n = \mathbf{x}$ in c_0 and since T is continuous

$$T(\mathbf{x}) = \lim_{n \to \infty} T(y_n) = \lim_{n \to \infty} \sum_{k=1}^{n} x_k a_k = \sum_{n=1}^{\infty} x_n a_n$$

We next show that $\mathbf{a} = \{a_n\} \in l^1$. To this end let

$$\gamma_m^{(n)} = \begin{cases} 1 & \text{if } m \leq n \text{ and } a_n \geq 0 \\ -1 & \text{if } m \leq n \text{ and } a_n < 0 \\ 0 & \text{if } m > n \end{cases}$$

Then
$$T(\gamma^{(n)}) = \sum_{m=1}^{\infty} \gamma_m^{(n)} a_m = \sum_{m=1}^{n} |a_m|$$

Now $\gamma^{(n)} \in c_0$ and $\|\gamma^{(n)}\| = 1$; so by Theorem 81.2

$$|T(\gamma^{(n)})| \leq \|T\|$$

Thus
$$\sum_{m=1}^{n} |a_m| \leq \|T\| \qquad \text{for } n = 1, 2, \ldots \qquad (82.1)$$

By Theorem 24.1, $a \in l^1$. Taking the limit in equation (82.1), we have

$$\|a\| = \sum_{n=1}^{\infty} |a_n| \leq \|T\|$$

On the other hand, if $x \in c_0$ and $\|x\| = 1$, then

$$|T(x)| = \left| \sum_{n=1}^{\infty} x_n a_n \right| \leq \sum_{n=1}^{\infty} |x_n| |a_n| \leq \sum_{n=1}^{\infty} |a_n| = \|a\|$$

By Theorem 81.2, $\|T\| \leq \|a\|$, and hence $\|T\| = \|a\|$. ∎

Exercises 82.2 and 82.3 show that $(l^1)^*$ may be identified with l^∞ and that $(l^2)^*$ may be identified with l^2.

Exercises

82.1 Let T be as in Theorem 82.3. Does there exist $x \in c_0$, $\|x\| = 1$, such that $T(x) = \|T\|$?

82.2 Let $T \in (l^1)^*$. Prove that there exists $a = \{a_n\} \in l^\infty$ such that
$$T(x) = \sum_{n=1}^{\infty} x_n a_n \qquad \text{for all } x = \{x_n\} \in l^1$$
and $\|T\| = \|a\|$. Does there exist $x \in l^1$, $\|x\| = 1$, such that $T(x) = \|T\|$?

82.3 Let $T \in (l^2)^*$. Prove that there exists $a = \{a_n\} \in l^2$ such that
$$T(x) = \sum_{n=1}^{\infty} x_n a_n \qquad \text{for all } x = \{x_n\} \in l^2$$
and $\|T\| = \|a\|$. Does there exist $x \in l^2$, $\|x\| = 1$, such that $T(x) = \|T\|$?

82.4 Show (by example) that there exists $T \in c^*$ such that for no sequence $a = \{a_n\}$ do we have
$$T(x) = \sum_{n=1}^{\infty} x_n a_n \qquad \text{for all } x = \{x_n\} \in c$$
(The space c is defined in Exercise 80.5.)

82.5 Let V and W be normed linear spaces. An *isometry* from V onto W is a linear transformation T from V onto W satisfying
$$\|T(x)\| = \|x\| \qquad \text{for all } x \in V.$$
  (a) Prove that if T is an isometry from V onto W, then T is a continuous vector space isomorphism of V onto W with continuous inverse.

Because of (a) if T is an isometry of V onto W, V and W are said to be *isometrically isomorphic*.

(b) Prove that if V and W are isometrically isomorphic normed linear spaces, then

    (i) If V is complete, then W is complete.

    (ii) V^* is isometrically isomorphic to W^*.

Does the converse of (ii) hold?

(c) Prove that the following pairs of normed linear spaces are isometrically isomorphic.

    (i) $\mathbf{R}^n$, $(\mathbf{R}^n)^*$

    (ii) $(c_0)^*$, l^1

    (iii) $(l^1)^*$, l^∞

    (iv) $(l^2)^*$, l^2

If we identify isometrically isomorphic normed linear spaces, we can state, for example, that the dual space of c_0 *is* l^1.

82.6 Let V and W be normed linear spaces with W complete. Prove that $\mathscr{L}(V, W)$ is a complete normed linear space. Deduce that the dual space of any normed linear space is complete.

83. Introduction to the Riesz Representation Theorem

In Example 80.4 we showed that if $\alpha \in BV[a, b]$, then the equation

$$T(f) = \int_a^b f \, d\alpha \qquad \text{for all } f \in C[a, b] \tag{83.1}$$

defines a continuous linear transformation from $C[a, b]$ into $\mathbf{R}$. That is, $T \in (C[a, b])^*$. It also follows from Example 80.4 and Theorem 81.3 that $\|T\| \leq V_a^b \alpha$ (verify). The Riesz representation theorem is the converse of the above: if $T \in (C[a, b])^*$, then there exists $\alpha \in BV[a, b]$ such that $T(f) = \int_a^b f \, d\alpha$. Moreover, we shall show that $\|T\| = V_a^b \alpha$. Thus the Riesz representation theorem states that $(C[a, b])^*$ may be identified with $BV[a, b]$. Linearity is one of the fundamental properties of the integral (Theorem 53.2). The Riesz representation theorem shows that linearity and continuity characterize the Riemann-Stieltjes integral with respect to continuous functions.

Let $T \in (C[a, b])^*$. To prove the Riesz representation theorem, we must define a function α on $[a, b]$, prove that $\alpha \in BV[a, b]$, and finally show that equation (83.1) holds. Consider the problem of defining α. If T satisfies equation (83.1) [and $\alpha(0) = 0$], we could recover α from this integral representation as follows. Let

$$f_x(y) = \begin{cases} 1 & a \leq y \leq x \\ 0 & x < y \leq b \end{cases}$$

(See Figure 83.1.) If α is right continuous at x, we have

$$\int_a^b f_x(y)\, d\alpha(y) = \int_0^x d\alpha = \alpha(x)$$

Thus, if the domain of T included the functions f_x, we could define $\alpha(x) = T(f_x)$. Unfortunately, f_x is not continuous (unless $x = b$), and thus f_x is not in the domain of T. The major part of the proof of the Riesz representation theorem consists of extending the domain of T to a subspace of $B[a, b]$ which includes the functions f_x. Once we prove that an extension T_2 of T to an appropriate subspace exists, we can define $\alpha(x) = T_2(f_x)$.

Note that f_x is the pointwise limit of a decreasing sequence $\{f_n\}$ of continuous functions on $[a, b]$. (See Figure 83.1.) We will prove that $\lim_{n \to \infty} T(f_n)$ exists. It then seems reasonable to define the extension T_2 by the equation $T_2(f_x) = \lim_{n \to \infty} T(f_n)$. This is exactly the method we shall employ.

In the remainder of this section we define some of the notation we will use in the proof of the Riesz representation theorem.

If f and g are functions on $[a, b]$, we write $f \geq g$ if $f(x) \geq g(x)$ for all x in $[a, b]$.

We let C_1 denote the set of all bounded functions f on $[a, b]$ such that f is the pointwise limit of a sequence $\{f_n\}$ in $C[a, b]$, where

$$f_n \geq f_{n+1} \qquad n = 1, 2, \ldots$$

As we noted above, C_1 contains the functions f_x. It is also clear that $C_1 \supset C[a, b]$.

The set C_1 is not quite a vector space, but we do have

(i) If $f, g \in C_1$, then $f + g \in C_1$.
(ii) If $f \in C_1$ and $c \geq 0$, then $cf \in C_1$. $\qquad\qquad$ (83.2)

To prove (ii), let $f \in C_1$ and let $c \geq 0$. Then there exists a decreasing sequence $\{f_n\}$ in $C[a, b]$ such that $\{f_n\}$ converges to f pointwise. Since $c \geq 0$, $\{cf_n\}$ is a

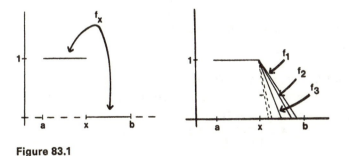

Figure 83.1

decreasing sequence in $C[a, b]$ such that $\{cf_n\}$ converges to cf pointwise. Therefore, $cf \in C_1$. Part (i) is proved in a similar way.

Since $-f_a \notin C_1$ (verify), C_1 is not closed under scalar multiplication, and thus C_1 is not a vector space. This difficulty necessitates enlarging our set of functions still further. Fortunately, this last extension is easy. We simply define

$$C_2 = \{f - g \mid f, g \in C_1\}$$

Using (83.2), it is easy to verify that C_2 is a vector space. Clearly C_2 contains C_1 and $C[a, b]$. Also $C_2 \subset B[a, b]$.

Functions such as f_x which assume only the values 0 or 1 play an important role in mathematics and are known as *characteristic functions*. In the remainder of this book we shall employ the following notation for a characteristic function. Let X be a set. We define

$$\chi_X(x) = \begin{cases} 1 & \text{if } x \in X \\ 0 & \text{if } x \notin X \end{cases}$$

Thus $f_x = \chi_{[a,x]}$.

One last bit of notation and we shall proceed to the proof of the Riesz representation theorem. If a is a real number, we define

$$\operatorname{sgn} a = \begin{cases} 1 & \text{if } a > 0 \\ 0 & \text{if } a = 0 \\ -1 & \text{if } a < 0 \end{cases}$$

It is then clear that $|a| = (\operatorname{sgn} a)a$ for all $a \in \mathbf{R}$.

Exercises

83.1 Verify that if T is defined by equation (83.1), then $\|T\| \leq V_a^b \alpha$.

83.2 Let $c \in [a, b]$. Define α on $[a, b]$ by the equation

$$\alpha(x) = \begin{cases} 0 & a \leq x \leq c \\ 1 & c < x \leq b \end{cases}$$

Define T by equation (83.1). Prove that $\|T\| = V_a^b \alpha$.

83.3 Prove (83.2)(ii).

83.4 Prove that $-f_a \notin C_1$.

83.5 Prove that $C_2 \supset C_1$ and $C_2 \subset B[a, b]$.

83.6 Let X and Y be sets. Prove that $\chi_{X \cap Y} = \chi_X \chi_Y$. Derive a formula for $\chi_{X \cup Y}$.

83.7 Let W be a subset of a vector space V such that

$$x + y \in W \quad \text{if } x, y \in W$$
$$cx \in W \quad \text{if } x \in W \text{ and } c \geq 0$$

Prove that

$$U = \{x - y \mid x, y \in W\}$$

is a subspace of V.

83.8 We say that f is *upper semi-continuous* (usc) *at c* if for every $\varepsilon > 0$, there exists $\delta > 0$ such that if $|x - c| < \delta$, then $f(x) < f(c) + \varepsilon$. We say that f is *usc on a set* $X \subset \mathbf{R}$ if f is usc at each point of X.

(a) Prove that f is usc on X if and only if $f^{-1}[(-\infty, t)]$ is open in X for every $t \in \mathbf{R}$.

(b) Prove that $f \in C_1$ if and only if f is a bounded, upper semi-continuous function on $[a, b]$.

84. Proof of the Riesz Representation Theorem

In this section we state and prove the Riesz representation theorem. We use the notation of Section 83.

Theorem 84.1 (Riesz Representation Theorem) Let $[a, b]$ be a closed interval and let $T \in (C[a, b])^*$. Then there exists $\alpha \in BV[a, b]$ such that

$$T(f) = \int_a^b f \, d\alpha \qquad \text{for all } f \in C[a, b]$$

and $\|T\| = V_a^b \alpha$.

Proof. Let $T \in (C[a, b])^*$. We begin by extending T to C_1. Let $f \in C_1$. Then there exists a decreasing sequence $\{f_n\}$ in $C[a, b]$ which converges pointwise to f. We first show that $\lim_{n \to \infty} T(f_n)$ exists.

Let $\phi_n = f_n - f_{n+1}$, $n = 1, 2, \ldots$. Let $e_n = \operatorname{sgn} T(\phi_n)$, $n = 1, 2, \ldots$. If $x \in [a, b]$, then

$$\left| \sum_{k=1}^n e_k \phi_k(x) \right| \le \sum_{k=1}^n \phi_k(x) = f_1(x) - f_n(x)$$

$$\le f_1(x) - f(x) \le \|f_1 - f\|$$

Thus

$$\left\| \sum_{k=1}^n e_k \phi_k \right\| \le \|f_1 - f\|$$

and

$$\sum_{k=1}^n |T(\phi_k)| = \sum_{k=1}^n e_k T(\phi_k) = T\left(\sum_{k=1}^n e_k \phi_k \right)$$

$$\le \|T\| \left\| \sum_{k=1}^n e_k \phi_k \right\| \le \|T\| \|f_1 - f\|$$

By Theorems 24.1 and 26.2, $\sum_{k=1}^\infty T(\phi_k)$ converges. Since

$$T(f_n) = T(f_1) - \sum_{k=1}^n T(\phi_k)$$

it follows that $\lim_{n \to \infty} T(f_n)$ exists. We define

$$T_1(f) = \lim_{n \to \infty} T(f_n) \qquad (84.1)$$

We next show that definition (84.1) is independent of the choice of the sequence $\{f_n\}$. That is, we show that if $\{f_n\}$ and $\{g_n\}$ are decreasing sequences in $C[a, b]$ with pointwise limit f, then $\lim_{n \to \infty} T(f_n) = \lim_{n \to \infty} T(g_n)$.
Let

$$F_n = f_n + \frac{1}{n} \qquad G_n = g_n + \frac{1}{n}$$

Then $\qquad \lim_{n \to \infty} T(F_n) = \lim_{n \to \infty} T(f_n) \qquad \lim_{n \to \infty} T(G_n) = \lim_{n \to \infty} T(g_n)$

Also $\{F_n\}$ and $\{G_n\}$ are decreasing sequences in $C[a, b]$ with pointwise limit f. Furthermore

$$F_n > f \qquad G_n > f \qquad \text{for } n = 1, 2, \ldots \qquad (84.2)$$

Let n_1 be a positive integer. We will show that there exists a positive integer m_1 such that $F_{n_1} > G_{m_1}$. Let

$$E_m = \{x \in [a, b] | F_{n_1}(x) > G_m(x)\} \qquad \text{for } m = 1, 2, \ldots$$

Since F_{n_1} and G_m are continuous, E_m is open for every positive integer m. Because of equation (84.2) and the fact that $\{G_m\}$ converges pointwise to f, the family $\{E_m\}$ covers $[a, b]$. Because $[a, b]$ is compact some finite subcollection of $\{E_m\}$ covers $[a, b]$. But

$$E_1 \subset E_2 \subset \cdots$$

so E_{m_1} covers $[a, b]$ for some positive integer m_1. From the definition of E_{m_1}, it follows that $F_{n_1} > G_{m_1}$.

Arguing as above, we construct a sequence such that

$$F_{n_1} > G_{m_1} > F_{n_2} > G_{m_2} > \cdots$$

Let $H_{2k} = G_{m_k}$ and $H_{2k-1} = F_{n_k}$. Then $\{H_k\}$ is a decreasing sequence in $C[a, b]$ with pointwise limit f. Therefore $\lim_{k \to \infty} T(H_k)$ exists. Now

$$\lim_{n \to \infty} T(f_n) = \lim_{n \to \infty} T(F_n) = \lim_{k \to \infty} T(F_{n_k}) = \lim_{k \to \infty} T(H_{2k-1})$$

$$= \lim_{k \to \infty} T(H_{2k}) = \lim_{k \to \infty} T(G_{m_k}) = \lim_{n \to \infty} T(G_n) = \lim_{n \to \infty} T(g_n)$$

Thus equation (84.1) is independent of the choice of the sequence $\{f_n\}$. It also follows by taking $f_n = f$ that $T_1(f) = T(f)$ if $f \in C[a, b]$.
It is easy to show that if $f, g \in C_1$ and $c \geq 0$, then

$$T_1(f + g) = T_1(f) + T_1(g) \qquad T_1(cf) = cT_1(f) \qquad (84.3)$$

Let $h \in C_2$. Then $h = f - g$, where $f, g \in C_1$. We define T_2 on C_2 by the equation

$$T_2(h) = T_1(f) - T_1(g) \tag{84.4}$$

We show that equation (84.4) is independent of the representation of h. That is, we show that if

$$f_1 - g_1 = h = f_2 - g_2$$

where $f_1, g_1, f_2, g_2 \in C_1$, then $T_1(f_1) - T_1(g_1) = T_1(f_2) - T_1(g_2)$. Since $f_1 + g_2 = f_2 + g_1$, by equation (84.3), $T_1(f_1) + T_1(g_2) = T_1(f_2) + T_1(g_1)$ and hence equation (84.4) is independent of the representation of h.

It follows easily from equations (84.3) and (84.4) that T_2 is a linear transformation on C_2 which agrees with T on $C[a, b]$.

We next show that T_2 is continuous on C_2 and $\|T_2\| = \|T\|$. Let $h \in C_2$, where $\|h\| \leq 1$. Then $h = f - g$, where $f, g \in C_1$. There exist decreasing sequences $\{f_n\}$ and $\{g_n\}$ in $C[a, b]$ which converge pointwise to f and g, respectively. Let $\phi_n = \min\{f_n, g_n + 1\}$ for $n = 1, 2, \ldots$. Then $\{\phi_n\}$ is a decreasing sequence in $C[a, b]$, and it follows from the inequality

$$g_n + 1 \geq g_n + h \geq g + h = f$$

that $\{\phi_n\}$ converges to f pointwise. Therefore

$$\lim_{n \to \infty} T(\phi_n) = T_1(f)$$

Let $\psi_n = \min\{g_n, \phi_n + 1\}$, $n = 1, 2, \ldots$. Then $\{\psi_n\}$ is a decreasing sequence in $C[a, b]$, and it follows from the inequality

$$\phi_n + 1 \geq f - h = g$$

that $\{\psi_n\}$ converges to g pointwise. Therefore $\lim_{n \to \infty} T(\psi_n) = T_1(g)$.

Since $\psi_n \leq \phi_n + 1$, it follows that

$$\psi_n - \phi_n \leq 1 \tag{84.5}$$

Now $\psi_n(x) = g_n(x)$ or $\psi_n(x) = \phi_n(x) + 1$. If $\psi_n(x) = g_n(x)$, then $\psi_n(x) - \phi_n(x) \geq g_n(x) - (g_n(x) + 1) = -1$. If $\psi_n(x) = \phi_n(x) + 1$, then $\psi_n(x) - \phi_n(x) = 1 \geq -1$. In any case

$$\psi_n - \phi_n \geq -1 \tag{84.6}$$

Combining equations (84.5) and (84.6), we have

$$\|\psi_n - \phi_n\| \leq 1$$

Therefore $\quad |T(\psi_n) - T(\phi_n)| = |T(\psi_n - \phi_n)| \leq \|T\| \tag{84.7}$

Taking limits in equation (84.7), we have

$$|T_1(g) - T_1(f)| \leq \|T\|$$

Thus $\qquad\qquad |T_2(h)| \le \|T\|$ for $\|h\| \le 1, h \in C_2$

By Theorems 80.1 and 81.3, T_2 is continuous on C_2 and $\|T_2\| \le \|T\|$. On the other hand,

$$\|T_2\| = \text{lub } \{|T_2(h)| \mid \|h\| \le 1, h \in C_2\}$$
$$\ge \text{lub } \{|T_2(h)| \mid \|h\| \le 1, h \in C[a, b]\}$$
$$= \text{lub } \{|T(h)| \mid \|h\| \le 1, h \in C[a, b]\}$$
$$= \|T\|$$

Therefore $\|T_2\| = \|T\|$.

We showed in Section 83, that if $c \in [a, b]$, then $\chi_{[a,c]} \in C_1$. Thus if $a \le c < d \le b$, then

$$\chi_{(c,d]} = \chi_{[a,d]} - \chi_{[a,c]} \in C_2$$

Define

$$\alpha(x) = T_2(\chi_{[a,x]}) \qquad \text{for all } x \in [a, b]$$

We next show that $\alpha \in BV[a, b]$ and $V_a^b\alpha \le \|T\|$.

Let $\{x_0, x_1, \ldots, x_n\}$ be a partition of $[a, b]$. Let $e_i = \text{sgn } T_2(\chi_{(x_{i-1},x_i]})$ for $i = 1, \ldots, n$. Now

$$\alpha(x_i) - \alpha(x_{i-1}) = T_2(\chi_{[a,x_i]}) - T_2(\chi_{[a,x_{i-1}]})$$
$$= T_2(\chi_{[a,x_i]} - \chi_{[a,x_{i-1}]})$$
$$= T_2(\chi_{(x_{i-1},x_i]})$$

Also $g = \sum_{i=1}^n e_i\chi_{(x_{i-1},x_i]} \in C_2$ and $\|g\| \le 1$. Thus

$$\sum_{i=1}^n |\alpha(x_i) - \alpha(x_{i-1})| = \sum_{i=1}^n |T_2(\chi_{(x_{i-1},x_i]})|$$
$$= \sum_{i=1}^n e_i T_2(\chi_{(x_{i-1},x_i]})$$
$$= T_2\left(\sum_{i=1}^n e_i\chi_{(x_{i-1},x_i]}\right)$$
$$= T_2(g) \le \|T_2\| = \|T\|$$

Therefore $\alpha \in BV[a, b]$, and

$$V_a^b\alpha \le \|T\| \qquad\qquad (84.8)$$

Finally, we show that $T(f) = \int_a^b f\, d\alpha$ for $f \in C[a, b]$ and $\|T\| = V_a^b\alpha$. Let $f \in C[a, b]$ and let $\varepsilon > 0$. Since f is uniformly continuous on $[a, b]$, there exists $\delta > 0$ such that if $|x - y| < \delta$, then $|f(x) - f(y)| < \varepsilon$. Let $P = \{y_0, y_1, \ldots, y_m\}$ be a partition of $[a, b]$ with max $\{|y_i - y_{i-1}| \mid 1 \le i \le m\} < \delta$. Let $P^* = \{x_0, x_1, \ldots, x_n\}$ be a partition of $[a, b]$ containing P and let

$t_i \in [x_{i-1}, x_i]$ for $i = 1, \ldots, n$. If $x \in [a, b]$, there exists a positive integer j, $1 \le j \le n$, such that $x \in (x_{j-1}, x_j]$. Now

$$\left| f(x) - \sum_{i=1}^{n} f(t_i)\chi_{(x_{i-1}, x_i]}(x) \right| = |f(x) - f(t_j)| < \varepsilon$$

since $|x - t_j| < \delta$. Thus $\| f - \sum_{i=1}^{n} f(t_i)\chi_{(x_{i-1}, x_i]} \| \le \varepsilon$. Therefore, if $\tau = \{t_1, \ldots, t_n\}$, we have

$$
\begin{aligned}
|T(f) - S(f, P^*, \tau)| &= \left| T(f) - \sum_{i=1}^{n} f(t_i)[\alpha(x_i) - \alpha(x_{i-1})] \right| \\
&= \left| T_2(f) - \sum_{i=1}^{n} f(t_i)T_2(\chi_{(x_{i-1}, x_i]}) \right| \\
&= \left| T_2[f - \sum_{i=1}^{n} f(t_i)\chi_{(x_{i-1}, x_i]}] \right| \\
&\le \|T_2\| \left\| f - \sum_{i=1}^{n} f(t_i)\chi_{(x_{i-1}, x_i]} \right\| \\
&\le \|T_2\|\varepsilon
\end{aligned}
$$

By Definition 53.1, $T(f) = \int_a^b f\, d\alpha$. By Example 80.4 and Theorem 81.3 $\|T\| \le V_a^b \alpha$ and combining this inequality with (84.8), we have $\|T\| = V_a^b \alpha$. ∎

Exercises

84.1 Prove equations (84.3).

84.2 Prove that T_2 is a linear transformation on C_2 which agrees with T on $C[a, b]$.

84.3 Prove that the function α constructed in the proof of the Riesz representation theorem is right continuous at every point of $[a, b)$.

84.4 We call a linear functional T in $(C[a, b])^*$ a *positive functional* if $T(f) \ge 0$ whenever $f \ge 0$ and $f \in C[a, b]$.
  (a) Let $T \in (C[a, b])^*$. Prove that T is a positive functional if and only if there exists an increasing function α on $[a, b]$ such that $T(f) = \int_a^b f\, d\alpha$ for all f in $C[a, b]$.
  (b) Let $T \in (C[a, b])^*$. Prove that there exist positive functionals U and V in $(C[a, b])^*$ such that $T = U - V$.

84.5 Let $g \in \mathscr{R}[a, b]$. Prove that if we define $T(f) = \int_a^b f(x)g(x)\, dx$, then $T \in (C[a, b])^*$. Apply Theorem 84.1 and describe the resulting α. What is $\|T\|$?

84.6 (a)* Let $T \in (C[-1, 1])^*$ and suppose also that $T(f \cdot g) = (T(f))(T(g))$, for all $f, g \in C[-1, 1]$. Prove that there exists $c \in [-1, 1]$ such that $T(f) = f(c)$ for $f \in C[-1, 1]$.
  (b)* Let $T \in (C[a, b])^*$ and suppose also that $T(f \cdot g) = (T(f))(T(g))$ for

all $f, g \in C[a, b]$. Prove that there exists $c \in [a, b]$ such that $T(f) = f(c)$ for $f \in C[a, b]$.

(c)* Prove that if T is a linear transformation from $C[a, b]$ into $\mathbf{R}$ satisfying $T(f \cdot g) = (T(f))(T(g))$ for all $f, g \in C[a, b]$, then $T \in (C[a, b])^*$, and hence by part (b), $T(f) = f(c)$ for some c in $[a, b]$ and for all $f \in C[a, b]$.

This exercise shows that the multiplicative linear transformations from $C[a, b]$ into $\mathbf{R}$ may be identified with the points of $[a, b]$.

XIV

The Lebesgue Integral

Consider the set S of rational numbers in $[0, 1]$. Since S is countable, we let $S = \{a_n\}_{n=1}^{\infty}$ be an enumeration of S. For $n = 1, 2, \ldots$, the function

$$f_n(x) = \begin{cases} 1 & \text{for } x = a_i \text{ with } i = 1, 2, 3, \ldots, n \\ 0 & \text{for all other } x \in [0, 1] \end{cases}$$

is Riemann integrable and

$$\int_0^1 f_n(x)\, dx = 0$$

Since $0 \le f_1(x) \le f_2(x) \le \cdots \le 1$ for all $x \in [0, 1]$, we let $f(x) = \lim_{n \to \infty} f_n(x)$ for each $x \in [0, 1]$. Clearly

$$f(x) = \begin{cases} 1 & \text{if } x \in S \\ 0 & \text{for all other } x \in [0, 1] \end{cases}$$

It seems reasonable that f should have integral 0, since each f_n has integral 0. However, since

$$\overline{\int_0^1} f(x)\, dx = 1 \qquad \underline{\int_0^1} f(x)\, dx = 0$$

f is not even Riemann integrable. This is an inadequacy of the Riemann and Riemann-Stieltjes integrals, which we now attempt to remedy.

Why is f not Riemann integrable? The fault lies not with the average used, namely $\alpha(x) = x$, but rather with the sets we averaged over, i.e., intervals. Using intervals is too restrictive. We now set a new goal. We wish to carry our concept of integral one more step, so that the class of functions which we can integrate is much larger than $\mathcal{R}[a, b]$ and so that the integral will not only retain the properties of the Riemann-Stieltjes integral, but will have additional properties which will allow us to deal with the integral of the limit of functions.

To keep the abstract flavor of this integral, we develop it from its rudimentary definitions and finish with examples of this integral on the real line.

85. The Extended Real Line

To accommodate our integration and measure theory, we now extend the real line **R** by adjoining two points, denoted ∞ and $-\infty$. The order relation $<$ and the operations of addition and multiplication are also extended in a natural way, with the exception of $0 \cdot \infty$ and $\infty - \infty$. We let $0 \cdot \infty = 0$, since we wish to ignore what happens on sets of measure zero. However, $\infty - \infty$ remains undefined since situations in which $\infty - \infty$ occur constitute major sources of difficulty in measure theory.

Definition 85.1

(i) $\tilde{\mathbf{R}} = \mathbf{R} \cup \{\infty, -\infty\}$.

For any $x \in \mathbf{R}$

(ii) $-\infty < x < \infty$ and $-\infty < \infty$. Also $<$ has its usual meaning on **R**.

(iii) $x + \infty = \infty + x = \infty + \infty = \infty$ and $x - \infty = -\infty + x = -\infty - \infty = -\infty$. Also $+$ has its usual meaning on **R** and $-$ has its usual meaning on **R**.

(iv) If $x > 0$, then $x \cdot \infty = \infty \cdot x = \infty \cdot \infty = (-\infty) \cdot (-\infty) = \infty$ and $x \cdot (-\infty) = (-\infty) \cdot x = \infty \cdot (-\infty) = (-\infty) \cdot \infty = -\infty$. $\cdot$ has its usual meaning on **R**.

(v) If $x < 0$, then $x \cdot \infty = \infty \cdot x = -\infty$ and $x \cdot (-\infty) = (-\infty) \cdot x = \infty$.

(vi) $0 \cdot \infty = \infty \cdot 0 = 0 \cdot (-\infty) = (-\infty) \cdot 0 = 0$.

Therefore, the extended operations $+$ and $\cdot$ are still commutative and associative, but the distributive law fails in $\tilde{\mathbf{R}}$. For example; $\infty \cdot (3 - 2) = \infty \cdot 1 = \infty$, but $(\infty \cdot 3) - (\infty \cdot 2) = \infty - \infty$ is undefined.

We shall sometimes refer to a function $f : x \to [-\infty, \infty]$ which takes on only real values as finite valued.

Definition 85.2 For $a, b \in \tilde{\mathbf{R}}$, $a < b$ define

$$[a, b] = \{x \in \tilde{\mathbf{R}} \mid a \le x \le b\}$$
$$(a, b] = \{x \in \tilde{\mathbf{R}} \mid a < x \le b\}$$
$$[a, b) = \{x \in \tilde{\mathbf{R}} \mid a \le x < b\}$$
$$(a, b) = \{x \in \tilde{\mathbf{R}} \mid a < x < b\}$$

Any of the above sets is called an *interval* of $\tilde{\mathbf{R}}$. The open sets in $\tilde{\mathbf{R}}$ are the open sets of $\tilde{\mathbf{R}}$ together with sets of the form $U \cup I_1 \cup I_2$, where U is open in $\mathbf{R}$ and I_1 and I_2 are intervals of $\tilde{\mathbf{R}}$ of the form $[-\infty, a)$ or $(a, \infty]$ for $a \in \tilde{\mathbf{R}}$.

Example. $(-1, 1) \cup (5, \infty]$ is an open set in $\tilde{\mathbf{R}}$. Also $[-\infty, \infty] = [-\infty, 1) \cup \varnothing \cup (0, \infty]$ is an open set in $\tilde{\mathbf{R}}$.

Note. $[-\infty, \infty] = \tilde{\mathbf{R}}$. We will use these notations interchangably.

Exercises

85.1 Show that the open sets of $\tilde{\mathbf{R}}$ form a topology for $\tilde{\mathbf{R}}$

85.2 Define a metric on $\tilde{\mathbf{R}}$, such that the metric topology is identical to the topology in Exercise 85.1.

86. σ-Algebras and Positive Measures

Definition 86.1 Let X be a set. A collection of subsets $\mathcal{M}$ of X is said to be a *σ-algebra in X* if

(i) $\varnothing \in \mathcal{M}$,

(ii) $A \in \mathcal{M}$ implies $A' \in \mathcal{M}$,

(iii) $A_n \in \mathcal{M}$ for $n = 1, 2, 3, \ldots$ implies $\bigcup_{n=1}^{\infty} A_n \in \mathcal{M}$.

Examples. Let X be any set and recall that $\mathcal{P}(X)$ denotes the collection of all subsets of X. It is easily seen that $\mathcal{P}(X)$ is a σ-algebra in X.

Another σ-algebra in X is $\mathcal{M} = \{\varnothing, X\}$, which also trivially satisfies Definition 86.1.

Note that any σ-algebra in X always lies between the σ-algebras $\{\varnothing, X\}$ and $\mathcal{P}(X)$.

Definition 86.2 Let $\mathcal{M}$ be a σ-algebra in X. A function μ from $\mathcal{M}$ into $[0, \infty]$ is called a *positive measure on $\mathcal{M}$* if

(i) For at least one set $A \in \mathcal{M}$, we have $\mu(A) < \infty$.

(ii) For $A_n \in \mathcal{M}, n = 1, 2, 3, \ldots$, and $A_i \cap A_j = \varnothing$ for each $i \neq j$, we have

$$\mu\left(\bigcup_{n=1}^{\infty} A_n\right) = \sum_{n=1}^{\infty} \mu(A_n) \ .$$

This last property is called the *countable additivity* of μ.

The property $A_i \cap A_j = \varnothing$ for each $i \neq j$ is called the *pairwise disjointness* of $\{A_n\}_{n=1}^{\infty}$. Alternatively, $\{A_n\}_{n=1}^{\infty}$ is said to be *pairwise disjoint*.

If $A_n \in \mathcal{M}$ and $\mu(A_n) < \infty$ for $n = 1, 2, \ldots$, $\sum_{n=1}^{\infty} \mu(A_n)$ is just the usual sum of an infinite series of real numbers. Since $\mu(A_n) \geq 0$ for $n = 1, 2, \ldots$, either the series converges or diverges to ∞. If $\mu(A_n) = \infty$ for some n, we define $\sum_{n=1}^{\infty} \mu(A_n) = \infty$.

Definition 86.3   Let X be a set, $\mathcal{M}$ be a σ-algebra in X and μ be a positive measure on $\mathcal{M}$, then we call the ordered triple $(X, \mathcal{M}, \mu)$ a *positive measure space*.

Example 86.4   Let X be any set. Let $\mathcal{M} = \mathscr{P}(X)$. For each $A \in \mathscr{P}(X)$ define

$$\mu(A) = \begin{cases} 0 & \text{if } A = \varnothing \\ \text{number of elements in } A & \text{if } A \text{ has a finite number of elements} \\ \infty & \text{if } A \text{ has an infinite number of elements} \end{cases}$$

We call μ *counting measure* on $\mathscr{P}(X)$. It may be verified (Exercise 86.1) that $(X, \mathcal{M}, \mu)$ is a positive measure space.

We now prove some of the basic properties of a positive measure space.

Theorem 86.5   Let $(X, \mathcal{M}, \mu)$ be any positive measure space. Then

(i) $\mu(\varnothing) = 0$.

(ii) If $A_1, A_2, \ldots, A_n$ are in $\mathcal{M}$ and $A_i \cap A_j = \varnothing$ for each $i \neq j$, then $\mu(\bigcup_{i=1}^{n} A_i) = \sum_{i=1}^{n} \mu(A_i)$.

(iii) If $A_n \in \mathcal{M}$ for $n = 1, 2, \ldots$, then $\bigcap_{n=1}^{\infty} A_n \in \mathcal{M}$. Also $\bigcap_{n=1}^{j} A_n \in \mathcal{M}$ for $j = 1, 2, \ldots$.

(iv) If A and B are in $\mathcal{M}$, then $A \backslash B \in \mathcal{M}$.

(v) If $A_1 \subset A_2 \subset A_3 \subset \cdots$ and each $A_n \in \mathcal{M}$, then $\mu(\bigcup_{n=1}^{\infty} A_n) = \lim_{n \to \infty} \mu(A_n)$

(vi) If A and B are in $\mathcal{M}$, and $A \subset B$, then $\mu(A) \leq \mu(B)$.

(vii) If $A_1 \supset A_2 \supset A_3 \supset \cdots$ with each $A_n \in \mathcal{M}$ and $\mu(A_1) < \infty$, then $\mu(\bigcap_{n=1}^{\infty} A_n) = \lim_{n \to \infty} \mu(A_n)$.

Proof.   (i) By Definition 86.2(i), there exists $A \in \mathcal{M}$ with $\mu(A) < \infty$. If $A_1 = A$, $A_2 = \varnothing$, $A_3 = \varnothing, \ldots$, then $A_n \in \mathcal{M}$ for $n = 1, 2, 3, \ldots$ and $A_i \cap A_j = \varnothing$ for $i \neq j$; therefore by Definition 86.2(ii), we have $\mu(A) = \mu(\bigcup_{n=1}^{\infty} A_n) = \sum_{n=1}^{\infty} \mu(A_n) < \infty$ which implies $\mu(\varnothing) = 0$, for otherwise, $\mu(\varnothing) > 0$ would give $\sum_{n=1}^{\infty} \mu(A_n) = \infty$.

(ii) If $A_1, A_2, \ldots, A_n$ are in $\mathcal{M}$ and $A_i \cap A_j = \varnothing$ for $i \neq j$, then let $A_{n+1} = \varnothing$, $A_{n+2} = \varnothing, \ldots$. By Definition 86.2(ii), we have

$$\mu\left(\bigcup_{i=1}^{n} A_i\right) = \mu\left(\bigcup_{i=1}^{\infty} A_i\right) = \sum_{i=1}^{\infty} \mu(A_i) = \sum_{i=1}^{n} \mu(A_i)$$

since $u(A_m) = 0$ for all $m \geq n + 1$.

(iii) If $A_n \in \mathcal{M}$ for $n = 1, 2, \ldots$, then $\bigcap_{n=1}^{\infty} = \left(\bigcup_{n=1}^{\infty} A'_n\right)'$ by DeMorgan's law. However, $A'_n \in \mathcal{M}$ for each n; so $\bigcup_{n=1}^{\infty} A'_n \in \mathcal{M}$, and therefore $\left(\bigcup_{n=1}^{\infty} A'_n\right)' \in \mathcal{M}$. Also $\bigcap_{n=1}^{j} A_n \in \mathcal{M}$ using (ii) and DeMorgan's law for $j = 1, 2, \ldots$.

(iv) If A and B are in $\mathcal{M}$, then $A \backslash B = A \cap B'$. Since $A \in \mathcal{M}$ and $B' \in \mathcal{M}$, we have by (iii) that $A \cap B' \in \mathcal{M}$.

(v) Define $B_1 = A_1$, $B_2 = A_2 \backslash A_1, \ldots$, $B_n = A_n \backslash A_{n-1}, \ldots$. By (iv), $B_n \in \mathcal{M}$ for each $n = 1, 2, \ldots$. Also it is easily seen that $B_i \cap B_j = \varnothing$ for each $i \neq j$. Therefore

$$\mu\left(\bigcup_{n=1}^{\infty} B_n\right) = \sum_{n=1}^{\infty} \mu(B_n)$$

Clearly $\bigcup_{n=1}^{\infty} B_n = \bigcup_{n=1}^{\infty} A_n$. But $\sum_{n=1}^{j} \mu(B_n) = \mu(\bigcup_{n=1}^{j} B_n) = \mu(A_j)$ for $j = 1, 2, \ldots$ by (ii). Therefore

$$\sum_{n=1}^{\infty} \mu(B_n) = \lim_{j \to \infty} \sum_{n=1}^{j} \mu(B_n) = \lim_{j \to \infty} \mu(A_j)$$

Thus
$$\mu\left(\bigcup_{n=1}^{\infty} A_n\right) = \lim_{n \to \infty} \mu(A_n).$$

(vi) $A \subset B$ implies that $B = A \cup (B \backslash A)$. Since $A \in \mathcal{M}$, $B \backslash A \in \mathcal{M}$, and $A \cap (B \backslash A) = \varnothing$, by (ii) we have

$$\mu(B) = \mu(A) + \mu(B \backslash A) \tag{86.1}$$

Since $\mu: \mathcal{M} \to [0, \infty]$, equation (86.1) implies that $\mu(A) \leq \mu(B)$.

(vii) If $A_1 \supset A_2 \supset \cdots$ and each $A_n \in \mathcal{M}$ and $\mu(A_1) < \infty$, we let $B_1 = A_1 \backslash A_2$, $B_2 = A_2 \backslash A_3, \ldots$, $B_n = A_n \backslash A_{n+1}, \ldots$. Then $B_n \in \mathcal{M}$ for all $n = 1, 2, \ldots$ and $B_i \cap B_j = \varnothing$ for each $i \neq j$. Now $A_1 = A_n \cup \left(\bigcup_{j=1}^{n-1} B_j\right)$ and $A_n \cap \left(\bigcup_{j=1}^{n-1} B_j\right) = \varnothing$; so

$$\mu(A_1) = \mu(A_n) + \mu\left(\bigcup_{j=1}^{n-1} B_j\right) \quad \text{for } n = 2, 3, \ldots \tag{86.2}$$

Since $A_n \subset A_1$, $\bigcup_{j=1}^{n-1} B_j \subset A_1$ and $\mu(A_1) < \infty$, we have $\mu(A_n) < \infty$ and $\mu(\bigcup_{j=1}^{n-1} B_j) < \infty$ by (vi). Therefore equation (86.2) gives

$$\mu\left(\bigcup_{j=1}^{n-1} B_j\right) = \mu(A_1) - \mu(A_n) \quad \text{for } n = 2, 3, \ldots \tag{86.3}$$

Similarly,
$$\bigcap_{n=1}^{\infty} A_n = A_1 \backslash \left(\bigcup_{n=1}^{\infty} B_n\right).$$

again
$$A_1 = \left(\bigcap_{n=1}^{\infty} A_n \right) \cup \left(\bigcup_{n=1}^{\infty} B_n \right)$$

$$\left(\bigcap_{n=1}^{\infty} A_n \right) \cap \left(\bigcup_{n=1}^{\infty} B_n \right) = \varnothing$$

$$\bigcap_{n=1}^{\infty} A_n \subset A_1 \quad \text{and} \quad \bigcup_{n=1}^{\infty} B_n \subset A_1$$

imply
$$\mu\left(\bigcap_{n=1}^{\infty} A_n \right) = \mu(A_1) - \mu\left(\bigcup_{n=1}^{\infty} B_n \right) = \mu(A_1) - \sum_{n=1}^{\infty} \mu(B_n)$$

$$= \mu(A_1) - \lim_{n\to\infty} \left(\sum_{j=1}^{n-1} \mu(B_j) \right)$$

$$= \lim_{n\to\infty} \left[\mu(A_1) - \sum_{j=1}^{n-1} \mu(B_j) \right]$$

$$= \lim_{n\to\infty} \left[\mu(A_1) - \mu\left(\bigcup_{j=1}^{n-1} B_j \right) \right]$$

$$= \lim_{n\to\infty} [\mu(A_1) - \mu(A_1) + \mu(A_n)] = \lim_{n\to\infty} \mu(A_n)$$

where we use equation (86.3) for the second to last equality. ∎

Property (ii) above is called the *finite additivity* of μ.

Example 86.6 In (vii) of Theorem 86.5 the assumption that $\mu(A_1) < \infty$ is necessary, for let $X = \mathbf{Z}$, the set of integers, let $\mathcal{M} = \mathcal{P}(\mathbf{Z})$, and let μ be counting measure on $\mathcal{M}$. If $A_1 = \{1, 2, 3, \ldots\}$, $A_2 = \{2, 3, 4, \ldots\}$, ..., $A_n = \{n, n+1, n+2, \ldots\}$, ..., then each $A_n \in \mathcal{M}$ and $A_1 \supset A_2 \supset \cdots$, but $\mu(A_n) = \infty$ for $n = 1, 2, \ldots$; so $\lim_{n\to\infty} \mu(A_n) = \infty$. However, $\mu(\bigcap_{n=1}^{\infty} A_n) = \mu(\varnothing) = 0$. So $\mu(\bigcap_{n=1}^{\infty} A_n) = 0 \neq \infty = \lim_{n\to\infty} \mu(A_n)$.

Exercises

86.1 Prove that if X is a set, then $\mathcal{P}(X)$ is a σ-algebra in X and that counting measure (Example 86.4) is a positive measure on $\mathcal{P}(X)$.

86.2 Use the measure space of Example 86.6 to give an example of each of parts (ii) to (vii) of Theorem 86.5.

86.3 Does there exist an *infinite* σ-algebra which has only *countably* many members?

86.4 Let $(X, \mathcal{M}, \mu)$ be a positive measure space. Prove that if $A_n \in \mathcal{M}$ for $n = 1, 2, \ldots$, then

$$\mu\left(\bigcup_{n=1}^{\infty} A_n \right) \leq \sum_{n=1}^{\infty} \mu(A_n)$$

(It is not assumed that the sequence $\{A_n\}$ is pairwise disjoint.)

86.5 Let X be a set. Let $\{\mathcal{M}_\alpha\}$ be a collection of σ-algebras in X. Prove that $\cap \, \mathcal{M}_\alpha$ is a σ-algebra in X. Give an example to show that the union of σ-algebras need not be a σ-algebra.

86.6 Prove that the sum $\sum_{n=1}^{\infty} \mu(A_n)$ which appears in Definition 86.2 is independent of the order in which the family of sets $\{A_n\}$ is indexed.

86.7 Let X be a set. When is the family of all countable subsets of X a σ-algebra?

86.8 If $(X, \mathcal{M}, \mu)$ is a measure space such that, whenever $A \subset B$, $B \in \mathcal{M}$, and $\mu(B) = 0$, we have $A \in \mathcal{M}$, we call $(X, \mathcal{M}, \mu)$ a *complete measure space*.

Let $(X, \mathcal{M}, \mu)$ be an arbitrary measure space. Let $\overline{\mathcal{M}}$ be the collection of all sets $S \cup A$, where $S \in \mathcal{M}$ and $A \subset B$, where $B \in \mathcal{M}$ and $\mu(B) = 0$. Define $\overline{\mu}$ on $\overline{\mathcal{M}}$ by $\overline{\mu}(S \cup A) = \mu(S)$.

   (a) Prove that $\overline{\mu}$ is well defined, that is, that the value of $\overline{\mu}$ does not depend on the representation $S \cup A$.

   (b) Prove that $\overline{\mathcal{M}}$ is a σ-algebra in X.

   (c) Prove that $\overline{\mu}$ is a measure on $\overline{\mathcal{M}}$.

   (d) Prove that $(X, \overline{\mathcal{M}}, \overline{\mu})$ is a complete measure space.

86.9 Let $\mathcal{M}$ be a σ-algebra in a set X.

   (a) Let $E \subset X$. Prove that

$$\mathcal{M}_E = \{A \cap E \mid A \in \mathcal{M}\}$$

is a σ-algebra in E. We call $\mathcal{M}_E$ a *relative σ-algebra*.

   (b) Suppose $(X, \mathcal{M}, \mu)$ is a positive measure space and let $E \in \mathcal{M}$. Prove that $(E, \mathcal{M}_E, \lambda)$, where $\lambda(A) = \mu\,(A \cap E)$ is a positive measure space. We call $(E, \mathcal{M}, \lambda)$ a *relative measure space*.

87. Measurable Functions

Definition 87.1 A function f from a set X, with a σ-algebra $\mathcal{M}$, into $[-\infty, \infty]$ is called an *extended real measurable function on X with respect to $\mathcal{M}$* if $f^{-1}(V) \in \mathcal{M}$ for every open set $V \subset [-\infty, \infty]$.

A function f which satisfies Definition 87.1 will be referred to as *$\mathcal{M}$-measurable* or just *measurable* if $\mathcal{M}$ is clear from the context.

Example. Let X be any set and $\mathcal{M} = \mathscr{P}(X)$. Then *every* function $f: X \to [-\infty, \infty]$ is measurable, since $f^{-1}(V) \in \mathscr{P}(X)$ for every V open in $[-\infty, \infty]$. We now wish to obtain some statements equivalent to Definition 87.1.

Theorem 87.2 Let X be a set and $\mathcal{M}$ be a σ-algebra in X and let f be a function from X into $[-\infty, \infty]$. Then the following are equivalent.

  (i) f is measurable.

  (ii) $\{x \mid f(x) > a\} \in \mathcal{M}$ for every $a \in \mathbf{R}$.

(iii) $\{x \mid f(x) < a\} \in \mathcal{M}$ for every $a \in \mathbf{R}$.
(iv) $\{x \mid f(x) \le a\} \in \mathcal{M}$ for every $a \in \mathbf{R}$.
 (v) $\{x \mid f(x) \ge a\} \in \mathcal{M}$ for every $a \in \mathbf{R}$.

Proof. (i) implies (ii) by Definition 87.1.

(ii) implies (iii). Assume (ii) holds. Then for any $a \in \mathbf{R}$

$$\{x \mid f(x) < a\} = X \setminus \bigcap_{n=1}^{\infty} \left\{x \mid f(x) > a - \frac{1}{n}\right\}$$

Since $\{x \mid f(x) > a - 1/n\} \in \mathcal{M}$ for each $n = 1, 2, \ldots$, then $\bigcap_{n=1}^{\infty} \{x \mid f(x) > a - 1/n\} \in \mathcal{M}$. Therefore, $X \setminus \bigcap_{n=1}^{\infty} \{x \mid f(x) > a - 1/n\} \in \mathcal{M}$, which gives (iii).

(iii) implies (iv). Assume (iii) holds. Then for any $a \in \mathbf{R}$ $\{x \mid f(x) \le a\} = \bigcap_{n=1}^{\infty} \{x \mid f(x) < a + 1/n\}$ which is in $\mathcal{M}$ since each $\{x \mid f(x) < a + 1/n\} \in \mathcal{M}$ by (iii), and $\mathcal{M}$ is closed under countable intersections.

(iv) implies (v). Assume (iv) holds. Then for any $a \in \mathbf{R}$, $\{x \mid f(x) \ge a\} = X \setminus \bigcup_{n=1}^{\infty} \{x \mid f(x) \le a - 1/n\}$ which is in $\mathcal{M}$ since $\{x \mid f(x) \le a - 1/n\} \in \mathcal{M}$ by (iv), and $\mathcal{M}$ is closed under countable unions and complementation in X.

(v) implies (i). Assume (v) holds. Let V be an open set in $[-\infty, \infty]$. Then V can be written uniquely as $\bigcup_{n=1}^{\infty} A_n$, where $A_i \cap A_j = \varnothing$ for each $i \ne j$, and A_n is of the form (a_n, b_n), $(a_n, \infty]$ or $[-\infty, a_n)$ for some $a_n, b_n \in \mathbf{\bar{R}}$. (These are the open intervals of $\mathbf{\bar{R}}$.) Therefore $f^{-1}(V) = \bigcup_{n=1}^{\infty} f^{-1}(A_n)$. But $\mathcal{M}$ is closed under countable unions; so to show f is measurable, it is sufficient to show that $f^{-1}(A) \in \mathcal{M}$ for all the open intervals A of $\mathbf{\bar{R}}$. We handle each type of open interval separately:

1. $f^{-1}((a, \infty]) = \bigcup_{n=1}^{\infty} \{x \mid f(x) \ge a + 1/n\} \in \mathcal{M}$ for any $a \in \mathbf{R}$.
2. $f^{-1}([-\infty, a)) = X \setminus \{x \mid f(x) \ge a\} \in \mathcal{M}$ for any $a \in \mathbf{R}$.
3. $f^{-1}((a, b)) = f^{-1}([-\infty, b)) \cap f^{-1}((a, \infty])$ by 1 and 2 for any $a, b \in \mathbf{R}$.
4. $f^{-1}((-\infty, \infty)) = \bigcup_{n=1}^{\infty} f^{-1}((-n, n)) \in \mathcal{M}$ by 3.
5. $f^{-1}((-\infty, a)) = f^{-1}([-\infty, a)) \cap f^{-1}((-\infty, \infty)) \in \mathcal{M}$ by 2 and 4 for any $a \in \mathbf{R}$.
6. $f^{-1}((a, \infty)) = f^{-1}((a, \infty]) \cap f^{-1}((-\infty, \infty)) \in \mathcal{M}$ by 1 and 4 for any $a \in \mathbf{R}$.

Items 1 to 6 cover all cases. ∎

We wish to show that the class of measurable functions on X with respect to $\mathcal{M}$ is closed under certain important operations. In the following theorems it is understood that X is a set and $\mathcal{M}$ is a σ-algebra in X.

Lemma 87.3 If $F : \mathbf{R} \to \mathbf{R}$ is continuous and $f : X \to \mathbf{R}$ is measurable, then $F \circ f$ is measurable.

Proof. Let U be an open subset of $\mathbf{R}$. Now $F^{-1}(U)$ is open since F is continuous. Since f is measurable $f^{-1}(F^{-1}(U)) \in \mathcal{M}$. Now $(F \circ f)^{-1}(U) = f^{-1}(F^{-1}(U)) \in \mathcal{M}$; hence $F \circ f$ is measurable. ■

Theorem 87.4 If f is a real-valued measurable function and c is a real number, then the functions cf, f^2, and $|f|$ are measurable. If further $f(x) \neq 0$ for $x \in X$, then $1/f$ is measurable.

Proof. The conclusions follow from Lemma 87.3 by taking $F(x)$ successively to be cx, x^2, $|x|$, and $1/x$. ■

Theorem 87.5 Let $\{f_n\}_{n=1}^{\infty}$ be any sequence of measurable functions on X. For each $x \in X$ define

$$g(x) = \text{lub } \{f_n(x) \mid n = 1, 2, \ldots\}$$

$$h(x) = \lim_{n \to \infty} \sup f_n(x)$$

$$k(x) = \text{glb } \{f_n(x) \mid n = 1, 2, \ldots\}$$

$$l(x) = \lim_{n \to \infty} \inf f_n(x)$$

Then g, h, k, and l are measurable functions on X.

Proof. First note that each of the above functions is well defined for each $x \in X$ and that each function takes X into $[-\infty, \infty]$.

For each $a \in \mathbf{R}$ we have

$$\{x \mid g(x) > a\} = \bigcup_{n=1}^{\infty} \{x \mid f_n(x) > a\} \in \mathcal{M}$$

$$\{x \mid h(x) \geq a\} = \bigcap_{m=1}^{\infty} \bigcup_{n=m}^{\infty} \left\{x \mid f_n(x) > a - \frac{1}{m}\right\} \in \mathcal{M}$$

$$\{x \mid k(x) < a\} = \bigcup_{n=1}^{\infty} \{x \mid f_n(x) < a\} \in \mathcal{M}$$

and $$\{x \mid l(x) \leq a\} = \bigcap_{m=1}^{\infty} \bigcup_{n=m}^{\infty} \left\{x \mid f_n(x) < a + \frac{1}{m}\right\} \in \mathcal{M}$$

Therefore by Theorem 87.2, g, h, k, and l are each measurable on X. ■

Corollary 87.6

(i) If f and g are measurable, then $\max \{f, g\}$ and $\min \{f, g\}$ are measurable.

(ii) For any function f define

$$f^+ = \max \{f, 0\} \qquad f^- = -\min \{f, 0\}$$

If f is measurable, then f^+ and f^- are measurable.

(iii) The pointwise limit of a convergent sequence of measurable functions is measurable.

The functions f^+ and f^- defined above will be used extensively in the next section. We note that $f = f^+ - f^-$ and $|f| = f^+ + f^-$.

After proving the following lemma, we will prove that sums and products of measurable functions are measurable.

Lemma 87.7 If f and g are measurable, then $\{x \mid f(x) < g(x)\} \in \mathcal{M}$.

Proof. If $A_r = \{x \mid f(x) < r\}$ and $B_r = \{x \mid g(x) > r\}$, then A_r, B_r, $A_r \cap B_r \in \mathcal{M}$. Now

$$\{x \mid f(x) < g(x)\} = \bigcup_{r \in Q} (A_r \cap B_r) \in \mathcal{M}$$

since $\mathcal{M}$ is closed under countable unions. ∎

Theorem 87.8 If f and g are real-valued measurable functions, then $f + g$ and fg are measurable.

Proof. We note that if a is a real number, then because of the equation

$$\{x \mid a - g(x) < r\} = \{x \mid a - r < g(x)\}$$

$a - g$ is measurable. Thus, for any number a, by Lemma 87.7,

$$\{x \mid f(x) + g(x) < a\} = \{x \mid f(x) < a - g(x)\} \in \mathcal{M}$$

Therefore $f + g$ is measurable.

By Theorem 87.4 and the result above, $(f + g)^2$ and $(f - g)^2$ are measurable. Thus

$$fg = \tfrac{1}{4}[(f + g)^2 - (f - g)^2]$$

is measurable. ∎

We now investigate a special class of functions on X.

Definition 87.9 A *simple function* is a function with finite range.

We shall be especially interested in real-valued, simple measurable functions on X. Suppose $\{c_1, \ldots, c_n\}$ is the set of distinct elements of the range of a real-valued, simple measurable function s on X. Let $A_i = s^{-1}(c_i)$ for $i = 1, \ldots, n$. Then $\{A_i\}$ is a pairwise disjoint family of sets in $\mathcal{M}$ whose union is X and s may be written

$$s = \sum_{i=1}^{n} c_i \chi_{A_i}$$

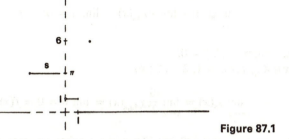

Figure 87.1

Example. Let $X = \mathbf{R}$, $A_1 = [0, 1]$, $A_2 = \{2\}$, $A_3 = [-3, -\frac{1}{2})$ and $c_1 = 1$, $c_2 = 6$, and $c_3 = \pi$. Then

$$s = \sum_{i=1}^{3} c_i \chi_{A_i} = 1 \cdot \chi_{[0,1]} + 6 \cdot \chi_{\{2\}} + \pi \chi_{[-3,-\frac{1}{2})}$$

is a simple function. (See Figure 87.1.)

Theorem 87.10 If f is any measurable function from X into $[0, \infty]$, then there exists a sequence $\{s_n\}_{n=1}^{\infty}$ of simple measurable functions such that

$$0 \le s_1 \le s_2 \le s_3 \le \cdots$$

and

$$\lim_{n \to \infty} s_n(x) = f(x) \qquad \text{for all } x \in X$$

Proof. Suppose $f: X \to [0, \infty]$ is measurable. For each $n = 1, 2, \ldots$ define the sets

$$A_{k,n} = f^{-1}\left(\left[\frac{k}{2^n}, \frac{k+1}{2^n}\right)\right) \qquad \text{for } k = 0, 1, 2, \ldots, n2^n - 1$$

and $\qquad A_n = f^{-1}([n, \infty])$

Then clearly for each $n = 1, 2, \ldots$, $X = \bigcup_{k=0}^{n2^n-1} A_{k,n} \cup A_n$ and A_n, $A_{1,n}$, $A_{2,n}, \ldots, A_{2^n-1,n}$ are mutually disjoint. By Theorem 87.2 each of these sets is in $\mathscr{M}$. Define

$$s_n = \sum_{k=0}^{n2^n-1} \frac{k}{2^n} \chi_{A_{k,n}} + n \chi_{A_n} \qquad \text{for } n = 1, 2, 3, \ldots$$

It is easy to show that each s_n is a measurable simple function (see Exercise 87.2). By construction $0 \le s_1 \le s_2 \le s_3 \le \cdots$.

We wish to show $\lim_{n \to \infty} s_n(x) = f(x)$ for each $x \in X$.

CASE 1.   Suppose $f(x) = \infty$.
Since $x \in A_n$ for $n = 1, 2, \ldots$,

$$\lim_{n \to \infty} s_n(x) = \lim_{n \to \infty} n\chi_{A_n}(x) = \lim_{n \to \infty} n = \infty = f(x)$$

CASE 2.   Suppose $f(x) = 0$.
Now $x \in A_{0,n}$ for $n = 1, 2, \ldots$; so

$$\lim_{n \to \infty} s_n(x) = \lim_{n \to \infty} \frac{0}{2^n}\chi_{A_{0,n}}(x) = \lim_{n \to \infty} 0 = 0 = f(x)$$

CASE 3.   Suppose $f(x) \in (0, \infty)$. Then there exists an integer N such that $f(x) < N$. By construction for all $n > N$, x is in exactly one $A_{k,n}$ for some $k = 0, 1, \ldots, n2^n - 1$ and also

$$0 \le f(x) - \frac{k}{2^n} < \frac{1}{2^n}$$

But $\qquad\qquad\qquad s_n(x) = \frac{k}{2^n}\chi_{A_{k,n}}(x) = \frac{k}{2^n}$

so for each $n > N$ we have $0 \le f(x) - s_n(x) < 1/2^n$. It follows that

$$\lim_{n \to \infty} s_n(x) = f(x)$$

By cases 1, 2, and 3, we have $\lim_{n \to \infty} s_n(x) = f(x)$ for all $x \in X$.   ∎

Exercises

In these exercises, unless specified otherwise, X is a set and $\mathcal{M}$ is a σ-algebra in X.

87.1   Deduce Corollary 87.6 from Theorem 87.5.

87.2   Prove that if $f: X \to [0, \infty]$ is bounded and measurable, then the sequence $\{s_n\}_{n=1}^{\infty}$ in Theorem 87.10 converges uniformly to f.

87.3   Let f be a real-valued function on X. Prove that f is a measurable function if and only if for each closed set $C \subset \mathbf{R}$, $f^{-1}(C) \in \mathcal{M}$. Show that this statement remains valid if *closed* is replaced by *compact*.

87.4   Let f be a real-valued measurable function and c be a real number. Prove that cf, $|f|$, and $1/f$ (provided $f(x) \ne 0$) are measurable directly from Theorem 87.2.

87.5   Let X be any set and $\mathcal{M} = \{\phi, X\}$. Prove that the class of measurable functions is exactly those functions which are constant on X.

87.6   Let X be a set and $\mathcal{M}$ be a σ-algebra in X.
(a)   Prove that a simple function $s = \sum_{i=1}^{n} c_i\chi_{A_i}$ on X, where the $\{c_i\}$

are distinct and not zero, is measurable with respect to $\mathcal{M}$ if and only if $A_1, A_2, \ldots, A_n$ are all elements of $\mathcal{M}$.

(b)   Prove that the sum of any two simple functions is a simple function.

87.7   Let X be a set, let $\mathcal{M}$ be a σ-algebra in X, and let f be a function from X into $[-\infty, \infty]$. Prove that the following are equivalent.
(a) f is measurable.
(b) $\{x \mid f(x) > r\} \in \mathcal{M}$ for every $r \in \mathbf{Q}$.
(c) $\{x \mid f(x) < r\} \in \mathcal{M}$ for every $r \in \mathbf{Q}$.
(d) $\{x \mid f(x) \leq r\} \in \mathcal{M}$ for every $r \in \mathbf{Q}$.
(e) $\{x \mid f(x) \geq r\} \in \mathcal{M}$ for every $r \in \mathbf{Q}$.

87.8   Let X be a set and $\mathcal{M}$ be a σ-algebra in X. Let $\{f_n\}_{n=1}^{\infty}$ be a sequence of measurable real-valued functions on X. Prove that the set $A = \{x \in X \mid \{f_n(x)\}_{n=1}^{\infty}$ converges in $\mathbf{R}\}$ is in $\mathcal{M}$.

87.9   Let f and g be measurable finite-valued functions on X with $g(x) \neq 0$ for all $x \in X$. Prove that f/g is measurable.

87.10   Let f and g be measurable functions on X. Prove that the sets
$$\{x \mid f(x) \leq g(x)\} \text{ and } \{x \mid f(x) = g(x)\}$$
are measurable.

87.11   Prove that if f is a real-valued measurable function, so is $|f|^c$, $c > 0$.

87.12   Let $\mathcal{M}$ be a σ-algebra in a set X and let $E \in \mathcal{M}$. Let $\mathcal{M}_E$ be the relative σ-algebra (see Exercise 86.9).
(a)   Let $f: X \to [-\infty, \infty]$ be $\mathcal{M}$-measurable.
Prove that $f|E$ is $\mathcal{M}_E$-measurable.
(b)   Let $f: E \to [-\infty, \infty]$ be $\mathcal{M}_E$-measurable.

Define $\qquad g(x) = \begin{cases} f(x) & x \in E \\ 0 & x \in X \backslash E \end{cases}$

Prove that g is $\mathcal{M}$-measurable.

87.13*   Prove the following theorem due to F. Riesz. Let $(X, \mathcal{M}, \mu)$ be a measure space and let f and $\{f_n\}$ be measurable functions on X such that for every $\delta > 0$,
$$\lim_{n \to \infty} \mu(\{x \mid |f(x) - f_n(x)| \geq \delta\}) = 0$$
Then there exists $S \in \mathcal{M}$ with $\mu(S) = 0$ and a subsequence $\{f_{n_k}\}$ such that $\lim_{k \to \infty} f_{n_k}(x) = f(x)$ for every $x \in X \backslash S$.

87.14*   Let $(X, \mathcal{M}, \mu)$ be a measure space with $\mu(X) < \infty$. Let $\{f_n\}$ be a sequence of measurable functions on X which converges to f on X. Prove that for any $\varepsilon > 0$, there exists a set $E \in \mathcal{M}$ with $\mu(E) < \varepsilon$ such that $\{f_n\}$ converges to f uniformly on $X \backslash E$.

88. Integration on Positive Measure Spaces

We now develop a theory of integration on any positive measure space $(X, \mathcal{M}, \mu)$. The simplicity of the development and the power of the results are the real beauty of this part of mathematics.

In the following X is a set, $\mathcal{M}$ is a σ-algebra in X, and μ is a positive measure on $\mathcal{M}$.

We begin by defining the integral of a nonnegative simple measurable function on X.

Definition 88.1 Let s be a nonnegative simple measurable function in X, $s = \sum_{i=1}^{n} c_i \chi_{A_i}$, where $A_i \in \mathcal{M}$ and $c_i \geq 0$ for $i = 1, \ldots, n$. We define *the integral of s over X with respect to μ* to be

$$\int_X s \, d\mu = \sum_{i=1}^{n} c_i \mu(A_i)$$

Notice that $\int_X s \, d\mu \geq 0$.

Consider $(\mathbf{P}, \mathscr{P}(\mathbf{P}), \mu)$, where μ is counting measure. Suppose $s = \{s_n\}$ is a nonnegative simple function on $\mathbf{P}$. If $s = \sum_{i=1}^{m} c_i \chi_{A_i}$, where $c_i > 0$ for $i = 1, \ldots, m$,

$$\int_{\mathbf{P}} s \, d\mu = \sum_{i=1}^{m} c_i \mu(A_i) = \sum_{n=1}^{\infty} s_n$$

This sum is finite if and only if for some N, $s_n = 0$ for $n \geq N$.

The reasoning for our choice of the number $\int_X s \, d\mu$ as our integral of s over X is clear, since we are averaging s over the set X and weighing the various disjoint subsets of X, where s takes on different values, by means of our measure μ. We used a similar method when integrating step functions over an interval $[a, b]$ with the Riemann integral.

We use the following definition to extend Definition 88.2 to the class of all nonnegative measurable functions on X.

Definition 88.2 Let f be a nonnegative extended real-valued function on X. We define the *integral of f over X with respect to μ* to be

$$\int_X f \, d\mu = \sup \left\{ \int_X s \, d\mu \mid s \text{ is a simple measurable function on } X \text{ with } 0 \leq s \leq f \right\}$$

Since $0 = \int_X 0 \, d\mu$ is in the set on the right, $\int_X f \, d\mu$ is well defined. It is easily verified (Exercise 88.1) that Definitions 88.1 and 88.2 are consistent.

Returning again to $(\mathbf{P}, \mathscr{P}(\mathbf{P}), \mu)$, where μ is counting measure, suppose

$\{a_n\}$ is a nonnegative sequence. First, suppose $\sum a_n = \infty$. Let M be any number. Choose N such that $M < \sum_{n=1}^{N} a_n$. Define

$$s_n = \begin{cases} a_n & n \leq N \\ 0 & n > N \end{cases}$$

Then s is a simple function, $s \leq a$; so

$$M < \sum_{n=1}^{N} a_n = \int_P s \, d\mu \leq \int_P a \, d\mu$$

It follows that $\int_P a \, d\mu = \infty = \sum_{n=1}^{\infty} a_n$.

Next, suppose $\sum_{n=1}^{\infty} a_n < \infty$. If $\{s_n\}$ is a simple function with $0 \leq s \leq a$, then

$$\int_P s \, d\mu = \sum_{n=1}^{\infty} s_n \leq \sum_{n=1}^{\infty} a_n$$

It follows that

$$\int_P a \, d\mu \leq \sum_{n=1}^{\infty} a_n \tag{88.1}$$

Let $\varepsilon > 0$. Choose N such that $\sum_{n=N+1}^{\infty} a_n < \varepsilon$. Let

$$s_n = \begin{cases} a_n & n \leq N \\ 0 & n > N \end{cases}$$

Then s is a simple function, $0 \leq s \leq a$, and

$$\sum_{n=1}^{\infty} a_n = \sum_{n=1}^{N} a_n + \sum_{n=N+1}^{\infty} a_n$$

$$< \int_P s \, d\mu + \varepsilon \leq \int_P a \, d\mu + \varepsilon$$

Since $\sum_{n=1}^{\infty} a_n < \int_P s \, d\mu + \varepsilon$ for every $\varepsilon > 0$, it follows that

$$\sum_{n=1}^{\infty} a_n \leq \int_P a \, d\mu$$

This inequality combined with (88.1) shows that

$$\sum_{n=1}^{\infty} a_n = \int_P a \, d\mu$$

Definition 88.3 Let f be a measurable function on X. Let f^+ and f^- be defined as in Corollary 87.6. Since f^+ and f^- are measurable, nonnegative functions, $\int_X f^+ \, d\mu$ and $\int_X f^- \, d\mu$ are defined. If at least one of the numbers $\int_X f^+ \, d\mu$ or $\int_X f^- \, d\mu$ is finite, we define the *integral of f over X with respect* *to* μ to be

$$\int_X f\, d\mu = \int_X f^+\, d\mu - \int_X f^-\, d\mu$$

If both $\int_X f^+\, d\mu$ and $\int_X f^-\, d\mu$ are ∞, we say that $\int_X f\, d\mu$ is not defined. If both $\int_X f^+\, d\mu$ and $\int_X f^-\, d\mu$ are finite, then $\int_X f\, d\mu$ is finite, and in this case we say that *f* is *Lebesgue integrable on X with respect to* μ, and we let $\mathscr{L}(X, \mathscr{M}, \mu)$ denote the set of all such *f*.

It is easily verified (Exercise 88.2) that Definitions 88.2 and 88.3 are consistent.

Again consider $(\mathbf{P}, \mathscr{P}(\mathbf{P}), \mu)$, where μ is counting measure. Let $\{a_n\}$ be an arbitrary sequence, then $a_n^+ = p_n$ and $a_n^- = q_n$, where p_n and q_n are defined in Section 26. The proof of Theorem 26.2 shows that $\sum_{n=1}^{\infty} p_n$ and $\sum_{n=1}^{\infty} q_n$ converge if and only if $\sum_{n=1}^{\infty} a_n$ converges absolutely. Since $\int_{\mathbf{P}} a^+\, d\mu = \sum_{n=1}^{\infty} p_n$ and $\int_{\mathbf{P}} a^-\, d\mu = \sum_{n=1}^{\infty} q_n$, it follows that $\{a_n\} \in \mathscr{L}(\mathbf{P}, \mathscr{P}(\mathbf{P}), \mu)$ if and only if $\sum_{n=1}^{\infty} a_n$ converges absolutely, and in this case we have

$$\int_{\mathbf{P}} a\, d\mu = \sum_{n=1}^{\infty} a_n$$

Theorem 88.4

(i) If $f, g \in \mathscr{L}(X, \mathscr{M}, \mu)$ and $f(x) \le g(x)$ for $x \in X$, then

$$\int_X f\, d\mu \le \int_X g\, d\mu$$

(ii) If *f* is measurable, $\mu(X) < \infty$, and for some $a, b \in \mathbf{R}$ we have $a \le f(x) \le b$ for $x \in X$, then

$$a\mu(X) \le \int_X f\, d\mu \le b\mu(X)$$

(iii) If *f* is measurable and bounded on *X* and if $\mu(X) < \infty$, then $f \in \mathscr{L}(X, \mathscr{M}, \mu)$.

(iv) If $f \in \mathscr{L}(X, \mathscr{M}, \mu)$ and $c \in \mathbf{R}$, then $cf \in \mathscr{L}(X, \mathscr{M}, \mu)$ and

$$\int_X cf\, d\mu = c \int_X f\, d\mu$$

(v) If $\mu(X) = 0$, then every measurable function *f* on *X* is in $\mathscr{L}(X, \mathscr{M}, \mu)$ and

$$\int_X f\, d\mu = 0$$

Proof. We prove (i) only. The other parts are proved similarly and are left as exercises (see Exercise 88.3).

Suppose $0 \le f(x) \le g(x)$ for $x \in X$. If *s* is a measurable simple function with $0 \le s \le f$, then also $0 \le s \le g$; hence

$$\int_X s \, d\mu \le \int_X g \, d\mu$$

Thus $\int_X g \, d\mu$ is an upper bound for the set of numbers

$\int_X s \, d\mu$ with $0 \le s \le f$. It follows that $\int_X f \, d\mu \le \int_X g \, d\mu$.

Let $f, g \in \mathscr{L}(X, \mathscr{M}, \mu)$ with $f(x) \le g(x)$ for $x \in X$. It is easily verified that $f^+ \le g^+$ and $f^- \ge g^-$. By the special case above

$$\int_X f^+ \, d\mu \le \int_X g^+ \, d\mu \qquad \int_X f^- \, d\mu \ge \int_X g^- \, d\mu$$

Therefore

$$\int_X f \, d\mu = \int_X f^+ \, d\mu - \int_X f^- \, d\mu \le \int_X g^+ \, d\mu - \int_X g^- \, d\mu = \int_X g \, d\mu \quad \blacksquare$$

If $f \in \mathscr{L}(X, \mathscr{M}, \mu)$, we sometimes wish to integrate f on a set $E \in \mathscr{M}$. This integral, denoted $\int_E f \, d\mu$, is defined to be $\int_X f\chi_E \, d\mu$. It is easy to show (see Exercise 88.4) that if $\int_X f \, d\mu$ is defined, then also $\int_X f\chi_E \, d\mu$ is defined so that $\int_E f \, d\mu$ is defined for each set $E \in \mathscr{M}$.

Our immediate goal is to prove the Lebesgue monotone convergence theorem (Theorem 88.6).

The proofs of other important theorems such as the linearity of the integral (Theorem 88.11) and other powerful convergence theorems (Theorems 88.13, 88.14, and 88.15) will depend on Theorem 88.6. First we must establish a lemma.

Lemma 88.5 Let s be a nonnegative measurable simple function. Let $\{E_n\}$ be a sequence of sets in $\mathscr{M}$ with $E_n \subset E_{n+1}$ for $n = 1, 2 \ldots$ and $\bigcup_{n=1}^{\infty} E_n = X$. Then

$$\lim_{n \to \infty} \int_{E_n} s \, d\mu = \int_X s \, d\mu$$

Proof. We show that $\lambda(E) = \int_E s \, d\mu$ defines a positive measure on $\mathscr{M}$. Since $\lambda(\varnothing) = 0$, we need only verify the countable additivity of λ.

Suppose $s = \sum_{i=1}^{m} c_i\chi_{A_i}$. For $E \in \mathscr{M}$ we have

$$\int_E s \, d\mu = \int_X s\chi_E \, d\mu = \int_X \left(\sum_{i=1}^{m} c_i\chi_{A_i} \right)\chi_E \, d\mu$$

$$= \int_X \left(\sum_{i=1}^{m} c_i\chi_{A_i \cap E} \right) d\mu = \sum_{i=1}^{m} c_i\mu(A_i \cap E) \cdot$$

Let $\{E_i\}$ be a pairwise disjoint sequence in $\mathscr{M}$. Letting $E = \bigcup_{i=1}^{\infty} E_i$, we have

$$
\begin{aligned}
\lambda(E) &= \int_E s \, d\mu = \sum_{i=1}^{m} c_i \mu(A_i \cap E) \\
&= \sum_{i=1}^{m} c_i \left[\mu \left(A_i \cap \left(\bigcup_{n=1}^{\infty} E_n \right) \right) \right] \\
&= \sum_{i=1}^{m} c_i \left[\sum_{n=1}^{\infty} \mu(A_i \cap E_n) \right] \\
&= \sum_{n=1}^{\infty} \left[\sum_{i=1}^{m} c_i \mu(A_i \cap E_n) \right] \\
&= \sum_{n=1}^{\infty} \int_{E_n} s \, d\mu = \sum_{n=1}^{\infty} \lambda(E_n)
\end{aligned}
$$

Thus λ is a positive measure on $\mathscr{M}$.

Let $\{E_n\}$ be a sequence in $\mathscr{M}$ with $E_n \subset E_{n+1}$ for $n = 1, 2, \ldots$ and $\bigcup_{n=1}^{\infty} E_n = X$. By Theorem 86.5(v),

$$
\lim_{n \to \infty} \int_{E_n} s \, d\mu = \lim_{n \to \infty} \lambda(E_n) = \lambda \left(\bigcup_{n=1}^{\infty} E_n \right) = \lambda(X) = \int_X s \, d\mu \qquad \blacksquare
$$

Theorem 88.6 (Lebesgue Monotone Convergence Theorem) Let $\{f_n\}$ be a sequence of nonnegative measurable functions on X such that $f_n(x) \leq f_{n+1}(x)$ for $n = 1, 2, \ldots$ and $x \in X$. Let $f(x) = \lim_{n \to \infty} f_n(x)$. Then f is measurable on X and

$$
\lim_{n \to \infty} \int_X f_n \, d\mu = \int_X f \, d\mu
$$

Proof. By Corollary 87.6(iii), f is measurable on X. Since

$$
\int_X f_n \, d\mu \leq \int_X f_{n+1} \, d\mu
$$

$\lim_{n \to \infty} \int_X f_n \, d\mu$ exists in $[0, \infty]$. Denote this limit by L.

Since $f_n \leq f$ for $n = 1, 2, \ldots$, we have $\int_X f_n \leq \int_X f$ for $n = 1, 2, \ldots$. Letting $n \to \infty$, we obtain

$$
L \leq \int_X f \tag{88.2}
$$

Let $t \in (0, 1)$ and let s be a simple measurable function with $0 \leq s \leq f$. Set

$$
E_n = \{x \in X \mid f_n(x) \geq ts(x)\}
$$

Then $E_n \subset E_{n+1}$ for $n = 1, 2, \ldots$ and $X = \bigcup_{n=1}^{\infty} E_n$. By Lemma 88.5,

$$\lim_{n \to \infty} \int_{E_n} ts \, d\mu = \int_X ts \, d\mu$$

Now

$$\int_X f_n \, d\mu \geq \int_{E_n} f_n \, d\mu \geq \int_{E_n} ts \, d\mu$$

and letting $n \to \infty$, we obtain

$$L = \lim_{n \to \infty} \int_X f_n \, d\mu \geq \lim_{n \to \infty} \int_{E_n} ts \, d\mu = \int_X ts \, d\mu = t \int_X s \, d\mu$$

Letting $t \to 1$, we obtain $L \geq \int_X s \, d\mu$. By the definition of $\int_X f \, d\mu$, we have

$$L \geq \int_X f \, d\mu$$

which, together with inequality (88.2), establishes the theorem. ∎

As an application of Theorem 88.6, let $\{a_{n,i}\}$ be a nonnegative double sequence. Let $f_n(m) = \sum_{i=1}^{n} a_{m,i}$. Consider the measure space $(\mathbf{P}, \mathscr{P}(\mathbf{P}), \mu)$, where μ is counting measure. Then $\{f_n\}$ is an increasing sequence of nonnegative functions on $\mathbf{P}$. Also, $f(m) = \lim_{n \to \infty} f_n(m) = \sum_{i=1}^{\infty} a_{m,i}$. By Theorem 88.6,

$$\lim_{n \to \infty} \int_{\mathbf{P}} f_n \, d\mu = \int_{\mathbf{P}} f \, d\mu \tag{88.3}$$

Now $\int_{\mathbf{P}} f \, d\mu = \sum_{m=1}^{\infty} \left(\sum_{i=1}^{\infty} a_{m,i} \right)$. Also,

$$\int_{\mathbf{P}} f_n \, d\mu = \sum_{m=1}^{\infty} \left(\sum_{i=1}^{n} a_{m,i} \right) = \sum_{i=1}^{n} \left(\sum_{m=1}^{\infty} a_{m,i} \right)$$

Thus equation (88.3) may be rewritten

$$\sum_{i=1}^{\infty} \left(\sum_{m=1}^{\infty} a_{m,i} \right) = \sum_{m=1}^{\infty} \left(\sum_{i=1}^{\infty} a_{m,i} \right)$$

We earlier established this result as Lemma 29.3.

Before establishing our next major result, which is essentially a generalization of Lemma 88.5, we must establish a special case of the linearity of the integral.

Lemma 88.7 If f and g are nonnegative measurable functions on X, then

$$\int_X (f + g) \, d\mu = \int_X f \, d\mu + \int_X g \, d\mu$$

Proof.   First suppose f and g are measurable simple functions. We may write

$$f = \sum_{i=1}^{n} a_i \chi_{A_i} \qquad g = \sum_{j=1}^{m} b_j \chi_{B_j}$$

where $\{A_i\}$ and $\{B_j\}$ are pairwise disjoint and $\cup A_i = X = \cup B_j$. Then $f + g = \sum_{i=1}^{n} \sum_{j=1}^{m} (a_i + b_j)\chi_{A_i \cap B_j}$. Therefore

$$\int_X (f + g)\, d\mu = \sum_{i=1}^{n} \sum_{j=1}^{m} (a_i + b_j)\mu(A_i \cap B_j)$$

$$= \sum_{i=1}^{n} a_i \sum_{j=1}^{m} \mu(A_i \cap B_j) + \sum_{j=1}^{m} b_j \sum_{i=1}^{m} \mu(A_i \cap B_j)$$

$$= \sum_{i=1}^{n} a_i \mu(A_i) + \sum_{j=1}^{m} b_j \mu(B_j) = \int_X f\, d\mu + \int_X g\, d\mu$$

Now suppose that f and g are nonnegative measurable functions on X. By Theorem 87.10 there exist sequences of simple measurable functions, $0 \leq s_1 \leq s_2 \leq \cdots$ and $0 \leq t_1 \leq t_2 \leq \cdots$ such that $\lim_{n\to\infty} s_n(x) = f(x)$ and $\lim_{n\to\infty} t_n(x) = g(x)$ for all $x \in X$. Clearly $\lim_{n\to\infty} (s_n + t_n)(x) = f(x) + g(x)$ for all $x \in X$; so by Theorem 88.6 and the special case above

$$\int_X (f + g)\, d\mu = \int_X \lim_{n\to\infty} (s_n + t_n)\, d\mu = \lim_{n\to\infty} \int_X (s_n + t_n)\, d\mu$$

$$= \lim_{n\to\infty} \left(\int_X s_n\, d\mu + \int_X t_n\, d\mu \right) = \lim_{n\to\infty} \int_X s_n\, d\mu + \lim_{n\to\infty} \int_X t_n\, d\mu$$

$$= \int_X (\lim_{n\to\infty} s_n)\, d\mu + \int_X (\lim_{n\to\infty} t_n)\, d\mu = \int_X f\, d\mu + \int_X g\, d\mu \qquad \blacksquare$$

Theorem 88.8   Let f be a nonnegative measurable function on X. Then

$$\lambda(E) = \int_E f\, d\mu$$

defines a positive measure on $\mathcal{M}$.

If $f \in \mathcal{L}(X, \mathcal{M}, \mu)$, $\lambda(E) = \int_E f\, d\mu$ defines a countably additive set function on $\mathcal{M}$.

Proof.   Let f be a nonnegative measurable function on X. Since $\lambda(\emptyset) = 0$, we need only verify the countable additivity of λ.

Let $\{E_i\}$ be a pairwise disjoint sequence in $\mathcal{M}$ and let $E = \bigcup_{i=1}^{\infty} E_i$. Define

$$f_n = \sum_{i=1}^{n} f \chi_{E_i}$$

Then $\{f_n\}$ is a sequence of nonnegative measurable functions on X such that

$f_n \leq f_{n+1}$ for $n = 1, 2, \ldots$. By Theorem 88.6, since $\lim_{n \to \infty} f_n = f\chi_E$,

$$\lim_{n \to \infty} \int_X f_n \, d\mu = \int_X f\chi_E \, d\mu$$

By Lemma 88.7

$$\int_X f_n \, d\mu = \int_X \left(\sum_{i=1}^n f\chi_{E_i} \right) d\mu = \sum_{i=1}^n \int_X f\chi_{E_i} \, d\mu = \sum_{i=1}^n \lambda(E_i)$$

Thus $\qquad \lambda(E) = \int_X f\chi_E \, d\mu = \lim_{n \to \infty} \sum_{i=1}^n \lambda(E_i) = \sum_{i=1}^\infty \lambda(E_i)$

If $f \in \mathscr{L}(X, \mathscr{M}, \mu)$, one shows that $\lambda(E) = \int_E f \, d\mu$ defines a countably additive set function on $\mathscr{M}$ by applying the result just established to f^+ and f^-. ∎

With modest additional hypotheses, a converse of Theorem 88.8 can be proved. This converse is a special case of a more general result known as the Lebesgue-Radon-Nikodym Theorem (see Hewitt and Stromberg, 1965, p. 315). One can show that if μ and λ are positive measures on the σ-algebra $\mathscr{M}$ satisfying $\mu(X) < \infty$ and $\lambda(X) < \infty$ and further $\mu(E) = 0$ implies $\lambda(E) = 0$, then there exists a nonnegative measurable function f on X such that

$$\lambda(E) = \int_E f \, d\mu$$

for every $E \in \mathscr{M}$.

In Section 57 we said that a property P holds almost everywhere if the set of points on the line where P does not hold is of measure zero. We now generalize this definition to an arbitrary measure space.

Definition 88.9 If a property P holds for every $x \in E \backslash A$ and if $\mu(A) = 0$ (where E, $A \in \mathscr{M}$, of course), then we say that P holds for *almost all $x \in E$ with respect to μ* or that P holds *μ-almost everywhere on E* or that P holds *μ-a.e. on E*.

An important example is given by the next theorem.

Theorem 88.10 If $f, g \in \mathscr{L}(X, \mathscr{M}, \mu)$ and $f = g$ μ-a.e. on X, then

$$\int_X f \, d\mu = \int_X g \, d\mu$$

Proof. Let

$$E = \{x \in X \mid f(x) \neq g(x)\}$$

By definition, $\mu(E) = 0$. Now

$$\int_{X\setminus E} f\, d\mu = \int_{X\setminus E} g\, d\mu \qquad \int_E f\, d\mu = 0 = \int_E g\, d\mu$$

Therefore, by Theorem 88.8,

$$\int_X f\, d\mu = \int_{X\setminus E} f\, d\mu + \int_E f\, d\mu = \int_{X\setminus E} g\, d\mu + \int_E g\, d\mu = \int_X g\, d\mu \quad \blacksquare$$

We next show that the integral is linear. Although it may appear that this fact should follow immediately from the definition it does not. Our proof depends directly on Theorem 88.4 and indirectly on Theorems 87.10 and 88.6.

Theorem 88.11 If $f, g \in \mathscr{L}(X, \mathscr{M}, \mu)$, then $f + g \in \mathscr{L}(X, \mathscr{M}, \mu)$ and

$$\int_X (f + g)\, d\mu = \int_X f\, d\mu + \int_X g\, d\mu$$

Proof. This result has been established (Lemma 88.7) for f and g nonnegative. Suppose $f \geq 0$ and $g \leq 0$. Let $A = \{x \in X \mid (f + g)(x) \geq 0\}$ and $B = \{x \in X \mid (f + g)(x) < 0\}$. Then $f + g, f$, and $-g$ are nonnegative on A. Therefore, by Theorem 88.4(iv) and Lemma 88.7,

$$\int_A f\, d\mu = \int_A (f + g)\, d\mu + \int_A (-g)\, d\mu = \int_A (f + g)\, d\mu - \int_A g\, d\mu \quad (88.4)$$

Similarly, $-(f + g), f$, and $-g$ are nonnegative on B. So

$$\int_B (-g)\, d\mu = \int_B f\, d\mu + \int_B -(f + g)\, d\mu$$

or

$$\int_B g\, d\mu = \int_B (f + g)\, d\mu - \int_B f\, d\mu \quad (88.5)$$

Adding equations (88.4) and (88.5), we get

$$\int_X (f + g)\, d\mu = \int_X f\, d\mu + \int_X g\, d\mu$$

Now let f and g be any functions in $\mathscr{L}(X, \mathscr{M}, \mu)$. Then $f = f^+ - f^-$ and $g = g^+ - g^-$ with f^+, f^-, g^+, and g^- in $\mathscr{L}(X, \mathscr{M}, \mu)$. By the above result and Lemma 88.7

$$\int_X (f + g)\, d\mu = \int_X (f^+ + g^+) + (-f^- - g^-)$$

$$= \int_X (f^+ + g^+)\, d\mu + \int_X -(f^- + g^-)\, d\mu$$

$$= \int_X (f^+ + g^+)\, d\mu - \int_X (f^- + g^-)\, d\mu$$

$$= \int_X f^+\, d\mu + \int_X g^+\, d\mu - \int_X f^-\, d\mu - \int_X g^-\, d\mu$$

$$= \int_X f^+\, d\mu + \int_X -f^-\, d\mu + \int_X g^+\, d\mu + \int_X -g^-\, d\mu$$

$$= \int_X (f^+ - f^-)\, d\mu + \int_X (g^+ - g^-)\, d\mu$$

$$= \int_X f\, d\mu + \int_X g\, d\mu \qquad\qquad\blacksquare$$

Theorem 88.11 together with Theorem 88.4(iv) shows that $\mathscr{L}(X, \mathscr{M}, \mu)$ is a real vector space and that

$$f \to \int_X f\, d\mu$$

is a linear functional on $\mathscr{L}(X, \mathscr{M}, \mu)$.

Theorem 88.12 $f \in \mathscr{L}(X, \mathscr{M}, \mu)$ if and only if $|f| \in \mathscr{L}(X, \mathscr{M}, \mu)$. In this case

$$\left| \int_X f\, d\mu \right| \le \int_X |f|\, d\mu$$

Proof. Suppose $f \in \mathscr{L}(X, \mathscr{M}, \mu)$. By definition $f^+, f^- \in \mathscr{L}(X, \mathscr{M}, \mu)$. By Theorem 88.11, $|f| = f^+ + f^- \in \mathscr{L}(X, \mathscr{M}, \mu)$. If $|f| \in \mathscr{L}(X, \mathscr{M}, \mu)$, since $f^+ \le |f|$, $\int_X f^+\, d\mu \le \int_X |f|\, d\mu < \infty$; so $f^+ \in \mathscr{L}(X, \mathscr{M}, \mu)$. Similarly, $f^- \in \mathscr{L}(X, \mathscr{M}, \mu)$. Therefore $f \in \mathscr{L}(X, \mathscr{M}, \mu)$.

Now $|\int_X f\, d\mu| = c \int_X f\, d\mu$ for $c = 1$ or $c = -1$. Since $cf \le |f|$, $|\int_X f\, d\mu| = c \int_X f\, d\mu = \int_X cf\, d\mu \le \int_X |f|\, d\mu$. $\blacksquare$

We next develop some powerful convergence theorems. That such theorems can be proved is a demonstration of the importance and usefulness of the Lebesgue integral.

Theorem 88.13 If $\{f_n\}_{n=1}^{\infty}$ is a sequence of nonnegative measurable functions and $f(x) = \sum_{n=1}^{\infty} f_n(x)$ for all $x \in X$, then

$$\int_X f\, d\mu = \sum_{n=1}^{\infty} \int_X f_n\, d\mu$$

Proof. The partial sums $s_m(x) = \sum_{n=1}^{m} f_n(x)$ form an increasing sequence of nonnegative measurable functions; so by Theorems 88.6 and 88.7 we have

$$\int_X f \, d\mu = \int_X \lim_{m \to \infty} s_m \, d\mu = \lim_{m \to \infty} \int_X s_m \, d\mu = \lim_{m \to \infty} \int_X \sum_{n=1}^{m} f_n \, d\mu$$

$$= \lim_{m \to \infty} \sum_{n=1}^{m} \int_X f_n \, d\mu = \sum_{n=1}^{\infty} \int_X f_n \, d\mu \qquad \blacksquare$$

Theorem 88.14 (Fatou's Lemma)   If $\{f_n\}_{n=1}^{\infty}$ is a sequence of nonnegative measurable functions, then

$$\int_X (\liminf_{n \to \infty} f_n) \, d\mu \leq \liminf_{n \to \infty} \int_X f_n \, d\mu$$

Proof.   By Theorem 87.5, $\liminf_{n \to \infty} f_n$ is measurable. For $n = 1, 2, \ldots$ and for each $x \in X$, let

$$p_n(x) = \text{glb} \{f_i(x) \mid i \geq n\}$$

Then p_n is measurable, $0 \leq p_1 \leq p_2 \leq \cdots$ on X, $p_n(x) \leq f_n(x)$ for all $x \in X$, and $\lim_{n \to \infty} p_n(x) = \liminf_{n \to \infty} f_n(x)$ for all $x \in X$. Using Theorem 88.10 and the above equations, we get

$$\int_X \liminf_{n \to \infty} f_n \, d\mu = \int_X \lim_{n \to \infty} p_n \, d\mu = \lim_{n \to \infty} \int_X p_n \, d\mu \leq \liminf_{n \to \infty} \int f_n \, d\mu \qquad \blacksquare$$

Now we state and prove one of the most powerful theorems of Lebesgue integration theory.

Theorem 88.15 (Lebesgue's Dominated Convergence Theorem)   Let $\{f_n\}_{n=1}^{\infty}$ be a sequence of measurable functions such that $\lim_{n \to \infty} f_n(x)$ exists for all $x \in X$. If there exists a function $g \in \mathcal{L}(X, \mathcal{M}, \mu)$ such that

$$|f_n(x)| \leq g(x) \qquad \text{for all } x \in X \text{ and all } n = 1, 2, \ldots$$

then $\lim_{n \to \infty} \int_X f_n \, d\mu = \int_X \lim_{n \to \infty} f_n \, d\mu$.

Proof.   Set $f(x) = \lim_{n \to \infty} f_n(x)$ for $x \in X$. Since $|f_n(x)| \leq g(x)$ for $x \in X$ and $n = 1, 2, \ldots$, we have $|f(x)| \leq g(x)$. By Theorem 88.4(i) $\int_X |f| \leq \int_X g$. Thus $|f| \in \mathcal{L}(X, \mathcal{M}, \mu)$. By Theorem 88.12 $f \in \mathcal{L}(X, \mathcal{M}, \mu)$ also. Similarly, $f_n \in \mathcal{L}(X, \mathcal{M}, \mu)$ for $n = 1, 2, \ldots$.

Now $f_n + g \geq 0$; so we may apply Fatou's lemma to obtain

$$\int_X \liminf_{n \to \infty} (f_n + g) \, d\mu \leq \liminf_{n \to \infty} \int_X (f_n + g) \, d\mu$$

But $$\liminf_{n \to \infty} (f_n + g) = \lim_{n \to \infty} f_n + g$$

so we have $$\int_X (\lim_{n \to \infty} f_n + g) \, d\mu \leq \liminf_{n \to \infty} \left(\int_X f_n \, d\mu + \int_X g \, d\mu \right)$$

This gives
$$\int_X \lim_{n\to\infty} f_n \, d\mu \le \liminf_{n\to\infty} \int_X f_n \, d\mu \qquad (88.6)$$

Since $g - f_n \ge 0$ on X, we see in a similar way that

$$\int_X (g - \lim_{n\to\infty} f_n) \, d\mu \le \liminf_{n\to\infty} \left(\int_X (g - f_n) \, d\mu \right)$$

so that
$$-\int_X (\lim_{n\to\infty} f_n) \, d\mu \le \liminf_{n\to\infty} \left(-\int_X f_n \, d\mu \right)$$

Therefore
$$\int_X (\lim_{n\to\infty} f_n) \, d\mu \ge \limsup_{n\to\infty} \int_X f_n \, d\mu \qquad (88.7)$$

Equations (88.6) and (88.7) combine to give

$$\limsup_{n\to\infty} \int_X f_n \, d\mu \le \int_X \lim_{n\to\infty} f_n \, d\mu \le \liminf_{n\to\infty} \int f_n \, d\mu$$

which implies $\int_X \lim_{n\to\infty} f_n \, d\mu = \lim_{n\to\infty} \int_X f_n \, d\mu$. ∎

We conclude this section by proving the following special result.

Theorem 88.16 Suppose f is measurable and $f(x) \ge 0$ for all $x \in X$. If $\int_X f \, d\mu = 0$, then $f = 0$ μ-a.e. on X.

Proof. If suffices to show that for the set $A = \{x \in X \mid f(x) \ne 0\}$, we have $\mu(A) = 0$. For then if $B = \{x \in X \mid f(x) = 0\}$, we would have $B' = A$; so $f = 0$ would be true μ-almost everywhere on X.

Let $A_n = \{x \in X \mid f(x) \ge 1/n\}$ for $n = 1, 2, 3, \ldots$. Each $A_n \in \mathcal{M}$ since f is measurable. Suppose $\mu(A_n) > 0$, then

$$\int_X f \, d\mu = \int_{A_n'} f \, d\mu + \int_{A_n} f \, d\mu \ge \int_{A_n'} f \, d\mu + \int_{A_n} \frac{1}{n} \, d\mu$$

But $f(x) \ge 0$ for all $x \in X$ implies $\int_{A_n'} f \, d\mu \ge 0$ so we have $\int_X f \, d\mu \ge \int_{A_n} (1/n) d\mu = (1/n) \mu(A_n) > 0$, a contradiction. Thus $\mu(A_n) = 0$ for $n = 1, 2, \ldots$. Now $A = \bigcup_{n=1}^\infty A_n$; so $0 \le \mu(A) = \mu(\bigcup_{n=1}^\infty A_n) \le \sum_{n=1}^\infty \mu(A_n) = 0$, and therefore $\mu(A) = 0$. ∎

Exercises

In these exercises, unless specified otherwise, $(X, \mathcal{M}, \mu)$ is a positive measure space.

88.1 Give the details of the argument that Definitions 88.1 and 88.2 are consistent for nonnegative simple measurable functions.

88.2 Show that Definitions 88.2 and 88.3 are consistent for nonnegative measurable functions.

88.3 Prove Theorem 88.4, parts (ii), (iii), (iv), and (v).

88.4 Prove that if $\int_X f \, d\mu$ is defined, then also $\int_X f \chi_E \, d\mu$ is defined for each $E \in \mathcal{M}$. Prove also that if $f \in \mathcal{L}(X, \mathcal{M}, \mu)$, then $f\chi_E \in \mathcal{L}(X, \mathcal{M}, \mu)$ for each $E \in \mathcal{M}$.

88.5 Let $(X, \mathcal{M}, \mu)$ be a positive measure space. Let $E \in \mathcal{M}$. Let $(E, \mathcal{M}_E, \lambda)$, where $\lambda(A) = \mu(A \cap E)$, be the relative measure space (see Exercise 86.9).

(a) Let $f \in \mathcal{L}(X, \mathcal{M}, \mu)$. Prove that $f | E \in \mathcal{L}(E, \mathcal{M}_E, \lambda)$ and

$$\int_E f | E \, d\lambda = \int_X f \chi_E \, d\mu$$

(b) Let $f \in \mathcal{L}(E, \mathcal{M}_E, \lambda)$. Define

$$g(x) = \begin{cases} f(x) & x \in E \\ 0 & x \in X \backslash E \end{cases}$$

Prove that $g \in \mathcal{L}(X, \mathcal{M}, \mu)$ and

$$\int_X g \, d\mu = \int_X g \chi_E \, d\mu = \int_E f \, d\lambda$$

88.6 Let $(X, \mathcal{M}, \mu)$ be a positive measure space with $\mu(X) < \infty$. Suppose $\lim_{n \to \infty} f_n(x) = f(x)$ for every $x \in X$ and for some M, $|f_n(x)| < M$ for $n = 1, 2, \ldots$ and for every $x \in X$. Prove that

$$\lim_{n \to \infty} \int_X f_n \, d\mu = \int_X f \, d\mu$$

88.7 Show that Theorems 88.6, 88.13, and 88.15 remain valid if *for all x* is replaced by *for almost all x with respect μ*.

88.8 Let $(X, \mathcal{M}, \mu)$ be any positive measure space. Suppose $f \in \mathcal{L}(X, \mathcal{M}, \mu)$. Prove that $f(x)$ must be finite μ-almost everywhere on X.

88.9 Let $X = \{1, 2, 3, 4, 5\}$, $\mathcal{M} = \mathcal{P}(X)$, and μ be counting measure. Let $E = \{1, 2\}$ and define $f_n = \chi_E$ if n is odd, $f_n = 1 - \chi_E$ if n is even. What is the relevance of this example to Fatou's lemma?

88.10 Let $(X, \mathcal{M}, \mu)$ be any positive measure space. Suppose $\mu(X) < \infty$ and $\{f_n\}_{n=1}^{\infty}$ is a sequence of bounded real measurable functions on X and $f_n \to f$ uniformly on X. Prove that

$$\lim_{n \to \infty} \int_X f_n \, d\mu = \int_X f \, d\mu$$

88.11 Show that the hypothesis "$\mu(X) < \infty$" cannot be omitted from Exercise 88.10.

88.12 Let $(X, \mathcal{M}, \mu)$ be any positive measure space. Suppose $f \in \mathcal{L}(X, \mathcal{M}, \mu)$. Prove that for every $\varepsilon > 0$ there exists a $\delta > 0$ such that

$$\int_E |f| \, d\mu < \varepsilon \qquad \text{whenever } \mu(E) < \delta \text{ and } E \in \mathcal{M}.$$

This property is sometimes referred to as the *uniform continuity of the integral.*

88.13 Let $(X, \mathcal{M}, \mu)$ be any positive measure space. Suppose $f_n: X \to [0, \infty]$ is measurable for $n = 1, 2, \ldots$; $f_1 \geq f_2 \geq f_3 \geq \cdots \geq 0$; $\lim_{n \to \infty} f_n(x) = f(x)$

for every x in X; and $f_1 \in \mathscr{L}(X, \mathscr{M}, \mu)$. Prove that

$$\lim_{n \to \infty} \int_X f_n\, f\mu = \int_X f\, d\mu$$

and show that this conclusion may fail if the hypothesis "$f_1 \in \mathscr{L}(X, \mathscr{M}, \mu)$" is omitted.

88.14 Suppose f is measurable and $f(x) = 0$ for almost all $x \in X$ with respect to μ. Prove that $\int_X f\, d\mu = 0$.

88.15 Suppose $\{f_n\}$ is a sequence of measurable functions, $\sum_{n=1}^{\infty} |f_n(x)|$ converges for all $x \in X$, and $\sum_{n=1}^{\infty} |f_n(x)| \in \mathscr{L}(X, \mathscr{M}, \mu)$. Prove that $\sum_{n=1}^{\infty} f_n(x)$ converges for all $x \in X$, $\sum_{n=1}^{\infty} f_n(x) \in \mathscr{L}(X, \mathscr{M}, \mu)$ and

$$\int_X \left(\sum_{n=1}^{\infty} f_n \right) d\mu = \sum_{n=1}^{\infty} \int_X f_n\, d\mu$$

88.16 Consider $(\mathbf{P}, \mathscr{P}(\mathbf{P}), \mu)$, where μ is counting measure. Let λ be any positive measure on $\mathscr{P}(\mathbf{P})$. Prove that there exists a nonnegative function f on $\mathbf{P}$ such that

$$\lambda(E) = \int_E f\, d\mu$$

for every $E \in \mathscr{P}(\mathbf{P})$.

89. Lebesgue Measure on R

We have developed an integral for certain measurable functions on any measure space $(X, \mathscr{M}, \mu)$. We have shown that some very powerful integral-limit theorems always hold for this integral (Theorems 88.6, 88.13, 88.14, 88.15). What does this have to do with the Riemann integral? The satisfying fact is that there is a σ-algebra $\mathscr{M}$ on $[a, b]$ and a positive measure m on $\mathscr{M}$ such that for *all* Riemann integrable functions f, we have $f \in \mathscr{L}([a, b], \mathscr{M}, m)$ and

$$\int_{[a,b]} f\, dm = \int_a^b f(x)\, dx$$

However, much more will be true. There will be many more functions than the Riemann integrable functions which we can integrate.

For example, returning to the function

$$f(x) = \begin{cases} 1 & \text{for } x \text{ a rational number in } [0, 1] \\ 0 & \text{for all other } x \in [0, 1] \end{cases}$$

discussed in the introduction to this chapter, we will find that we can integrate f. Recall that f was the pointwise limit of the functions

$$f_n(x) = \begin{cases} 1 & \text{for } x = a_i,\ i = 1, 2, \ldots, n \\ 0 & \text{for all other } x \in [0, 1] \end{cases}$$

where $\{a_n\}$ is the set of rational numbers in $[0, 1]$. Since each function f_n is m-measurable, Theorem 88.6 tells us f is m-measurable and

$$0 = \lim_{n \to \infty} \int_0^1 f_n(x)\, dx = \lim_{n \to \infty} \int_{[0,1]} f_n\, dm$$

$$= \int_{[0,1]} (\lim_{n \to \infty} f_n)\, dm = \int_{[0,1]} f\, dm$$

Since f is *not* Riemann integrable, we see that we have extended the class of integrable functions and obtained a much more useful class of integrable functions.

We will actually construct a σ-algebra of **R** and a measure m on this σ-algebra such that, in a very natural way, we may restrict m and the σ-algebra to any closed interval $[a, b]$ and get the desired results.

Definition 89.1 Let U be any nonempty open subset of **R**. U can be written uniquely as a countable union of pairwise disjoint open intervals (see Exercise 39.8); so

$$U = \bigcup_{n \in J} I_n \qquad \text{with } I_n = (a_n, b_n) \text{ for some } a_n, b_n \in \tilde{\mathbf{R}}$$

(We recall that intervals such as $(1, \infty)$ and $(-\infty, \infty)$ are open in **R**.) As a convention we will always assume these open intervals to be indexed by $J = \{1, 2, \ldots, n\}$ if the collection is finite or $J = \{1, 2, 3, 4, \ldots\}$ if the collection of intervals making up U is infinite.

We define

$$m^*(U) = \sum_{n \in J} (b_n - a_n) \tag{89.1}$$

Since $b_n - a_n \in [0, \infty]$ for each $n \in J$ and J is countable, there is no ambiguity in the order of the summation of equation (89.1). (See Theorem 29.7.)

Examples.   (a) If $U = (1, 2)$, then $J = \{1\}$, $I_1 = (1, 2) = (a_1, b_1)$ and

$$m^*(U) = \sum_{n \in J} (b_n - a_n) = b_1 - a_1 = 2 - 1 = 1$$

(b) If $V = (-\infty, \infty)$, then $J = \{1\}$, $I_1 = (-\infty, \infty) = (a_1, b_1)$ and

$$m^*(V) = \sum_{n \in J} (b_n - a_n) = b_1 - a_1 = \infty - (-\infty) = \infty$$

(c) $W = \bigcup_{n=0}^{\infty} (n - 1/2^n, n + 1/2^n)$, we have

$$W = (-1, 1) \cup (1 - \tfrac{1}{2}, 1 + \tfrac{1}{2}) \cup (2 - \tfrac{1}{4}, 2 + \tfrac{1}{4}) \cup \cdots$$
$$= (-1, \tfrac{3}{2}) \cup (2 - \tfrac{1}{4}, 2 + \tfrac{1}{4}) \cup (3 - \tfrac{1}{9}, 3 + \tfrac{1}{9}) \cup \cdots$$

as the unique pairwise disjoint union of open intervals; so $J = \{1, 2, 3, \ldots\}$

with $J_1 = (-1, \frac{3}{2})$, and $J_n = (n - 1/2^n, n + 1/2^n)$ for $n = 2, 3, 4, \ldots$. Therefore

$$m^*(W) = \sum_{n \in J} (b_n - a_n) = \left(\frac{3}{2} - (-1)\right) + \sum_{n=2}^{\infty} \left(n + \frac{1}{2^n}\right) - \left(n - \frac{1}{2^n}\right)$$

$$= \frac{5}{2} + \sum_{n=2}^{\infty} \frac{2}{2^n} = \frac{5}{2} + 2 \sum_{n=2}^{\infty} \frac{1}{2^n} = \frac{5}{2} + 2\left(\frac{1}{2}\right) = \frac{7}{2}$$

It turns out that m^* is countably additive on the collection of all open intervals of **R**, but unfortunately this collection is not a σ-algebra. We now extend the definition of m^* to a function on $\mathscr{P}(\mathbf{R})$.

Definition 89.2 For each $S \subset \mathbf{R}$ we define

$$m(S) = \inf \left\{ \sum_{n=1}^{\infty} m^*(I_n) \middle| \text{ where each } I_n \text{ is an open interval} \right.$$

$$\left. \text{or } \varnothing \text{ in } \mathbf{R} \text{ and } S \subset \bigcup_{n=1}^{\infty} I_n \right\}$$

Notice that there is no stipulation that the I_n be pairwise disjoint and also that m is a function from $\mathscr{P}(\mathbf{R})$ into $[0, \infty]$.

$\mathscr{P}(\mathbf{R})$ is a σ-algebra in **R** and $m: \mathscr{P}(\mathbf{R}) \to [0, \infty]$, however, m is not countably additive on $\mathscr{P}(\mathbf{R})$. This is the m we want; however, the σ-algebra on which it is a measure lies somewhere between the collection of all open sets in **R** and the σ-algebra $\mathscr{P}(\mathbf{R})$.

Theorem 89.3 If V is open in **R**, then $m(V) = m^*(V)$.

Proof. If $V = \varnothing$, it is clear that $m(V) = m^*(V) = 0$. Suppose $V \neq \varnothing$. By Definition 89.1, $V = \bigcup_{n \in J} I_n$, where the I_n are pairwise disjoint open intervals and $m^*(V) = \sum_{n \in J} m^*(I_n)$. If J is infinite, then clearly $V \subset \bigcup_{n=1}^{\infty} I_n$; so by Definition 89.2, we have

$$m(V) \leq \sum_{n=1}^{\infty} m^*(I_n) = \sum_{n \in J} m^*(I_n) = m^*(V)$$

If J is finite, i.e., $J = \{1, 2, \ldots, l\}$, let $I_k = \varnothing$ for $k > l$. Then $V \subset \bigcup_{n=1}^{\infty} I_n$; so by Definition 89.2, we have $m(V) \leq \sum_{n=1}^{\infty} m^*(I_n) = \sum_{n=1}^{l} m^*(I_n) = \sum_{n \in J} m^*(I_n) = m^*(V)$. Therefore

$$m(V) \leq m^*(V) \qquad \text{for all open } V \text{ in } \mathbf{R} \qquad (89.2)$$

If $m^*(V) = \infty$, then it is clear that $m^*(V) = m(V)$. So suppose $m^*(V) < \infty$, then $V = \bigcup_{n \in J} I_n$ with $I_n = (a_n, b_n)$ and $a_n, b_n \in \mathbf{R}$ (both finite) and the I_n pairwise disjoint. Now suppose $\{\sigma_n\}_{n=1}^{\infty}$ is any collection of open intervals or $\varnothing$ such that $V \subset \bigcup_{n=1}^{\infty} \sigma_n$. If J is finite, it is clear that $m^*(V) = \sum_{n \in J} (b_n - $

$a_n) = \sum_{n=1}^{l} (b_n - a_n) \le \sum_{n=1}^{\infty} m^*(\sigma_n)$ (verify). But if $J = \{1, 2, \ldots\}$ we have that

$$\sum_{n=1}^{l} (b_n - a_n) \le \sum_{n=1}^{\infty} m^*(\sigma_n) \quad \text{for } l = 1, 2, \ldots$$

Therefore $\quad m^*(V) = \sum_{n \in J} (b_n - a_n) = \sum_{n=1}^{\infty} (b_n - a_n) = \lim_{l \to \infty} \sum_{n=1}^{l} (b_n - a_n)$

$$\le \sum_{n=1}^{\infty} m^*(\sigma_n)$$

We conclude, since $m(V)$ is the inf of the sums $\sum_{n=1}^{\infty} m^*(\sigma_n)$, that

$$m^*(V) \le m(V) \tag{89.3}$$

Equations (89.2) and (89.3) give us that $m^*(V) = m(V)$ for all V open in **R**. ■

Theorem 89.4 If $A_n \in \mathscr{P}(\mathbf{R})$ for $n = 1, 2, \ldots$, then

$$m\left(\bigcup_{n=1}^{\infty} A_n\right) \le \sum_{n=1}^{\infty} m(A_n)$$

Proof. By Definition 89.2, it is clear that for any sets $A \subset B$ in **R** we have

$$m(A) \le m(B)$$

Since $A_k \subset \bigcup_{n=1}^{\infty} A_n$, if $m(A_k) = \infty$ for some k, it follows that $\infty = m(A_k) \le m(\bigcup_{n=1}^{\infty} A_n)$; therefore,

$$m\left(\bigcup_{n=1}^{\infty} A_n\right) = \infty = \sum_{n=1}^{\infty} m(A_n)$$

Therefore, we assume $m(A_n) < \infty$ for $n = 1, 2, \ldots$. Let $\varepsilon > 0$. By Definition 89.2 for each positive integer n we may choose a covering $\{I_k^{(n)}\}_{k=1}^{\infty}$ of A_n such that $\sum_{k=1}^{\infty} m(I_k^{(n)}) \le m(A_n) + \varepsilon/2^n$. Then $\bigcup_{n=1}^{\infty} \bigcup_{k=1}^{\infty} I_k^{(n)}$ is a countable covering of $\bigcup_{n=1}^{\infty} A_n$; so by Definition 89.2, we have

$$m\left(\bigcup_{n=1}^{\infty} A_n\right) \le \sum_{n=1}^{\infty} \sum_{k=1}^{\infty} m(I_k^{(n)}) \le \sum_{n=1}^{\infty} \left(m(A_n) + \frac{\varepsilon}{2^n}\right) = \sum_{n=1}^{\infty} m(A_n) + \varepsilon$$

Since $\varepsilon > 0$ was arbitrary, we have

$$m\left(\bigcup_{n=1}^{\infty} A_n\right) \le \sum_{n=1}^{\infty} m(A_n) \qquad\qquad ■$$

We have constructed a function $m: \mathscr{P}(\mathbf{R}) \to [0, \infty]$ with the following properties:

(i) $m(\varnothing) = 0$.

(ii) $m(A) \leq m(B)$ if $A \subset B \subset \mathbf{R}$.

(iii) $m(\bigcup_{n=1}^{\infty} A_n) \leq \sum_{n=1}^{\infty} m(A_n)$ for any sequence $\{A_n\}_{n=1}^{\infty}$ of subsets of $\mathbf{R}$.

We now return to the abstract setting, because the development of a σ-algebra on which m is countably additive does not depend on $\mathbf{R}$, but only on the properties stated above.

Definition 89.5 Let X be a set. A function μ from $\mathscr{P}(X)$ into $[0, \infty]$ is called an *outer measure* if

(i) $\mu(\varnothing) = 0$.

(ii) $\mu(A) \leq \mu(B)$ whenever $A \subset B \subset X$.

(iii) $\mu(\bigcup_{n=1}^{\infty} A_n) \leq \sum_{n=1}^{\infty} \mu(A_n)$ for all sequences $\{A_n\}_{n=1}^{\infty}$ of subsets of X.

This last property is called the *countable subadditivity* of μ. We note that countable additivity demands the pairwise disjointness of the sets involved, while countable subadditivity does not.

The following definition is due to Carathéodory and is the key to defining a useful σ-algebra on which μ will be a positive measure.

Definition 89.6 Let X be a set and let μ be an outer measure on $\mathscr{P}(X)$. A subset S of X is said to be *μ-measurable* if

$$\mu(T) = \mu(S \cap T) + \mu(S' \cap T) \qquad \text{for all } T \subset X$$

Let $\mathscr{M}_\mu$ denote the family of all μ-measurable subsets of X.

Since $T = (S \cap T) \cup (S' \cap T)$, it follows from the subadditivity of μ that

$$\mu(T) \leq \mu(S \cap T) + \mu(S' \cap T)$$

Thus the equation in Definition 89.6 holds precisely when

$$\mu(T) \geq \mu(S \cap T) + \mu(S' \cap T)$$

A subset S of X is not μ-measurable if there exists a set $T \subset X$ with

$$\mu(T) > \mu(S \cap T) + \mu(S' \cap T)$$

Notice that in this case μ is not even finitely additive on the disjoint pair $\{S \cap T, S' \cap T\}$. It can be shown (see Hewitt and Stromberg, 1965) that there exists a non-m-measurable set $S \subset \mathbf{R}$, where m is given in Definition 89.2. Thus m is not even finitely additive on $\mathscr{P}(\mathbf{R})$. Again, we see that we must restrict m to a σ-algebra which is properly contained in $\mathscr{P}(\mathbf{R})$.

We now proceed to show that $\mathscr{M}_\mu$ is a σ-algebra in X and μ is a positive measure on $\mathscr{M}_\mu$.

Theorem 89.7   Every subset S of X such that $\mu(S) = 0$ is in $\mathcal{M}_\mu$.

Proof. Let $S \subset X$ and $\mu(S) = 0$. Let T be any subset of X. Then by Definition 89.5(ii), since $S \cap T \subset S$, we have $0 \le \mu(S \cap T) \le \mu(S) = 0$, which implies $\mu(S \cap T) = 0$.

By (iii) and (ii) of Definition 89.5, we have

$$\mu(T) \le \mu(S \cap T) + \mu(S' \cap T) = \mu(S' \cap T) \le \mu(T)$$

Therefore, it follows that

$$\mu(T) = \mu(S \cap T) + \mu(S' \cap T)$$

so $S \in \mathcal{M}_\mu$. ∎

Theorem 89.8   If $S \in \mathcal{M}_\mu$, then $S' \in \mathcal{M}_\mu$.

Proof. Trivial. ∎

Theorem 89.9   If A and B are in $\mathcal{M}_\mu$, then $A \cap B'$ is in $\mathcal{M}_\mu$.

Proof. It suffices to prove that if $E \subset A \cap B'$ and $F \subset (A \cap B')'$, then $\mu(E \cup F) = \mu(E) + \mu(F)$.

Now $F = (F \cap B) \cup (F \cap B')$ and $B \in \mathcal{M}_\mu$; so

$$\mu(E) + \mu(F) = \mu(E) + \mu((F \cap B) \cup (F \cap B'))$$
$$= \mu(E) + \mu(F \cap B) + \mu(F \cap B') \tag{89.4}$$

Since $E \subset A$ and $F \cap B' \subset A'$ and $A \in \mathcal{M}_\mu$, we have

$$\mu(E) + \mu(F \cap B') + \mu(F \cap B) = \mu(E \cup (F \cap B')) + \mu(F \cap B) \tag{89.5}$$

Again $E \cup (F \cap B') \subset B'$ and $F \cap B \subset B$; so

$$\mu(E \cup (F \cap B')) + \mu(F \cap B) = \mu(E \cup (F \cap B') \cup (F \cap B))$$
$$= \mu(E \cup F) \tag{89.6}$$

Combining equations (89.4), (89.5), and (89.6)

$$\mu(E) + \mu(F) = \mu(E \cup F) \qquad\qquad ∎$$

Lemma 89.10   Let $\{A_n\}_{n=1}^\infty$ be a sequence of pairwise disjoint sets in $\mathcal{M}_\mu$. Then

$$\mu(T) = \sum_{n=1}^\infty \mu(A_n \cap T) + \mu\left(\left(\bigcup_{n=1}^\infty A_n\right)' \cap T\right) \qquad \text{for all } T \subset X \tag{89.7}$$

Proof. By Definition 89.5(iii), we have $\mu(T) \le \sum_{n=1}^\infty \mu(T \cap A_n) + \mu(T \cap (\bigcup_{n=1}^\infty A_n)')$. If $\mu(T) = \infty$, then equation (89.7) follows immediately.

Therefore suppose that $\mu(T) < \infty$. We first show that

$$\mu(T) = \sum_{n=1}^{p} \mu(T \cap A_n) + \mu\left(T \cap \left(\bigcup_{n=1}^{p} A_n\right)'\right) \qquad \text{for all } p \in \mathbf{P} \quad (89.8)$$

We use induction on p. For $p = 1$, equation (89.8) becomes $\mu(T) = \mu(T \cap A_1) + \mu(T \cap A_1')$. But this follows since $A_1 \in \mathscr{M}_\mu$. Suppose that equation (89.8) is true for a positive integer p and all $T \subset X$. Since $A_{p+1} \in \mathscr{M}_\mu$, we have

$$\mu(T) = \mu(T \cap A_{p+1}) + \mu(T \cap A_{p+1}') = \mu(T \cap A_{p+1})$$

$$+ \sum_{n=1}^{p} \mu(T \cap A_{p+1}' \cap A_n) + \mu\left(T \cap A_{p+1}' \cap \left(\bigcup_{n=1}^{p} A_n\right)'\right) \quad (89.9)$$

The last equality follows from the inductive hypothesis applied to the set $T \cap A_{p+1}'$.

Now since $A_n \geq A_{p+1}'$ for $n \neq p + 1$, equation (89.9) can be written

$$\mu(T) = \mu(T \cap A_{p+1}) + \sum_{n=1}^{p} \mu(T \cap A_n) + \mu\left(T \cap A_{p+1}' \cap \left(\bigcup_{n=1}^{p} A_n\right)'\right)$$

$$= \sum_{n=1}^{p+1} \mu(T \cap A_n) + \mu\left(T \cap \left(\bigcup_{n=1}^{p+1} A_n\right)'\right)$$

which is equation (89.8) for $p + 1$. The sequence of numbers $\{\mu(T \cap (\bigcup_{n=1}^{p} A_n)')\}_{p=1}^{\infty}$ is a nonincreasing sequence bounded below by the number $\mu(T \cap (\bigcup_{n=1}^{\infty} A_n)')$. Thus it has a limit, which is greater than or equal to $\mu(T \cap (\bigcup_{n=1}^{\infty} A_n)')$. Taking limits in the equality (89.8), we get

$$\mu(T) = \lim_{p \to \infty} \sum_{n=1}^{p} \mu(T \cap A_n) + \lim_{p \to \infty} \mu\left(T \cap \left(\bigcup_{n=1}^{p} A_n\right)'\right)$$

$$\geq \sum_{n=1}^{\infty} \mu(T \cap A_n) + \mu\left(T \cap \left(\bigcup_{n=1}^{\infty} A_n\right)'\right)$$

Since the reverse inequality has already been established, this completes the proof. ∎

Theorem 89.11 $\mathscr{M}_\mu$ is a σ-algebra in X and μ is a positive measure on $\mathscr{M}_\mu$ [i.e., $(X, \mathscr{M}_\mu, \mu)$ is a positive measure space].

Proof. Let $\{A_n\}_{n=1}^{\infty}$ be any sequence of sets in $\mathscr{M}_\mu$. Then

$$\bigcup_{n=1}^{\infty} A_n = A_1 \cup (A_2 \cap A_1') \cup (A_3 \cap A_2' \cap A_1') \cup \cdots$$

$$\cup (A_n \cap A_{n-1}' \cap \cdots \cap A_1') \cup \cdots$$

By Theorem 89.9, each set of the form $B_n = (A_n \cap A_{n-1}' \cap \cdots \cap A_1')$ is in $\mathscr{M}_\mu$. Furthermore, the sets B_n are pairwise disjoint. Let $T \subset X$. By Lemma

89.10 and the countable subadditivity of μ, we have

$$\mu(T) = \sum_{n=1}^{\infty} \mu(T \cap B_n) + \mu\left(T \cap \left(\bigcup_{n=1}^{\infty} B_n\right)'\right)$$

$$\geq \mu\left(T \cap \left(\bigcup_{n=1}^{\infty} B_n\right)\right) + \mu\left(T \cap \left(\bigcup_{n=1}^{\infty} B_n\right)'\right)$$

By countable subadditivity, we have

$$\mu(T) \leq \mu\left(T \cap \left(\bigcup_{n=1}^{\infty} B_n\right)\right) + \mu\left(T \cap \left(\bigcup_{n=1}^{\infty} B_n\right)'\right)$$

and so

$$\mu(T) = \mu\left(T + \left(\bigcup_{n=1}^{\infty} B_n\right)\right) + \mu\left(T \cap \left(\bigcup_{n=1}^{\infty} B_n\right)'\right)$$

This implies $\bigcup_{n=1}^{\infty} B_n = \bigcup_{n=1}^{\infty} A_n$ is in $\mathcal{M}_\mu$. This along with Theorem 89.8 implies that $\mathcal{M}_\mu$ is a σ-algebra in X. Now let $\{B_n\}_{n=1}^{\infty}$ be an arbitrary pairwise disjoint collection of elements of $\mathcal{M}_\mu$. We have just shown that $\bigcup_{n=1}^{\infty} B_n \in \mathcal{M}_\mu$. Letting $T = \bigcup_{k=1}^{\infty} B_k$ in Lemma 89.10, we have

$$\mu\left(\bigcup_{k=1}^{\infty} B_k\right) = \sum_{n=1}^{\infty} \mu\left(B_n \cap \left(\bigcup_{k=1}^{\infty} B_k\right)\right) + \mu\left(\left(\bigcup_{n=1}^{\infty} B_n\right)' \cap \left(\bigcup_{k=1}^{\infty} B_k\right)\right)$$

$$= \sum_{n=1}^{\infty} \mu(B_n) + \mu(\varnothing) = \sum_{k=1}^{\infty} \mu(B_k)$$

Thus μ is countably additive on the σ-algebra $\mathcal{M}_\mu$, and so μ is a positive measure on $\mathcal{M}_\mu$. ∎

We finally return to $\mathbf{R}$ and the outer measure m on $\mathscr{P}(\mathbf{R})$ (Definition 89.2). By Theorem 89.11 there is a σ-algebra $\mathcal{M}_m$ in $\mathbf{R}$ such that m is a positive measure on $\mathcal{M}_m$.

We also know, by definition, that $A \in \mathcal{M}_m$ if and only if $m(T) = m(T \cap A) + m(T \cap A')$ for all $T \subset \mathbf{R}$.

Definition 89.12 $\mathcal{M}_m$ is called the σ-algebra of *Lebesgue measurable sets* in $\mathbf{R}$.

We wish to show that every open set $U \subset \mathbf{R}$ is in $\mathcal{M}_m$. However, we must first prove some preliminary facts about m.

Theorem 89.13   Let T be any subset of $\mathbf{R}$ and let U be an open subset of $\mathbf{R}$. Then

(i) $m(T) = \inf \{m(V) \mid V \text{ is open in } \mathbf{R} \text{ and } T \subset V\}$.
(ii) $m(U) = \sup \{m(K) \mid K \text{ compact in } \mathbf{R} \text{ and } K \subset U\}$.
(iii) $m(U) = \sup \{m(V) \mid V \text{ is open and } \overline{V} \text{ is compact in } \mathbf{R} \text{ with } \overline{V} \subset U\}$.

Proof. (i) Let

$$m^*(T) = \inf \{m(V) \mid V \text{ open and } T \subset V\} \tag{89.10}$$

By Definition 89.2

$$m(T) = \inf \left\{ \sum_{n=1}^{\infty} m(I_n) \middle| \text{ where each } I_n \text{ is an open interval} \right.$$

$$\left. \text{or } \varnothing \text{ and } T \subset \bigcup_{n=1}^{\infty} I_n \right\} \tag{89.11}$$

Suppose I_n is an open interval or $\varnothing$ for $n = 1, 2, \ldots$ and $T \subset \bigcup_{n=1}^{\infty} I_n$. Then $V = \bigcup_{n=1}^{\infty} I_n$ is open in $\mathbf{R}$ and $T \subset V$. By Theorem 89.4, we have $m(V) = m(\bigcup_{n=1}^{\infty} I_n) \leq \sum_{n=1}^{\infty} m(I_n)$. Therefore, we have shown for every sum $\sum_{n=1}^{\infty} m(I_n)$ on the right side of equation (89.11), there is a number $m(V)$ on the right side of equation (89.10) with $m(V) \leq \sum_{n=1}^{\infty} m(I_n)$. Therefore

$$m^*(T) \leq m(T) \tag{89.12}$$

Suppose V is an open set in $\mathbf{R}$ with $T \subset V$. Using Definition 89.1, $V = \bigcup_{n \in J} I_n$, where the I_n are pairwise disjoint open intervals and $m(V) = \sum_{n \in J} m(I_n)$. If J is finite, say $J = \{1, 2, \ldots, l\}$, define $I_n = \varnothing$ for all $n > l$. Then for either J finite or infinite, we have $T \subset \bigcup_{n=1}^{\infty} I_n = \bigcup_{n \in J} I_n$ and $m(V) = \sum_{n \in J} m(I_n) = \sum_{n=1}^{\infty} m(I_n)$. Therefore for each number $m(V)$ on the right side of equation (89.10), there is a number $\sum_{n=1}^{\infty} m(I_n)$ on the right side of equation (89.11) such that $m(V) = \sum_{n=1}^{\infty} m(I_n)$. This implies that

$$m(T) \leq m^*(T) \tag{89.13}$$

By equations (89.12) and (89.13) we have $m(T) = m^*(T)$ for any $T \subset \mathbf{R}$.

(ii) Let U be any open set in $\mathbf{R}$. Then $U = \bigcup_{n \in J} I_n$, where the I_n are disjoint open intervals and

$$m(U) = \sum_{n \in J} m(I_n) \tag{89.14}$$

CASE 1. Suppose $m(U) < \infty$. Let $\varepsilon > 0$. Then equation (89.14) implies there is some positive integer N such that

$$m(U) - \sum_{n=1}^{N} m(I_n) < \frac{\varepsilon}{2}$$

Since $I_n = (a_n, b_n)$ with $a_n, b_n \in \mathbf{R}$ for $n = 1, 2, \ldots, N$, we may let

$$K_n = \left[a_n + \frac{\varepsilon}{2^{n+2}}, b - \frac{\varepsilon}{2^{n+2}} \right] \quad \text{for } n = 1, 2, \ldots, N$$

(Note that some K_n may be void.) Then each K_n is compact, and so $\bigcup_{n=1}^{N} K_n = K$ is compact. Since $K \subset U$, we have

$$0 \leq m(U) - m(K) = m(U) - \sum_{n=1}^{N} m(I_n) + \sum_{n=1}^{N} m(I_n) - m(K)$$

$$< \frac{\varepsilon}{2} + \sum_{n=1}^{N} m(I_n) - \sum_{n=1}^{N} m\left(\left[a_n + \frac{\varepsilon}{2^{n+2}}, b_n - \frac{\varepsilon}{2^{n+2}}\right]\right)$$

$$\leq \frac{\varepsilon}{2} + \sum_{n=1}^{N} 2\frac{\varepsilon}{2^{n+2}} < \frac{\varepsilon}{2} + \frac{\varepsilon}{2} = \varepsilon \qquad (89.15)$$

Since $\varepsilon > 0$ was arbitrary, equation (89.15) implies that $m(U) \leq \sup \{m(K) \mid K$ compact and $K \subset U\}$. Also, since $K \subset U$ implies $m(K) \leq m(U)$, the opposite inequality is clear.

CASE 2.   Suppose $m(U) = \infty$. Then U contains an interval of the form $(-\infty, b)$ or (a, ∞) for some a or $b \in \mathbf{R}$. Suppose $(a, \infty) \subset U$ for some $a \in \mathbf{R}$. Let $K_n = [a + 1, a + n]$ for $n = 1, 2, \ldots$. Each K_n is compact and $K_n \subset U$ and $\sup \{m(K_n) \mid n = 1, 2, \ldots\} = \infty$; so (ii) follows.

   A similar argument applies if $(-\infty, b) \subset U$ for some $b \in \mathbf{R}$. Cases 1 and 2 complete the proof of (ii).

   (iii) Let U be any open set in $\mathbf{R}$. Then $U = \bigcup_{n \in J} I_n$, where the I_n are disjoint open intervals.

CASE 1.   Suppose $m(U) < \infty$. The proof of case 1(ii) applies if we replace K_n by $\overline{U}_n$ where either U_n is void or

$$U_n = \left(a_n + \frac{\varepsilon}{2^{n+2}}, b - \frac{\varepsilon}{2^{n+2}}\right).$$

In any case $\overline{U}_n$ is compact. Letting $U = \bigcup_{n=1}^{N} U_n$, we see that (iii) follows.

CASE 2.   Suppose $m(U) = \infty$. We may again modify (ii), case 2, letting $U_n = (a + 1, a + n)$ and (iii) follows.   ∎

Theorem 89.14   If U is any open subset of $\mathbf{R}$, then $U \in \mathcal{M}_m$.

   Proof.   Let $T \subset \mathbf{R}$. We must show that $m(T) = m(T \cap U) + m(T \cap U')$. By the subadditivity of m, $m(T) \leq m(T \cap U) + m(T \cap U')$. The reverse inequality is obvious if $m(T) = \infty$. Therefore, assume $m(T) < \infty$. Let $\varepsilon > 0$. By Theorem 89.13(i) there is an open set V with $T \subset V$ such that $m(V) < m(T) + \varepsilon/2$. By the same theorem there is an open set H such that $V \cap U' \subset H$ and $m(H) < m(V \cap U') + \varepsilon/4$. Since $V \cap U$ is open, we may use Theorem 89.13(iii) to get an open set W such that $\overline{W} \subset V \cap U$ and $m(W) + \varepsilon/4 > m(V \cap U)$. Now let $W^* = V \cap H \cap (\overline{W})'$. Then W and W^* are disjoint open sets. Since $V \cap U'$ is a subset of each of the sets V, H, and $(\overline{W})'$, it follows that $V \cap U' \subset W^* \subset H$, and so

$$0 \le m(W^*) - m(V \cap U') \le m(H) - m(V \cap U') < \frac{\varepsilon}{4}$$

Therefore $|m(W) + m(W^*) - [m(V \cap U) + m(V \cap U')]|$

$$\le |m(W^*) - m(V \cap U')| + |m(W) - m(V \cap U)|$$

$$< \frac{\varepsilon}{4} + \frac{\varepsilon}{4} = \frac{\varepsilon}{2}$$

Combining this with the fact that $W \cup W^* \subset V$ and the fact that m is additive on disjoint open sets by Definition 89.1, we have

$$m(T) + \varepsilon > m(V) + \frac{\varepsilon}{2} \ge m(W \cup W^*) + \frac{\varepsilon}{2}$$

$$= m(W) + m(W^*) + \frac{\varepsilon}{2} > m(V \cap U) + m(V \cap U')$$

$$\ge m(T \cap U) + m(T \cap U')$$

Since $\varepsilon > 0$ was arbitrary,

$$m(T) \ge m(T \cap U) + m(T \cap U') \qquad \blacksquare$$

Theorem 89.14 states that all open sets in $\mathbf{R}$ are in $\mathcal{M}_m$; so we may conclude, since $\mathcal{M}_m$ is a σ-algebra, that all countable unions and intersections of open or closed sets in $\mathbf{R}$ are in $\mathcal{M}_m$. We have now reached our first goal, the construction of a σ-algebra in $\mathbf{R}$ with a measure corresponding to length on intervals.

Exercises

89.1 Let $A \in \mathcal{M}_m$ and $r \in \mathbf{R}$. Prove that each set $\{x + r \,|\, x \in A\}$, $\{-x \,|\, x \in A\}$, and $\{xr \,|\, x \in A\}$ is in $\mathcal{M}_m$.

89.2 Prove that $(\mathbf{R}, \mathcal{M}_m, m)$ is a complete measure space (see Exercise 86.7 for the definition of *complete measure space*).

89.3 We say that a set $E \subset [a, b]$ is a *perfect* subset of $[a, b]$ if E is closed and every point in E is an accumulation point of E.
  (a) Prove that the Cantor set is a perfect subset of $[0, 1]$ of Lebesgue measure zero.
  (b)* Prove that for any $\delta > 0$ there exists a nowhere dense perfect subset of $[0, 1]$ having Lebesgue measure greater than $1 - \delta$.

89.4 Let $E \subset \mathbf{R}$. Prove that there exists a G_δ set $S \supset E$ such that $m(S) = m(E)$. (G_δ set is defined in Exercise 47.5.)

89.5 Prove that a subset E of $\mathbf{R}$ is m-measurable if and only if $E \cap [-n, n]$ is measurable for $n = 1, 2, \ldots$.

89.6* Prove that a subset E of $\mathbf{R}$ is m-measurable if and only if for every $\varepsilon > 0$, there exists a closed set $F \subset E$ such that $m(E \backslash F) < \varepsilon$.

89.7 A subset E of $\mathbf{R}$ is said to be an F_σ set if E is a countable union of closed sets. The collection of *Borel sets* $\mathscr{B}$ is the least σ-algebra in $\mathbf{R}$ containing the closed sets. Prove

(a) $\mathscr{B}$ is the least σ-algebra in $\mathbf{R}$ containing the open sets.

(b) Every Borel set is m-measurable.

(c) Every F_σ set and every G_δ set is in $\mathscr{B}$.

(d)* A subset E of $\mathbf{R}$ is m-measurable if and only if $E = B \cup S$, where $B \in \mathscr{B}$ and $m(S) = 0$.

89.8 Let E be a subset of $\mathbf{R}$. Prove that the following are equivalent.

(a) E is m-measurable.

(b) For every $\varepsilon > 0$, there exists an open set $U \supset E$ such that $m(U \backslash E) < \varepsilon$.

(c) There exists an F_σ set B and a set S of measure zero such that $A = B \cup S$.

(d) There exists a G_δ set B and a set S of measure zero such that $A = B \backslash S$.

89.9* Let f be a differentiable function on $[a, b]$. Prove that f' is m-measurable in $[a, b]$.

89.10 Let X and Y be (not necessarily Lebesgue measurable) subsets of $\mathbf{R}$ with glb $\{|x - y| \,|\, x \in X, y \in Y\} > 0$. Prove that $m(X \cup Y) = m(X) + m(Y)$.

89.11 Prove that a continuous function on an m-measurable subset E of $\mathbf{R}$ is m-measurable.

89.12 (a) Let f be an increasing function on an m-measurable subset E of $\mathbf{R}$. Prove that f is m-measurable.

(b) Let g be an m-measurable function on an m-measurable subset E of $\mathbf{R}$ and suppose that f is an increasing function on a set containing the range of g. Prove that $f \circ g$ is m-measurable.

89.13 Let E be a bounded, m-measurable subset of $\mathbf{R}$ and let f be an m-measurable function on E. Prove that for every $\varepsilon > 0$, there exists an m-measurable subset A of E such that $m(A) < \varepsilon$ and f is bounded on $E \backslash A$.

90. Lebesgue Measure on $[a, b]$

Let $[a, b]$ be any closed interval in $\mathbf{R}$. We have the σ-algebra $\mathscr{M}_m$ in $\mathbf{R}$ and the positive measure m on $\mathscr{M}_m$ (see 89.12 through 89.14).

Definition 90.1 Let $[a, b]$ be a closed interval in $\mathbf{R}$. Define

$$\mathscr{M}_m[a, b] = \{S \cap [a, b] \mid S \in \mathscr{M}_m\}$$

Now $[a, b] \in \mathscr{M}_m$ since $[a, b]'$ is open in $\mathbf{R}$. Therefore, $S \cap [a, b] \in \mathscr{M}_m$ for all $S \in \mathscr{M}_m$. Since $\mathscr{M}_m$ is a σ-algebra in $\mathbf{R}$, we conclude that $\mathscr{M}_m[a, b]$ is

σ-algebra in the set $[a, b]$ (verify). Also since m is a positive measure on $\mathcal{M}_m$, m is a positive measure on $\mathcal{M}_m[a,b]$ (verify).

Definition 90.2   The positive measure space $([a, b], \mathcal{M}_m[a, b], m)$ is called the *Lebesgue measure on* $[a, b]$.

In Section 57 we introduced the concept of a subset of $[a, b]$ having measure zero. It is now clear from the definition of m that a subset A of $[a, b]$ has measure zero if and only if $A \in \mathcal{M}_m[a, b]$ and $m(A) = 0$ (see Theorem 89.7). From Section 57 we may therefore conclude that every countable subset of $[a, b]$ is of m measure zero and the Cantor set in $[0, 1]$ is of m measure zero.

Recall that a property P is said to hold almost everywhere on $[a, b]$ if the set of points where P does not hold has measure zero. Therefore, we have that m-almost everywhere coincides with the concept of almost everywhere.

$([a, b], \mathcal{M}_m[a, b], m)$ is a positive measure space; so by Section 88 we have an integral $\int_{[a,b]} f \, dm$ defined on a subclass of the class of all $\mathcal{M}_m[a, b]$-measurable functions on $[a, b]$. Thus the theorems of Section 88 are valid for this integral. We call this integral the *Lebesgue integral on* $[a, b]$, and we call the functions in $\mathcal{L}([a, b], \mathcal{M}_m[a, b], m)$ (Definition 88.3), the *Lebesgue integrable functions on* $[a, b]$, denoted $\mathcal{L}[a, b]$.

We now fill in the important link between the Riemann integral and the Lebesgue integral on $[a, b]$.

Theorem 90.3   If $f \in \mathcal{R}[a, b]$, then $f \in \mathcal{L}[a, b]$ and

$$\int_{[a,b]} f \, dm = \int_a^b f(x) \, dx$$

Proof.   Suppose $f \in \mathcal{R}[a, b]$. To prove that $f \in \mathcal{L}[a, b]$, it suffices (Theorem 88.4(iii)) to show that f is measurable.

Let V be an open subset of $\mathbf{R}$. Let

$$X = \{x \in [a, b] \mid f \text{ is continuous at } x\}$$

For each $x \in X \cap f^{-1}(V)$ there exists an open interval U_x such that $x \in U_x \subset f^{-1}(V)$. Now $U = \cup \{U_x \mid x \in X \cap f^{-1}(V)\}$ is open and thus measurable by Theorem 89.14 and Definition 86.1. By Theorem 58.5, $X' \cap [a, b]$ has measure zero and since

$$f^{-1}(V)\backslash U \subset X' \cap [a, b]$$

it follows that $f^{-1}(V)\backslash U$ also has measure zero. By Theorem 89.7, $f^{-1}(V)\backslash U$ is measurable. Thus $f^{-1}(V) = [f^{-1}(V)\backslash U] \cup U$ is measurable by Theorem 86.5(ii). Therefore f is measurable and $f \in \mathcal{L}[a, b]$.

Now we wish to show that $\int_a^b f(x)\, dx = \int_{[a,b]} f\, dm$.

First assume that $f \in \mathscr{R}[a, b]$ and $f(x) \geq 0$ for all x in $[a, b]$. Let $P = \{x_0, x_1, \ldots, x_n\}$ be a partition of $[a, b]$. Let

$$s = \sum_{i=1}^{n-1} m_i \chi_{[x_{i-1}, x_i)} + m_n \chi_{[x_{n-1}, x_n]}$$

where $m_i = \text{glb}\,\{f(x) \mid x \in [x_{i-1}, x_i)\}$ for $i = 1, 2, \ldots, n$.

Since each of the intervals $[x_{i-1}, x_i)$ for $i = 1, 2, \ldots, n - 1$ and $[x_{n-1}, x_n]$ is measurable, s is a simple measurable function on $[a, b]$. Now $0 \leq s \leq f$; so by Definition 88.2, we have

$$L(f, P) = \sum_{i=1}^{n} m_i(x_i - x_{i-1}) = \int_{[a,b]} s\, dm \leq \int_{[a,b]} f\, dm$$

for every partition P of $[a, b]$. By Definition 51.3, we have

$$\int_a^b f(x)\, dx \leq \int_{[a,b]} f\, dm$$

Let $f \in \mathscr{L}[a, b]$. Since f is bounded, there exists a number M such that $0 \leq f(x) + M$ for all x in $[a, b]$. By the above argument, we have

$$\int_a^b (f(x) + M)\, dx \leq \int_{[a,b]} (f + M)\, dm$$

which reduces to

$$\int_a^b f(x) \leq \int_{[a,b]} f\, dm$$

Repeating the argument above for $-f$, we have

$$\int_a^b -f(x)\, dx \leq \int_{[a,b]} (-f)\, dm$$

which reduces to

$$\int_a^b f(x)\, dx \geq \int_{[a,b]} f\, dm$$

Therefore, $\int_a^b f(x)\, dx = \int_{[a,b]} f\, dm$ for all $f \in \mathscr{R}[a, b]$. ∎

Exercises

90.1 Define $f_n(x) = n\chi_{(0, 1/n)}$ on $[0, 1]$. Prove that $\{f_n\}$ converges pointwise to $f(x) = 0$ on $[0, 1]$ and that each function f_n and f is in $\mathscr{L}[0, 1]$, yet

$$\lim_{n \to \infty} \int_{[0,1]} f_n \neq \int_{[0,1]} f$$

Explain why each convergence theorem (Theorems 88.6, 88.13, and 88.15) does not apply here.

90.2 Let n be a positive integer. Select the unique positive integer i such that $2^{i-1} \le n \le 2^i - 1$. Let $r = n - 2^{i-1}$.

Let $f_n = \chi_{[2^{i-1}+(r/2^{i-1}),\, 2^{i-1}+((r+1)/2^{i-1})]}$ on [0, 1].

Let $f(x) = 0$ on [0, 1]. Prove that

(a) $f_n, f \in \mathscr{L}[0, 1]$ for $n = 1, 2, \ldots$.

(b) $\lim_{n\to\infty} \int_{[0,1]} f_n = \int_{[0,1]} f$.

(c) $\lim_{n\to\infty} f_n(x)$ does not exist for any $x \in [0, 1]$.

Prove that there exists a subsequence $\{f_{n_k}\}$ of $\{f_n\}$ such that $\{f_{n_k}\}$ converges to f m-almost everywhere.

90.3* Let $f \in \mathscr{L}[a, b]$. Define

$$F(x) = \int_{[a,x]} f \qquad x \in [a, b]$$

Prove that F is absolutely continuous, uniformly continuous, and of bounded variation on [a, b]. (The definition of absolute continuity is given in Exercise 56.1.)

90.4 (a) Prove that $\mathscr{L}[a, b]$ is a vector space if we define operations pointwise.

(b) Prove that for $f \in \mathscr{L}[a, b]$

$$\|f\| = \int_{[a,b]} |f| \, dm$$

defines a norm on the vector space $\mathscr{L}[a, b]$, except that $\|f\| = 0$ implies $f = 0$ is replaced by $\|f\| = 0$ implies $f = 0$ m-almost everywhere. The space $\mathscr{L}[a, b]$ with this norm is often denoted $\mathscr{L}^1[a, b]$.

(c) Let $g \in \mathscr{L}[a, b]$, where $g = h$ m-almost everywhere and h is bounded. (In this case, g is said to be *essentially bounded*.) Define

$$T(f) = \int_{[a,b]} fg \qquad f \in \mathscr{L}[a, b]$$

Prove that T is a continuous linear functional on $\mathscr{L}[a, b]$. (Actually, every continuous linear function on $\mathscr{L}[a, b]$ is of this form, but the proof is rather involved. (See Hewitt and Stromberg, 1965.)

90.5 (a)* Suppose $f, |f| \in \mathscr{R}[c, b]$ for each c satisfying $a < c < b$ and that the improper Riemann integral $\int_{a^+}^b |f|$ converges.

Prove that $f \in \mathscr{L}[a, b]$ and that

$$\int_{[a,b]} f = \int_{a^+}^b f$$

(b) Use (a) to compute $\int_{[0,1]} f$, where

$$f(x) = \begin{cases} 1/\sqrt{x} & 0 < x \le 1 \\ 0 & x = 0 \end{cases}$$

90.6 State and prove a result similar to that of Exercise 90.5(a) for improper Riemann integrals of the first kind.

90.7 Let f be a bounded function on [a, b]. Fix a nonnegative integer n. Let $\Delta = (b - a)/2^n$ and $x_i = a + i\Delta$ for $i = 0, 1, \ldots, 2^n$. Define

$$m_i = \text{glb}\,\{f(x)|x \in (x_{i-1}, x_i)\}$$
$$M_i = \text{lub}\,\{f(x)|x \in (x_{i-1}, x_i)\}$$

and for $x \in [a, b]$, define

$$L_n(x) = \begin{cases} m_i & x \in (x_{i-1}, x_i) \\ m_0 & \text{otherwise} \end{cases}$$

$$U_n(x) = \begin{cases} M_i & x \in (x_{i-1}, x_i) \\ M_0 & \text{otherwise} \end{cases}$$

(a) Prove that $L_n(x) \leq L_{n+1}(x) \leq U_{n+1}(x) \leq U_n(x)$ for $n = 1, 2, \dots$.

(b) Prove that L_n and U_n are measurable functions for $n = 1, 2, \dots$.

(c) Prove that $\{L_n\}$ and $\{U_n\}$ converge pointwise to measurable functions (hereafter, denoted L and U, respectively).

(d) Prove that $f \in \mathscr{R}[a, b]$ if and only if $L(x) = U(x)$ m-almost everywhere on $[a, b]$.

(e) Prove that if $x \neq x_i$ for any i corresponding to any n, then $L(x) = U(x)$ if and only if f is continuous at x.

90.8 Let f be a bounded, measurable function on $[a, b]$. Let $(c, d) \supset f([a, b])$. Let

$$P: c = y_0 < y_1 < \cdots < y_n = d$$

be a partition of $[c, d]$. Let $E_i = f^{-1}([y_{i-1}, y_i))$, $m_i = \text{glb}\,E_i$, and $M_i = \text{lub}\,E_i$. Define

$$L(f, P) = \sum_{i=1}^{n} m_i m(E_i) \qquad U(f, P) = \sum_{i=1}^{n} M_i m(E_i)$$

(a) Prove that the set

$$\{L(f, P)|\ P \text{ is a partition of } [c, d]\}$$

is bounded above and that the corresponding set for $U(f, P)$ is bounded below.

(b) Prove that if P and S are partitions of $[c, d]$ with $S \supset P$, then

$$L(f, P) \leq L(f, S) \leq U(f, S) \leq U(f, P)$$

and that for any partitions P and S of $[c, d]$, we have

$$L(f, S) \leq U(f, P)$$

Define

$$\underline{\mathscr{L}}_a^b f = \text{lub}\,\{L(f, P)\}$$
$$\overline{\mathscr{T}}_a^b f = \text{glb}\,\{U(f, P)\}.$$

(c) Prove that $\underline{\mathscr{L}}_a^b f = \int_{[a,b]} f = \overline{\mathscr{T}}_a^b f$.

90.9* Let $F(x, y)$ be a function on $[a, b] \times (c, d)$ and suppose that for each $x \in [a, b]$, the function $g_x(y) = F(x, y)$ is differentiable on (c, d). Letting $D_2 F(x, y) = g_x'(y)$ suppose further than for some M, $|D_2 F(x, y)| < M$ for all $x \in [a, b]$ and $y \in (c, d)$. Suppose also that for each $y \in (c, d)$ the function $f_y(x) = F(x, y)$ is m-measurable in $[a, b]$. Set $G(y) = \int_{[a,b]} F(x, y)\, dm(x)$ and prove that $G'(y) = \int_{[a,b]} D_2 F(x, y)\, dm(x)$.

91.   The Hilbert Spaces $\mathscr{L}^2(X, \mathscr{M}, \mu)$

We complete our text with a discussion of a large class of Hilbert spaces.

Definition 91.1   Let $(X, \mathscr{M}, \mu)$ be a positive measure space and define $\mathscr{L}^2(X, \mathscr{M}, \mu) = \{f \mid f$ is a real measurable function on X and $\int_X |f|^2 \, d\mu < \infty$, with the identification that $f = g$ if and only if $f = g$ almost everywhere on $X\}$.

We shall prove that $\mathscr{L}^2(X, \mathscr{M}, \mu)$ is an inner product space and is complete, i.e., that $\mathscr{L}^2(X, \mathscr{M}, \mu)$ is a Hilbert space.

Lemma 91.2   If $f, g \in \mathscr{L}^2(X, \mathscr{M}, \mu)$, then $fg \in \mathscr{L}(X, \mathscr{M}, \mu)$.

Proof.   By Theorems 87.4 and 87.8 $|fg|$ is measurable. Let

$$A = \{x \in X \mid |f(x)| \geq |g(x)|\}$$

Then $A \in \mathscr{M}$ and

$$0 \leq \int_X |fg| \, d\mu = \int_A |fg| \, d\mu + \int_{A'} |fg| \, d\mu \leq \int_A |f|^2 \, d\mu + \int_{A'} |g|^2 \, d\mu$$

$$\leq \int_X |f|^2 \, d\mu + \int_X |g|^2 \, d\mu < \infty$$

Therefore, $fg \in \mathscr{L}(X, \mathscr{M}, \mu)$.   ∎

Theorem 91.3 $\mathscr{L}^2(X, \mathscr{M}, \mu)$ is an inner product space with inner product

$$(f, g) = \int_X fg \, d\mu \qquad \text{for all } f, g \in \mathscr{L}^2(X, \mathscr{M}, \mu)$$

Proof.   Using Lemma 91.2 and the properties of the integral everything follows as long as we can show that $\mathscr{L}^2(X, \mathscr{M}, \mu)$ is a vector space. The only difficulty is in proving that if $f, g \in \mathscr{L}^2(X, \mathscr{M}, \mu)$, then $f + g \in \mathscr{L}^2(X, \mathscr{M}, \mu)$. But using Lemma 91.2 again,

$$\int_X |f + g|^2 \, d\mu \leq \int_X |f|^2 \, d\mu + 2 \int_X |fg| \, d\mu + \int_X |g|^2 \, d\mu < \infty$$

Note that with our identification, $(f, f) = 0$ if and only if $f = 0$ almost everywhere on X.   ∎

We also note that in $\mathscr{L}^2(X, \mathscr{M}, \mu)$ we have $\|f\| = (\int_X |f|^2 \, d\mu)^{1/2}$ and that the Cauchy-Schwarz inequality can be written as $\int_X |fg| \, d\mu \leq \|f\| \cdot \|g\|$ for all $f, g \in \mathscr{L}^2(X, \mathscr{M}, \mu)$.

Theorem 91.4 $\mathscr{L}^2(X, \mathscr{M}, \mu)$ is a Hilbert space.

Proof. We must show that $\mathscr{L}^2(X, \mathscr{M}, \mu)$ is complete. Let $\{f_n\}$ be a Cauchy sequence in $\mathscr{L}^2(X, \mathscr{M}, \mu)$. There is a subsequence $\{f_{n_j}\}_{j=1}^{\infty}$ with $n_1 < n_2 < \cdots$ such that

$$\|f_{n_{j+1}} - f_{n_j}\| < \frac{1}{2^j} \qquad \text{for } j = 1, 2, \ldots \qquad (91.2)$$

Set $\qquad g_k = \sum_{j=1}^{k} |f_{n_{j+1}} - f_{n_j}| \qquad$ and $\qquad g = \sum_{j=1}^{\infty} |f_{n_{j+1}} - f_{n_j}|$

By equation (91.2) and using the fact that the triangle inequality holds for $\|\cdot\|$, we see that $\|g_k\| < 1$ for $k = 1, 2, \ldots$. Now using Theorem 88.14 (Fatou's lemma), we have

$$\|g\|^2 = \int_X \left| \sum_{j=1}^{\infty} |f_{n_{j+1}} - f_{n_j}| \right|^2 d\mu = \int_X \liminf_{k \to \infty} \left(\sum_{j=1}^{k} |f_{n_{j+1}} - f_{n_j}| \right)^2 d\mu$$

$$\leq \liminf_{k \to \infty} \int_X \left(\sum_{j=1}^{k} |f_{n_{j+1}} - f_{n_j}| \right)^2 d\mu = \liminf_{k \to \infty} \|g_k\|^2 \leq 1$$

Therefore $\|g\| \leq 1$. This implies, in particular, that $g(x) < \infty$ a.e. on X; so that the series

$$f_{n_1}(x) + \sum_{j=1}^{\infty} [f_{n_{j+1}}(x) - f_{n_j}(x)] \qquad (91.3)$$

converges absolutely for almost all $x \in X$. Denote the sum in equation (91.3) by $f(x)$, for all those x at which (91.3) converges and set $f(x) = 0$ on the remaining set of measure 0. Since

$$f_{n_1} + \sum_{j=1}^{k-1} (f_{n_{j+1}} - f_{n_j}) = f_{n_k}$$

we see that

$$f(x) = \lim_{k \to \infty} f_{n_k}(x)$$

for almost all x in X.

We now prove that f is the limit of $\{f_n\}$ in $\mathscr{L}^2(X, \mathscr{M}, \mu)$. Let $\varepsilon > 0$. There exists an N such that $\|f_n - f_m\| < \varepsilon$ for all $m, n \geq N$. For every $m \geq N$, Fatou's lemma shows that

$$\int_X |f - f_m|^2 d\mu \leq \liminf_{j \to \infty} \int_X |f_{n_j} - f_m|^2 d\mu \leq \varepsilon^2 \qquad (91.4)$$

From equation (91.4) we conclude that

$$h = f - f_m \in \mathscr{L}^2(X, \mathscr{M}, \mu)$$

Therefore, $f = f_m + h \in \mathscr{L}^2(X, \mathscr{M}, \mu)$. Equation (91.4) also gives us that $\|f - f_m\| \to 0$ as $m \to \infty$. ∎

We turn to a familiar example, $([a, b], \mathscr{M}_m[a, b], m)$, Lebesgue measure on $[a, b]$. (See Definition 90.2.) We shall be interested in $\mathscr{L}^2([a, b], \mathscr{M}_m[a, b], m)$, which we shall abbreviate to $\mathscr{L}^2[a, b]$.

By Theorem 90.3, $\mathscr{R}[a, a + 2\pi] \subset \mathscr{L}^2[a, a + 2\pi]$, so that the trigonometric set $\{\phi_n\}$ is a subset of $\mathscr{L}^2[a, a + 2\pi]$. Since the Riemann and Lebesgue integrals agree for Riemann integrable functions, $\{\phi_n\}$ is an orthonormal set in $\mathscr{L}^2[a, a + 2\pi]$. Therefore, we may develop Fourier series in $\mathscr{L}^2[a, a + 2\pi]$ relative to $\{\phi_n\}$. In fact, throughout the remainder of this section, we will assume that all Fourier series are relative to this orthonormal set. Our immediate goal will be to show that the trigonometric set is a complete orthonormal set in $\mathscr{L}^2[a, a + 2\pi]$. Along the way we shall derive a result (Theorem 91.6) which is interesting in its own right. The proof that $\{\phi_n\}$ is complete in $\mathscr{L}^2[a, a + 2\pi]$ (Theorem 91.7) parallels the proof of Theorem 78.5.

Theorem 91.5 If $f \in \mathscr{L}^2[a, b]$ and $\varepsilon > 0$, there exists a simple function $h \in \mathscr{L}^2[a, b]$ such that $\|f - h\| < \varepsilon$.

Proof. First suppose that f is nonnegative on $[a, b]$ and $\varepsilon > 0$. By Theorem 87.10 there exists an increasing sequence $\{s_n\}$ of nonnegative simple functions such that $\lim_{n \to \infty} s_n(x) = f(x)$. Now $|f(x) - s_n(x)|^2 \le [f(x)]^2$ and since $f^2 \in \mathscr{L}[a, b]$, we may use the Lebesgue dominated convergence theorem (Theorem 88.15) to conclude that

$$\lim_{n \to \infty} \int_{[a,b]} |f - s_n|^2 = \int_{[a,b]} \lim_{n \to \infty} |f - s_n|^2 = 0$$

In particular, there exists n such that $\|f - s_n\| < \varepsilon$. For an arbitrary $f \in \mathscr{L}^2[a, b]$ and $\varepsilon > 0$, choose simple functions s and t, as above, such that $\|f^+ - s\| < \varepsilon/2$ and $\|f^- - t\| < \varepsilon/2$. Then $h = s - t$ is a simple function and

$$\|f - h\| = \|(f^+ - f^-) - (s - t)\| \le \|f^+ - s\| + \|f^- - t\| < \varepsilon \quad \blacksquare$$

It follows from the next result that the completion, in the sense of Theorem 46.7, of $\mathscr{R}[a, b]$ relative to the norm $\|f\| = (\int f^2)^{\frac{1}{2}}$, is $\mathscr{L}^2[a, b]$. Similarly (see Exercises 91.4 and 91.5), it can be shown that the completion of $\mathscr{R}[a, b]$ relative to the norm $\|f\| = \int |f|$ is $\mathscr{L}[a, b]$.

Theorem 91.6 If $f \in \mathscr{L}^2[a, b]$ and $\varepsilon > 0$, there exists a step function $g \in \mathscr{R}[a, b]$ such that $\|f - g\| < \varepsilon$.

Proof. First assume that $f = \chi_E$, a characteristic function, where E is a measurable subset of $[a, b]$. Let $\varepsilon > 0$. By Theorem 89.13(i) there exists an open set U such that $E \subset U$ and $m(U) < m(E) + \varepsilon^2/2$. We may write

$U = \bigcup_{n=1}^{\infty} U_n$, where $\{U_n\}$ is a collection of pairwise disjoint open intervals. Since $m(U) = \sum_{n=1}^{\infty} m(U_n)$ and $m(U)$ is finite, the series $\sum_{n=1}^{\infty} m(U_n)$ converges. Therefore, there exists N such that $\sum_{n=N+1}^{\infty} m(U_n) < \varepsilon^2/2$. Let $V = \bigcup_{n=1}^{N} U_n$. Then χ_V is a step function. We will show that $\|\chi_V - \chi_E\| < \varepsilon$.

First, note that since $V\backslash E \subset U\backslash E$, $m(V\backslash E) \leq m(U\backslash E) < \varepsilon^2/2$. Also, since $E\backslash V \subset \bigcup_{n=N+1}^{\infty} U_n$, $m(E\backslash V) < \varepsilon^2/2$. If $t = |\chi_V - \chi_E|^2$, then

$$\int_{[a,b]} t = \int_{[a,b]\backslash(V\cup E)} t + \int_{V\backslash E} t + \int_{E\backslash V} t + \int_{E\cap V} t$$

$$= \int_{V\backslash E} t + \int_{E\backslash V} t = \int_{V\backslash E} 1 + \int_{E\backslash V} 1 = m(V\backslash E) + m(E\backslash V) < \varepsilon^2$$

Therefore, $\|\chi_V - \chi_E\| < \varepsilon$.

Now let f be an arbitrary function in $\mathscr{L}^2[a, b]$ and let $\varepsilon > 0$. By Theorem 91.5 there exists a simple function $h \in \mathscr{L}^2[a, b]$ such that $\|f - h\| < \varepsilon/2$. We may write $h = \sum_{k=1}^{n} c_k \chi_{E_k}$, where $c_k \neq 0$ for $k = 1, \ldots, n$. By the special case above, there exist step functions g_k such that $\|g_k - \chi_{E_k}\| < \varepsilon/(2n|c_k|)$. Now $g = \sum_{k=1}^{n} c_k g_k$ is a step function and

$$\|f - g\| \leq \|f - h\| + \|h - g\| = \|f - h\| + \left\| \sum_{k=1}^{n} c_k \chi_{E_k} - \sum_{k=1}^{n} c_k g_k \right\|$$

$$\leq \|f - h\| + \sum_{k=1}^{n} |c_k| \|\chi_{E_k} - g_k\| < \frac{\varepsilon}{2} + \sum_{k=1}^{n} |c_k| \frac{\varepsilon}{2n|c_k|} = \varepsilon \quad \blacksquare$$

We can now prove that the trigonometric set is complete in $\mathscr{L}^2[a, a + 2\pi]$.

Theorem 91.7 The trigonometric set $\{\phi_n\}$ is a complete orthonormal set in $\mathscr{L}^2[a, a + 2\pi]$.

Proof. Let $f \in \mathscr{L}^2[a, a + 2\pi]$ and let $\{s_n\}$ be the sequence of partial sums of the Fourier series of f. Let $\varepsilon > 0$. By Theorem 91.6 there exists a step function $g \in \mathscr{R}[a, a + 2\pi]$ such that

$$\|f - g\| < \frac{\varepsilon}{2}$$

Let $\{t_n\}$ be the sequence of partial sums of the Fourier series of g. By Theorem 78.5, $\lim_{n\to\infty} \|g - t_n\| = 0$. Thus, there exists a positive integer N such that if $n \geq N$,

$$\|g - t_n\| < \frac{\varepsilon}{2}$$

By Theorem 75.5,

$$\|f - s_n\| \leq \|f - t_n\|$$

Thus, if $n \geq N$,

$$\|f - s_n\| \leq \|f - t_n\| \leq \|f - g\| + \|g - t_n\| < \frac{\varepsilon}{2} + \frac{\varepsilon}{2} = \varepsilon \quad \blacksquare$$

Corollary 91.8 Let $f, g \in \mathscr{L}^2[a, a + 2\pi]$. Suppose

$$\frac{a_0}{2} + \sum_{n=1}^{\infty} (a_n \cos nx + b_n \sin nx)$$

is the Fourier series of f and

$$\frac{c_0}{2} + \sum_{n=1}^{\infty} (c_n \cos nx + d_n \sin nx)$$

is the Fourier series of g. Then

(i) $\dfrac{a_0 c_0}{2} + \displaystyle\sum_{n=1}^{\infty} (a_n c_n + b_n d_n) = \dfrac{1}{\pi} \int_a^{a+2\pi} fg$

(ii) (Parseval's theorem).

$$\frac{a_0^2}{2} + \sum_{n=1}^{\infty} (a_n^2 + b_n^2) = \frac{1}{\pi} \int_a^{a+2\pi} f^2$$

(iii) If

$$\int_a^{a+2\pi} f(x) \cos nx \, dm = 0 = \int_a^{a+2\pi} f(x) \sin nx \, dm \qquad \text{for } n = 0, 1, 2, \dots$$

then $f = 0$ almost everywhere in $[a, a + 2\pi]$.

(iv) If f and g have the same Fourier series, then $f = g$ almost everywhere in $[a, a + 2\pi]$.

Proof.   The proof follows immediately from Theorems 91.7 and 75.10. $\blacksquare$

It follows from Bessel's inequality (Theorem 75.6) that if $f \in \mathscr{L}^2[a, a + 2\pi]$ and $a_n = (f, \phi_n)$, then $\{a_n\} \in l^2$. The converse is known as the Riesz-Fischer theorem (Theorem 91.9) and answers affirmatively a question raised in Section 78: If $\{a_n\} \in l^2$, does there exist a function f in $\mathscr{L}^2[a, a + 2\pi]$ such that $(f, \phi_n) = a_n$?

Theorem 91.9 (Riesz-Fischer Theorem)   Suppose $\{a_n\} \in l^2$, then there exists an $f \in \mathscr{L}^2[a, a + 2\pi]$ such that $a_n = (f, \phi_n)$ for all n.

Proof.   We look at the sequence of functions

$$f_k = \sum_{n=0}^{k} a_n \phi_n \qquad \text{for } k = 1, 2, 3, \dots$$

By Theorems 75.3 and 75.4, we have $\|f_n - f_m\|^2 = \sum_{k=m+1}^{n} a_k^2$ for all $n > m$, and this implies that $\{f_k\}$ is a Cauchy sequence in $\mathscr{L}^2[a, a + 2\pi]$. Since $\mathscr{L}^2[a, a + 2\pi]$ is complete (Theorem 91.4), we conclude that there exists $f \in \mathscr{L}^2[a, a + 2\pi]$ such that $\{f_k\}$ converges to f; in fact, clearly $f = \sum_{n=0}^{\infty} a_n \phi_n$. Now $(f, \phi_n) = (\sum_{k=0}^{\infty} a_k \phi_k, \phi_n) = \sum_{k=0}^{\infty} (a_k \phi_k, \phi_n) = a_n$ for $n = 0, 1, 2, \ldots$ by Corollary 75.2. ∎

If we define $T: \mathscr{L}^2[a, a + 2\pi] \to l^2$ by $T(f) = \{(f, \phi_n)\}$, Bessel's inequality (Theorem 75.6) assures us that the image of T is contained in l^2. Parseval's theorem (Corollary 91.8(ii)) tells us that T preserves norms, $\|f\| = \|T(f)\|$. In particular, T is one to one. The Riesz-Fischer theorem (Theorem 91.9) states that T is onto. Thus, as Hilbert spaces, $\mathscr{L}^2[a, a + 2\pi]$ and l^2 are, in a sense, identical.

Exercises

Throughout these exercises, $(X, \mathscr{M}, \mu)$ denote a positive measure space.

91.1   Prove that if $f_n \to f$ in $\mathscr{L}^2(X, \mathscr{M}, \mu)$, there exists a subsequence $\{f_{n_k}\}$ such that $f_{n_k} \to f$ almost everywhere.

91.2   Do Exercise 91.1 with $\mathscr{L}^2(X, \mathscr{M}, \mu)$ replaced by $\mathscr{L}(X, \mathscr{M}, \mu)$.

91.3*   Prove that $\mathscr{L}^2[a, b] \subset \mathscr{L}[a, b]$ and that containment is proper.

91.4*   Prove that $\mathscr{L}(X, \mathscr{M}, \mu)$ is complete in the norm $\|f\| = \int_X |f|$.

91.5*   Show that the completion of $\mathscr{R}[a, b]$ in the norm $\|f\| = \int_a^b |f|$ is $\mathscr{L}[a, b]$.

91.6   Prove that if $f \in \mathscr{L}^2[a, b]$ and $\varepsilon > 0$, there exists $h \in CP[a, b]$ with $\|f - h\| < \varepsilon$. Show that this statement remains valid if "$CP[a, b]$" is replaced by "polynomial."

91.7   Prove that if $f \in \mathscr{L}[a, b]$ and $\varepsilon > 0$, there exists a step function h with $\|f - h\| < \varepsilon$. Show that this statement remains valid if "step-function" is replaced by either "$CP[a, b]$" or "polynomial."

91.8   Give the details of the argument which shows that if we define $T: \mathscr{L}^2(X, \mathscr{M}, \mu) \to l^2$ by $T(f) = \{(f, \phi_n)\}$, then T is one to one and onto and $\|f\| = \|T(f)\|$.

91.9*   Prove the following version of the Riemann-Lebesgue lemma. If $f \in \mathscr{L}^2[a, a + 2\pi]$, then $\lim_{n \to \infty} (f, \phi_n) = 0$.

91.10   (a)   Let $g \in \mathscr{L}^2[a, a + 2\pi]$. Prove that $T(f) = (f, g)$ defines a continuous linear functional on $\mathscr{L}^2[a, a + 2\pi]$. (That is, $T \in \mathscr{L}^2[a, a + 2\pi]^*$.) Prove that $\|T\| \leq \|g\|$.

   The remainder of this exercise is devoted to showing that if $T \in \mathscr{L}^2[a, a + 2\pi]^*$, then T is of the form $T(f) = (f, g)$ for some $g \in \mathscr{L}^2[a, a + 2\pi]$ and $\|T\| = \|g\|$, so that, in effect, the dual space $\mathscr{L}^2[a, a + 2\pi]^*$ is $\mathscr{L}^2[a, a + 2\pi]$ itself. Assume $T \in \mathscr{L}^2[a, a + 2\pi]^*$.

(b)* Set $x_n = T(\phi_n)$. Show that if $f \in \mathscr{L}^2[a, a + 2\pi]$, then $T(f) = \sum_{n=1}^{\infty}$ $(f, \phi_n)x_n$. The convergence of the series is part of the conclusion.

(c) By considering the functions $h_n = \sum_{i=1}^{n} x_i\phi_i$ and the inequality $|T(h_n)| \le \|T\| \|h_n\|$, prove that $\{x_n\} \in l^2$ and $\|\{x_n\}\| \le \|T\|$.

(d) Use the Riesz-Fisher theorem (Theorem 91.9) to obtain $g \in \mathscr{L}^2$ $[a, a + 2\pi]$ with $x_n = (g, \phi_n)$. Prove that $T(f) = (f, g)$.

(e)* Prove that $\|T\| = \|g\|$.

91.11 (a) Prove that if $\{a_n\} \in l^1$, there exists $f \in \mathscr{L}[a, a + 2\pi]$ such that $(f, \phi_n) = a_n$ for $n = 1, 2, \ldots$.

(b) Prove that there exists $f \in \mathscr{L}[a, a + 2\pi]$ such that $\{(f, \phi_n)\} \notin l^1$.

91.12* Prove that if $f \in \mathscr{L}[a, b]$ and $\varepsilon > 0$, there exists an m-measurable set E with $m([a, b]\backslash E) < \varepsilon$ such that $f|E$ is continuous.

Appendix: Vector Spaces

In this brief appendix we summarize some of the key definitions and theorems concerning vector spaces which are used in this text.

Definition. Let F be a field. A *vector space V over a field F* is the pair (V, F) together with operations $+ : V \times V \to V$ and $\cdot : F \times V \to V$ satisfying

1. $x + y = y + x$ for all $x, y \in V$.
2. $x + (y + z) = (x + y) + z$ for all $x, y, z \in V$.
3. There exists $\theta \in V$ such that $x + \theta = x$ for all $x \in V$.
4. For each $x \in V$, there exists $y \in V$ such that $x + y = \theta$.
5. $c(dx) = (cd)x$ for all $c, d \in F$; $x \in V$.
6. $1x = x$ for all $x \in V$, where 1 is the multiplicative identity for F.
7. $c(x + y) = cx + cy$ for all $c \in F$; $x, y \in V$.
8. $(c + d)x = cx + dx$ for all $c, d \in F$; $x \in V$.

We call V the set of *vectors* and the operation $+ : V \times V \to V$ *vector addition*. We call F the set of *scalars* and the operation $\cdot : F \times V \to V$ *scalar multiplication*. When the field F is understood, we speak of the vector space V. We also note that a vector space V is always nonempty since $\theta \in V$. In case $F = \mathbf{R}$, we call V a *real vector space*. Throughout the remainder of this appendix we shall assume that all vector spaces are real vector spaces.

If X is a set and $\mathscr{F}_X$ is the set of real-valued functions on X, then $\mathscr{F}_X$ is a vector space if we define vector addition and scalar multiplication pointwise:

$$(f + g)(x) = f(x) + g(x) \qquad f, g \in \mathscr{F}_X; x \in X$$

$$(cf)(x) = cf(x) \qquad c \in \mathbf{R}; f \in \mathscr{F}_X$$

Most of the vector spaces considered in this text are subsets of $\mathscr{F}_X$ for some set X. For example, if we let $X = \{1, 2, \ldots, n\}$, $\mathscr{F}_X$ is the vector space $\mathbf{R}^n$. It can be shown that any (real) vector space is essentially a subset of $\mathscr{F}_X$ for some set X.

Definition. Let V be a vector space and $X \subset V$. We say that X is *linearly independent* if whenever $\{x_1, \ldots, x_n\}$ is a finite subset of X and for scalars $c_1, \ldots, c_n$ we have $c_1 x_1 + \cdots + c_n x_n = \theta$, it follows that

$$0 = c_1 = c_2 = \cdots = c_n$$

The set X is *linearly dependent* if it is not linearly independent, that is, if there exists a finite subset $\{x_1, \ldots, x_n\}$ of X and scalars $c_1, \ldots, c_n$, not all zero, such that $c_1 x_1 + \cdots + c_n x_n = \theta$.

Definition. Let $W, X \subset V$. If for every element $w \in W$, there exist $x_1, \ldots, x_n \in X$ and $c_1, \ldots, c_n \in \mathbf{R}$ such that

$$w = c_1 x_1 + \cdots + c_n x_n$$

we say that X *spans* W.

Definition. A subset $X \subset V$ is a *basis* for V if X is a linearly independent set and X spans V.

Theorem. A subset $X \subset V$ is a basis for V if and only if every element $v \in V$ may be written uniquely as

$$v = c_1 x_1 + \cdots + c_n x_n$$

where $c_1, \ldots, c_n \in \mathbf{R}$ and $x_1, \ldots, x_n \in X$.

Definition. A vector space V is *finite dimensional* if $V = \{\theta\}$ or if there exists a basis X for V, where X is a finite set.

Theorem. If V is a finite dimensional vector space and $V \neq \{\theta\}$, then every basis for V is finite and any two bases have the same number of elements.

Definition. If V is a finite dimensional vector space, $V \neq \{\theta\}$, and X is a basis for V, we define the *dimension* of V to be the number of elements in X. If $V = \{\theta\}$, we define the dimension of V to be zero.

Theorem. The (vector space) dimension of $\mathbf{R}^n$ is n.

Theorem. Let $x_1, \ldots, x_n \in \mathbf{R}^n$. The following are equivalent.

(i) $\{x_1, \ldots, x_n\}$ is a basis for $\mathbf{R}^n$.

(ii) $\{x_1, \ldots, x_n\}$ spans $\mathbf{R}^n$.

(iii) $\{x_1, \ldots, x_n\}$ is a linearly independent set.

Definition. If $W \subset V$, we say that W is a *subspace* of V if W itself is a vector space.

Theorem. W is a subspace of V if and only if whenever $x, y \in W$ and $c \in \mathbf{R}$, $x + cy \in W$.

Throughout the remainder of this Appendix V_1 and V_2 will denote real vector spaces.

Definition. A *linear transformation* T from V_1 to V_2 is a function $T: V_1 \to V_2$ satisfying $T(x + cy) = (Tx) + c(Ty)$ for $x, y \in V_1$ and $c \in \mathbf{R}$.

Definition. A *linear functional* on V_1 is a linear transformation from V_1 to $\mathbf{R}$, where $\mathbf{R}$ is considered a vector space over itself.

Definition. Let $\mathscr{L}'(V_1, V_2)$ denote the set of linear transformations from V_1 to V_2. If $T, U \in \mathscr{L}'(V_1, V_2)$, $c \in \mathbf{R}$, and $v \in V_1$, we define

$$(T + U)(v) = (Tv) + (Uv)$$

$$(cT)(v) = c(Tv)$$

Theorem. $\mathscr{L}'(V_1, V_2)$ with vector addition and scalar multiplication defined as above is a real vector space.

Theorem. If $T \in \mathscr{L}'(V_1, V_2)$ and $S \in \mathscr{L}'(V_2, V_3)$, then the composition $S \circ T \in \mathscr{L}'(V_1, V_3)$. (Of course, V_3 denotes a real vector space.)

The composition $S \circ T$ is ordinarily written ST.

Definition. If $T \in \mathscr{L}'(V_1, V_2)$ and T is one to one and onto, we call T a (*vector space*) *isomorphism* and we say that V_1 and V_2 are *isomorphic* (*vector spaces*).

Theorem. Let V be a vector space. V is isomorphic to $\mathbf{R}^n$ if and only if V is n-dimensional.

Definition. If $T \in \mathscr{L}'(V_1, V_2)$ and there exists $S \in \mathscr{L}'(V_2, V_1)$ such that TS is the identity on V_2 and ST is the identity on V_1, we say that T is *invertible*. In this case we denote S by T^{-1}.

Theorem. Let $T \in \mathscr{L}'(\mathbf{R}^n, \mathbf{R}^n)$. The following are equivalent:

(i) T is an isomorphism.

(ii) T is one to one.

(iii) T is onto.

(iv) T is invertible.

Definition. If $T, U \in \mathscr{L}'(V_1, V_1)$ and there exists an invertible $S \in \mathscr{L}'(V_1, V_1)$ such that $T = S^{-1}US$, we say that T and U are *similar* linear transformations.

Definition. Suppose that $T \in \mathscr{L}'(V_1, V_1)$ and there exists a nonzero vector x and a scalar c such that $Tx = cx$. We call c an *eigenvalue* and x an *eigenvector* of T.

Good references on linear algebra are Halmos: *Finite-Dimensional Vector Spaces* (1974), Herstein (1975), Hoffman and Kunze (1971), and Strang (1976).

References

Articles

R. P. Boas, Estimating remainders, *Math. Mag.*, **51** (1978), 83–89.

R. P. Boas, Partial sums of infinite series, and how they grow, *Amer. Math. Mo.*, **84** (1977), 237–258.

A. M. Bruckner, Derivatives: Why they allude classification, *Math. Mag.*, **49** (1976), 5–11.

A. M. Bruckner and J. Leonard, Derivatives, *Amer. Math. Mo.*, **73** (4) part II (1966), 24–56.

P. Calabrese, A note on the alternating series, *Amer. Math. Mo.*, **69** (1962), 215–217. Reprinted in T. Apostol et al., *Selected Papers on Calculus*, 352–353.

L. Carleson, On convergence and growth of partial sums of Fourier series, *Acta Math.*, **116** (1966), 137–157.

C. Goffman and D. Waterman, Some aspects of Fourier series, *Amer. Math. Mo.*, **77** (1970), 119–134.

T. H. Hildebrant, Definitions of Stieltjes integrals of the Riemann type, *Amer. Math. Mo.*, **45** (1938), 265–278.

R. Johnsonbaugh, Another proof of an estimate for e, *Amer. Math. Mo.*, **81** (1974), 1011–1012. Reprinted in T. Apostol et al., *Selected Papers on Precalculus*, 107–108.

R. Johnsonbaugh, Compact and connected spaces, *Math. Mag.*, **50** (1977), 24–25.

H. Kestelman, Riemann integration of limit functions, *Amer. Math. Mo.*, **77** (1970), 182–187.

E. R. Lorch, Continuity and Baire functions, *Amer. Math. Mo.*, **78** (1971), 748–770.

E. Schenkman, The intervals of convergence of some power series, *Amer. Math. Mo.*, **78** (1971), 890–892.

J. A. Shohat, Definite integrals and Riemann sums, *Amer. Math. Mo.*, **46** (1939), 538–545.

J. A. Shohat, On a certain transformation of infinite series, *Amer. Math. Mo.*, **40** (1933), 226–229. Reprinted in T. Apostol et al., *Selected Papers on Calculus*, 342–345.

E. Tolsted, An elementary derivation of the Cauchy, Hölder, and Minkowski inequalities from Young's inequality, *Math. Mag.*, **37** (1964), 2–12.

Books

T. M. Apostol, *Mathematical Analysis*, 2d ed., Addison-Wesley, Reading, Mass., 1974.

T. M. Apostol et al. (eds.), *Selected Papers on Calculus*, Mathematical Association of America, 1969.

T. M. Apostol et al. (eds.), *Selected Papers on Precalculus*, Mathematical Association of America, 1977.

R. P. Boas, *A Primer of Real Functions*, Carus Math. Monograph No. 13, Wiley, New York, 1960.

G. Chrystal, *Textbook of Algebra*, vols. I and II, 7th ed., Chelsea, New York, 1964.

H. G. Eggleston, *Elementary Real Analysis*, Cambridge University Press, Cambridge, 1962.

L. M. Graves, *The Theory of Functions of Real Variables*, 2d ed., McGraw-Hill, New York, 1956.

P. R. Halmos, *Finite-Dimensional Vector Spaces*, 2nd ed., Springer-Verlag, New York, 1974.

P. R. Halmos, *Naive Set Theory*, Springer-Verlag, New York, 1974.

G. H. Hardy, *A Course of Pure Mathematics*, 10th ed., Cambridge University Press, Cambridge, 1952.

G. H. Hardy, *Divergent Series*, Oxford University Press, Oxford, 1949.

G. H. Hardy and W. Rogosinski, *Fourier Series*, 2d ed., Cambridge University Press, Cambridge, 1950.

I. N. Herstein, *Topics in Algebra*, 2d ed., Xerox, Lexington, Mass., 1975.

E. Hewitt, *Numbers, Series, and Integrals*, Springer-Verlag, New York, to appear.

E. Hewitt and K. Stromberg, *Real and Abstract Analysis*, Springer-Verlag, New York, 1965.

K. Hoffman and R. Kunze, *Linear Algebra*, 2d ed., Prentice-Hall, Englewood Cliffs, N.J., 1971.

W. Hurewicz and H. Wallman, *Dimension Theory*, Princeton University Press, Princeton, 1941.

I. Kaplansky, *Set Theory and Metric Spaces*, 2d ed., Chelsea, New York, 1977.

J. L. Kelley, *General Topology*, Van Nostrand, New York, 1955.

K. Knopp, *Theory and Application of Infinite Series*, Blackie, London, 1951.

E. Landau, *Foundations of Analysis*, 2d ed., Chelsea, New York, 1960.

R. Larsen, *Functional Analysis, An Introduction*, Marcel Dekker, New York, 1973.

K. O. May, *Index of the American Mathematical Monthly*, Mathematical Association of America, 1977.

J. Olmsted, *The Real Number System*, Appleton-Century-Crofts, New York, 1962.

A. L. Rabenstein, *Introduction to Ordinary Differential Equations*, 2d (enlarged) ed., Academic Press, New York, 1972.

K. A. Ross, *Elementary Analysis: The Theory of Calculus*, Springer-Verlag, New York, 1980.

H. L. Royden, *Real Analysis*, Macmillan, New York, 1963.

W. Rudin, *Principles of Mathematical Analysis*, 3d ed., McGraw-Hill, New York, 1976.

W. Rudin, *Real and Complex Analysis*, 2d ed., McGraw-Hill, New York, 1974.

G. Strang, *Linear Algebra and Its Applications*, Academic Press, New York, 1976.

K. Stromberg, *An Introduction to Classical Real Analysis*, Prindle, Weber, & Schmidt, Boston, 1981.

R. Wheedon and A. Zygmund, *Measure and Integration*, Marcel Dekker, New York, 1977.

A. Zygmund, *Trigonometric Series*, 2d ed., Cambridge University Press, Cambridge, 1968.

Hints to Selected Exercises

Section 5

5.9 Argue by contradiction. Then there exist $c, d \in (a, b)$ with $c < d$ and $f(c) \geq f(d)$. Now consider
$$\text{lub } \{x \mid a < x < d \text{ and } f(x) \geq f(d)\}$$

Section 6

6.6 Suppose $S(n)$ is false for some n and consider the smallest such n (Theorem 6.10).

Section 7

7.9 $\mathbf{R}'$ possesses a set of positive integers $\mathbf{P}'$ by virtue of Definition 6.3. Let $\mathbf{P}' = \{1', 2', \ldots\}$. We may define
$$\mathbf{Q}' = \left\{ \frac{p'}{q'} \bigg| q' \in \mathbf{P}', p' \in \mathbf{P}' \cup -\mathbf{P}' \cup \{0\} \right\}$$
If we define $f(p/q) = p'/q'$, f maps $\mathbf{Q}$ onto $\mathbf{Q}'$.
If $x \in \mathbf{R}$, define
$$f(x) = \text{lub } \{f(r) \mid r < x, r \in \mathbf{Q}\}$$
Show that f has the desired properties.

Section 8

8.4 Use Exercise 8.3.

Section 9

9.7 Use induction on n.

9.8 Use Exercise 9.7.

9.9(c) Use Exercise 9.8.

9.10 Suppose the plane can be written as $\bigcup_{n=1}^{\infty} L_n$, where each L_n is a straight line. Show that there exists a real number t such that the line $L: y = t$ is different from L_n for every n. Consider $L \cap L_n$, $n = 1, 2, \ldots$.

Section 16

16.11 Set $\delta_n = n^{1/n} - 1$. Then $n = (1 + \delta_n)^n \geq \frac{1}{2}n(n-1)\delta_n^2$, $n \geq 2$. Thus $\delta_n^2 \leq 2/(n-1)$, $n \geq 2$. Now show that $\lim_{n \to \infty} \delta_n = 0$.

16.16(a) Divide both sides of the equation by a_{n+1}.

Section 17

17.3 Use Exercise 17.2 and Theorems 17.4(vi) and 14.3.

Section 18

18.2 Consider the bounded and unbounded cases separately.

18.3 Argue by contradiction.

Section 22

22.4 $\dfrac{1}{n(n+1)} = \dfrac{1}{n} - \dfrac{1}{n+1}$

Section 23

23.2 See Exercise 22.4.

Section 24

24.3 Compare the series $\sum_{n=1}^{\infty} 1/n^2$ and $\sum_{n=1}^{\infty} 1/n(n+1)$.

24.9 Consider $s_{2n} - s_n$.

Section 25

25.5(d)   Use 25.5(c).

Section 26

26.1(h)   Assume that $\sum_{n=1}^{\infty} (n^{1/n} - 1)$ converges and then use Exercise 24.9 and inequality (16.1).

26.1(i)   Expand $(1 + 1/n)^n$ by the binomial theorem and show that $e - (1 + 1/n)^n > 1/2n$, $n \geq 2$.

26.1(j)   Show that

$$\frac{1}{2n} \leq \frac{1 \cdot 3 \cdot 5 \cdots (2n - 1)}{2 \cdot 4 \cdot 6 \cdots (2n)} \leq \frac{1}{\sqrt{n + 1}}$$

26.6   Consider the series $\sum_{n=1}^{\infty} p_n$ and $\sum_{n=1}^{\infty} q_n$ in conjunction with Exercise 22.8, where p_n and q_n are defined as in Theorem 26.2.

26.10   Let a_1 be the greatest of $0, 1, 2, \ldots, 9$ such that $a_1/10 \leq x$. Let a_2 be the greatest of $0, 1, 2, \ldots, 9$ such that $(a_1/10) + (a_2/10^2) \leq x$. Continue in this way and then prove that $\sum_{n=1}^{\infty} a_n/10^n = x$.

26.11   To derive the right-most inequality, write

$$R_n = \frac{1}{(n + 1)!} \left(1 + \frac{1}{n + 2} + \frac{1}{(n + 3)(n + 2)} + \cdots \right)$$

and compare the resulting series with

$$1 + \frac{1}{n + 2} + \frac{1}{(n + 2)^2} + \cdots$$

Section 29

29.11(c)   Use part (b).

29.13(a)   Consider the sets $|X|_n = \{|x| \mid x \in X \text{ and } |x| > 1/n\}$.

Section 33

33.6   Use Exercise 32.5 and Theorem 32.2.

Section 36

36.6   Examine the proof of Theorem 36.1.

36.7   Complete the square.

Section 37

37.11 Suppose $\lim_{k \to \infty} a_n^{(k)} = a_n$ for every positive integer n. Choose a positive integer M such that $\sum_{n=M+1}^{\infty} 2/2^n < \varepsilon/2$. Show that there exists a positive integer N such that if $k \geq N$, then

$$\sum_{n=1}^{M} \frac{|a_n^{(k)} - a_n|}{2^n} < \frac{\varepsilon}{2}$$

Deduce that if $k \geq N$, then $d(a^{(k)}, a) < \varepsilon$. The proof of the converse is similar to the proof of Theorem 37.3.

Section 40

40.14(d) Use (c) and Theorem 40.5.

40.16(a) Use Exercise 37.11.

40.16(b) Use Theorem 40.2.

Section 42

42.2 Let $y \in X'$. For each $x \in X$, there exist open sets U_x and V_x such that $U_x \cap V_x$ is empty, $x \in U_x$, and $y \in V_x$. The collection $\{U_x \cap X \mid x \in X\}$ is an open cover of X and so has a finite subcover $\{U_{x_1} \cap X, \ldots, U_{x_n} \cap X\}$. Prove that

$$y \in V_{x_1} \cap \cdots \cap V_{x_n} \subset X'$$

42.12 Argue by contradiction. Investigate the minimum of the continuous function $g(x) = d(f(x), x)$.

Section 43

43.3 For $n \in \mathbf{P}$, cover M by a finite collection of open balls of radius $1/n$. Choose a point in each ball and let A_n be the set of points obtained. Now let $X = \bigcup_{n=1}^{\infty} A_n$.

Section 44

44.6(e) See Exercise 35.7.

44.6(f) Replace M by M^* using (e). Now use Exercise 40.16.

44.6(g) Use (f) and Exercise 43.3.

44.7 Use Exercise 43.4.

Section 47

47.6(c)   If such an f exists, we have $\mathbf{Q} = \bigcap_{n=1}^{\infty} \{x \mid \omega_f(x) < 1/n\}$ which contradicts Exercise 47.5.

47.7(b)   Show that G_ε is open for every $\varepsilon > 0$. Then $\bigcap_{n=1}^{\infty} G_{1/n}$ is nonempty by Theorem 47.2.

47.8   Suppose we can write $[a, b]$ as the disjoint union of a collection of closed intervals of length less than $b - a$. Prove that the collection is countable so that we may write $[a, b] = \bigcup_{n=1}^{\infty} [a_n, b_n]$. Let $X = (\{a_n\} \cup \{b_n\}) \backslash \{a, b\}$. Prove that X is a compact set with no isolated points. Use Corollary 47.8 to obtain a contradiction.

47.9   Set $X_n = \{x \in \mathbf{R} \mid f_x \text{ is of degree } n\}$ and $Y_m = \{y \in \mathbf{R} \mid f_y \text{ is of degree } m\}$. Then $\mathbf{R} = \bigcup_{n=0}^{\infty} X_n = \bigcup_{m=0}^{\infty} Y_m$ implies $X_k^\circ \neq \varnothing \neq Y_l^\circ$ for some integers k and l by the Baire category theorem.

Section 53

53.3   To prove that (c) implies (a), for each positive integer n, choose a partition P_n^* such that if $P^{**}, P^* \supset P_n^*$, then

$$|S(f, P^{**}, T^{**}) - S(f, P^*, T^*)| < \frac{1}{n}$$

Let $P_n = P_1^* \cup \cdots \cup P_n^*$. Choose points T_n in P_n. Let $a_n = S(f, P_n, T_n)$. Prove that $\{a_n\}$ is a Cauchy sequence. Now let $I = \lim_{n \to \infty} a_n$.

Section 54

54.10   Use Exercise 54.8(a), Theorem 54.8, and Exercise 33.6.

Section 56

56.5(a)   Let $F(t) = \int_{c-t}^{c+t} f - t[f(c + t) + f(c - t)]$ and $G(t) = F(t) - (t/h)^3 F(h)$. Apply Rolle's theorem to G on $[0, h]$ to obtain

$$0 = -\eta[f'(c + \eta) - f'(c - \eta)] - (3\eta^2/h^3)F(h)$$

Now apply the mean-value theorem to f' on $[c - \eta, c + \eta]$.

56.6(a)   Let $F(t) = \int_{c-t}^{c+t} f - (t/3)[f(c + t) + 4f(c) + f(c - t)]$ and $G(t) = F(t) - (t/h)^5 F(h)$. Apply Rolle's theorem to G, G', and G'', then the mean-value theorem as in Exercise 56.5(a).

56.10   Write

$$f(\tfrac{1}{2}) = \int_0^1 f + \int_0^{\frac{1}{2}} tf'(t)\, dt - \int_{\frac{1}{2}}^1 (1 - t)f'(t)\, dt$$

Section 58

58.2 It suffices to prove that if $f \in \mathcal{R}[a, b]$ and $f(x) = 0$ almost everywhere in $[a, b]$, then $\int_a^b f = 0$. Apply Theorem 51.8 to the functions max $\{f, 0\}$ and $-\min\{f, 0\}$.

Section 60

60.8 Let $\varepsilon > 0$ and consider the sets $U_n = \{x \in M \mid |f(x) - f_n(x)| < \varepsilon\}$.
60.10 Use Theorem 60.7, Corollary 60.5, and Exercise 60.7.

Section 61

61.4 See Kestelman (1970).

Section 62

62.4 Use Exercise 60.8.

Section 63

63.5 Argue by contradiction using Theorem 63.1(iv) and Rolle's theorem.
63.6 Use Exercise 63.5.
63.7(c) Compute the derivative of $f(x)/(1 + x)^a$.

Section 64

64.3 Use Theorems 29.9 and 64.3.
64.4 Imitate the proof of Theorem 64.1.

Section 66

66.7 Consider the derivative of $e^{-ct} T(t)$.
66.8 Consider the derivative of $e^{-x} f(x)$.

Section 68

68.9(e) Use Wallis' product given in Exercise 68.8.

Section 69

69.8 Imitate the proof of Theorem 45.7.

69.11(b) Imitate the proof of Theorem 46.7.

Section 72

72.6(i) Use (g) and (h) and argue by induction.

Section 74

74.5(d) See Exercise 66.7 and Exercise 68.12.

Section 75

75.6 Let $x \in V$. By Bessel's inequality $\sum_{n=1}^{\infty} (x, x_n)^2$ converges. Now use the generalized Pythagorean theorem and completeness to deduce that $\sum_{n=1}^{\infty} (x, x_n)x_n$ converges. Let $y = \sum_{n=1}^{\infty} (x, x_n)x_n$. Now show that $x - y$ is orthogonal to x_n for every positive integer n.

75.8 Use Theorem 75.5.

Section 76

76.7 Use Exercise 64.4 and Theorem 76.3(i).

Section 77

77.6 Use Exercise 77.5 and imitate the proof of Fejér's theorem.

77.7 Let X be the set of polynomials with rational coefficients. Now use Weierstrass' theorem.

Section 78

78.7(a) Approximate the series $\sum_{n=1}^{\infty} 1/n^3$ by a partial sum and compare the result to π^3/m.

78.8(a) Use Theorem 36.3.

Section 80

80.3 Consider a basis for $\mathbf{R}^n$.

80.6 By the Weierstrass approximation theorem there exist polynomials p such that $p(2)$ is large, but $||p||$ is small.

Section 81

81.8   Use Weierstrass' theorem, Corollary 77.10.

Section 84

84.6(a) Let $f \in C[-1,.1]$ with $\|f\| \le 1$. Then $\|f^n\| \le 1$, and thus $|T(f^n)| \le \|T\|$. Therefore, $|T(f)| \le \|T\|^{1/n}$. Deduce $\|T\| \le 1$. Let $g(x) = x$ for $x \in [-1, 1]$. Define $c = T(g)$. Show that $c \in [-1, 1]$. Now show that if p is a polynomial, then $T(p) = p(c)$. Finally, use the Weierstrass approximation theorem to show that $T(f) = f(c)$ for all $f \in C[-1, 1]$.

(b) Let ϕ be a linear function from $[a, b]$ onto $[-1, 1]$. Define $S(f) = T(f \circ \phi)$. Prove that $S \in (C[-1, 1])^*$ and $S(f \cdot g) = (Sf)(Sg)$ for $f, g \in C[-1, 1]$. Use 84.6(a) to deduce the conclusion.

(c) If T is not continuous, there exists $f \in C[a, b]$ with $\|f\| \le 1$ such that $|T(f)| > 1$. Define $g = f/(T(f) - f)$ and note that $g \in C[a, b]$. Show that $f - gT(f) + f \cdot g = 0$ and deduce that $T(f) = 0$ which is a contradiction.

Section 87

87.13 Choose positive integers $n_1 < n_2 < \cdots$ such that
$$\mu(\{x \mid |f(x) - f_{n_k}(x)| \ge 1/(k+1)\}) < 1/2^{k+1}.$$
Let
$$A_l = \bigcup_{k=l}^{\infty} \{x \mid |f(x) - f_{n_k}(x)| \ge 1/k\}$$
Set $S = \bigcap_{l=1}^{\infty} A_l$.

87.14 Let $F_{i,j} = \{x \mid |f_n(x) - f(x)| < 1/i$ for $n \ge j\}$. Show that for each i, $\lim_{l \to \infty} \mu(F_{i,j}) = \mu(X)$. For each i, choose a positive integer j such that
$$\mu(X \setminus F_{i,j_i}) < \frac{\varepsilon}{2^i}$$
Let $E = X \setminus \bigcap_{i=1}^{\infty} F_{i,j_i}$.

Section 89

89.3(b) Choose n so that $1/(n-2) < \delta$. Imitate the construction of the Cantor set. First, instead of removing the middle third, remove the middle section of length $1/n$. From each of the two remaining intervals remove the middle section of length $1/n^2$ so that an additional $2/n^2$ is removed. Continue inductively.

89.6 If the condition holds, choose closed sets $F_n \subset E$ such that $m(E \setminus F_n) < 1/n$ for $n = 1, 2, \ldots$. Let $A = \cup F_n$. Then $E = A \cup (E \setminus A)$. Show that A and $E \setminus A$ are m-measurable. If E is m-measurable, use Exercise 89.5 and Theorem 89.13(ii) to obtain a closed set $F_n \subset E_n = E \cap [-n, n]$ with $m(E_n \backslash F_n) < \varepsilon/2^n$. Take $F = \cap F_n$.

89.7(d)   For the case when E is m-measurable, use Exercise 89.6.

89.9   Consider the functions

$$f_n(x) = \frac{f(x + 1/n) - f(x)}{1/n}$$

Section 90

90.3   Use Exercise 56.1 and Exercise 88.12.

90.5(a)   Use Theorem 88.6.

90.9   Fix y. Set

$$f_n(x) = \frac{F(x, y + 1/n) - F(x, y)}{1/n}$$

and apply Theorem 88.15.

Section 91

91.3   Write $[a, b] = \{x \mid |f(x)| > 1\} \cup \{x \mid |f(x)| \le 1\}$.

91.4   Examine the proof of Theorem 91.4.

91.5   Use Exercise 91.3.

91.9   First obtain $g \in \mathscr{L}^2[a, a + 2\pi] \cap \mathscr{R}[a, a + 2\pi]$ with $\int_a^{a+2\pi} |f - g|$ small. Then apply Theorem 75.8 to g.

91.10(b)   Use Theorem 91.7 to write $f = \sum_{n=1}^{\infty} (f, \phi_n)\phi_n$.
   (e)   Let $n \to \infty$ in part (c).

91.12   Use Exercises 91.2 and 91.7 to produce a sequence of continuous functions $\{f_n\}$ which converges pointwise almost everywhere to f. Now use Exercise 87.14.

Index

A

Abel's lemma, 92
Abel's limit theorem, 263
Abel's summability, 265
Abel's test, 91, 262
Abel's theorem, 263
Absolute continuity, 228
Absolute convergence, 81, 96
  of improper Riemann-Stieltjes
    integrals, 240
Absolute value, 14
Accumulation point, 102, 131
Addition, 10
Additive identity, 10
Algebraic number, 32
Almost everywhere properties, 233, 375
Alternating series, 80
  test, 80
Archimedean ordering, 19

B

Baire category theorem, 168
Banach algebra, 343
Banach space, 283
Banach's theorem, 165
Bessel's inequality, 305

Binary operation, 9
Binomial series, 261
Binomial theorem, 20
Bolzano-Weierstrass theorem,
  for metric spaces, 150
  for reals, 58
Bonnet's theorem, 229
Boundary of a set, 136
Bounded:
  above, 14
  below, 14
  function, 145
  linear transformation, 337
  sequence, 45
  set, 151
  variation, 214

C

c_0, 120
  metric, 120
  norm, 284
C^n, 185
C^∞, 185
Cantor set, 232
Cartesian product, 5
Cauchy condition, 60
  for functions, 107

421

Errata

Page	Line	Correction
2	−12	$x \in \mathcal{A}$ should be $x \in A$
6	−13	"said to an *extension*" should be "said to be an *extension*"
10	−3	Just before line 3 from the bottom, insert the line "It should be noted that $x = y$ if and only if $x - y = 0$."
29	−1	$\min\{f^{-1}(x)\}$ should be $\min\{f^{-1}(\{x\})\}$
30	−6	$f(n, m)$ should be $f((n, m))$
47	−3	$c_n \le L + \varepsilon$ should be $c_n < L + \varepsilon$
52	−11	Leave out the word "recently"
56	11	"are increasing" should be "are strictly increasing"
84	12	Replace by $$L = \lim_{n \to \infty} \left| \frac{a_{n+1}}{a_n} \right| \text{ exists} \qquad (L = \infty \text{ is allowed})$$
94	−5	$\le \left(\sum\limits_{n=1}^{\infty} a_{m,n} \right)$ should be $\le \sum\limits_{m=1}^{\infty} \left(\sum\limits_{n=1}^{\infty} a_{m,n} \right)$
108	−4	$x > M \ (x < M)$ should be $x > M \ (x < M)$ and $x \in X$
110	5	"*at a* if" should be "*at* $a \in X$ if"
151	5	"Let $x \in M$." should be "Let $x \in C$."
153	4	"in M_2" should be "in $f(M_1)$"
158	12	"f is continuous" should be "f is (uniformly) continuous"
169	1	"a complete" should be "a nonempty complete"
169	6	"a compact" should be "a nonempty compact"
174	−12	"for all $x \in R$" should be "for all $x \in R$ (where $x^0 = 1$)"
199		In Theorem 51.13, all R_α should be $\mathcal{R}_\alpha$
226	7	"F is continuous" should be "F is (uniformly) continuous"
230	8	"f is continuous" should be "f is bounded and continuous"
232	−8	2^n should be 2^{n-1}
239	−11	Insert "and say that $\int_{-\infty}^{\infty} f \, d\alpha$ converges." before line −11
251	5	Change $\left\{ \begin{array}{c} 0 \\ 1 \end{array} \right.$ to $\left\{ \begin{array}{c} 1 \\ 0 \end{array} \right.$
251	7	Change $\left\{ \begin{array}{c} 0 \\ 1 \end{array} \right.$ to $\left\{ \begin{array}{c} 1 \\ 0 \end{array} \right.$
257	7	h_m should be $a + h_m$
270	12	$e\alpha$ should be e^α
303	−10	"f is continuous" should be "f is (uniformly) continuous"
333	14	"the terms of" should be "the terms $\{na_n\}$ of"
411	2	5.9 should be 5.8

A CATALOG OF SELECTED
DOVER BOOKS
IN SCIENCE AND MATHEMATICS

Mathematics

FUNCTIONAL ANALYSIS (Second Corrected Edition), George Bachman and Lawrence Narici. Excellent treatment of subject geared toward students with background in linear algebra, advanced calculus, physics and engineering. Text covers introduction to inner-product spaces, normed, metric spaces, and topological spaces; complete orthonormal sets, the Hahn-Banach Theorem and its consequences, and many other related subjects. 1966 ed. 544pp. $6^{1}/_{8}$ x $9^{1}/_{4}$. 0-486-40251-7

DIFFERENTIAL MANIFOLDS, Antoni A. Kosinski. Introductory text for advanced undergraduates and graduate students presents systematic study of the topological structure of smooth manifolds, starting with elements of theory and concluding with method of surgery. 1993 edition. 288pp. $5^{3}/_{8}$ x $8^{1}/_{2}$. 0-486-46244-7

VECTOR AND TENSOR ANALYSIS WITH APPLICATIONS, A. I. Borisenko and I. E. Tarapov. Concise introduction. Worked-out problems, solutions, exercises. 257pp. $5^{3}/_{8}$ x $8^{1}/_{4}$. 0-486-63833-2

AN INTRODUCTION TO ORDINARY DIFFERENTIAL EQUATIONS, Earl A. Coddington. A thorough and systematic first course in elementary differential equations for undergraduates in mathematics and science, with many exercises and problems (with answers). Index. 304pp. $5^{3}/_{8}$ x $8^{1}/_{2}$. 0-486-65942-9

FOURIER SERIES AND ORTHOGONAL FUNCTIONS, Harry F. Davis. An incisive text combining theory and practical example to introduce Fourier series, orthogonal functions and applications of the Fourier method to boundary-value problems. 570 exercises. Answers and notes. 416pp. $5^{3}/_{8}$ x $8^{1}/_{2}$. 0-486-65973-9

COMPUTABILITY AND UNSOLVABILITY, Martin Davis. Classic graduate-level introduction to theory of computability, usually referred to as theory of recurrent functions. New preface and appendix. 288pp. $5^{3}/_{8}$ x $8^{1}/_{2}$. 0-486-61471-9

AN INTRODUCTION TO MATHEMATICAL ANALYSIS, Robert A. Rankin. Dealing chiefly with functions of a single real variable, this text by a distinguished educator introduces limits, continuity, differentiability, integration, convergence of infinite series, double series, and infinite products. 1963 edition. 624pp. $5^{3}/_{8}$ x $8^{1}/_{2}$. 0-486-46251-X

METHODS OF NUMERICAL INTEGRATION (SECOND EDITION), Philip J. Davis and Philip Rabinowitz. Requiring only a background in calculus, this text covers approximate integration over finite and infinite intervals, error analysis, approximate integration in two or more dimensions, and automatic integration. 1984 edition. 624pp. $5^{3}/_{8}$ x $8^{1}/_{2}$. 0-486-45339-1

INTRODUCTION TO LINEAR ALGEBRA AND DIFFERENTIAL EQUATIONS, John W. Dettman. Excellent text covers complex numbers, determinants, orthonormal bases, Laplace transforms, much more. Exercises with solutions. Undergraduate level. 416pp. $5^{3}/_{8}$ x $8^{1}/_{2}$. 0-486-65191-6

RIEMANN'S ZETA FUNCTION, H. M. Edwards. Superb, high-level study of landmark 1859 publication entitled "On the Number of Primes Less Than a Given Magnitude" traces developments in mathematical theory that it inspired. xiv+315pp. $5^{3}/_{8}$ x $8^{1}/_{2}$. 0-486-41740-9

CALCULUS OF VARIATIONS WITH APPLICATIONS, George M. Ewing. Applications-oriented introduction to variational theory develops insight and promotes understanding of specialized books, research papers. Suitable for advanced undergraduate/graduate students as primary, supplementary text. 352pp. 5³/₈ x 8¹/₂. 0-486-64856-7

MATHEMATICIAN'S DELIGHT, W. W. Sawyer. "Recommended with confidence" by *The Times Literary Supplement,* this lively survey was written by a renowned teacher. It starts with arithmetic and algebra, gradually proceeding to trigonometry and calculus. 1943 edition. 240pp. 5³/₈ x 8¹/₂. 0-486-46240-4

ADVANCED EUCLIDEAN GEOMETRY, Roger A. Johnson. This classic text explores the geometry of the triangle and the circle, concentrating on extensions of Euclidean theory, and examining in detail many relatively recent theorems. 1929 edition. 336pp. 5³/₈ x 8¹/₂. 0-486-46237-4

COUNTEREXAMPLES IN ANALYSIS, Bernard R. Gelbaum and John M. H. Olmsted. These counterexamples deal mostly with the part of analysis known as "real variables." The first half covers the real number system, and the second half encompasses higher dimensions. 1962 edition. xxiv+198pp. 5³/₈ x 8¹/₂. 0-486-42875-3

CATASTROPHE THEORY FOR SCIENTISTS AND ENGINEERS, Robert Gilmore. Advanced-level treatment describes mathematics of theory grounded in the work of Poincaré, R. Thom, other mathematicians. Also important applications to problems in mathematics, physics, chemistry and engineering. 1981 edition. References. 28 tables. 397 black-and-white illustrations. xvii + 666pp. 6¹/₈ x 9¹/₄. 0-486-67539-4

COMPLEX VARIABLES: Second Edition, Robert B. Ash and W. P. Novinger. Suitable for advanced undergraduates and graduate students, this newly revised treatment covers Cauchy theorem and its applications, analytic functions, and the prime number theorem. Numerous problems and solutions. 2004 edition. 224pp. 6¹/₂ x 9¹/₄. 0-486-46250-1

NUMERICAL METHODS FOR SCIENTISTS AND ENGINEERS, Richard Hamming. Classic text stresses frequency approach in coverage of algorithms, polynomial approximation, Fourier approximation, exponential approximation, other topics. Revised and enlarged 2nd edition. 721pp. 5³/₈ x 8¹/₂. 0-486-65241-6

INTRODUCTION TO NUMERICAL ANALYSIS (2nd Edition), F. B. Hildebrand. Classic, fundamental treatment covers computation, approximation, interpolation, numerical differentiation and integration, other topics. 150 new problems. 669pp. 5³/₈ x 8¹/₂. 0-486-65363-3

MARKOV PROCESSES AND POTENTIAL THEORY, Robert M. Blumenthal and Ronald K. Getoor. This graduate-level text explores the relationship between Markov processes and potential theory in terms of excessive functions, multiplicative functionals and subprocesses, additive functionals and their potentials, and dual processes. 1968 edition. 320pp. 5³/₈ x 8¹/₂. 0-486-46263-3

ABSTRACT SETS AND FINITE ORDINALS: An Introduction to the Study of Set Theory, G. B. Keene. This text unites logical and philosophical aspects of set theory in a manner intelligible to mathematicians without training in formal logic and to logicians without a mathematical background. 1961 edition. 112pp. 5³/₈ x 8¹/₂. 0-486-46249-8

A TREATISE ON ELECTRICITY AND MAGNETISM, James Clerk Maxwell. Important foundation work of modern physics. Brings to final form Maxwell's theory of electromagnetism and rigorously derives his general equations of field theory. 1,084pp. 5⅜ x 8½. Two-vol. set.   Vol. I: 0-486-60636-8   Vol. II: 0-486-60637-6

MATHEMATICS FOR PHYSICISTS, Philippe Dennery and Andre Krzywicki. Superb text provides math needed to understand today's more advanced topics in physics and engineering. Theory of functions of a complex variable, linear vector spaces, much more. Problems. 1967 edition. 400pp. 6½ x 9¼.   0-486-69193-4

INTRODUCTION TO QUANTUM MECHANICS WITH APPLICATIONS TO CHEMISTRY, Linus Pauling & E. Bright Wilson, Jr. Classic undergraduate text by Nobel Prize winner applies quantum mechanics to chemical and physical problems. Numerous tables and figures enhance the text. Chapter bibliographies. Appendices. Index. 468pp. 5⅜ x 8½.   0-486-64871-0

METHODS OF THERMODYNAMICS, Howard Reiss. Outstanding text focuses on physical technique of thermodynamics, typical problem areas of understanding, and significance and use of thermodynamic potential. 1965 edition. 238pp. 5⅜ x 8½.
0-486-69445-3

THE ELECTROMAGNETIC FIELD, Albert Shadowitz. Comprehensive under- graduate text covers basics of electric and magnetic fields, builds up to electromagnetic theory. Also related topics, including relativity. Over 900 problems. 768pp. 5⅜ x 8¼.
0-486-65660-8

GREAT EXPERIMENTS IN PHYSICS: FIRSTHAND ACCOUNTS FROM GALILEO TO EINSTEIN, Morris H. Shamos (ed.). 25 crucial discoveries: Newton's laws of motion, Chadwick's study of the neutron, Hertz on electromagnetic waves, more. Original accounts clearly annotated. 370pp. 5⅜ x 8½.   0-486-25346-5

EINSTEIN'S LEGACY, Julian Schwinger. A Nobel Laureate relates fascinating story of Einstein and development of relativity theory in well-illustrated, nontechnical volume. Subjects include meaning of time, paradoxes of space travel, gravity and its effect on light, non-Euclidean geometry and curving of space-time, impact of radio astronomy and space-age discoveries, and more. 189 b/w illustrations. xiv+250pp. 8⅜ x 9¼.   0-486-41974-6

THE VARIATIONAL PRINCIPLES OF MECHANICS, Cornelius Lanczos. Philosophic, less formalistic approach to analytical mechanics offers model of clear, scholarly exposition at graduate level with coverage of basics, calculus of variations, principle of virtual work, equations of motion, more. 418pp. 5⅜ x 8½.   0-486-65067-7

Paperbound unless otherwise indicated. Available at your book dealer, online at www.doverpublications. com, or by writing to Dept. GI, Dover Publications, Inc., 31 East 2nd Street, Mineola, NY 11501. For current price information or for free catalogues (please indicate field of interest), write to Dover Publications or log on to www.doverpublications.com and see every Dover book in print. Dover publishes more than 400 books each year on science, elementary and advanced mathematics, biology, music, art, literary history, social sciences, and other areas.